AF837932

Modeling of the Whole Lifecycle Dioxins for Municipal Solid Waste Incineration

This book by leading experts consolidates information on the numerical simulation and AI-driven modeling of multiple points of dioxin concentration in the municipal solid waste incineration (MSWI) process to provide readers with the skills to develop safer and more efficient waste management practices.

The authors of *Modeling of the Whole Lifecycle Dioxins for Municipal Solid Waste Incineration: Numerical Simulation and AI-Driven Method* explore the complex interplay between municipal solid waste (MSW) management and environmental health, specifically focusing on the generation and production of dioxins during incineration. This comprehensive work amalgamates theoretical insights and practical applications, leveraging numerical simulations and AI techniques to offer innovative modeling solutions. It addresses the critical problem of dioxin generation and emissions from MSWI through a structured approach integrating exhaustive research data, case studies, and advanced computational methodologies, providing an integrated model accounting for various operational parameters and their effects on emissions. It both discusses the existing theoretical frameworks and empirical studies and applies advanced modeling techniques to yield practical insights. Readers will find clear representations of simulation processes alongside discussions on optimization and control mechanisms. This facilitates an understanding of the underlying mechanisms of dioxin formation, enabling targeted solutions that are informed by both experimental results and advanced simulation techniques. This ultimately furnishes practitioners with robust tools to better understand and mitigate the risks associated with dioxin emissions so they can contribute to safer and more efficient waste management practices.

This is an essential resource for engineers, policy-makers, and researchers involved in sustainable waste management and environmental protection. In particular, environmental engineers, MSWI plant operators, and researchers will benefit from the integrated theoretical and practical approaches and discussions of real-world scenarios and applications.

Modeling of the Whole Lifecycle Dioxins for Municipal Solid Waste Incineration

Numerical Simulation and AI-Driven Method

Jian Tang
Loai Aljerf
Junfei Qiao

CRC Press
Taylor & Francis Group
Boca Raton London New York

CRC Press is an imprint of the
Taylor & Francis Group, an **informa** business

First edition published 2027
by CRC Press
2385 NW Executive Center Drive, Suite 320, Boca Raton FL 33431

and by CRC Press
4 Park Square, Milton Park, Abingdon, Oxon, OX14 4RN

CRC Press is an imprint of Taylor & Francis Group, LLC

© 2027 Jian Tang, Loai Aljerf, and Junfei Qiao

Reasonable efforts have been made to publish reliable data and information, but the author and publisher cannot assume responsibility for the validity of all materials or the consequences of their use. The authors and publishers have attempted to trace the copyright holders of all material reproduced in this publication and apologize to copyright holders if permission to publish in this form has not been obtained. If any copyright material has not been acknowledged please write and let us know so we may rectify in any future reprint.

Except as permitted under U.S. Copyright Law, no part of this book may be reprinted, reproduced, transmitted, or utilized in any form by any electronic, mechanical, or other means, now known or hereafter invented, including photocopying, microfilming, and recording, or in any information storage or retrieval system, without written permission from the publishers.

For permission to photocopy or use material electronically from this work, access www.copyright.com or contact the Copyright Clearance Center, Inc. (CCC), 222 Rosewood Drive, Danvers, MA 01923, 978-750-8400. For works that are not available on CCC please contact mpkbookspermissions@tandf.co.uk

For Product Safety Concerns and Information please contact our EU representative GPSR@taylorandfrancis.com. Taylor & Francis Verlag GmbH, Kaufingerstraße 24, 80331 München, Germany.

Trademark notice: Product or corporate names may be trademarks or registered trademarks and are used only for identification and explanation without intent to infringe.

ISBN: 978-1-041-06183-0 (hbk)
ISBN: 978-1-041-06184-7 (pbk)
ISBN: 978-1-003-63419-5 (ebk)

DOI: 10.1201/9781003634195

Typeset in Times
by SPi Technologies India Pvt Ltd (Straive)

Contents

PART I Dioxins Characteristics Analysis of in MSWI Process

PART III Dioxins with AI Driven Detection Modeling and Risk Warning

About the Authors

Jian Tang is a professor at Beijing University of Technology and Deputy Director of Beijing Laboratory of Smart Environmental Protection. He earned his PhD from Northeastern University, Shenyang, China, in 2012. He has authorized over 50 US/Chinese invention patents and obtained over 30 software copyrights. He is a senior member of IEEE.

Loai Aljerf is a professor at Damascus University. He has created partnerships and memberships in prestigious organizations, including the Advances in Cleaner Production Network, the German Chemical Society e.V. (GDCh), the Asian Chemical Society, and the American University of Beirut.

Junfei Qiao is a professor with the Beijing University of Technology and Director of Beijing Laboratory of Smart Environmental Protection, Beijing, China. He received his PhD from Northeastern University, Shenyang, China in 1998. He is a member of several government bodies and director of several national committees in China. He is a senior member of IEEE.

Preface

Dioxins (DXNs), a group of persistent organic pollutants, pose significant threats to human health and the environment. Municipal solid waste incineration (MSWI) has emerged as a major source of DXN emissions, raising concerns about its impact on ecological sustainability. While advancements in air pollution control technologies have significantly reduced emissions, the need for accurate, real-time DXN monitoring remains a critical challenge in the quest for optimal pollution control of MSWI processes.

This book provides a comprehensive exploration of numerical simulation and AI-driven modeling techniques for the precise and comprehensive characterization of multiple points of DXN concentration in MSWI processes. It delves into the intricate mechanisms governing DXN generation, adsorption, and emission at different locations, providing a solid foundation for understanding the complex interplay of various factors influencing DXN behavior in MSWI's whole process.

The book's core focus lies in addressing the challenges of DXN characteristics analysis for an actual MSWI power plant, DXN concentration numerical simulation modeling for multiple points, accurate AI-based DXN detection modeling, and risk warning. For the first focus, it includes MSWI process prescription for DXN, investigation of DXN emission characteristics during the complete maintenance operating period, and assessment of DXN formation and emission characteristics for one year. Second, it includes non-manipulated variable analysis of solid-phase combustion in the furnace and unveiling DXN concentration dynamics based on a whole-process simulation model in terms of multiple points. Finally, it includes multi-scale modeling using Generative Adversarial Network (GAN) and deep forest regression for DXN emission concentration, a hybrid simulator-based mechanism, and data-driven modeling for multi-demand DXN emissions concentration, multimodal data-driven interpretable deep modeling approach of DXN generation concentration, AI-based tree modeling for multi-point DXN concentrations, and a virtual sample generation method based on a generative adversarial fuzzy neural network for DXN emission risk warning.

By bridging the gap between theoretical understanding and practical application, the book offers a comprehensive framework for developing robust and interpretable models for DXN emission control. The insights provided empower engineers and researchers to develop more effective strategies for minimizing DXN emissions, ultimately promoting environmental sustainability and safeguarding human health.

This book is intended for a wide audience, including:

- Environmental Engineers: Professionals involved in the design, operation, and optimization of MSWI plants.
- Researchers: Academics working on modeling and controlling pollutants in industrial processes.
- Students: Individuals pursuing higher education in environmental engineering, data science, and related fields.

The book offers valuable insights into the latest advancements in numerical simulation and AI-driven modeling, equipping readers with the knowledge and tools necessary to address the challenges of DXN emission control. By understanding the principles and methodologies presented, readers can contribute to the development of sustainable waste management practices and protect the environment for future generations.

Thank you to the National Natural Science Foundation of China (62073006), Smart Environmental Protection Beijing Laboratory, Beijing University of Technology, and Beijing Institute of Artificial Intelligence for their long-term support. Thank you to the colleagues and graduate students at the Beijing Laboratory of Smart Environmental Protection, especially Jiakun Chen, Heng Xia, Canlin Cui, and other students who participated in the drafting of this book.

We would also like to extend our gratitude to both experts and scholars in the field of MSWI. It is through your inspiration and encouragement that the content of this book has been significantly enriched.

1 Introduction

1.1 RESEARCH STATUS OF ARTIFICIAL INTELLIGENCE (AI) ON ENVIRONMENTAL SCIENCE AND ENGINEERING

Humanity is currently confronted with global pollution issues such as climate change, ozone layer depletion, and acid rain, alongside local pollution challenges like water, air, and soil contamination. These problems have made the sustainable development of the ecological environment a significant global challenge [1]. According to the World Health Organization, environmental pollution is responsible for 23% of global mortality, making it one of the most significant threats to human health worldwide. Among various pollutants, air pollution poses the most serious risk. As a result, pollution continues to be a critical environmental crisis that countries around the world must address. It is evident that achieving sustainable development requires implementing strategies such as the circular economy, green technologies, and policy regulations. These measures aim to strike a balance that meets present needs without compromising the well-being of future generations [2]. Currently, using the municipal solid waste incineration (MSWI) process as an example, artificial intelligence (AI) technology is transforming the pollution control system. By leveraging intelligent monitoring, predictive modeling, process optimization, and precise traceability, AI is shifting the pollution control paradigm from an "experience-driven" approach to a "data-driven" methodology [3].

Environmental Science and Engineering is an interdisciplinary field that studies the interactions between human activities and the natural environment. Its goal is to address pollution issues and achieve sustainable development of the ecological environment through engineering and technological solutions. The scope of research in this field encompasses both environmental science and environmental engineering. The following text discusses pollutant modeling, small sample modeling, fundamental AI concepts, explainable AI algorithms, tree-based ensemble learning, fuzzy neural network modeling, and virtual sample generation. These topics are systematically elaborated, progressing from surface-level concepts to more in-depth analyses.

1.1.1 Pollutant Modeling Description in Municipal Solid Waste Incineration (MSWI) Process

The development of the global economy and the growth of urban populations have resulted in an annual increase of 8%–10% in the production of municipal solid waste (MSW) [1]. The common methods for treating MSW are landfill and composting. Landfilling offers several advantages, including mature technology, straightforward operation, large treatment capacity, and low investment and operating costs. However, it also has significant drawbacks, such as limited waste reduction effectiveness, low resource utilization rates, extensive land use, and potential risks of groundwater and soil contamination. On the other hand, composting has the benefits of microbial

DOI: 10.1201/9781003634195-1

sterilization of MSW and the production of organic fertilizers. Nevertheless, it comes with challenges, including low fertilizer efficiency, strong odor emissions, and the risk of soil pollution [4]. Clearly, there is a need for more advanced MSW processing equipment and technologies. MSWI technology is a critical solution to address the inherent drawbacks of traditional waste treatment methods, promote sustainable social development, and achieve strategic goals related to waste-to-energy (WTE) [5]. In addition to fulfilling the requirements for harmlessness, waste reduction, and resource utilization, MSWI offers several advantages, including rapid processing speed, reliable operation, and established technology. Furthermore, its service life typically exceeds 20 years.

The essence of MSWI lies in its ability to decompose organic matter in waste into simple inorganic substances through combustion in a high-temperature, oxygen-rich environment. This process effectively reduces the volume of MSW while simultaneously recovering usable energy [3]. Therefore, MSWI not only effectively mitigates methane emissions associated with landfill technology but also reduces CO_2 emissions by recovering heat energy instead of relying on coal-fired power generation, thereby achieving a dual carbon reduction effect. Research has demonstrated that MSWI technology can significantly reduce waste volume by up to 70%, decrease capacity by as much as 90%, and recycle energy at rates of up to 19%. Its potential economic and environmental benefits have been acknowledged by numerous developing countries. Consequently, the MSWI process plays a crucial role in promoting low-carbon initiatives, environmental protection, and sustainable energy development [6], making it a cornerstone industry in the advancement of ecological civilization and the circular economy [7].

MSWI technology represents the most advanced waste treatment method available today; however, it involves complex interactions among numerous sequential and parallel processes, with physical and chemical mechanisms that remain unclear. Consequently, MSWI is a complex dynamic system with spatiotemporal dimensions, characterized by significant uncertainty, nonlinearity, strong coupling, and non-stationarity. Due to these dynamic characteristics, the MSWI process inevitably generates exhaust gases, wastewater, and residues, prompting environmental protection agencies worldwide to classify MSWI as a source of pollution emissions. The pollutants emitted by the MSWI process include nitrogen oxides (NO and NO_2), carbon monoxide (CO), carbon dioxide (CO_2), nitrous oxide (N_2O), sulfur dioxide (SO_2), hydrochloric acid (HCl), particulate matter, DXN, and other flue gas pollutants. The excessive emissions of these pollutants have led to the "not in my back yard (NIMBY)" effect regarding MSWI plants [3], where public perception and regulatory pressures may hinder their acceptance and operation. Consequently, addressing the environmental pollution issues associated with the MSWI process has become a key area of focus for research in both academic research and industry applications.

The MSWI process is characterized by multiple stages, various influencing factors, and complex mechanisms [8, 9]. Operational experience has demonstrated that adopting the "3T + E" principle can effectively ensure the decomposition and combustion of harmful substances in flue gas [10]. This principle stipulates that the furnace temperature must be greater than 850°C, the flue gas residence time should exceed 2 seconds, there must be sufficient turbulence intensity in the flue gas, and an appropriate excess air coefficient should be maintained. Under this principle, researchers in developed countries identify the main manipulated variables as the MSW feed rate, grate speed, air volume ratio, and temperature. The primary controlled variables include combustion line, furnace temperature, oxygen content in the flue gas, and steam flow rate. Based on these variables, they develop an automatic combustion control (ACC) system tailored to their specific national conditions [11–14]. However, a critical prerequisite for the long-term stable operation of such systems is the consistent composition and calorific value of the MSW being processed. Based on statistics from the World Bank [15] and data provided by Gu and Yamada [16, 17], it is evident that MSW in developed countries primarily consists of classified combustible components, which exhibit relatively small fluctuations in calorific value. In contrast, developing countries such as China are still in the process of gradually improving and promoting their classification policies and corresponding management systems for MSW. This has led to high uncertainty, lower calorific values, and greater variability in the composition of collected MSW in these regions [3]. Therefore, ACC systems developed in advanced countries may not be directly applicable to the operational control of MSWI processes in developing countries [18].

Currently, MSWI power plants in developing countries such as China primarily rely on domain experts (i.e., knowledge workers) to assess operating conditions based on their understanding of mechanisms and experiential knowledge. They employ a differentiated manual operation approach tailored to various scenarios, effectively functioning as a manual control mode that incorporates intelligent autonomous behavior [19]. This approach relies on multi-modal data, including structured process data, unstructured images and text, and even language communication, to adjust manipulation variables according to the assessed scenario needs. Fundamentally, this process is grounded in an understanding of the MSWI mechanism, the accumulation of operational experience, and the recognition of variations in working conditions. Essentially, this is an optimization control mode based on the embodied intelligence of domain experts, following the "perception—prediction—control" process, as illustrated in Figure 1.1.

Essentially, this is an embodied intelligent mode summarized in combination with the actual situation of many strong interferences existing in the MSWI dynamic system, which can be briefly described as follows. Based on multi-modal information such as on-site flame images, time serial process data, and operation record, domain experts, in accordance with their own "brain models," use on-site time serial process data and flame videos to monitor in real time easily measurable process parameters such as furnace temperature, oxygen content in flue gas, and emission concentration of conventional flue gas pollutants, as well as predict difficult-to-measure process parameters such as emission concentration of trace pollutants. Based on the process mechanisms mastered through professional education and training and the long-term accumulated experience and knowledge, after reasoning and understanding the current dynamic operating conditions and treatment priorities, the control behavior in the solid waste combustion, waste heat exchange, and flue gas purification stages is empirically regulated.

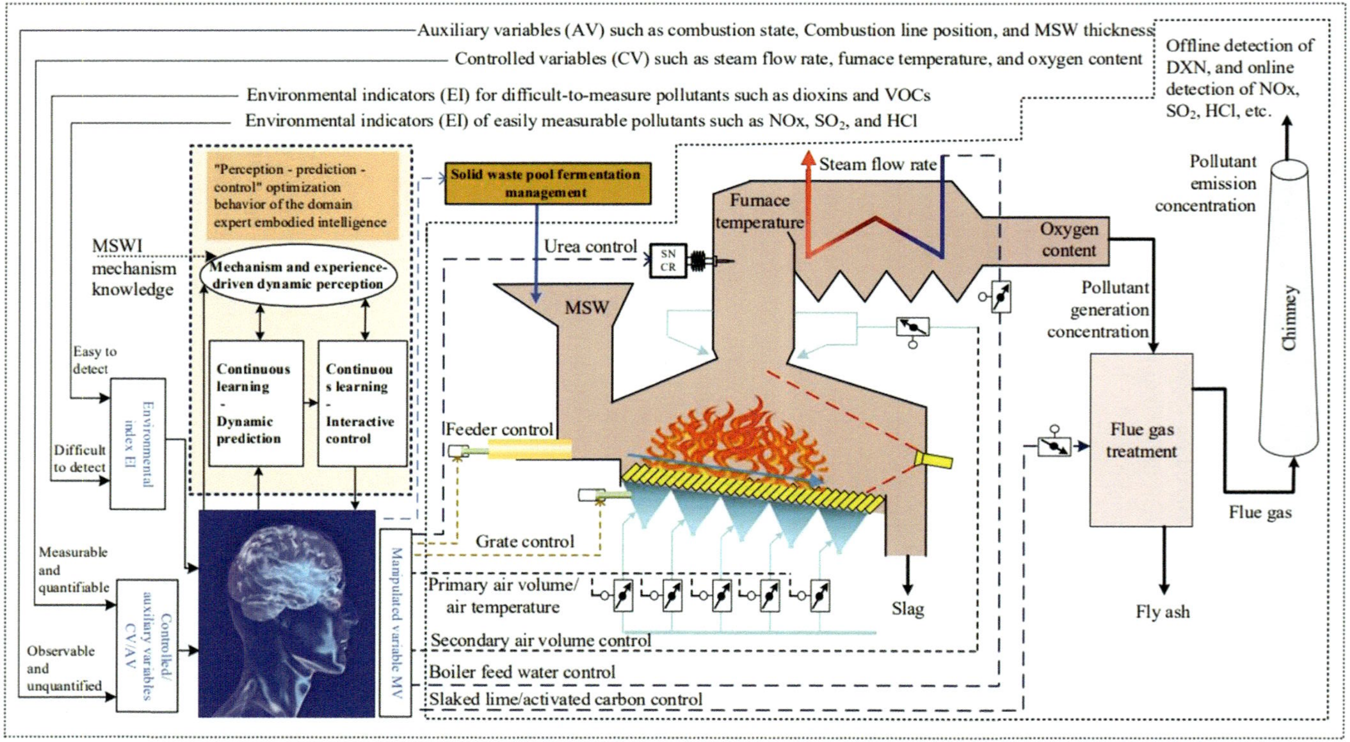

FIGURE 1.1 Optimization control mode of "perception—prediction—control" based on the behavior of the domain expert embodied intelligence.

With the continuous development of AI, the embodied intelligence models can be employed to replace the domain experts' process of "perception—prediction—control" in order to achieve intelligent control. Obviously, the primary perceptual function is the ability to perceive and measure difficult-to-measure pollutants such as dioxins.

DXN, which includes polychlorinated dibenzo-*p*-dioxins (PCDDs) and polychlorinated dibenzofurans (PCDFs), is a significant contributor to the "NIMBY" effect by MSWI power plants [20, 21]. This is primarily due to their strong toxicity and high persistence in the environment, which raises serious concerns regarding public health and environmental safety. The presence of these compounds in emissions can lead to increased regulatory scrutiny and public opposition, further complicating the operation and acceptance of MSWI facilities. DXN exhibits extremely high thermal stability and bioaccumulation potential, which can harm the reproductive, immune, and endocrine systems, and may even pose a cancer risk [22]. Statistics indicate that approximately 37.6% of global anthropogenic DXN emissions originate from the MSWI process [23]. The formation mechanism of dioxins spans multiple temperature zones and multiphase states, and is highly sensitive to operational disturbances. Currently, achieving stable and reliable online monitoring remains a challenge, making it particularly difficult to predict and control trace pollutants in the MSWI process [24].

The commonly used offline direct detection methods for DXN include high-resolution gas chromatography coupled with high-resolution mass spectrometry (HRGC/HRMS), biological detection methods, and immunoassays [25, 26]. The typical procedure involves continuous online sampling of flue gas for 4–6 hours, followed by offline analytical testing in the laboratory, which can take nearly a week. This approach has several disadvantages, including a long turnaround time, significant lag in results, and high costs [27]. The method proposed in reference [28] reduces operational time and lowers the consumption and cost of harmful solvents; however, online real-time detection has not yet been achieved. Reference [29] developed a device for the continuous sampling of exhaust gases containing DXN. Currently, the primary method for monitoring DXN emissions in industrial settings and environmental protection agencies remains this offline direct detection method. This approach is characterized by high complexity, significant lag times (ranging from weeks to months), and reliance on laboratory analysis.

To overcome the bottleneck of lengthy offline direct detection cycles, some studies have explored the use of chlorine-containing precursors and other indicators or associated substances that exhibit a strong correlation with DXN for online indirect detection [30]. This approach involves selecting indicators or associated substances that can be measured in real-time within flue gas and constructing a mapping model based on their concentration changes to indirectly infer the emission levels of DXN. By utilizing the online detection results of these indicators or associated substances, combined with their empirical or statistical correlations with DXN concentration, it becomes possible to achieve indirect online detection of DXN emission concentrations.

Another method is real-time soft sensing detection, which utilizes process variables that are easy to collect during the MSWI process (such as temperature, oxygen content, airflow, load, and pressure) combined with mechanistic knowledge or data-driven approaches to establish a mapping relationship between inputs and outputs. This technique enables the online real-time measurement of DXN discharge concentrations. Compared to offline direct detection methods and online indirect detection methods, real-time soft sensing detection offers significant advantages in terms of real-time capability, cost-effectiveness, maintainability, and engineering applicability. As a result, it has gradually emerged as one of the mainstream technologies for achieving real-time measurement of DXN emission concentrations.

However, actual industrial detection of DXN concentrations faces limitations such as high costs and time delays, making labeled real samples (true input and true output) used for modeling DXN emission concentrations in the MSWI process quite scarce. Consequently, DXN soft sensing modeling essentially becomes a small-sample modeling problem.

1.1.2 Small Sample Modeling

In practical industrial applications, a common challenge faced in soft sensor modeling for difficult-to-measure parameters, such as product quality and environmental indicators, is how to obtain sufficient and comprehensive modeling samples that encompass a wide range of operating conditions [31]. Specifically, the implicit knowledge derived from process data obtained during the early stages of production is limited, which makes it challenging to support the inference of sample distribution [32]. Despite the continuous expansion of modern industry, the available process information remains extremely scarce. Research has shown that sufficient and comprehensive modeling samples are essential for developing effective learning models [33]. In process industries, operational indicators such as product quality and trace pollutant concentrations are often difficult to directly measure on-site due to time constraints or economic costs. As a result, these modeling samples can typically only be obtained during the experimental design phase or shortly after the process is shut down and restarted. Otherwise, acquiring them may come at the expense of the company's economic benefits or require a longer waiting period. Consequently, constructing robust data-driven soft sensing models for industrial applications using scarce, incomplete, and sparse small sample data remains a significant challenge [34]. This task necessitates the adoption of learning algorithms specifically designed for small sample scenarios.

When the sample size for engineering applications is less than 50 and for academic research is less than 30, the associated machine learning problems are referred to as small sample learning problems [35]. Furthermore, the literature [36] articulates this concept as follows:

$$\alpha = \frac{n_{\text{sample}}}{p_{\text{feature}}} \tag{1.1}$$

where α is the ratio of the sample size n_{sample} to feature number p_{feature} with range $\{2, 5, 10\}$.

Clearly, a dataset that is too small α cannot provide adequate support for constructing reliable learning models. In the context of classification problems, the Literature [37] has investigated the interrelationships among classification error, training sample size, input feature dimension, and the complexity of classification algorithms. For certain typical classifiers, the Literature [38] elucidates the underlying reasons for the necessity of having sufficient and comprehensive training samples. It specifically focuses on studying classifier performance at $\alpha \leq 1$, particularly examining the generalization performance of linear classifiers when n_{sample} falls below p_{feature}.

The features after dimensionality reduction are denoted as $p_{\text{feature\_redu}}$, and a new metric is defined as follows:

$$\alpha_{\text{redu}} = \frac{n_{\text{sample}}}{p_{\text{feature\_redu}}} \tag{1.2}$$

If the value α_{redu} obtained after dimensionality reduction still fails to meet the requirements for building a learning model with robust generation performance, alternative methods must be employed to address the shortage of training samples.

To ensure the operational safety of the whole industrial process, as well as high production quality and efficiency, the actual industrial process often functions in a compromise stable state. This may even involve sacrificing economic considerations to prioritize safety, leading to prolonged operation in a suboptimal state. Consequently, the operating conditions, as represented by the operating data, tend to remain relatively constant [39]. Therefore, even with the collection of a substantial amount of process data, the range of operating condition fluctuations it encompasses remains limited. This results in a restricted number of representative samples, which primarily include conventional operating condition data while lacking potential multiple operating condition data. Consequently, these data do not adequately represent the complete distribution necessary for the expected modeling sample space. This article identifies this issue as poor completeness of sample distribution, which leads to models that are applicable only to stable operating processes and struggle to adapt to actual processes with continually changing working conditions [40]. Clearly, there is a need to employ modeling algorithms that can expand the sample set for modeling purposes.

Clearly, the mechanistic knowledge contained in the limited sample data used to construct models for complex industrial processes and difficult-to-measure operational indices is relatively insufficient. The primary reasons are as follows: first, the scarcity of sample size; second, the incomplete sample distribution; and finally, unclear mechanisms that complicate the understanding of domain knowledge. In such cases, interpretable AI algorithms are essential for developing reliable soft sensing detection and warning models based on small sample data.

1.1.3 Basic AI Concepts

AI aims to develop systems that can simulate human intelligent behavior, enabling machines to learn, reason, make decisions, and solve problems [41]. The primary objective is to achieve intelligent behaviors such as perception, understanding, thinking, and action by integrating algorithms and data, thereby extending the boundaries of human capabilities. Essentially, AI is a technology that integrates multiple interdisciplinary fields, including computer science, mathematics, psychology, and neuroscience. It spans various application areas such as robotics, natural language processing, image recognition, and industrial process control.

As a highly intelligent organism on Earth, the human brain comprises approximately 86 billion neurons, which facilitate abstract thinking, language creation, knowledge inheritance, and the construction of civilization through complex neural networks. This intricate system endows humans with the ability to perceive, regulate emotions, and creatively solve problems—capabilities that far exceed those of other organisms. In contrast, AI seeks to replicate human information processing patterns by simulating neural connections, such as those found in deep learning neural networks. As a result, AI has now achieved intelligence that surpasses domain experts in specific tasks. So far, the development of AI has gone through multiple stages, evolving from theoretical exploration to technological breakthroughs and progressing from specialized weak AI to general strong AI. These stages can be outlined as follows:

(1) The Initial Exploration Stage (1950s–1960s): The Turing Test, proposed in 1950, laid the conceptual foundation for AI, while the Dartmouth conference in 1956 marked its official inception. During this period, researchers aimed to simulate human thinking using symbolic rules. However, progress was slow due to limitations in computing power. Despite these challenges, this stage provided a crucial theoretical framework for subsequent research.
(2) The Cold Winter Period of the 1970s: AI research entered a low point during this stage, primarily due to unresolved technical bottlenecks and unmet high expectations. The symbolic processing theory from the initial stage struggled to handle fuzzy logic, which hindered its applicability in real-world scenarios. Additionally, the computing power and data volume required for AI research and application were inadequate to meet the demands of the field.
(3) The Recovery and Expert System Stage (1980s): This stage benefited from advancements in computer technology, leading to the emergence of expert systems that could simulate human expert decision-making. These systems achieved initial success in fields such as medical diagnosis and financial analysis. However, their reliance on manually encoded knowledge limited their generalization ability, making it challenging for them to adapt to dynamic environments.
(4) The Rising Stage of Machine Learning (1990s–2000s): Progress during this period was driven by factors such as statistical learning methods, increased data volumes, and enhanced computing power. A milestone that exemplified

AI's capability to outperform human experts in specific tasks was IBM's "Deep Blue" defeating the Chess champion in 1997. Key machine learning techniques from this era include support vector machines (SVM) and random forests (RF), which reduced reliance on artificial rules by emphasizing data-driven pattern modeling.

(5) The Deep Learning Revolution (2010s–Present): This stage is marked by significant milestones, including the success of the convolutional neural network (CNN) AlexNet in winning the ImageNet competition in 2012 and AlphaGo's victory over the world Go champion in 2016. The recent rise of generative AI and large language model (LLM) technologies signifies the golden age of AI, expanding the boundaries of content creation and interactive applications.

From the current development perspective, the core elements of AI include computing power, algorithms, and data. Computing power enables the training of complex models, algorithms determine the efficiency of learning, and data serves as the foundation for model optimization. Together, these three components collaborate to extract patterns from vast amounts of information and facilitate autonomous optimization. Presently, AI is transitioning from specialized weak AI to the pursuit of general strong AI.

The application of AI in industrial process modeling has progressed through the following three stages:

(1) Technology Foundation Stage (1980s–2010s): During this period, early AI technologies such as expert systems and neural networks were employed to develop specialized small models for rule inference and simple predictions within industrial processes. Following the widespread adoption of programmable logic controller (PLC), distributed control system (DCS), and manufacturing execution system (MES) systems, these technologies played a crucial role in automating industrial production processes.
(2) Technological Breakthrough Stage (2011–2020): This stage marked a significant advancement with the breakthrough of deep learning in image recognition in 2012. This development enabled AI to model complex multimodal industrial data, including images and speech, using machine learning models. Consequently, applications such as equipment fault prediction and production quality detection became feasible, enhancing the capabilities of industrial process modeling.
(3) Large-Scale Modeling Stage (2021–Present): In this current phase, parameter scales have reached billions, and large-scale industrial models that integrate industry knowledge are being developed. These models are beginning to explore optimization parameter modeling for multi-modal interactions, further advancing the effectiveness and efficiency of AI in industrial processes.

In the future, AI will understand the physical world through continuous learning, constructing a world model that can generate semantically and geometrically consistent three-dimensional environments. It will learn physical laws, simulate environmental dynamics, deduce future states, and evaluate behavioral consequences to achieve autonomous decision-making. For modeling trace pollutants such as DXN in processes similar to MSWI, it is clearly necessary to employ interpretable AI algorithms.

1.1.4 Explainable AI Algorithm

Explainable AI (XAI) refers to the use of technological methods to make the decision-making processes of AI systems transparent, enabling users to comprehend their internal logic and algorithmic framework [42]. The core objective of XAI is to address the traditional "black box" problem associated with AI. By enhancing transparency, XAI aims to meet the transparency requirements of certain fields, ensuring that stakeholders can trust and effectively interact with AI systems.

So far, XAI has progressed through multiple stages, specifically:

(1) Early Stage (Before the 1980s): During this period, XAI primarily consisted of rule-based expert systems that relied on manually written knowledge bases. While these systems offered strong interpretability, they struggled with scalability, making it difficult to adapt to more complex scenarios.
(2) Machine Learning Era (After 2000): In this stage, XAI evolved into data-driven models such as decision tree (DT) and logistic regression. These models provided partial interpretability, particularly for analyzing feature importance, allowing users to gain insights into how specific input features contributed to the model's decisions.
(3) Deep Learning Stage (After 2012): With the rise of complex neural networks as mainstream tools, the demand for interpretability increased significantly. This led to the development of various interpretability tools, such as LIME (Local Interpretable Model-agnostic Explanations) and SHAP (SHapley Additive exPlanations). Notably, the concept of XAI was formally proposed by DARPA in 2016, emphasizing the need for transparent AI systems in increasingly complex applications.

The XAI algorithms can be divided into four categories, described in detail as follows:

(1) Model-Based Approach: This category includes inherently interpretable models such as DTs, rule lists, and linear models. These models are designed to have clear and transparent decision-making processes, allowing users to directly understand how decisions are made based on the input features.
(2) Post-Hoc Explanation Methods: This category encompasses techniques such as LIME and SHAP, which provide explanations for predictions made by black box models after they have been trained. These methods aim to clarify the prediction results through local approximation or feature importance analysis.

    The LIME method explains black box models through local linear approximation. Its core steps include generating perturbation samples around the instance to be explained, using the original model to predict these perturbed samples, and then training an interpretable linear model (like a DT) weighted by sample similarity to approximate the behavior of the original model.

Based on the Shapley value from game theory, SHAP provides interpretability by calculating the marginal contribution of each feature to the predicted outcomes. It adheres to axioms such as local accuracy, absence, and consistency, enabling it to globally quantify feature importance in a consistent manner. However, it can be computationally intensive for complex models.

(3) Proxy-Based Approach: This approach involves training an interpretable proxy model, such as a DT or a set of rules, to approximate the behavior of a complex model. The essence of this method is to simplify complex models by using these proxy models, which can provide more intuitive and understandable explanations of the predictions made by the original model. By creating a simpler representation, users can gain insights into how the complex model operates without needing to delve into its intricate details.

(4) Mixed Methods: This category of methods is characterized by the integration of multiple techniques to enhance interpretability. For example, one could combine deep learning models with rule-based engines or incorporate attention mechanisms to improve the transparency of the model's decision-making process. These mixed approaches leverage the strengths of different methodologies to create more interpretable and comprehensible AI systems, allowing for a better understanding of complex interactions within the models.

The selection of XAI algorithms for industrial processes typically requires careful consideration of various factors, including data characteristics, model complexity, and specific business scenarios. Details are as follows.

(1) Data Characteristics: From the perspective of post-hoc interpretability in XAI, LIME is preferred for small sample sizes, while SHAP is more suitable for larger datasets. Regarding data types, temporal data lends itself well to global interpretation using SHAP, whereas image data is best interpreted locally with LIME. In the context of model-based XAI, DT algorithms with strong interpretability are advisable when dealing with data quality issues.

(2) Model Complexity: If a simple model is sufficient, inherent interpretability methods such as linear regression and DTs should be employed. However, when working with complex models, it may be necessary to combine post-hoc interpretation techniques like LIME and SHAP to achieve both interpretability and performance.

(3) Specific Business Scenarios: In high-risk decision-making contexts, such as chemical reaction control, SHAP is crucial for providing global interpretations that inform critical decisions. For real-time inspections, such as quality checks on production lines, LIME is preferable due to its rapid generation of local interpretations. Additionally, SHAP can be utilized to evaluate the coupling effects of interactive features during process optimization.

Therefore, LIME is well-suited for real-time scenarios requiring quick interpretations, SHAP is ideal for high-risk decisions necessitating global insights, and DTs are effective for modeling problems with clear rules.

Clearly, for complex industrial processes characterized by distinct causal relationships yet lacking transparency, adopting various enhanced modeling algorithms based on DTs is the preferred choice.

1.1.5 Tree-Based Ensemble Learning

Research has demonstrated that ensemble modeling methods, which integrate multiple sub-models, can enhance the generalization, effectiveness, and credibility of predictive models. The initial ensemble modeling approach originated from neural network ensembles, where a finite number of neural networks are utilized to address the same problem. For a given input sample, the outputs of each neural network within the ensemble collectively determine the final output [43]. This collaborative approach helps to mitigate individual model biases, leading to more robust and accurate predictions.

The construction of integrated modeling can be divided into two main steps: the development of sub-models and the merging of these sub-models. The generalization error of neural network ensemble models can be expressed as the difference between the average generalization error of the sub-models and the average ambiguity of the sub-models, which can be understood as the degree of difference among individual learners. This degree of difference among sub-models can be achieved by employing various topological structures and training datasets. Common methods for obtaining different training datasets include: resampling the training samples (e.g., splitting the training samples into different subsets), manipulating the input features (e.g., creating different subsets of input features), altering the output targets (e.g., applicable to multi-type problems), and injecting randomness into the learning algorithm (e.g., using different initial weights for learning algorithms applied to the same training set). These strategies help to enhance the diversity of the sub-models, thereby improving the overall performance of the ensemble model.

In theory, ensemble learning enhances the generalization ability of the overall model and significantly reduces the risk of overfitting associated with individual base learners [44, 45]. Ensemble learners have the advantage of achieving an optimal approximation of the model's objectives. The primary reason for the widespread adoption of ensemble learning is its capacity to increase the overall diversity of the model. This increased diversity enables the model to explore a broader range of data correlations, resulting in strong generalization performance when applied to various datasets. However, existing methods also come with certain limitations, including high computational costs, reduced interpretability, and insufficient generalization performance in some cases. These challenges highlight the need for further advancements in ensemble learning techniques to improve their efficiency and applicability across different contexts.

The DT is a fundamental model in machine learning characterized by internal nodes, leaf nodes, and directed edges. It operates in three main steps: first, it applies a minimum loss function to segment the sample space; second, it employs an induction strategy to construct the decision-making model; and third, it utilizes this model to predict unseen samples. The DT algorithm is often referred to as a "white-box model" because its decision-making process can be easily interpreted as a set of rules. However, the split criterion based on "qualitative" information used in classification tasks may not be well-suited for the "quantitative" nature of regression problems. To address this limitation, the CART (classification and regression trees) model is commonly employed, as it utilizes split criteria based on both the Gini index and squared error. In particular, the squared error minimization criterion is widely used in constructing binary DTs for industrial process modeling, as it effectively captures the relationships between input features and output variables in regression scenarios.

The accuracy of a single-tree model is generally lower compared to that of a tree ensemble that utilizes voting decisions. While the strategy of combining multiple models has gained widespread attention, the diversity of training data continues to pose a significant challenge to the advancement of ensemble learning. Bootstrap sampling, commonly referred to as bagging, is often employed to address this issue by generating subsets from the original dataset, thereby enhancing diversity among the models. Although the bagging model can achieve strong generalization performance, it still exhibits some inherent bias. Consequently, the performance of the bagging model may be inferior to that of other techniques, such as AdaBoost and similar methods, which are designed to reduce bias and improve overall predictive accuracy. This highlights the ongoing need for innovative approaches to enhance the effectiveness of ensemble learning methods.

The ensemble DT structure is primarily divided into two types: parallel forests (PF) and cascade forests (CF). Both PF and CF methods can be utilized to construct process models with learners based on the CART framework. Among the methods based on the PF structure, the RF algorithm is the most representative and widely used due to its robust ability to handle noise and effectively model nonlinear data [46]. RF is developed by combining bagging techniques with the random subspace method, where multiple training subsets are generated through bootstrap sampling. Additionally, random feature selection is applied to each training subset, allowing each CART tree to grow to its maximum potential without overfitting to any particular subset of the data. The predictions from each individual CART tree are then aggregated using a "simple average" strategy, resulting in enhanced predictive accuracy and improved generalization performance across various datasets. Its versatility and resilience make it a popular choice for various applications in machine learning and data analysis [47].

1.1.6 Fuzzy Neural Network Modeling

Artificial neural networks (ANNs) are mathematical models inspired by the principles of biological neural networks. By understanding and abstracting the structure of the human brain and its response mechanisms to external stimuli, ANNs utilize network topology as a theoretical basis to simulate the information processing capabilities of the brain's nervous system. ANNs models are characterized by several key features, including parallel distributed processing capabilities, high fault tolerance, intelligence, and self-learning ability. ANNs effectively combine information processing and storage, utilizing unique methods of knowledge representation along with adaptive learning mechanisms. This allows them to learn from data, recognize patterns, and make predictions, making them powerful tools for solving complex problems across various domains in machine learning and AI.

An ANN typically consists of three main layers: an input layer, one or more hidden layers, and an output layer. The input layer directly receives raw data or preprocessed feature data. Its primary function is to pass external inputs to the subsequent hidden or output layers without performing any computations. The hidden layer is responsible for achieving complex feature learning and nonlinear mapping of the original data. It transforms the input data into high-order features that capture intricate patterns and relationships. These learned features are essential for supporting the completion of tasks such as classification and regression in the network's later stages. The output layer takes the high-order features generated by the hidden layer and converts them into the final output format required for specific tasks, such as class labels in classification problems, continuous values in regression tasks, or probabilities in probabilistic tasks. Together, these layers enable ANNs to learn from data, model complex relationships, and make accurate predictions or classifications based on the input provided.

ANNs do have several disadvantages, particularly when it comes to modeling complex industrial processes using small sample sizes. The first one is the high overfitting risk. ANNs can easily overfit to the training data, especially when sample sizes are small. This occurs because the model may learn to memorize noise or specific details from the training samples rather than general patterns. As a result, the network's performance on unseen data can suffer significantly, leading to poor generalization. The second one is the unstable training. When working with insufficient data, the weight adjustment process during training can be heavily influenced by individual samples. This can result in large fluctuations in the model's parameters, making it difficult for the network to converge to an optimal solution. Such instability can hinder the learning process and lead to inconsistent results. The third one is the limited generalization performance. The lack of sufficient orthogonality and completeness in small sample datasets means that the features learned by the model may not adequately represent the underlying data distribution. This insufficiency can lead to a significant decrease in generalization performance, as the model struggles to capture the essential characteristics necessary for making accurate predictions on new data.

These challenges highlight the importance of using larger, more representative datasets and potentially exploring alternative modeling approaches or regularization techniques to mitigate overfitting and improve the robustness of ANNs when applied to complex industrial processes. In practical, complex scenarios, there is often a significant amount of uncertain information, such as that encountered in natural disasters, weather forecasts, human investment behavior, and advancements in science and technology [48]. Practical experience suggests that the human brain is adept at transforming this uncertain information from complex systems into precise, reliable, and deterministic knowledge through reasoning.

To effectively describe and analyze uncertainty in such systems, a fuzzy reasoning method based on fuzzy sets and fuzzy logic was first proposed in 1965 [49, 50]. This innovative approach allows for the representation of data that is not strictly binary (true or false), accommodating degrees of truth that are more aligned with real-world complexities. The ability of fuzzy reasoning methods to handle data uncertainty and perform approximate calculations has gained significant attention across various disciplines. These include engineering, where it aids in control systems and decision-making processes; medicine, for diagnostic systems and patient management; chemistry, particularly in reaction modeling; computational biology, for analyzing biological data; and pattern recognition, where it enhances classification tasks in noisy environments. The versatility and effectiveness of fuzzy logic in managing uncertainty make it a valuable tool for researchers and practitioners in these fields [51].

A fuzzy neural network (FNN) is an intelligent algorithm that integrates the strengths of fuzzy logic and neural networks, allowing for effective modeling of complex systems by fuzzifying uncertain information and leveraging the learning capabilities of neural networks. The essence of an FNN lies in mapping precise inputs to fuzzy sets within the [0, 1] interval through membership functions, which can take various shapes, such as triangular or Gaussian.

Typically, an FNN consists of four layers. The input layer receives raw data from external sources. The fuzzification layer transforms the input data into a fuzzy language set. This transformation often employs Gaussian basis functions to calculate the membership degrees of each input value concerning the defined fuzzy sets. The fuzzy inference layer performs fuzzy inference by calculating the strength of the rule matches through node combinations. The output here reflects the degree to which the input signals align with the established fuzzy rules. The clarity layer converts the fuzzy outputs into crisp, clear values using a weight-averaging method, effectively defuzzifying the results.

FNNs commonly utilize the backpropagation (BP) algorithm to optimize their parameters, updating weights and biases by minimizing the mean square error (MSE) between predicted outputs and actual values. This optimization process enhances the network's predictive accuracy. One of the significant advantages of FNNs is their ability to automatically extract features relevant to industrial processes through the combination of fuzzy rule layers and neural network learning capabilities. They are particularly well-suited for predictive modeling of key parameters such as furnace temperature and pollution emission concentration in the MSWI process. When faced with a shortage of modeling samples, FNNs can enhance the model's generalization ability by using fuzzy logic to effectively process and interpret fuzzy information, thus improving performance in uncertain environments.

1.1.7 Virtual Sample Generation

Virtual sample generation (VSG) technology has its roots in the field of image recognition, where researchers generate effective virtual image samples by leveraging prior knowledge to enhance the accuracy of model recognition. The core principle of VSG involves extracting latent information from the limitations of small-sample data to create an appropriate number of virtual samples. This process expands the modeling sample size and improves the performance of model soft measurements. The spatial distribution relationship between real samples and virtual samples suggests that the essence of VSG lies in achieving sample augmentation by "filling" the information gaps and voids within the overall data space. Research results demonstrated that the VSG method, which incorporates prior knowledge, functions as a regularization strategy, wherein the prior knowledge is integrated into a regularization matrix [52]. Furthermore, the VSG method based on prior knowledge has been successfully applied in various fields, including text recognition and noise source identification. However, acquiring domain-specific prior knowledge can be both challenging and time-consuming. This is especially true for complex industrial processes, which often require extensive experience and insights from domain experts. Effective prior knowledge that directly informs the generation of virtual samples necessitates a comprehensive understanding of the entire process system. In reality, most domain experts are only able to gather limited local knowledge about specific processes. As a result, researchers across various fields have increasingly focused on methods for extracting knowledge from small samples to generate virtual samples.

The general definition of virtual samples [53] is as follows:

Definition: Let $\{\boldsymbol{x}_l, y_l\}$ be the real sample, where, $\boldsymbol{x}, y \in \mathrm{R}^n, l = 1, \cdots, k$, k is the number of real samples; Based on prior knowledge Know, a transformation $\{T, f(T)\}$ is used to generate a new sample $\{T\boldsymbol{x}, f(T)y\}$, namely:

$$\left.\begin{matrix}\{\boldsymbol{x}_l, y_l\} \\ \text{Know}\end{matrix}\right\} \xrightarrow{\{T, f(T)\}} \{T\boldsymbol{x}, f(T)y\} \tag{1.3}$$

The new sample $\{T\boldsymbol{x}, f(T)y\}$ referred to here is termed a virtual sample. The methods employed for transformation may vary depending on the specific application context. Based on the definition provided, this book presents the following inference:

Inference: Given the real sample dataset $\{\boldsymbol{x}_l, y_l\}_{l=1}^{k}$, the process of generating the virtual sample dataset $\{\boldsymbol{x}_{l'}^{\mathrm{VS}}, y_{l'}^{\mathrm{VS}}\}_{l'=1}^{k^{\mathrm{VS}}}$ through appropriate transformations can be expressed as:

$$\begin{cases} \{\boldsymbol{x}_l\}_{l=1}^{k} \xrightarrow{T} \{\boldsymbol{x}_{l'}^{\mathrm{VS}}\}_{l''=1}^{k^{\mathrm{VS}}}, & \boldsymbol{x}_{\mathrm{low}}^{\mathrm{VS}} \le \boldsymbol{x}_{l'}^{\mathrm{VS}} \le \boldsymbol{x}_{\mathrm{high}}^{\mathrm{VS}} \\ \left.\begin{matrix}\{\boldsymbol{x}_{l'}^{\mathrm{VS}}\}_{l''=1}^{k^{\mathrm{VS}}} \\ \{y_l\}_{l=1}^{k}\end{matrix}\right\} \xrightarrow{f(T)} \{y_{l'}^{\mathrm{VS}}\}_{l''=1}^{k^{\mathrm{VS}}}, & y_{\mathrm{low}}^{\mathrm{VS}} \le y_{l'}^{\mathrm{VS}} \le y_{\mathrm{high}}^{\mathrm{VS}} \end{cases} \tag{1.4}$$

where, $\boldsymbol{x}_{\mathrm{low}}^{\mathrm{VS}}$ and $\boldsymbol{x}_{\mathrm{high}}^{\mathrm{VS}}$, $y_{\mathrm{low}}^{\mathrm{VS}}$ and $y_{\mathrm{high}}^{\mathrm{VS}}$ respectively represent the lower and upper limits of the input and output values of the virtual sample.

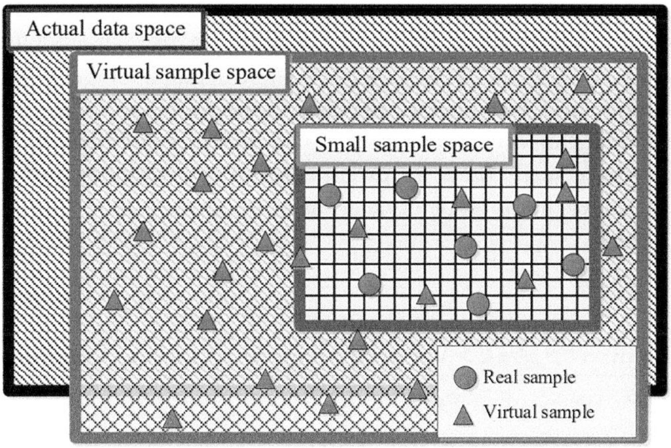

FIGURE 1.2 Relationship between virtual samples and real samples.

The relationship between virtual samples, real samples, virtual sample space, small sample space, and actual data space is shown in Figure 1.2.

Figure 1.2 illustrates that the small sample space can only provide a one-sided representation of the actual data space, leading to several issues: an incomplete coverage of the actual data space, resulting in areas with information gaps; the presence of information gaps between samples; and uneven distribution of samples within the actual data space. Consequently, many scholars studying VSG aim to make the virtual sample space as close as possible to the actual data space. However, regardless of the VSG method employed, it is inevitable that some virtual samples will not conform to the true characteristics and distribution of the actual data. Such samples can be detrimental to model training and may result in a decline in generalization performance.

To generate reasonable virtual samples, at least four key issues need to be addressed:

(1) Strategy for Generating Virtual Samples: This involves deciding on approaches such as utilizing prior knowledge, perturbing the original samples, or adding noise to the input data.
(2) Input for Virtual Samples: This includes selecting methods for generating inputs, such as employing prior knowledge techniques, randomly selecting points within the super domain of real sample input [54], functionalizing virtual populations [55], implementing information dispersion techniques based on real sample input intervals [56], using interval kernel density estimation techniques, applying mega trend dispersion functions [57], utilizing membership functions based on fuzzy datasets [58], incorporating group VSG techniques [59], trend dispersion methods based on fuzzy theory [60], Gaussian distribution techniques, and generative adversarial networks (GANs).

(3) Output for Virtual Samples: This involves determining how to produce outputs, which may include average neural network outputs, information dispersion techniques based on actual sample output intervals, genetic algorithms [61], and semi-supervised collaborative training.
(4) Determining the Number of Virtual Samples: The optimal number of virtual samples is typically based on experimental data. However, finding a theoretical framework to guide or determine its optimal value remains an open challenge.

1.2 RESEARCH STATUS OF DIOXINS (DXN) MECHANISM

1.2.1 Brief Overview of DXN

Since the DXN emissions from the MSWI process first drew significant attention in 1977 [62], DXN has become a central issue in both international regulation and academic research. Dioxin-like chemicals (DLCs) represent a broad class of compounds, primarily including PCDDs, polychlorinated dibenzofurans (PCDFs), and coplanar polychlorinated biphenyls (Co-PCBs). Among these, PCDDs and PCDFs are often collectively referred to as DXN. In the context of MSWI, DXN are commonly referred to as DXN, and their basic molecular structures are illustrated in Figure 1.3.

As illustrated in Figure 1.3, the basic skeletons of PCDDs and PCDFs consist of two benzene rings linked by two oxygen bridges and one oxygen bridge, respectively. Each benzene ring can be substituted with 0–4 chlorine atoms, where both the number and the positions of these substitutions determine the variety of their congener groups and isomers. Statistical data indicate that PCDDs comprise 75 isomers, while PCDFs comprise 135 isomers. It is generally recognized that DXN containing 1–3 chlorine atoms exhibits low or negligible toxicity, whereas those containing 4–8 chlorine atoms possess significant toxic potential. In particular, the 17 congeners that are chlorine-substituted at all the 2, 3, 7, and 8 positions exhibit the highest toxicity and thus represent the primary focus of current research and monitoring efforts, as detailed in Table 1.1.

The international toxic equivalency factor (I-TEF), established by the North Atlantic Treaty Organization (NATO), is used to characterize the relative toxicity of each PCDD/F congener that is chlorine-substituted at all the 2,3,7,8 positions,

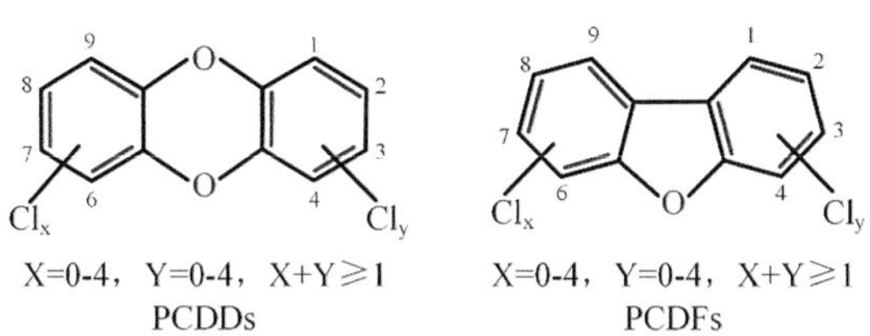

FIGURE 1.3 Molecular structure of DXN.

TABLE 1.1
The 17 Congeners and Their International Toxic Equivalency Factors (I-TEFs)

Category	I-TEFs	Category	I-TEFs
2,3,7,8-TCDD	1	2,3,7,8-TCDF	0.1
1,2,3,7,8-PeCDD	0.5	1,2,3,7,8-PeCDF	0.5
1,2,3,4,7,8-HxCDD	0.1	2,3,4,7,8-PeCDF	0.05
1,2,3,6,7,8-HxCDD	0.1	1,2,3,4,7,8-HxCDF	0.1
1,2,3,7,8,9-HxCDD	0.1	1,2,3,6,7,8-HxCDF	0.1
1,2,3,4,6,7,8-HpCDD	0.01	2,3,4,6,7,8-HxCDF	0.1
OCDD	0.001	1,2,3,7,8,9-HxCDF	0.1
—	—	1,2,3,4,6,7,8-HpCDF	0.01
—	—	1,2,3,4,7,8,9-HpCDF	0.01
—	—	OCDF	0.001

compared to 2,3,7,8-tetrachlorodibenzo-*p*-dioxin (2,3,7,8-TCDD). It is important to note that the relative toxicity of these congeners decreases as the number of chlorine substitutions at non-critical positions (e.g., positions 1, 4, 6, and 9) within the molecule increases. Consequently, the I-TEF values exhibit a declining trend with increasing degree of chlorination.

To comprehensively assess the overall toxicity of DXN in a sample, the international toxic equivalency quantity (I-TEQ) is calculated. This is achieved by multiplying the concentration of each of the 17 toxic isomers by its corresponding I-TEF and then summing the products. The calculation formula is as follows:

$$\mathrm{TEQ}_{(\mathrm{PCDD/Fs})} = \sum_{i=1}^{17} \left(c_i \times \mathrm{TEF}_i \right) \tag{1.5}$$

where c_i is the concentration of the i-th 2,3,7,8-chlorine-substituted DXN congener in ng/m^3, and TEF_i is the corresponding toxic equivalency factor for the i-th congener.

The European Union stipulates an emission limit for DXN in the flue gas from MSWI processes at 0.1 ng I-TEQ/Nm^3. This limit has been adopted as an environmental emission standard by several developing countries, including China.

1.2.2 Theoretical Mechanism Research

Owing to its complex molecular structure and significant bioaccumulative toxicity, the DXN emission control has remained a pivotal scientific challenge in the MSWI process. Numerous studies have demonstrated that DXN is not directly derived from the organic components present in raw MSW. Instead, it is generated through a series of pyrolysis, recombination, and catalytic reactions during the multi-stage MSWI process. Therefore, an in-depth elucidation of its formation and transformation mechanisms constitutes the foundation for achieving source suppression and whole-lifecycle emission reduction. Building upon this foundation, this section will systematically review the current research status and key influencing factors pertaining to the decomposition, re-synthesis, adsorption, and emission mechanisms of DXN.

1.2.2.1 Formation and Decomposition Mechanisms of DXN

In a typical MSWI plant employing a mechanical grate system, MSW undergoes four sequential stages along the grate: drying, volatiles release, fixed carbon combustion, and burnout. This process primarily occurs within the solid-phase combustion zone, as illustrated in Figure 1.4.

During this stage, trace amounts of DXN inherently present in the MSW are released via pyrolysis. Concurrently, the incomplete combustion of some organic materials under local oxygen-deficient conditions generates a variety of DXN precursors, often referred to as products of incomplete combustion (PICs). Typical PICs include chlorobenzenes, chlorophenols, polycyclic aromatic hydrocarbons (PAHs), and polychlorinated biphenyls (PCBs). The primary formation mechanisms of these precursors are elaborated below.

In local reducing zones within the furnace, the complex organic components of MSW undergo pyrolysis to yield low-molecular-weight aliphatic hydrocarbons (such as ethylene, butene, and acetylene). These compounds serve as critical precursors for the formation of aromatic structures. These aliphatic hydrocarbons undergo a series of reactions in a high-temperature environment, including bond cleavage, polymerization, dehydrogenation, and cyclization. These processes progressively lead to the formation of benzene ring structures. A representative formation pathway is illustrated in Figure 1.5.

With increasing temperature and reaction progress, the generated benzene ring structures can form more complex PAHs through intermolecular condensation and dehydrogenation reactions. In chlorine-rich atmospheres, PAHs can undergo further

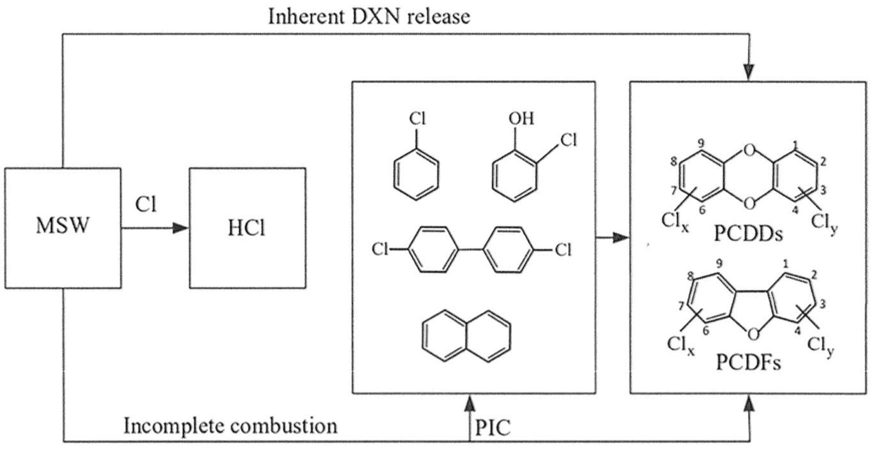

FIGURE 1.4 Schematic of DXN pyrolytic release and precursor formation in the solid-phase combustion zone.

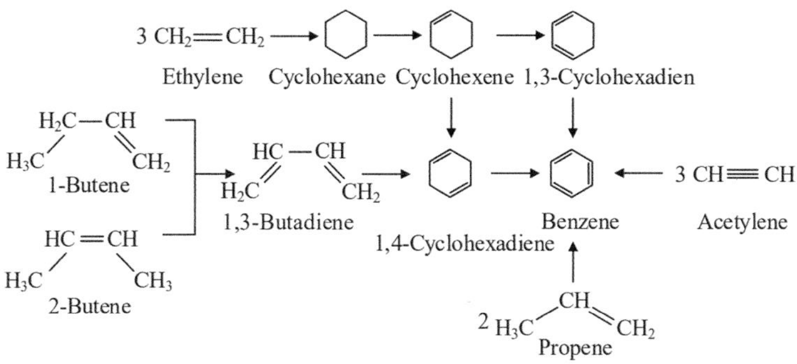

FIGURE 1.5 Mechanism of benzene formation from aliphatic hydrocarbons.

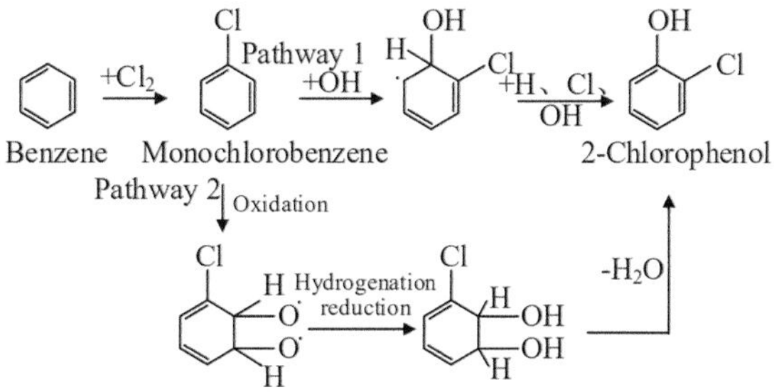

FIGURE 1.6 Mechanism of chlorophenol formation from benzene and chlorobenzene.

chlorination or oxidation reactions to form chlorinated polycyclic aromatic hydrocarbons (Cl-PAHs), thereby providing a stable aromatic framework for the subsequent formation of DXN.

Simultaneously, in oxidative atmospheres, benzene can undergo electrophilic substitution to form chlorobenzene. At higher temperatures or under free radical conditions, chlorobenzene can be further oxidized to yield chlorophenols. The primary reaction pathways for these transformations are illustrated in Figure 1.6.

Furthermore, under oxygen-deficient or high-temperature reducing conditions, chlorobenzene or chlorophenol molecules can undergo radical coupling and dechlorination-condensation reactions to form PCBs. These PCBs share a basic structural similarity with PCDFs.

Ultimately, these precursors, along with any incompletely decomposed dioxin molecules, enter the gas-phase combustion zone of the furnace. There, at high temperatures between 800°C and 1,000°C, they are oxidized and decomposed into CO_2 and HCl, thereby completing the DXN destruction process, as depicted in Figure 1.7.

1.2.2.2 Re-synthesis Mechanisms of DXN

Multiple re-synthesis pathways for DXN exist along the flue gas cooling section. Early research predominantly focused on high-temperature gas-phase pathways. However, under most actual operating conditions, it is widely acknowledged that

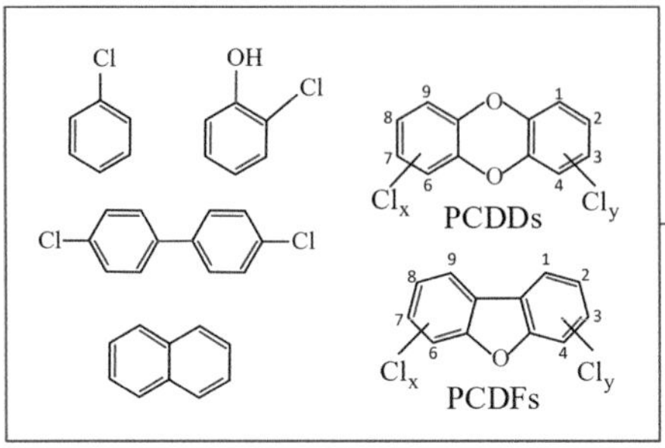

FIGURE 1.7 Schematic of high-temperature DXN decomposition.

once the flue gas enters medium-to-low-temperature zones containing fly ash, surface-catalyzed precursor transformation pathways and *de novo* synthesis pathways are considered the predominant sources of DXN re-synthesis. Accordingly, the following subsections sequentially review the current research status in terms of three key pathways, i.e., high-temperature gas-phase reactions, precursor catalytic reactions, and *de novo* synthesis reactions.

1.2.2.2.1 High-Temperature Gas-Phase Reactions

High-temperature gas-phase reactions represent one of the primary pathways for DXN re-synthesis in the MSWI process. This pathway specifically refers to the transformation reactions of chlorinated aromatic compounds under high-temperature gas-phase conditions. This reaction typically occurs in the high-temperature flue gas zone between 500°C and 800°C, which corresponds to the superheater heat exchange stage of the waste heat boiler.

In early research, Choudhry and Hutzinger [63] systematically investigated the formation mechanisms of DXN from a chemical perspective. They proposed that PCBs could generate PCDFs; that the pyrolysis of chlorophenols, phenoxyacetic acids, and their salts, as well as phenyl-2-phenyl ketones, yields PCDDs; and that the pyrolysis of PCDEs and chlorobenzenes produces mixtures of DXN. Subsequently, Ritter et al. [64] elucidated that PCBs, PCDEs, and PCDFs can yield PCDDs through thermal transformation under high-temperature gas-phase conditions. Building upon this foundation, Ballschmiter et al. [65] summarized the principal pathways for DXN formation in flue gas from the perspective of classical organic chemistry. These include: (1) cyclization of PCBs; (2) cyclization of PCDEs; (3) chlorination of dibenzofuran (DF); and (4) dechlorination of octachlorodibenzofuran (OCDF). However, Stanmore's review [66] has suggested that high-temperature gas-phase reactions may not constitute the dominant source of DXN in incinerators [67], and their actual contribution requires further in-depth assessment.

Furthermore, utilizing isotopic tracing experiments, Wikström et al. [68] demonstrated that gaseous chlorine species are the critical driver for DXN formation via high-temperature gas-phase reactions. Within the temperature range of 500°C–800°C, molecular chlorine (Cl_2) and HCl are activated via radical chain reactions to generate chlorine radicals, which subsequently react with organic intermediates to form DXN.

Research indicates that within this temperature interval, high-temperature gas-phase formation accounts for the majority of DXN generation in the superheater section of the waste heat boiler, with minimal contributions from solid- and liquid-phase pathways.

Current research suggests that the oxidative coupling of chlorophenols is considered the predominant pathway in high-temperature gas-phase reactions. In addition, DXN can also be formed from other precursors such as chlorobenzenes, hydroquinones, PCBs, and PAHs. In summary, this process is essentially one of coupling and cyclization reactions involving chlorinated aromatic radicals under oxidative and chlorinating conditions. This encompasses four major mechanistic types, which are detailed as follows.

1.2.2.2.1.1 Mechanism for DXN Formation from Chlorophenols Weber and Hagenmaier [69] observed that introducing O_2 into 2-monochlorophenol (2-MCP) significantly enhanced its conversion to DXN. Based on this finding, they proposed a formation mechanism that has gained wide acceptance within the academic community. This work established chlorophenols as one of the primary precursors for DXN generation under high-temperature gas-phase conditions [70].

The mechanisms for DXN formation from 2-MCP are illustrated in Figures 1.8 and 1.9, respectively.

Under normal operating conditions, the concentration of chlorophenols in flue gas typically ranges from 800 to 46,600 ng/Nm^3, which is 2–3 orders of magnitude

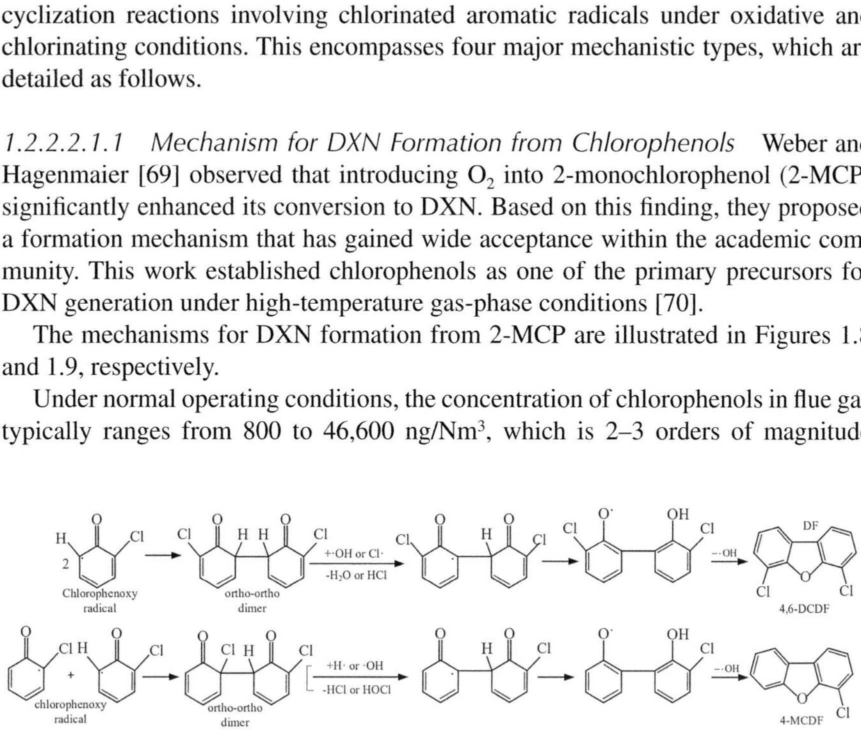

FIGURE 1.8 Mechanism of PCDFs formation from 2-MCP.

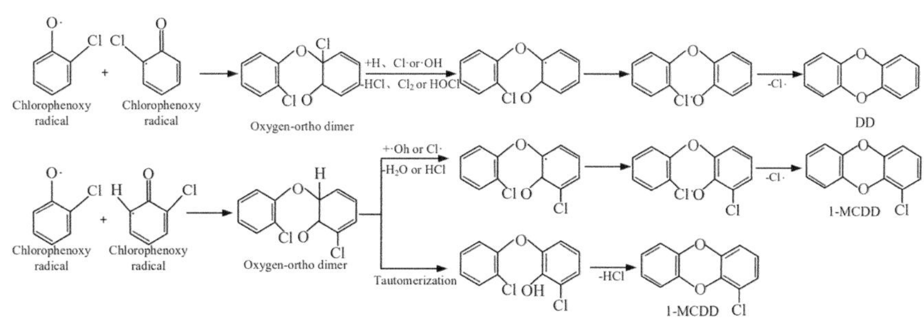

FIGURE 1.9 Mechanism of PCDDs formation from 2-MCP.

higher than that of DXN [71]. Among these, 2,4,6-trichlorophenol (2,4,6-TrCP) and pentachlorophenol (PCP) are present at the highest levels [72]. Chlorophenols originate from multiple pathways, including the oxidation of chlorobenzenes, the decomposition of oxygenated PAHs (at approximately 250°C [73]), the cyclization of small organic molecules (between 275°C and 350°C [74]), and the recombination of carbon-based radicals (between 150°C and 400°C [75]). The mechanism for their transformation into DXN generally proceeds in three stages: (1) formation of chlorophenoxy radicals from chlorophenols; (2) oxidative coupling of chlorophenoxy radicals to yield intermediates; and (3) cyclization of the intermediates to form DXN. A detailed description follows.

(a) Formation of Chlorophenoxy Radicals from Chlorophenols: The primary pathways for this transformation can be categorized into three types, as illustrated in Figure 1.10.

As illustrated in Figure 1.10, the first type involves a bond cleavage reaction of the O – H bond under pyrolytic conditions [76]. The second type is an oxidation reaction. This is attributable to the fact that the bond energy of the hydroxyl O – H bond is lower than that of the aromatic C – H bond; in oxygen-rich atmospheres, molecular oxygen can abstract the hydroxyl hydrogen atom, thereby generating a phenoxy radical and H_2O [77]. The third type is a chlorination-substitution reaction. The rationale here is that when a chlorine radical substitutes for the hydrogen atom in the hydroxyl group, a chlorophenoxy radical and HCl are produced. Studies indicate that these three mechanistic types can occur in parallel under high-temperature flue gas conditions, with their relative contributions determined by the prevailing redox conditions and chlorine source concentrations.

In addition to chlorophenols, hydroquinone compounds (e.g., 1,4-dihydroxybenzene and 1,2-dihydroxybenzene) are also important potential precursors for DXN formation. These compounds can be produced from the thermal degradation of biomass such as lignin and cellulose and are capable of generating phenoxy radicals via radical pathways. Under both oxygen-rich and oxygen-deficient conditions, the mechanism for DXN formation from 1,4-dihydroxybenzene is similar to that from chlorophenols [78]. The key distinction lies in the formation of the phenoxy radicals, as illustrated in Figure 1.11.

(b) Following their formation, Chlorophenoxy radicals undergo oxidative coupling reactions in an oxidizing atmosphere [79]. Due to the resonance effect, the reactive sites of phenoxy radicals can be distributed across the

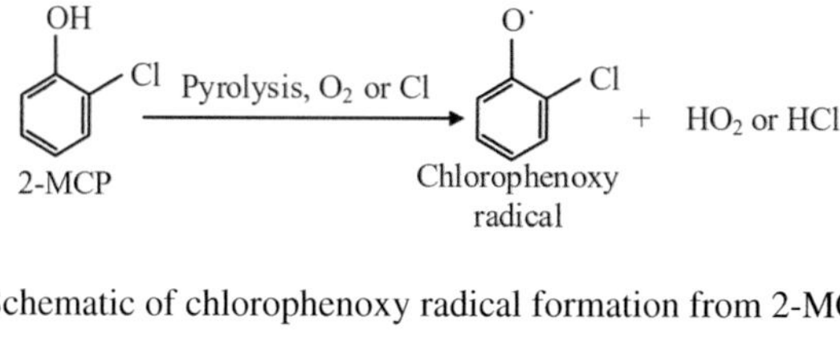

FIGURE 1.10 Schematic of chlorophenoxy radical formation from 2-MCP.

oxygen atom (i.e., an oxygen-centered radical) and the *ortho* and *para* carbon atoms, corresponding to the three resonance contributors depicted in Figure 1.12. Consequently, the coupling of these radicals can proceed through three distinct orientations, leading to the six classes of dimers illustrated in Figure 1.13.

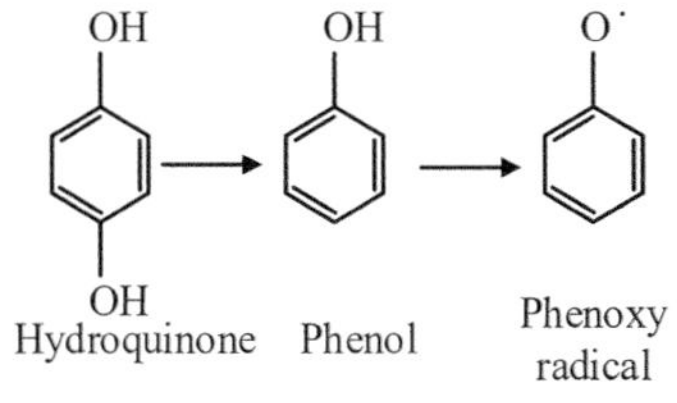

FIGURE 1.11 Process of phenoxy radical formation from 1,4-dihydroxybenzene.

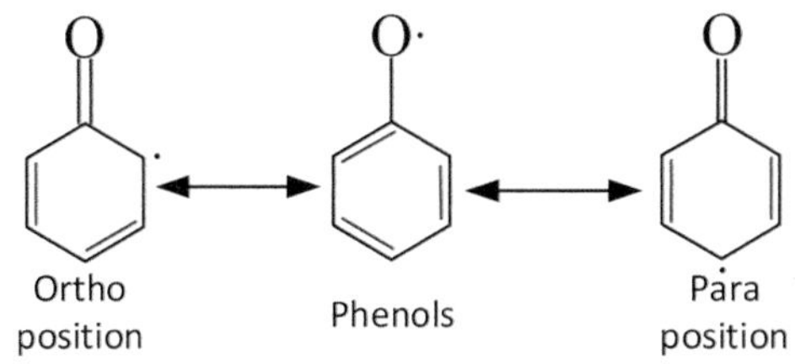

FIGURE 1.12 Resonance structures of the phenoxy radical.

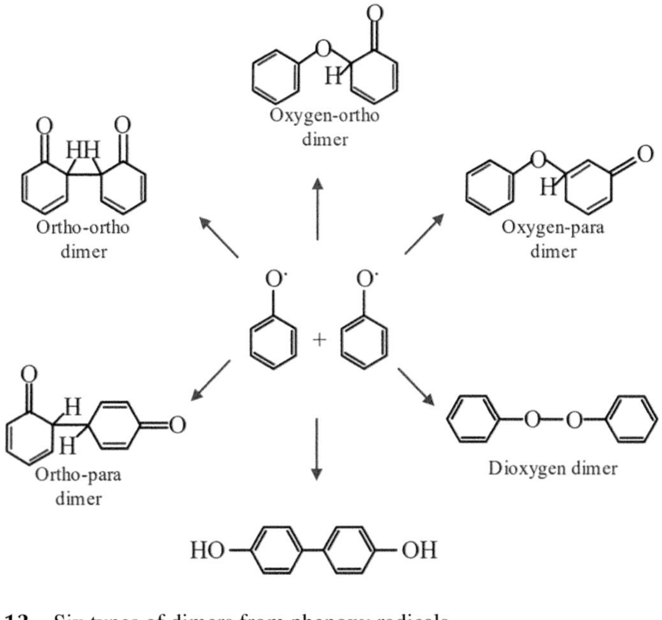

FIGURE 1.13 Six types of dimers from phenoxy radicals.

Based on charge density calculations, Janoschek et al. [80] reported a spin density distribution of approximately 41.3% on the oxygen atom, 39.1% on the *ortho* carbons, and 19.6% on the *para* carbon of the phenoxy radical. Molecular orbital theory posits that sites with higher spin density are more prone to coupling, suggesting that *para-para* dimers should be dominant [81]. However, experimental results show that *ortho-ortho* dimers prevail [82]. Subsequently, *ortho-ortho* dimers and oxygen-*ortho* dimers can respectively transform into dihydroxybiphenyls (DOHBs) and *ortho*-phenoxyphenols (POPs), which serve as the direct precursors for PCDFs and PCDDs, respectively (Figure 1.14). As the degree of chlorination increases, the tendency to form polychlorinated POPs is enhanced.

(c) During the cyclization stage, the DOHB and POP intermediates undergo intramolecular dehydration and oxidative rearrangement reactions to yield DXN. The formation of PCDFs primarily originates from the DOHB structure generated by the coupling of two radicals. The dehydration of DOHB can yield a limited amount of dibenzofuran (DF). Notably, the conversion rate of DF is significantly enhanced when a methyl group capable of abstracting phenolic hydrogen is introduced. This process is schematically illustrated in Figure 1.15.

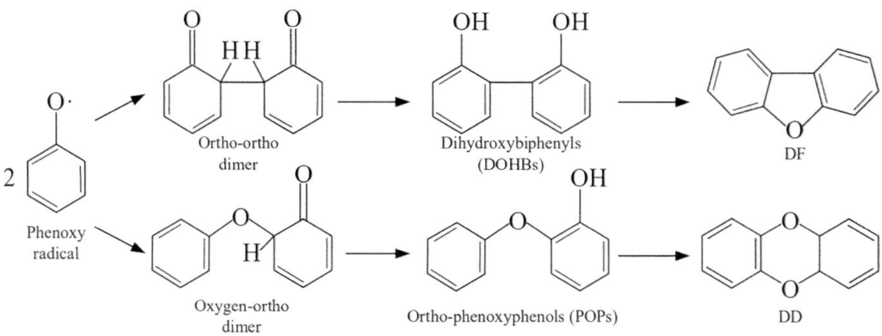

FIGURE 1.14 Schematic of DOHB and POP formation from phenoxy radicals.

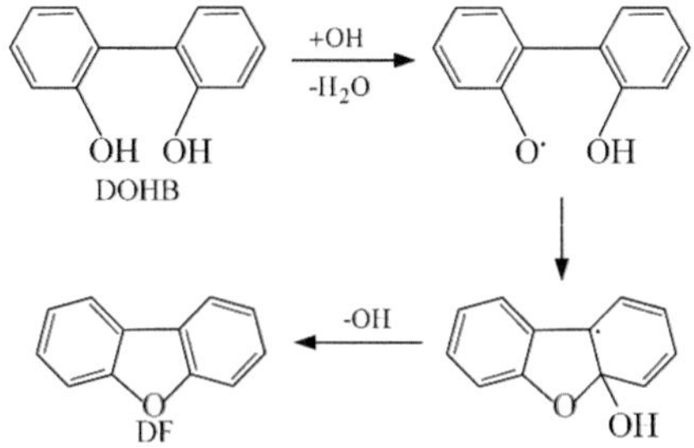

FIGURE 1.15 Schematic of DOHB transformation to DF.

In contrast, the formation of PCDDs involves three coupling pathways: radical/radical, molecule/radical, and molecule/molecule, with the radical/radical mechanism predominating [83]. Furthermore, during chlorinated condensation, at least one of the two chlorophenols forming PCDDs must possess an *ortho*-chlorine substituent, while at least one of the two forming PCDFs must retain an *ortho*-hydrogen.

1.2.2.2.1.2 Mechanism for DXN Formation from Benzene and Chlorobenzene Benzene and chlorobenzene compounds can also generate DXN under high-temperature oxidative conditions. The underlying mechanism bears similarities to the chlorophenol pathways but involves distinct precursors and key intermediates. Research indicates that in oxygen-rich atmospheres, chlorobenzene is more prone to undergo oxidative coupling reactions, yielding PCDFs and PCBs. Conversely, under reducing or oxygen-deficient conditions, chlorination reactions predominate, leading mainly to highly chlorinated benzenes [84].

Taking monochlorobenzene (MCB) as an example, under oxygen-rich conditions, it can undergo oxidation to yield 4-monochlorodibenzofuran (4-MCDF) and 2,2′-dichlorobiphenyl (2,2′-DCB). The corresponding reaction pathways proceed as follows. First, heating induces the homolytic cleavage of the weakest C – Cl bond in MCB, generating a chlorine radical and a chlorophenyl radical. Subsequently, this chlorophenyl radical can follow two distinct fates. On one hand, it can undergo radical coupling with another MCB molecule to form 2,2′-DCB. On the other hand, in the oxygen-rich atmosphere, it can be oxidized to a chlorophenoxy radical. Finally, this chlorophenoxy radical reacts further with MCB and undergoes intramolecular cyclization to yield 4-MCDF. Consequently, under oxidative conditions, chlorobenzenes with varying degrees of chlorination can all serve as precursors participating in the pathways described above. The resulting product profile is predominantly PCDFs, with PCDDs typically constituting less than 10% of the total [85].

1.2.2.2.1.3 Mechanism for DXN Formation from Polychlorinated Biphenyls (PCBs) Under high-temperature gas-phase conditions, the oxidative cyclization of PCBs is widely recognized as a significant pathway for DXN re-synthesis, with PCDFs being the primary products [86]. Based on the reaction mechanism, the potential formation of specific PCDF isomers can be predicted from the structural features of the parent PCBs [87]. Research indicates that the PCDD/PCDF formation ratio is closely correlated with reaction temperature, the position of hydroxyl groups in chlorophenols, and oxygen concentration. Bani et al. [88] proposed a nonlinear kinetic model that considers both chlorine atom count and process conditions to characterize the complex kinetic behavior of PCB incineration systems. In isothermal flow reactor experiments designed to minimize the influence of wall catalysis, researchers investigated the pyrolysis and oxidative reaction characteristics of 4,4′-dichlorobiphenyl (4,4′-DCB) as a model compound over a temperature range of 400°C–800°C. The experimental results demonstrated that the primary product formed was 3,7-dichlorodibenzofuran (3,7-DCDF) [89]. The conversion of 4,4′-DCB to 3,7-DCDF proceeds through three distinct stages. Firstly, a hydrogen atom is abstracted from the biphenyl ring by molecular oxygen, forming a highly reactive peroxide-bound intermediate complex. Subsequently, homolytic cleavage of the

O – O bond within this complex triggers a radical rearrangement, yielding a phenoxy-type intermediate. Finally, this intermediate undergoes intramolecular cyclization to form 3,7-DCDF.

1.2.2.2.1.4 Mechanism for DXN Formation from Polycyclic Aromatic Hydrocarbons (PAHs) In high-temperature gas-phase reactions, certain PAHs can serve as precursors for DXN re-synthesis. Among these, anthracene is one of the most abundant PAHs in the atmosphere, with a concentration of approximately 40 ng/m^3. In developing countries such as China, anthracene accounts for about 12.5% of the total PAH concentration [90, 91]. Due to its molecular structure and reactivity, anthracene is widely regarded as an important potential precursor for dibenzofuran (DF) and dibenzo-*p*-dioxin (DD) [92].

The oxidative transformation of anthracene can be simplified into a three-step sequence. First, a hydrogen abstraction process generates an anthryl radical. Subsequently, this radical is oxidized to anthrone, which is further converted to xanthone and xanthene. Finally, under high-temperature oxidative conditions, these oxygenated intermediates undergo dehydration and cyclization reactions to yield DF and DD, respectively. Notably, xanthone can serve as a precursor for both DF and DD, whereas xanthene leads exclusively to DD. It is important to note that the schematic presents a simplified mechanism; in actual reaction systems, both the hydrogen abstraction and oxidation steps involve the participation of radicals (such as $^{\cdot}OH$, $^{\cdot}O$, etc.).

1.2.2.2.2 Catalytic Reactions of Precursors

Precursor molecules undergo heterogeneous reactions on the surface of fly ash, catalyzed by metal oxides, leading to the formation of DXN. The key factors influencing this process include the type of precursor, the type/oxidation state of the catalyst, and the reaction temperature.

Regarding precursors, typical potential precursors include chlorophenols, chlorobenzenes, and PCBs. The conversion of chlorophenols to DXN is analogous to Ullmann-type reactions, representing a metal-catalyzed aromatic coupling reaction. Under oxygen-deficient conditions, approximately 0.5% of 2-monochlorophenol (2-MCP) is converted to DXN [93]. In the synthesis experiments conducted by Cains et al. using chlorophenols as precursors, only PCDDs were detected in the products [94]. Tuppurainen et al. further elucidated that the mechanisms for precursor-derived PCDDs and PCDFs are distinct [95]. PCDDs primarily originate from surface-catalyzed chlorophenol coupling reactions, followed by multiple steps such as ring closure. In contrast, PCDFs are more likely generated via ring-closure reactions of chlorobenzenes and PCBs under metal catalysis. Therefore, from a mechanistic propensity perspective, the precursor catalytic pathway favors the formation of PCDDs over PCDFs.

Regarding the necessity of catalysis, experimental evidence has demonstrated that 2-MCP can also be catalytically oxidized to form DXN on SiO_2, but its yield is approximately one order of magnitude lower than that achieved with metal oxide catalysts [96]. This finding demonstrates that even in the absence of transition metals, fly ash particles still exhibit a certain catalytic effect on DXN formation, but the catalytic effect is significantly more pronounced with transition metal oxides [97, 98]. Among these, cupric oxide (CuO) exhibits a higher yield in catalyzing DXN formation compared to other metal oxides, and surpasses cuprous oxide (Cu_2O) and elemental copper [99]. Oxides such as CuO, ZnO, MnO_2, TiO_2, and Co_3O_4 generally demonstrate a promotional effect, whereas others, including CaO, Al_2O_3, Fe_2O_3, and PbO, are frequently reported as inhibitors [97].

Regarding temperature effects, in experiments on the catalytic formation of DXN from 2,4,6-trichlorophenol (2,4,6-TrCP) and 2,3,4,6-tetrachlorophenol (2,3,4,6-TeCP), the yield of PCDDs peaks at approximately 300°C, whereas that of PCDFs peaks at around 350°C. Notably, the maximum yield of PCDFs is significantly lower than that of PCDDs [100]. Karasek et al. found that the optimum temperature for DXN formation from 2,4,6-TrCP was approximately 300°C, while the generation from pentachlorophenol (PCP)-containing fly ash was notably higher within the temperature range of 250°C–350°C [101]. Furthermore, PCP was found to predominantly generate octachlorodibenzo-p-dioxin (OCDD) at 280°C, accompanied by smaller amounts of 1,2,3,4,6,7,8- and 1,2,3,4,6,7,9-heptachlorodibenzo-*p*-dioxins (1,2,3,4,6,7,8-/1,2,3,4,6,7,9-HpCDD) [97]. These findings collectively underscore the decisive influence of the temperature window on both the reaction selectivity and final yield.

In summary, under metal oxide catalysis and at appropriate temperatures, the heterogeneous pathway for DXN formation from precursors can be conceptualized as three sequential stages: (1) adsorption of precursors onto the catalyst surface, (2) condensation of precursors to yield adsorbed DXN, and (3) desorption of DXN.

The following discussion utilizes 2-MCP as a representative precursor to elucidate the key steps and intermediate evolution within these three stages.

1.2.2.2.2.1 Adsorption of Precursors onto the Catalyst Surface The chemical adsorption of 2-MCP on the metal oxide surface can occur via two distinct pathways [102], the mechanisms of which are illustrated in Figure 1.16.

As shown in Figure 1.16, there are two pathways. Details are as follows.

In Pathway 1, 2-MCP loses a water molecule (dehydration) under the influence of the metal catalyst surface and adsorbs onto it, forming an initial intermediate. This intermediate undergoes transfer of a radical electron, followed by the elimination of an OH group, leading to a chlorophenate-type intermediate. This chlorophenate

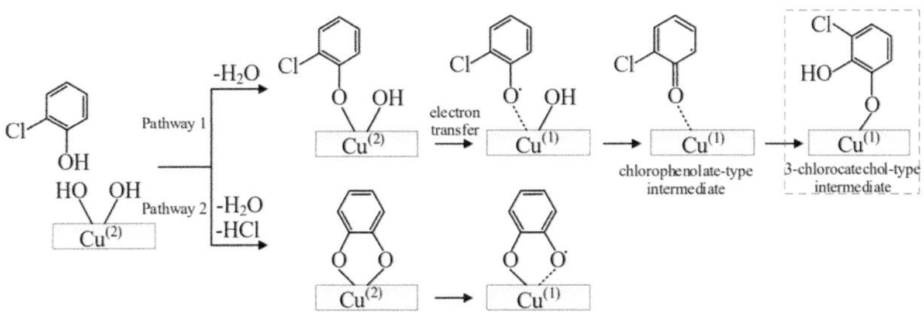

FIGURE 1.16 Schematic of the 2-MCP adsorption process on catalyst surfaces.

species represents the form of 2-MCP adsorbed on the catalyst according to the Langmuir-Hinshelwood (LH) mechanism. Within the Eley-Rideal (ER) mechanism framework, after a radical electron transfer, the *ortho* position of the benzene ring binds to the metal catalyst. This is followed by cleavage of the metal-oxygen bond and transfer of a hydrogen atom, resulting in the formation of a 3-chlorocatechol-type intermediate.

In Pathway 2, 2-MCP loses both H_2O and HCl (dehydrochlorination) under the influence of the metal catalyst surface and adsorbs to form an initial intermediate. This intermediate subsequently undergoes a radical electron transfer, resulting in the formation of a paramagnetic intermediate. It is generally accepted that Pathway 1 is favored at lower temperatures, whereas Pathway 2 predominates under higher temperature conditions [103].

1.2.2.2.2.2 Condensation of Precursors to Yield Adsorbed DXN During the condensation stage, the activated precursors adsorbed on the catalyst surface evolve along two typical pathways, as depicted in Figure 1.17.

As depicted in Figure 1.17, under the LH mechanism, two chlorophenoxy radicals in the chlorophenate form, adsorbed on the metal oxide surface, undergo *ortho*-coupling to form a ketone-ketone bridged intermediate. This intermediate subsequently undergoes rearrangement and cyclization to yield adsorbed 4,6-dichlorodibenzofuran (4,6-DCDF). Under the ER mechanism, surface-adsorbed 3-chlorocatechol interacts heterogeneously with gaseous 2-MCP. This interaction, accompanied by the elimination of two HCl molecules, leads to the formation of an adsorbed precursor structure for DD.

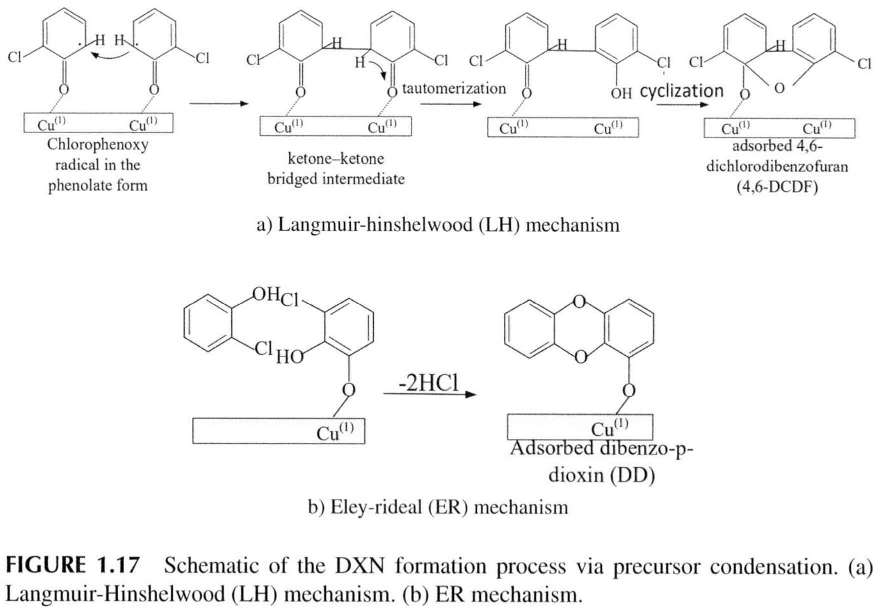

FIGURE 1.17 Schematic of the DXN formation process via precursor condensation. (a) Langmuir-Hinshelwood (LH) mechanism. (b) ER mechanism.

1.2.2.2.2.3 Desorption of DXN The DXN desorption process is illustrated in Figure 1.18.

As illustrated in Figure 1.18, within the LH mechanism, adsorbed 4,6-DCDF desorbs into the gas phase via cleavage of the carbon-oxygen bond linking it to the copper site on the catalyst surface. In the ER mechanism, the adsorbed DD intermediate can evolve through two distinct pathways: it may either desorb directly into the gas phase under reducing conditions or, under oxygen-deficient conditions, undergo further chlorination to form other PCDD congeners prior to desorption.

1.2.2.2.3 De Novo Synthesis Reactions

Stieglitz and Vogg first proposed the *de novo* synthesis mechanism, positing that elements such as carbon, hydrogen, oxygen, and chlorine can undergo a series of elementary reactions on fly ash surfaces to directly form DXN [104]. This process typically occurs during the flue gas cooling stage (200°C–400°C) and is considered one of the primary pathways for DXN re-synthesis in the post-combustion zone. Subsequent research has further demonstrated a significant correlation between the concentration of entrained fly ash in the gas phase and DXN levels [105]. Building upon this foundation, Gullett et al. delineated the *de novo* synthesis process into three consecutive stages [106]. First, during the combustion stage, chlorine sources in the flue gas are released as HCl, which then reacts with O_2, catalyzed by metal oxides via the Deacon reaction, to form Cl_2. Subsequently, the generated Cl_2 undergoes substitution or addition reactions with the carbonaceous matrix on the fly ash surface,

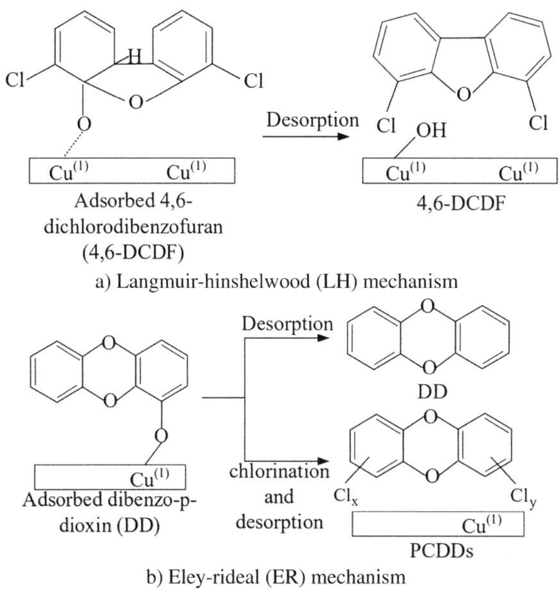

FIGURE 1.18 Schematic of the DXN desorption process. (a) Langmuir-Hinshelwood (LH) mechanism. (b) ER mechanism.

yielding typical precursors such as chlorobenzenes, chlorophenols, and PCBs. Finally, these precursors, catalyzed by metal oxides, undergo oxidative condensation and intramolecular cyclization reactions to form DXN with a bicyclic aromatic framework.

Further advancing the understanding, Zhao et al. have summarized three principal pathways for the *de novo* synthesis reactions [107], as outlined in Figure 1.19.

As depicted in Figure 1.19, the *de novo* synthesis reactions proceed via three distinct routes. The first one involves the structural decomposition of residual carbon, which yields benzene that subsequently undergoes chlorination to form chlorophenols. The second one is that the inorganic chlorides are transferred to the macromolecular structure of residual carbon under the action of transition metal chlorides. The residual carbon is then disrupted via gasification reactions with oxygen, leading to the formation of aliphatic precursors [108], which are subsequently converted to chlorophenols. The third one is that the pre-existing phenols are chlorinated to yield chlorophenols, which are ultimately catalytically converted to DXN. The Cl_2 required for the chlorination reactions in routes (1) and (3) can be supplied by the Deacon reaction.

Overall, *de novo* synthesis is a multiphase coupled process involving gaseous chlorine sources, a carbonaceous matrix, and metal catalysts. Studies show that factors such as temperature, residence time, carbon and chlorine content, particle size, catalyst type, and reaction atmosphere all significantly influence DXN formation. The following sections will review research related to these key influencing factors separately.

1.2.2.2.3.1 Temperature The temperature of the MISW Process is widely regarded as the most critical parameter influencing DXN formation [109]. Research related to *de novo* synthesis often employs fixed-bed experimental setups, with the reaction temperature typically set within the range of 270°C–400°C to simulate the typical temperature conditions of the cooling zone in incineration flue gas. It indicates that the yield of DXN from *de novo* synthesis reactions typically peaks within the temperature range of 300°C–370°C. In a comparative study of fly ash from

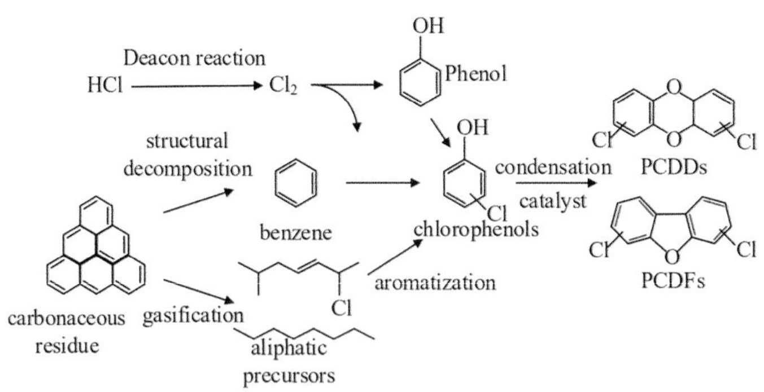

FIGURE 1.19 Mechanism of de novo synthesis.

different sources, Chang and Huang found that the optimal temperatures for DXN formation varied with the ash type [110]: approximately 350°C for boiler ash, 300°C for cyclone ash, and 250°C for electrostatic precipitator ash. Furthermore, Schwarz and Stieglitz [111] and Zhang et al. [112] noted a decrease in the average chlorination level of DXN with increasing temperature [111, 112]. Multiple *de novo* synthesis experiments also demonstrated that when the temperature exceeds 300°C, the PCDF/PCDD ratio increases significantly [110, 111, 113]. Thus, elevated temperatures appear to favor the formation of PCDFs [104].

1.2.2.2.3.2 Time (Residence Time) Compared to the transformation pathways involving chlorophenol precursors, which typically require only a few minutes at 300°C, *de novo* synthesis generally necessitates a longer residence time. Experiments by Weber indicated that a period of approximately 30 minutes is required to observe significant DXN generation via the *de novo* route [114]. Furthermore, another study conducted at 300°C with a 30-minute reaction time revealed that the formation rate of PCDFs was higher than that of PCDDs. This dynamic is reflected in the observed increase of the PCDD/PCDF ratio from 0.5 to 3 over the course of the reaction.

1.2.2.2.3.3 Carbon The carbon matrix serves a dual role in the *de novo* synthesis process, acting as both the primary reaction substrate and a catalytic support. Under oxygen- and chlorine-containing atmospheres within the 200°C–400°C range and in the presence of metal sites, its surface undergoes concurrent and/or alternating processes of oxidation and chlorination. These processes are accompanied by bond cleavage and rearrangement, progressively leading to the formation of aromatic frameworks and oxygen-/chlorine-containing functional groups. These, in turn, serve as the foundation for generating various DXN precursors. Specifically, the former provides the necessary carbon source for DXN formation, while the latter, by forming a synergistic interface with metal oxides, facilitates the oxidation and chlorination reactions. By comparing four fly ash samples (three from MSWI plants and one from a coal-fired power plant) on the potential of different carbon sources in promoting carbon gasification and DXN generation from native carbon, it showed that the gasification rate of native carbon in MSW fly ash was at least an order of magnitude higher than that of pure carbon, indicating the presence of a significant catalytic mechanism within the fly ash. In contrast, coal fly ash exhibited low activity in both DXN generation and carbon gasification. Furthermore, research by Chang and Huang [110] revealed a positive correlation between DXN formation yield and carbon content within the range of 0 to 5% in the flue gas.

1.2.2.2.3.4 Chlorine The chlorination pathway plays a crucial role in *de novo* synthesis and is one of the essential prerequisites for DXN formation. Literature reported that the chlorine content in fly ash ranges from 0.20 to 34.27 mg/g and indicated that the non-extractable organic chlorine (NEOCI) within fly ash serves as the primary chlorine source for DXN generation [115]. It is hypothesized that the activation and catalytic role of chlorine may be more critical than its mere abundance.

1.2.2.2.3.5 Catalyst The catalytic role of transition metal chlorides or oxides is paramount in *de novo* synthesis reactions. Among them, $CuCl_2$ is widely regarded as the most potent catalyst [116, 117]. It shows that the dynamic changes and kinetic modeling of copper during the *de novo* process elucidate the relationship between different chemical forms of copper and DXN formation. Furthermore, Luijk et al. conducted *de novo* synthesis studies at 300°C by varying the mass fraction of $CuCl_2$ (0%, 0.1%, 0.5%, 1.0%, 5.0%) [118]. Their results showed that the PCDDs/PCDFs content ratio decreased from 33 to 0.2. This indicates an inverse correlation between $CuCl_2$ mass fraction and PCDDs yield, and a positive correlation with PCDFs yield. Research in the Literature [119] demonstrated that heating fly ash to 300°C induces significant changes in the chemical speciation of copper, i.e., approximately 45% as CuCl, 29% as Cu, and 26% as Cu_2O. When the temperature was increased to 400°C, the distribution shifted to 61%, 20%, and 19%, respectively. Taking $CuCl_2$ as an example, it is reduced to CuCl during the chlorination process, simultaneously forming a C–Cl bond. The reaction is as follows:

$$2CuCl_2 + R?\ H \rightarrow 2CuCl + R?\ Cl + HCl \tag{1.6}$$

In addition to copper, chlorides or oxides of other metals, including Ca, Cd, Fe, Hg, Mg, Mn, Ni, Pb, Sn, Zn, K, and Cr, also exhibit varying degrees of catalytic activity in *de novo* synthesis reactions [120–124]. Stieglitz et al. investigated the catalytic role of $FeCl_3$ using a model fly ash [125]. Their results showed that at 350°C, the formation yields of PCDDs and PCDFs were 3,369 and 6,594 ng/g of fly ash, respectively. Through a comparative analysis of the elemental composition of fly ash and chlorinated aromatics, Öberg et al. suggested that Cr and Ni may facilitate the progress of the *de novo* synthesis [124].

Studies indicate significant variations in the catalytic activity among different metals. For instance, the catalytic activity of $ZnCl_2$ is approximately 5%–10% of that of $CuCl_2$ [120]. Fujimori et al. conducted comprehensive testing on 11 distinct samples and ranked their catalytic capability [121], which are as follows:

$$\begin{aligned} &CuCl_2 \cdot 2H_2O > Cu_2(OH)_3Cl > FeCl_3 \cdot 6H_2O > FeCl_2 \cdot 4H_2O \\ &> CuO > Fe_2O_3 > PbCl_2 > \text{Blank} > ZnCl_2 > PbO > ZnO \end{aligned} \tag{1.7}$$

1.2.2.2.3.6 Reaction Atmosphere *De novo* synthesis is highly sensitive to the gas-phase composition. Gaseous species such as O_2, H_2O, HCl, and Cl_2 can significantly alter both the reaction rate and product distribution. Their roles can be conceptualized within a framework of three coupled pathways, i.e., oxygen supply, catalyst regeneration, and chlorination.

Firstly, oxygen plays a multifaceted role in de novo synthesis. It serves a dual function. The first function is to oxidize the carbonaceous matrix, thereby providing oxygen-containing functional groups for precursor formation. The second function is to re-oxidize reduced metal sites while facilitating the conversion of HCl to Cl_2 through the Deacon reaction. Experimental evidence strongly supports the critical influence of oxygen concentration. As O_2 levels increased from 1% to 10%, Vogg et al. observed a dramatic rise in PCDDs (from 738 to 8,708 ng/g) and PCDFs (from 1,255 to 4,361 ng/g) in fly ash [126]. Similarly, Addink and Olie reported a marked increase in total DXN concentration with elevated oxygen content [127]. These findings collectively underscore that oxygen acts as a key driver for de novo synthesis, and that reducing its concentration can, to a certain extent, suppress DXN formation.

Secondly, moisture exhibits a dual influence on DXN formation. On one hand, as a product of the Deacon reaction, an elevated H_2O partial pressure shifts the equilibrium, thereby suppressing Cl_2 generation. On the other hand, H_2O can modify the surface acidity of catalyst supports, alter active sites, and influence the hydrolysis/rearrangement of chlorinated aromatics. Consequently, depending on the specific system conditions, the presence of water can either promote or inhibit DXN formation.

Finally, in most cases, HCl itself cannot directly chlorinate aromatic hydrocarbons. It must first be converted into Cl_2, either via the Deacon reaction or through redox cycles involving metals/metal salts, to become effective. Experimental results demonstrate that increasing the Cl_2 content in flue gas elevates the yield of highly chlorinated DXN congeners, thereby raising the total DXN output [128]. This clearly indicates that Cl_2 serves as the key active chlorine source responsible for chlorination in de novo synthesis.

1.2.2.3 Adsorption Mechanism of DXN

During the flue gas cleaning stage, the combined technology of activated carbon injection (ACI) followed by baghouse filtration has emerged as one of the cornerstone methods for DXN removal, owing to its operational simplicity, cost-effectiveness, and superior adsorption performance. The adsorption of DXN via this technology constitutes a complex multiphase mass transfer process, which is commonly summarized by researchers into the following two consecutive stages.

(1) In-duct Adsorption (Entrained-Flow Adsorption): Following injection into the flue gas duct by high-pressure gas, activated carbon (AC) particles become entrained in the turbulent gas stream. During this transport, DXN molecules in the flue gas are primarily adsorbed onto the AC particles. Additionally, a fraction of finer AC particles may deposit on the duct walls, contributing a secondary adsorption effect on DXN.

(2) Filter-Cake Adsorption: The flue gas, now laden with AC, enters the baghouse filter where particulate matter (including fly ash and AC) is intercepted by the filter bags, gradually accumulating to form an adsorptive filter cake on their surface. This cake functions analogously to a fixed-bed reactor, requiring the flue gas to permeate through it. Unsaturated AC within the cake layer continues to adsorb residual DXN from the gas stream over an extended contact time. Theoretically, adsorption efficiency increases with cake thickness. Consequently, a key operational strategy for enhancing overall DXN removal is to appropriately reduce the bag cleaning frequency, provided that the resulting increase in system pressure drop remains within permissible limits.

To quantitatively describe the aforementioned two-stage process, literature attributes the overall DXN removal efficiency to the combined effects of entrained-flow and filter-cake adsorption [129]. It proposes the following calculation models for the entrained-flow adsorption efficiency η_1 and the filter-cake adsorption efficiency η_2.

$$\eta_1 = 1 - \exp\left(-k_1\left(C_A + \frac{C_F}{60}\right)t_t A_s\right) \tag{1.8}$$

$$\eta_2 = 1 - \exp\left[-k_2\left(\rho_B \times 10^3\right)\left(f_A + \frac{f_F}{60}\right)t_f A_s f_s\right] \tag{1.9}$$

where k_1 and k_2 represent the adsorption coefficients for the entrained-flow and filter-cake stages, respectively; C_A is the injection rate of the adsorbent; C_F denotes the fly ash concentration in the flue gas; t_t is the contact time from the adsorbent injection point to the filter cake; A_s indicates the number of adsorption sites per unit mass of adsorbent at the operating temperature (the number of adsorption sites on fly ash is approximately 1/60 of that on AC [130]); ρ_B is the bulk density of the filter cake; f_A and f_F are the mass fractions of AC and fly ash within the cake, respectively; t_f is the gas contact time within the filter cake; and f_s represents the fraction of unoccupied adsorption sites.

Consequently, the overall DXN removal efficiency of the combined AC and baghouse filter system is related to the removal efficiencies achieved in the entrained-flow and filter-cake stages by the following expression:

$$1 - \eta_g = \left(1 - \eta_1\right)\left(1 - \eta_2\right) \tag{1.10}$$

Fundamentally, the adsorption of DXN onto AC is a process of molecular accumulation on the carbon surface. The efficiency of this process is governed by the combined influence of three factors, i.e., the physicochemical properties of the adsorbent, the characteristics of DXN itself, and the process operating parameters. Ultimately, the adsorption capacity is fundamentally determined by the intrinsic properties of both the adsorbent and the adsorbate.

Regarding the adsorbent, the pore structure of AC is considered a critical factor influencing its adsorption capacity for DXN [131], particularly the surface area contributed by mesopores and macropores. As for the adsorbate, the degree of chlorination of DXN molecules constitutes another key factor. It has been observed that the adsorption capacity approximately doubles with each additional chlorine substituent on the benzene rings [132]. This manifests as an increase in DXN removal efficiency with the number of chlorine substitutions, a trend that approximates a power function of the corresponding vapor pressure [133]. Although adsorption capacity varies with the degree of chlorination, congeners with the same number of chlorine substituents (e.g., the 17 toxic DXN) exhibit similar adsorption strengths [131, 134].

At the industrial process operational level, several parameters directly modulate adsorption efficacy. Since adsorption is an exothermic process, an increase in temperature under a given equilibrium pressure exerts an inhibitory effect, leading to a decline in DXN removal rate with rising temperature [134]. In contrast, prolonging the adsorption contact time or increasing the ACI rate can substantially enhance DXN removal [135]. Experimental work by Everaert et al. demonstrated an optimal DXN removal efficiency at ACI rates between 50 and 80 mg/Nm^3 [129]. Further supporting this, Chang et al. indicated that when the specific surface area of AC exceeds 1,050 m^2/kg, and the injection rate is greater than 70 mg/Nm^3, an adsorption efficiency above 85% can be achieved [136]. Under these conditions, the DXN concentration at the baghouse outlet can meet the emission limit of 0.1 ng I-TEQ/Nm^3. To quantitatively describe the combined influence of these factors, researchers have also pursued adsorption kinetics studies. For instance, Li et al. experimentally determined the adsorption isotherms and kinetics of DXN on AC [137, 138]. They found that an intraparticle diffusion model based on the Langmuir isotherm provided an excellent fit for the breakthrough curves.

Building upon the mechanistic understanding and parametric studies described above, the dual-baghouse filter system has emerged to further enhance cleaning efficiency and optimize operational costs. Research indicates that this configuration can achieve a DXN removal efficiency of up to 99.3%. Compared to a single-stage system, it reduces DXN emissions by a factor of more than four while concurrently decreasing AC consumption by approximately 40% [139, 140], representing substantial economic and environmental benefits.

1.2.2.4 Memory Effect of DXN

This section provides a detailed description of the impact of the memory effect on DXN emissions. It may occur at various locations within an MSWI plant.

1.2.2.4.1 Definition of the Memory Effect

The memory effect is defined as a delayed emission phenomenon associated with certain persistent organic pollutants (POPs). This effect is predominantly particulate-related. Particles are retained on the inner walls of flue gas ducts and the air pollution control (APC) system through mechanisms such as impaction, interception, or gravitational settling [141]. Under specific temperature ranges, prolonged exposure may induce significant desorption of adsorbed DXN from these captured particles, thereby elevating DXN concentrations in the final emissions. The memory effect primarily targets specific POPs, such as PAHs, PCBs, and DXN, which resist chemical or biological degradation and thus persist in the captured particulate phase [142, 143].

Specifically, the memory effect manifests as two fundamental phenomena [144]. The first one is the *adsorption-based memory effect*, which refers to fly ash particles deposited on the inner walls of flue gas ducts or the APC system acting as an adsorption substrate [143, 145, 146]. Under prolonged exposure within specific temperature ranges, previously adsorbed DXN desorbs and is released into the flue gas stream. The second one is the *de novo synthesis-based memory effect*, where residual carbon within the ash continuously generates DXN via the *de novo* synthesis mechanism, facilitated by catalysts. This effect is often triggered by non-steady-state operating conditions, such as incinerator startup or incomplete combustion, accompanied by high DXN emissions.

These memory effect mechanisms do not operate in isolation. They predominantly occur within key components of the flue gas treatment system. Due to significant variations in temperature, flow conditions, particulate loading, and catalyst types, different APC devices exhibit distinct sensitivities and manifestations of the DXN memory effect. As the core stages of the MSWI, the flue gas cleaning process, the baghouse filter, and the selective catalytic reduction (SCR) unit are of particular concern. Conducting an in-depth analysis of the DXN memory effect within these two devices is therefore crucial for achieving stable, compliant emissions from MSWI systems.

1.2.2.4.2 Baghouse Filter and the Memory Effect

The baghouse filter is a critical component within the flue gas treatment system, primarily responsible for the highly efficient removal of particulate matter, including fly ash and injected AC. AC tends to preferentially adsorb low-chlorinated DXN congeners, which can lead to the gradual accumulation of highly chlorinated homologs on the aged filter surface. These accumulated species may subsequently be released under conditions of lower gaseous DXN concentration. The memory effect becomes particularly pronounced during non-steady-state operational periods, such as incinerator startup or shutdown. Under these conditions, the DXN concentration at the baghouse outlet can even exceed that at the inlet. Supporting this observation, Li et al. [147] reported that in a baghouse filter after 49 months of service, the DXN concentration measured at the stack was higher than at the inlet, manifesting as an increase in the gas-phase DXN concentration.

Regarding the memory effect in baghouse filters under varying operating conditions, there are two primary viewpoints. The first one is the *low-temperature conditions (approximately 165°C)*. Under these conditions, the amount of DXN released from adhesion to duct walls or the baghouse filter itself is minimal. Consequently, the observed memory effect is likely attributable to the aging of the baghouse filter. The second viewpoint is the h*igh-temperature conditions (225°C–230°C)*. This temperature range aligns with the optimal window for *de novo* synthesis of DXN. The substantial accumulation of fly ash particles deposited on the filter provides ideal reaction sites for the catalytic formation and subsequent adsorption of DXN, resulting in a pronounced memory effect within the baghouse. DXN formed and adsorbed on these fly ash particles can be gradually released into the flue gas, leading to an increase in DXN concentration at the outlet by approximately one order of magnitude [148].

1.2.2.4.3 Memory Effect in the Selective Catalytic Reduction (SCR) Unit

The SCR unit, while primarily employed for flue gas denitrification, can also serve as a potential site for DXN reformation. This is evidenced by observations that the I-TEQ value of DXN at the SCR outlet may exceed that at the inlet. This phenomenon is primarily attributed to two interrelated processes occurring on the SCR catalyst surface, i.e., chlorination reactions and the associated memory effect. The details are shown as follows.

(1) Catalytic reformation and chlorination: Studies indicate that upon passage through the SCR unit, concentrations of specific low-chlorinated PCDF and PCDD congeners exhibit elevated concentrations. For instance, levels

of 1,2,3,7,8-pentachlorodibenzofuran (1,2,3,7,8-PeCDF), 2,3,7,8-TCDD, and 1,2,3,4,7,8-hexachlorodibenzofuran (1,2,3,4,7,8-HxCDF) have been observed to increase by 69.3%, 64.4%, and 48.7%, respectively. Given their relatively high toxic equivalency factors (TEFs), this marked increase directly elevates the total I-TEQ value of DXN at the SCR outlet. This selective regeneration phenomenon suggests that chemical transformation reactions are operative within the SCR unit. Specifically, at its operating temperature, active components of the catalyst may utilize gaseous chlorine sources to catalyze sequential chlorination reactions of non-DXN precursors or lower-chlorinated DXN congeners. This catalytic conversion alters the congener profile of outlet PCDFs and PCDDs, thereby enhancing the generation of more toxic species.

(2) Load fluctuations and the memory effect: DXN concentrations within the SCR unit exhibit sensitivity to fluctuations in incinerator load. Research indicates that when the load increases from 80% to 100%, the DXN concentration at the SCR outlet rises to 1.38 ng/Nm^3 [149]. This non-steady-state transition activates the memory effect, which primarily manifests in two ways. Firstly, the increased load may promote incomplete combustion within the furnace, significantly altering the distribution and concentration of DXN and its precursors entering the SCR unit. Secondly, interactions between fluctuating parameters (e.g., temperature and flow velocity) and contaminants accumulated within the SCR system further promote catalytic reformation and desorption of DXN. The synergistic action of these factors intensifies the memory effect, ultimately leading to elevated DXN concentrations at the outlet.

1.2.3 Discussion and Analysis

Advances in the mechanistic understanding of DXN formation provide a crucial foundation for its emission reduction and control. However, from the perspective of control science, a substantial gap remains between the current theoretical framework and the engineering imperatives of real-time detection, feedback control, and optimization-driven reduction. This section presents a systematic analysis of the key limitations inherent in current research on DXN formation mechanisms.

1.2.3.1 Lack of Systematic Theoretical Research on Whole-Lifecycle Mechanisms

The formation and transformation of DXN span multiple stages, each governed by a complex interplay of factors including fluctuating feedstock composition, chemical kinetics, temperature field distributions, and operational states of equipment. While existing research has yielded substantial mechanistic insights into individual stages (furnace combustion, flue gas cooling, and flue gas cleaning), the systemic challenges persist. These include a lack of unified model parameters, inconsistent boundary conditions, and, most critically, a poor understanding of the coupling relationships between successive stages. Furthermore, the MSWI process is characterized by inherent dynamic properties such as strong perturbations, high variability, abrupt process transitions, and rapid load changes. In contrast, the majority of current mechanistic studies on DXN are based on steady-state or quasi-steady-state assumptions. This fundamental mismatch hinders the ability of existing models to accurately capture the transient responses, dynamic evolution, and cross-stage transfer characteristics of real-world systems.

Consequently, a critical deficiency remains: the absence of a cross-stage, multi-scale, and dynamically resolvable theoretical framework for the whole-lifecycle DXN process. This theoretical gap not only limits a holistic understanding of the DXN "formation-decomposition-reformation-adsorption" lifecycle but also hinders the development of emission optimization and plant-wide synergistic control strategies that are fundamentally guided and constrained by mechanistic principles.

1.2.3.2 Incomplete Mechanistic Understanding of the Memory Effect

The memory effect lies at the heart of the lagged, episodic, and periodic fluctuations observed in DXN emissions. Its formation spans multiple stages of the DXN lifecycle, including adsorption, accumulation, redistribution, and release, and is intrinsically linked to the dynamic operating conditions of the MSWI process. Although existing research has confirmed its role in shaping emission patterns, a deep mechanistic understanding of its underlying drivers remains limited. Crucially, a comprehensive theoretical framework capable of systematically explaining its cross-stage characteristics is still lacking. The current deficiencies are primarily manifested in two aspects. Firstly, a lack of microkinetic understanding hinders quantitative analysis of the long-term retention and release behaviors of DXN on various substrates (e.g., the filter-cake surface on bags, deposits on heat exchange surfaces, ash layers on duct walls) under operational disturbances such as temperature fluctuations, load changes, and process transitions. Secondly, within the temperature window associated with the memory effect, multiple reaction mechanisms coexist and compete. These mechanisms include adsorption-desorption, *de novo* synthesis, and thermal decomposition. However, current knowledge of their interrelationships remains largely at the macroscopic trend level, lacking experimental evidence and theoretical explanations that can elucidate their relative contributions, modes of coupling, and dynamic evolution in response to varying operating conditions. This incompleteness in mechanistic understanding renders the memory effect a phenomenon that is observable in practice yet difficult to deconvolute in terms of pathways and contributions.

In summary, the current understanding of the memory effect has yet to advance beyond the qualitative or semi-quantitative stage. A systematic mechanistic framework capable of explaining its cross-scale and cross-stage behavior remains to be established. Future research must prioritize the in-depth elucidation of both its microscopic reaction pathways and macroscopic evolution chains. This endeavor is essential to lay a solid mechanistic foundation for achieving dynamic monitoring and emission reduction optimization of DXN.

1.2.3.3 Lack of Integrated Multiscale Visualization for Whole-Lifecycle Mechanisms

Integrated multiscale visualization studies of the DXN "formation-decomposition-reformation-adsorption" whole lifecycle play an indispensable role in elucidating its complex mechanistic framework. At the microscopic level, they enable researchers to directly reveal key reaction sites, evolutionary pathways, and the logical connections between different stages. At the macroscopic level, such visualization provides critical support for developing interpretable whole-lifecycle models and formulating effective emission reduction strategies. However, within the context of whole-lifecycle DXN mechanism research, the means and application of integrated multiscale visualization remain significantly underdeveloped. Current studies predominantly rely on microscopic computational methods, such as density functional theory (DFT) and molecular dynamics (MD), to deduce DXN reaction pathways, potential energy surfaces, and catalytic mechanisms. While valuable, these computational results are often confined to isolated, idealized models. A critical gap persists in translating these findings into a dynamic macro-level representation that corresponds to, and can be validated against, the complex realities of actual MSWI processes. Concurrently, some studies have attempted to construct numerical simulation models using software such as Aspen Plus. Yet, these models generally lack the capability to visually elucidate the influence of key operational parameters on DXN concentration. These parameters include furnace temperature fields, velocity fields, air flow regulation, and feedstock composition. While fluid simulation software with advanced visualization capabilities excels at depicting macroscopic-scale flow fields, temperature distributions, and heat/mass transfer processes, it cannot map the intricate chemical pathways underpinning the full DXN lifecycle. The "formation-decomposition-reformation-adsorption" sequence and other critical reactions rely on a multitude of kinetic parameters that are difficult to obtain and are not reliably implementable within general-purpose fluid dynamics software.

In summary, research on DXN mechanisms still lacks a systematic visualization framework capable of coupling micro-level reactions with macro-level behaviors. The absence of this cross-scale visualization linkage not only constrains the validation and refinement of mechanistic models but also hinders the development of mechanism-informed optimization and control strategies. Addressing this gap represents a pressing fundamental challenge that must be overcome to advance DXN emission control research.

1.2.3.4 Lack of an Effective Coupling Mechanism between Theoretical Research and Industrial Practice

A significant contextual gap persists between current research on DXN formation mechanisms and the demands of real-time detection and emission control in industrial MSWI settings. The MSWI process itself is characterized by high volatility and substantial uncertainties. Factors such as load variations, fluctuations in chlorine-containing components, oxygen disturbances, equipment aging, and frequent operational transitions collectively impact the actual formation and transformation of DXN, rendering the system non-steady-state, strongly coupled, and multiscale.

In contrast, the majority of contemporary DXN mechanistic studies are conducted under controlled, homogeneous, and quasi-steady-state laboratory conditions. While this approach successfully elucidates key reaction pathways, catalytic patterns, and critical temperature windows, the resultant findings often fail to maintain their consistency when confronted with the complexities of industrial scenarios. This discrepancy leads to significant deviations of model predictions upon their migration to real-world operating conditions, and even a complete breakdown at times.

In summary, research on DXN formation mechanisms still lacks a robust coupling mechanism capable of systematically calibrating experimental findings and numerical simulation results against actual industrial behaviors. This deficiency impedes the ability of mechanistic models to adequately support critical engineering needs such as operational optimization, control strategy design, and emission early warning. Consequently, bridging this gap represents a pivotal bottleneck that must be resolved to translate DXN emission control from theoretical research into practical engineering applications.

1.3 RESEARCH STATUS OF DIOXINS NUMERICAL SIMULATION

Numerical simulation provides a quantifiable and traceable research tool for investigating the formation, transformation, and emission processes of DXN. This section systematically reviews the progress of DXN numerical simulation across different stages, encompassing studies within the incinerator, through the cleaning stages, and those targeting the whole lifecycle.

1.3.1 Numerical Simulation within the Incinerator

The earliest numerical simulation modeling of DXN dates back to 1983, the first simplified kinetic model (ST model) was proposed based on the mechanism of DXN formation from phenolic compounds. This model did not incorporate the crucial radical coupling mechanism, resulting in predictions that were at least five orders of magnitude lower than values observed in actual MSWI process. This discrepancy highlighted the significant limitations of early models in terms of both accuracy and mechanistic completeness. Subsequently, researchers have engaged in successive refinements and extensions of the "ST model." For instance, Sidhu et al. enhanced the model's ability to fit actual emission data by adjusting key reaction activation energies [150]. Huang et al., through comparative experiments, demonstrated that the original "ST model" underpredicted experimental results by at least four orders of magnitude [151]. They hypothesized the existence of gas-phase reaction pathways for DXN formation not accounted for in the model. Building on this discrepancy, Babushok et al. revised the reaction processes involving chlorine species and hydroxyl radicals, leading to significant improvements in the model's predictions [152]. Despite these improvements, a notable gap between model outputs and actual emissions persisted. Further analysis of the residual model deviation led Khachatryan et al. to infer the contribution of additional [153], unknown non-gas-phase reaction pathways to PCDF formation.

In addition to the ongoing extensions of the "ST model," other researchers have proposed a variety of kinetic models specifically for furnace combustion. For example, Gerasimov constructed a kinetic model for DXN formation applicable to coal combustion processes [154], based on *de novo* synthesis reaction rates provided by Altwicker et al. [155]. This model enabled both qualitative description and quantitative estimation of DXN formation profiles. Recognizing that the development of existing kinetic models often requires the integration of empirical knowledge and struggles to differentiate among the 17 toxic DXN congeners, Stanmore developed separate kinetic models for homolog- and isomer-specific reactions [156]. Their approach succeeded in confining the discrepancy between model predictions and actual measurements to within 20%.

In summary, the development of DXN kinetic models for incinerator applications has undergone an evolutionary trajectory from simplified gas-phase models toward an integration of microscopic computational methods with macro-scale kinetic modeling. While this evolution marks significant progress, the field continues to face substantial challenges, including limited predictive accuracy and incomplete mechanistic coverage.

1.3.2 Numerical Simulation in the Flue Gas Cleaning Stage

Numerical simulation research targeting the cleaning stage primarily focuses on the AC adsorption process. For instance, Li et al. experimentally determined the Langmuir isotherm parameters for various types of commercial AC [138], laying the groundwork for subsequent adsorption modeling. Separately, Prisciandaro et al. investigated the adsorption thermodynamic properties of six PCDD congeners [157], with their simulated adsorption profiles showing high consistency with other experimental data. The core objective of such studies is to establish high-fidelity adsorption equilibrium models to guide the optimized control of ACI rates.

1.3.3 Numerical Simulation Research for the Whole Lifecycle

With a deepened understanding of DXN full-process mechanisms, research has shifted toward constructing integrated numerical simulation models that encompass all stages of the MSWI process. Chen et al. employed the MATLAB Simulink toolbox to simulate the degradation of DXN under electron beam irradiation [158], thereby identifying the optimal degradation conditions. In a separate approach, Sun et al. coupled the kinetic model proposed by Stanmore [156] with the FLIC software [159]. Based on user-defined HCl profiles, their model simulated the concentration distribution trends of DXN along the grate movement direction during solid-phase combustion. However, practical validation of this model remains challenging. Demonstrating the feasibility of process simulation software in DXN research, Tungalag et al. utilized the Fortran subroutine capability within Aspen Plus to model the formation process of 2,3,7,8-TCDD during MSW pyrolysis [160].

Building upon this foundation, Xia et al. developed a comprehensive numerical simulation model (CNSM) for the whole lifecycle of DXN [161]. By encompassing multiple stages (precursor formation, high-temperature pyrolysis, re-synthesis, AC adsorption), the model describes the evolution of DXN from combustion to emission in the MSWI process, providing a theoretical basis for online detection and optimization control. The specific description is as follows. The CNSM employs a multi-software coupling strategy to effectively overcome the functional limitations of any single software in simulating solid-phase combustion, gas-solid reactions, and adsorption processes. Specifically, the FLIC software simulates the solid-phase combustion process on the grate and provides accurate flue gas composition data to Aspen Plus. Aspen Adsorption simulates the adsorption behavior of DXN onto AC and calculates the adsorption removal efficiency. Aspen Plus serves as the integration platform, responsible for coupling all modules and performing steady-state simulations of the whole MSWI process, thereby completing the whole-lifecycle simulation of DXN.

1.3.4 Discussion and Analysis

Although existing numerical simulation studies on DXN have made considerable progress in kinetic modeling, software coupling, and whole-lifecycle simulation, significant shortcomings persist within the current research framework from the perspective of control science. These limitations hinder the potential application of these models in areas such as online detection, feedback control, and emission reduction optimization. This section presents an analysis centered on the following four critical issues.

1.3.4.1 Lack of a Visualization Mechanism for DXN Concentration Evolution in Industrial Process-Oriented Numerical Simulation

While notable progress has been made in developing multi-stage coupled DXN numerical simulation models for characterizing specific actual MSWI processes, the current simulation framework still lacks systematic visualization capabilities. This deficiency hinders the direct application of simulation results to subsequent engineering tasks such as mechanistic interpretation, real-time detection, and emission optimization. A core manifestation of this shortcoming is the absence of dynamic visualization that elucidates the causal relationships between key process variables and DXN concentration at different zones. Existing models with visualization predominantly focus on displaying steady-state results or cross-sectional distributions of physical quantities such as temperature fields, flow fields, or particulate matter concentrations. Evidently, none of these are trace pollutants. However, the formation and transport of DXN exhibit distinct temporal and stage-specific characteristics. Current research fails to provide spatiotemporal visualization of the evolution of critical process variables like temperature, precursor concentrations, and residence time. Consequently, existing numerical simulations fall short of offering a systematic, process-level understanding, thereby limiting their value in guiding practical process adjustments and emission reduction strategies for MSWI operations.

In summary, existing DXN numerical simulation tools tailored for practical MSWI processes predominantly focus on "result presentation," lacking the "process visualization" capability essential for mechanistic analysis and emission optimization. This deficiency constrains the utility of such numerical simulations in critical applications, including identifying DXN formation causes, pinpointing process anomalies, and guiding emission reduction strategies. To enhance the efficacy of the simulation framework, there is a pressing need to develop advanced visualization techniques that are higher-dimensional, more dynamic, and possess a stronger ability to map underlying mechanisms.

1.3.4.2 Numerical Simulation Models for the Mechanistically Unclear Memory Effect Remain Incomplete

In essence, the memory effect is a spatiotemporal uncertain process characterized by "long-term accumulation and transient release" across multiple temporal and spatial scales. In contrast, existing DXN numerical simulation frameworks designed for practical industrial processes are primarily constructed around instantaneous reactions or steady-state conditions, making them ill-equipped to capture these core dynamic spatiotemporal characteristics. Current numerical simulation platforms lack the capability to model the temporal accumulation of adsorbates within internal equipment structures. Consequently, prevailing models typically resort to simplifications such as constant adsorption coefficients, fixed removal efficiencies, or isothermal equilibrium relationships. These approximations treat adsorption as an instantaneous event, thereby failing to reproduce critical features of DXN behavior during long-term operation, including gradual enrichment, spatial redistribution, and carrier saturation. Evidently, existing numerical simulations lack a mechanism for simulating the dynamic release of DXN under non-steady-state operating conditions. In an actual MSWI power plant, the memory effect is often triggered by transient events such as temperature jumps, load disturbances, filter cleaning operations, or startup/shutdown cycles. However, most current numerical simulation frameworks cannot describe the instantaneous release behavior induced by such disturbances.

In summary, current DXN numerical simulation still lacks a unified functional framework capable of depicting the complete dynamic chain of "accumulation-release" across multiple spatiotemporal scales, particularly in the context of specific industrial sites. This research gap renders existing numerical simulations inadequate for supporting memory effect-related tasks such as real-time detection, risk early warning, feedback control, and operational optimization. Addressing this deficiency constitutes a critical technological shortfall that must be resolved to advance DXN simulation tools from the research stage to a state of engineering applicability.

1.3.4.3 Integrated Whole-Lifecycle Numerical Simulation Models Based on a Single Platform Remain Underdeveloped

Current DXN whole-lifecycle numerical simulations are predominantly constructed through the discontinuous coupling of multiple specialized software packages, with each related MSWI process stage implemented by a distinct, mature numerical simulation platform. While this "modular assembly" approach can cover key stages (formation, decomposition, re-synthesis, and adsorption) of DXN evolution in MSWI, it presents several inherent drawbacks in terms of consistency and continuity. These limitations are particularly critical for modeling trace pollutants characterized by cross-temperature-zone behavior and tightly coupled multi-mechanism interactions.

Discrepancies in numerical schemes, underlying physical assumptions, and solution strategies among the different software packages hinder the enforcement of unified mass and energy conservation across stages. This directly compromises the overall credibility and traceability of the simulation. Furthermore, inconsistencies in spatial discretization, time-stepping methods, and iteration algorithms further undermine the stability of the coupled results. Consequently, such whole-lifecycle models exhibit significant shortcomings in convergence, reproducibility, and parameter sensitivity analysis.

In summary, current DXN whole-lifecycle numerical simulation modeling struggles to provide a unified description of the multi-stage, multi-physical-field, and multi-mechanism coupled behaviors inherent to the MSWI process. It is evident that constructing an integrated, whole-lifecycle numerical simulation model based on a single, unified platform is a critical prerequisite for advancing DXN simulation models from "research-oriented analytical tools" to "engineering decision-making tools." Furthermore, such a model constitutes an essential technical foundation for future advancements in high-precision real-time detection, emission early warning, disturbance analysis, and emission reduction control strategies.

1.3.4.4 Lack of Numerical Simulation Models for Regionalized and Multi-Furnace-Type Practical Processes

Current DXN numerical simulation studies designed for actual MSWI processes are predominantly based on industrial scenarios involving a single region and a specific furnace type. While capable of reflecting the emission behavior of that particular furnace under specific operating conditions, these models struggle to adapt to the significant variations observed across different regions, including MSW composition, furnace configuration, and flue gas cleaning technologies. Consequently, the engineering generalizability of existing numerical simulation models is limited. The physicochemical properties of MSW (moisture content, proportion of combustibles, chlorine content, and inorganic composition) exhibit substantial regional disparities. The lack of adjustable parametric inputs to account for these regional characteristics prevents numerical simulation models from accurately capturing the impact of compositional variations on combustion processes and DXN concentration evolution trends. This, in turn, constrains the generalization capability of such numerical simulations. Furthermore, MSWI facilities employing different furnace types (e.g., grate, fluidized bed) differ markedly in combustion organization, flow structures, residence time distributions, and temperature windows. Since the formation mechanisms of DXN are highly dependent on system architecture and operating conditions, these differences are critical. Existing numerical simulation models are primarily constructed for a single furnace type or a fixed set of flue gas cleaning equipment. Therefore, they cannot readily encompass multiple equipment types or their combined process configurations, lacking the transferability needed for direct application to other MSWI power plants.

In summary, the absence of a generalized framework capable of adapting to varying regional MSW characteristics and multiple equipment types results in significant shortcomings in the engineering portability, cross-plant generalizability, and regional adaptability of current DXN numerical simulation models. This constitutes a key factor constraining their advancement from isolated case studies to broader, scalable applications.

1.4 RESEARCH STATUS OF DIOXINS CONCENTRATION DETECTION AND EARLY WARNING

This chapter provides a comprehensive review of research advances in DXN concentration detection and early-warning. It introduces, in sequence, the detection methods for DXN emission concentration, generation concentration, and whole-lifecycle concentration, followed by the current state of development in emission concentration early-warning technologies. This structured presentation aims to offer a holistic overview of the primary research directions within the current DXN concentration detection and early-warning framework.

1.4.1 Current Research Status on Detection of DXN Emission Concentration

Based on the temporal characteristics of the detection cycle, DXN emission concentration detection methods are categorized in this chapter into three classes, i.e., offline direct detection, online indirect detection, and real-time soft sensing, as illustrated in Figure 1.20.

1.4.1.1 Offline Direct Detection

Offline direct detection represents the earliest established and most mature technical framework for DXN concentration analysis. It primarily encompasses multiple analytical pathways such as chromatographic methods, laser mass spectrometry, biological methods, and immunoassay methods. This section provides a concise overview of the technical procedures, advantages, and limitations associated with these typical offline direct detection methods.

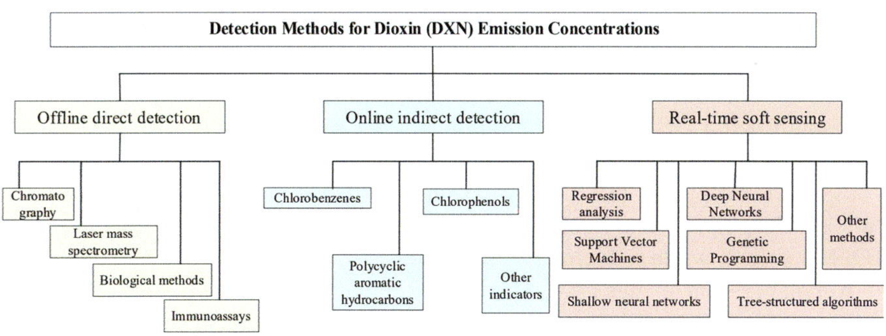

FIGURE 1.20 Classification of DXN emission concentration detection methods.

1.4.1.1.1 Chromatography

Chromatography is currently the internationally recognized gold-standard technique for DXN detection, with HRGC/HRMS being the most representative method. This technique relies on a rigorous laboratory workflow involving multiple steps: on-site sample collection, solvent extraction, cleaning, isotopic internal standard calibration, and quantification via high-resolution mass spectrometry. Owing to its exceptional resolution and selectivity, chromatography enables the accurate determination of individual DXN congeners, serving as the benchmark for TEQ assessment. However, its application is challenged by several factors, including dependence on complex sample pretreatment, expensive isotopically labeled standards, and high-precision instrumentation. Consequently, the results are susceptible to operator skill, complex matrix effects, and instrumental performance. More critically, the large size of HRGC/HRMS systems, coupled with their requirement for specialized laboratory environments and trained personnel, precludes their deployment at industrial sites. This confines the method primarily to high-precision analytical tasks under controlled laboratory conditions.

1.4.1.1.2 Laser Mass Spectrometry

Laser mass spectrometry employs laser irradiation to selectively ionize target molecules while suppressing background ion signals, followed by mass analysis using time-of-flight (TOF) mass spectrometers. Its theoretical foundation lies in the characteristic narrow-band spectral structures unique to different DXN congeners [162]. Driven by laboratory research, laser mass spectrometry has demonstrated growing potential for the real-time detection of DXN and its precursors. Tsuruga et al. developed a vacuum ultraviolet single-photon ionization ion trap time-of-flight mass spectrometry (VUV-SPI-IT-TOFMS) system for the long-term stable monitoring of PeCDF [163]. Subsequently, Morii et al. utilized vacuum ultraviolet photoionization ion trap time-of-flight mass spectrometry (VUV-IT-TOFMS) to achieve online detection of PeCDF while simultaneously measuring related trichlorobenzenes (TrCBz) [164]. Heger et al. redesigned and constructed a resonance-enhanced multiphoton ionization time-of-flight mass spectrometry (REMPI-TOFMS) instrument [165]. By fixing the laser wavelength at 266 or 248 nm, it enabled highly sensitive detection of PAHs, achieving a detection limit as low as 10 parts-per-trillion (ppt) for naphthalene. Furthermore, the team conducted chlorobenzene detection experiments on an industrial hazardous waste incinerator and, in conjunction with gas chromatography-mass spectrometry (GC-MS) analysis, established a DXN concentration detection model [166]. Hafner et al. developed a novel supersonic jet inlet system integrated with a nozzle-skimmer and a fused silica capillary [167]. This design provided samples with excellent adiabatic cooling, significantly enhancing the sensitivity and selectivity of the REMPI-TOFMS system while reducing fragmentation.

Although recent studies have extended the application of laser mass spectrometry to the online rapid detection of DXN precursors, its core technical workflow remains complex, involving on-site sampling, sample enrichment/extraction, low-energy laser excitation, high-energy laser ionization, and detection by high-resolution TOFMS. Compared to chromatographic methods, laser mass spectrometry offers advantages including rapid response, high sensitivity, strong temporal resolution, and the capability for parallel analysis of multiple congeners [168, 169]. However, this method relies on precise spectral databases, requiring prior knowledge of the characteristic spectral structures of target congeners. Furthermore, it imposes stringent demands on laser stability, optical path alignment, and background interference suppression.

1.4.1.1.3 Biological Methods

Biological methods constitute a class of offline direct detection approaches that assess the DXN TEQ based on the response intensity of biological receptors or biochemical reactions. Their common characteristics include relatively short analysis cycles, lower costs, and applicability for the rapid screening of large sample sets. In contrast to chromatographic and laser mass spectrometry methods, which are capable of distinguishing individual DXN congeners, biological methods focus on the quantification of bulk toxicity. This makes them suitable for pollution level assessment and trend analysis. Three representative biological assays are briefly introduced herein.

1.4.1.1.3.1 Enzyme Induction Bioassay This bioassay quantifies the TEQ of a sample by measuring the change in enzymatic activity induced upon activation of the aryl hydrocarbon receptor (AhR) by DXN. The procedure encompasses several steps: sample collection, DXN extraction and separation, culture medium preparation, receptor activation reaction, and fluorescence intensity measurement [170]. The primary advantages of this method include a short analysis cycle, low cost, and the capability for high-throughput rapid screening of samples. However, a significant limitation is its inability to distinguish the specific concentrations of individual DXN congeners in the sample. Consequently, it is primarily used for the rough estimation of total toxicity and exhibits lower quantitative precision compared to chromatographic methods.

1.4.1.1.3.2 Enzyme Immunoassay (EIA) The EIA quantifies sample TEQ by leveraging the high-affinity binding of monoclonal or polyclonal antibodies (typically derived from mice or rabbits) to DXN congeners through a competitive inhibition reaction [171]. Key procedural steps involve DXN extraction and cleaning, antibody preparation, the competitive reaction itself, optical density measurement, and subsequent toxicity calculation [172]. This method offers distinct advantages such as straightforward operation, ease of implementation, and a short analysis cycle. Relevant studies indicate that its accuracy surpasses that of enzyme induction bioassays [173]. However, notable limitations persist. Firstly, EIA cannot provide concentration information for individual DXN congeners. Secondly, its performance is heavily reliant on the calibration of standard curves and the spectroscopic system used for readout. This dependence can significantly increase testing complexity when handling large sample batches.

1.4.1.1.3.3 Luciferase Reporter Gene Assay The luciferase reporter gene assay achieves DXN TEQ detection by genetically engineering a specific mammalian cell line to integrate a construct containing a DXN-responsive promoter (e.g., from a cytochrome P450 gene) linked to the firefly luciferase reporter gene into its chromosomes. This genetic modification enables the synthesis of luciferase enzyme in proportion to the DXN concentration, which subsequently catalyzes a substrate to produce a measurable luminescent signal [174]. A standard analytical workflow for this assay comprises sample collection, DXN extraction, the cellular bioassay (involving reporter cell incubation and luciferase expression), and subsequent luminescence measurement. Capitalizing on genetic engineering, this method offers distinct advantages, including high sensitivity, strong specificity, and a relatively short analysis cycle (approximately 24 hours). Studies have demonstrated that its accuracy surpasses that of enzyme induction bioassays [175, 176]. The literature indicates its significant potential for application in fields such as food, soil, and water environmental monitoring, though its standardized implementation and broader engineering application require further development in specific regional contexts [26]. Despite its prominence in rapid toxicity screening, the assay results only reflect the overall TEQ of a sample and cannot discriminate the specific concentrations of individual DXN congeners. Furthermore, the performance of this bioassay is sensitive to variations in cell culture conditions, solvent residues, and environmental perturbations, which may compromise the stability and reproducibility of the measurements.

1.4.1.1.4 Immunoassay Methods

Immunoassay methods primarily rely on the specific recognition between antibodies and DXN or its structural analogs to achieve detection. These methods can be classified into two categories. The first one utilizes artificially prepared, DXN-specific antibodies to recognize particular DXN structures via antigen-antibody binding. The second one quantifies the overall sample TEQ based on the signaling cascade initiated upon activation of the AhR by DXN. A typical immunoassay workflow involves sample collection, DXN extraction and cleaning, preparation of antibodies or receptor systems, the antigen-antibody binding reaction, detection of precipitates or fluorescent signals, and final TEQ calculation. Immunoassays offer distinct advantages such as relatively straightforward operational procedures, lower instrumental requirements, and more controllable costs compared to other chemical analysis methods, making them valuable for the rapid screening and semi-quantitative analysis of DXN. However, several significant limitations exist: (1) The antibody preparation process is complex and highly dependent on technical conditions, with substantial variability in specificity among different antibody batches; (2) It is difficult to distinguish the specific concentrations of individual DXN congeners, often yielding only an overall TEQ value; (3) The spectrum of detectable DXN types is limited by the available antibody repertoire, making it challenging to cover all 17 highly toxic congeners. These constraints position immunoassays in engineering applications as more suitable for auxiliary screening rather than precise quantitative analysis.

1.4.1.1.5 Summary

As evidenced above, chromatography, laser mass spectrometry, biological methods, and immunoassays constitute the most mature suite of offline direct detection techniques for DXN. While they enable high-precision quantitative analysis or toxic equivalency estimation, significant limitations persist. Although laser mass spectrometry, biological methods, and some immunoassays have markedly reduced the analysis cycle, i.e., from weeks down to 3–4 days or even within 24 hours. These methods remain fundamentally reliant on offline procedures such as "sample collection, extraction, cleaning, and laboratory analysis." Consequently, they cannot achieve the online detection mode desired for MSWI processes. The shared core limitations of these techniques thus manifest in three key aspects: complex operational workflows, a strong dependence on laboratory conditions and operator expertise, and an inherent inability to respond at second- or minute-level timescales to the operational fluctuations inherent in actual MSWI processes.

1.4.1.2 Online Indirect Detection

To overcome the bottleneck of long analysis cycles inherent in offline direct detection, research efforts have explored online indirect detection. This approach utilizes a proxy substance that exhibits a high correlation with DXN. The proxy substances include chlorinated precursors or other indicators/related compounds. The core strategy involves selecting these real-time measurable indicators/associates from flue gas and constructing mapping models based on their concentration variations to indirectly infer the DXN emission concentration [30]. Literature categorizes such indicators/associated compounds into four classes [177]: small-molecule inorganic gases (e.g., CO, Cl_2, and HCl), chlorinated precursors (e.g., chlorobenzenes, chlorophenols, and PCBs), non-chlorinated precursors (e.g., PAHs, ethylene, and propylene), and semi-volatile or low-volatility organic chlorinated compounds. Online indirect detection of DXN emission concentration is achieved by combining the real-time detection results of these indicators or associates with empirical or statistical mappings established between their concentrations and DXN concentration.

The following section will provide a review of commonly used indicators or associated compounds in online indirect detection methods.

1.4.1.2.1 Chlorobenzene (CBz)

Chlorobenzene compounds represent one of the most extensively studied and, in terms of engineering application, the most mature class of indicators for DXN concentration. A strong correlation between hexachlorobenzene (HCB) and DXN was first identified in the flue gas from hazardous waste incineration [178]. Subsequently, Pandelova et al. confirmed similar characteristics for pentachlorobenzene (PeCBz) in the flue gas from the co-combustion of MSW and coal [179]. Although Literature [180] indicates that PCBs may exhibit an even stronger correlation with DXN; chlorobenzenes hold a distinct advantage in practical engineering detection. This advantage stems from their typically higher concentrations in flue gas, better suitability for online detection, and the availability of more mature detection technologies.

Research indicates that the high correlation observed between CBz and DXN emissions stems from their shared involvement in the *de novo* synthesis pathway, where they utilize common reaction intermediates. The correlation coefficient between the two can be as high as 0.98 [181]. Building on this robust relationship, researchers have endeavored to construct high-precision mapping models linking chlorobenzene concentrations to DXN emission levels. For instance, Literature [182] employed partial least squares regression to develop a multivariate regression model.

Numerous studies are currently focused on enhancing the online detection accuracy and temporal resolution of CBz. For instance, Literature [183, 184] developed mass spectrometry systems capable of real-time CBz detection. In a significant advancement, Literature [185, 186] conducted research on DXN online indirect detection based on chlorobenzenes using resonance-enhanced multiphoton ionization (REMPI) coupled with time-of-flight mass spectrometry (TOFMS), achieving further improvements in detection precision and response speed. Notably, Literature [187] proposed an online detection method that reduced the CBz measurement cycle to 0.5 hours. This method underwent a three-month industrial test, demonstrating excellent real-time performance. Further pushing the boundaries of Literature [188] implemented online detection of 1,2,4-Trichlorobenzene (1,2,4-TrCBz) using laser absorption spectroscopy-tunable ionization time-of-flight mass spectrometry (LASTI-TOFMS) and subsequently achieved indirect detection of PCDD/F emission concentrations in flue gas. The experimental results revealed that 1,2,4-TrCBz concentrations were approximately three orders of magnitude higher than the corresponding I-TEQ values. The correlation coefficients between 1,2,4-TrCBz and I-TEQ under different operating conditions were 0.867, 0.953, and 0.944, respectively, and the established linear prediction model showed high consistency with the measured values. A particularly notable application in a more realistic operational setting is documented in [177]. This study collected concurrent CBz and DXN concentration data from multiple typical locations within a circulating fluidized bed (CFB)-based MSWI process. By constructing a multi-point mapping model across these zones, the study successfully reduced the detection lag time for DXN concentrations at multiple points to 0.5 hours.

1.4.1.2.2 Polycyclic Aromatic Hydrocarbons (PAH)

PAH is ubiquitous incomplete combustion products in incineration flue gas, characterized by relatively high concentrations and favorable online detectability [189]. Consequently, they are regarded as another class of promising indicator compounds for DXN. Studies have shown an association between PAH formation pathways and those of DXN precursors, with their concentration trends exhibiting significant consistency [190, 191]. From a mechanistic analysis perspective, Literature [192] obtained concentration profiles of PAH and DXN precursors by offline analysis of simulated flue gas generated from the pyrolysis of Polyvinyl Chloride (PVC) under nitrogen conditions. These profiles were subsequently used to estimate the DXN concentration range. Further advancing the field, a mass spectrometry system capable of real-time PAH detection was developed.

In summary, PAH offer advantages including ready detectability, rapid response, and relatively high concentrations. Research conducted under both laboratory and select engineering conditions has demonstrated a significant correlation with DXN concentrations. However, their generation and emission concentrations are susceptible to influences such as operational fluctuations of the furnace. Consequently, the stability and reliability of PAH as DXN indicators require further validation under a broader range of actual MSWI operating conditions.

1.4.1.2.3 Chlorophenols (CP)

CP, another potential indicator class, serve as key precursors in DXN formation and possess strong suitability for online detection. Literature [193] analyzed the correlation between gas-phase CP and particle-phase DXN, subsequently validating the feasibility of using CP for DXN prediction by constructing a partial least squares (PLS) model. Expanding on this, combustion tests involving four typical types of industrial solid waste and specifically investigating the correlation between CP and DXN in both a fluidized-bed incinerator and a rotary kiln-grate furnace system were conducted. The results demonstrated the applicability of CP as a DXN indicator across various incineration conditions.

The inherent characteristics of the MSWI process can lead to variability in the correlation between indicator/associated compounds and DXN. Building on this, Literature [194, 195] proposed the use of a combined "dual-indicator" online indirect detection framework employing both CBz and CP to enhance the robustness of DXN concentration detections. However, research in Literature [196] indicates that CP demonstrates weaker modeling performance compared to CBz as an indicator. The primary reason is that CP is more susceptible to the influence of various factors, including furnace temperature fluctuations, variations in chlorine components, and metal catalysis effects. This susceptibility results in significant variability in the CP-DXN correlation across different incineration operating conditions.

In conclusion, while CP possess a mechanistic justification for serving as DXN indicators, the stability of their correlation is inferior to that of CBz. Therefore, CPs are more suitable as auxiliary or supplementary indicators and cannot serve as a standalone primary indicator for the engineering implementation of online indirect DXN detection.

1.4.1.2.4 Other Indicators

Beyond chlorobenzenes, PAHs, and chlorophenols, several other potential indicators for online indirect DXN detection have been explored. For instance, Literature [30] indicated a certain degree of correlation between low-volatile organohalogen compounds (LVOH) and DXN emissions. Building on this, Literature [197] developed an online detection system capable of automatic sampling and analysis at hourly intervals for low-volatile organic chlorine compounds (LVOCl), and established a mapping relationship with DXN concentration as described below:

$$\rho_{\mathrm{DXN}} = 0.00154\rho_{\mathrm{LVOCl}}^{1.34} \tag{1.11}$$

However, this system is hampered by drawbacks, including a complex structure and demanding operational and maintenance requirements, which limit its widespread adoption in engineering applications.

In summary, indicators such as LVOH and LVOCl offer novel avenues for the online indirect detection of DXN. However, challenges, including insufficient correlation strength and stability, coupled with the high maintenance costs of the corresponding detection systems, remain. Consequently, research in this area is still at an exploratory stage, and significant further validation and optimization are required before engineering deployment can be considered.

1.4.1.2.5 Summary

As outlined above, online indirect detection approaches based on indicator compounds provide a crucial strategy to overcome the bottleneck of direct online DXN detection. Various substances (e.g., CBz, PAHs, CP, and LVOH) have demonstrated different degrees of correlation with DXN in laboratory and pilot-scale tests. Among these, CBz is considered the most promising candidate for engineering implementation, owing to its similar formation mechanism, high correlation coefficient, and the relatively mature state of its online detection technology. In contrast, the correlations of PAHs and CP with DXN are more susceptible to influences from operating conditions, furnace temperature, combustion completeness, and metal catalysis. Consequently, their stability and overall modeling performance are generally inferior to those of CBz, rendering them better suited as supplementary indicators. Furthermore, while novel indicators such as LVOH and LVOCl offer alternative perspectives for observing DXN emission trends, their associated detection systems are often structurally complex and difficult to maintain, currently falling short of meeting the demands for continuous industrial operation.

While online indirect detection methods offer distinct advantages in improving detection frequency and reducing response time, they are fundamentally constrained by several key limitations. Foremost, the correlation between indicator/associated compounds and DXN is highly dependent on specific factors inherent to the MSWI system, including furnace type, combustion state fluctuations, flue gas treatment processes, and fly ash characteristics. This dependency severely limits the direct transferability of established mapping models across different MSWI facilities. Furthermore, most of these methods rely on complex analytical systems, such as mass spectrometers, laser-based instruments, or automated sampling units. Consequently, they suffer from high engineering implementation costs and significant challenges in maintaining stable, long-term operation. Therefore, although feasible in experimental setups and demonstration cases, online indirect detection faces critical challenges for widespread industrial application in real-world MSWI processes. These include unstable correlations, insufficient model generalizability, and high system maintenance complexity. Substantial further development and refinement are imperative to advance these methods toward robust engineering implementation.

1.4.1.3 Real-Time Soft Sensing

The real-time soft sensing method described here utilizes easily accessible process variables (e.g., temperature, oxygen level, air flow rate, load, and pressure) during MSWI and integrates them with either mechanistic knowledge or data-driven methodologies to establish a mapping relationship between inputs and outputs. This enables the online, real-time detection of DXN emission concentrations. Compared to offline direct detection and online indirect detection methods, real-time soft sensing offers distinct advantages in terms of real-time capability, cost, maintainability, and engineering applicability. Consequently, it is increasingly emerging as a mainstream technology for achieving real-time detection of DXN emission concentrations.

To systematically delineate the developmental trajectory of real-time soft sensing methods for DXN and to elucidate the distinctions among these approaches, existing research is categorized herein into six classes: Regression Analysis, SVM, shallow neural networks, deep neural networks, genetic programming, tree-structured algorithms, and other methods. A review and analysis of the performance advantages and limitations associated with each category are subsequently presented.

1.4.1.3.1 Regression Analysis Methods

Regression analysis methods operate by first establishing a correlation or functional relationship between measurable variables and the DXN. Subsequently, model parameters are fitted using historical data to achieve rapid estimation of the DXN. These methods are characterized by advantages such as model simplicity, computational efficiency, and favorable interpretability. Current research in this area can be broadly categorized into two groups: models based on univariate relationships and those based on multivariate relationships.

1.4.1.3.1.1 Univariate Regression Models

Early-stage research typically focused on a single characteristic variable within the MSWI process to explore its correlation with DXN emission concentration. For instance, Hasberg et al. proposed a mapping relationship between the maximum flue gas temperature and DXN concentration [198], as follows:

$$\mathrm{DXN} = A_1 e^{\left(-T_p / B_1\right)} \tag{1.12}$$

where T_p represents the maximum temperature of the emitted flue gas, $A_1 = 5.2 \times 10^7$, and $B_1 = 40.6$. Furthermore, the researchers proposed a mapping relationship between PCDD and CO concentration as follows:

$$\mathrm{PCDDs} = \left(\frac{\mathrm{CO}}{k_{\mathrm{CO}}}\right)^2 \tag{1.13}$$

where k_{CO} is a constant. The study indicated that when the O_2 concentration deviates from the 7%–9% range, both CO and DXN emission concentrations exhibit an upward trend.

Furthermore, based on an analysis of 100 samples collected from 17 incinerators, Tillman et al. investigated the impact of the excess air ratio (EA) on the formation of DXN and furans. They proposed the following empirical correlation for operating conditions where EA exceeds 80% [199]:

$$\mathrm{DXN} = 0.0376\mathrm{EA} - 3.305 \tag{1.14}$$

Other studies have further validated significant correlations between specific univariate parameters and DXN emission concentrations in the MSWI process. For instance, it found a significant correlation between CO and DXN under stable operating conditions in an MSWI process based on a fluidized-bed. Furthermore, Literature [144] noted that DXN levels correlate more strongly with the CO concentration measured 3–4 hours prior to the DXN sampling point. This observation provides the first experimental evidence in the literature revealing the inherent memory effect of DXN, which evidently increases the difficulty of implementing real-time soft sensing methods.

1.4.1.3.1.2 Multivariate Regression Models

With advancing research, the focus has shifted toward employing multivariate regression models to delineate the inherently coupled, multi-factor mechanisms underlying DXN emissions. Specifically, for the Modular and Waterwall furnace types, a model for DXN emission concentration under a representative operating condition of 7% O_2 content was expressed as follows:

$$\begin{cases} \text{DXN} = 2670.2 - 1.37\text{TEMP} + 100.06\text{CO} \\ \text{DXN} = 4754.6 - 5.14\text{TEMP} + 103.41\text{CO} \end{cases} \tag{1.15}$$

where TEMP denotes the secondary combustion chamber temperature in the Modular furnace or the internal temperature in the Waterwall furnace.

Further extending this work, the Literature [200] developed a DXN emission concentration model applicable to three furnace types (e.g., Modular, Waterwall, and RDF) under the same operating conditions, as expressed below:

$$\begin{cases} \text{DXN} = -423.6119 + 0.0288\text{FLOW} + 2.6141\text{CO} + 505.6183\text{ESP} - 68.1576\text{DRY} \\ \text{DXN} = 148.9954 + 2.2666\text{CO} - 0.1565\text{HCL} + 445.3134\text{ESP} \\ \text{DXN} = -1.2271 + 0.0101\text{TEMPC} + 0.0014\text{CO} \end{cases} \tag{1.16}$$

where FLOW denotes the flue gas flow rate at the stack, a parameter correlated with both the flue gas residence time and the EA; ESP and DRY are binary variables that take a value of 1 if an electrostatic precipitator (ESP) and a dry fabric filter, respectively; 1 is employed as the end-of-pipe treatment devices in the MSWI system, and 0 otherwise; and TEMPC represents the temperature of the upper combustion chamber.

The study further indicated that feedback control on DXN emissions could be implemented by regulating the feeding rate and steam output, proposing the control model shown below [200]:

$$\begin{cases} \text{DXN} = 346.4892 - 12.9031\text{STEAM} + 0.2369\text{TEMPC} \\ \text{DXN} = 1245.5399 - 84.8730\text{FEED} - 1.1067\text{HCL} \\ \text{DXN} = 0.1305 + 0.0016\text{TEMPC} + 0.0076\text{HCL} \end{cases} \tag{1.17}$$

where FEED denotes the hourly feeding rate of the Modular furnace, and STEAM represents the hourly steam production rate of the Waterwall furnace.

Furthermore, Chang et al. developed a multivariate regression model for three categories of incinerators based on empirical data collected from multiple testing projects in North America [201]. The results indicate that the constructed model possesses reasonable statistical significance and explanatory power.

In summary, as the earliest adopted soft sensing modeling approach for DXN emission concentration, regression analysis has played a crucial role in elucidating key influencing factors and verifying the statistical coupling patterns between process variables and DXN. However, inherent characteristics (e.g., high nonlinearity, multivariable coupling, and time hysteresis) of DXN formation pose significant challenges for regression analysis. Due to its foundation in statistical averaging techniques, this method struggles to maintain robust soft sensing performance across multiple operating conditions and in response to dynamic process changes.

1.4.1.3.2 Support Vector Machines (SVM)

SVM, with their flexible kernel functions, can effectively model nonlinear mapping relationships between inputs and outputs without substantially increasing model complexity. This characteristic makes them particularly suitable for data-driven modeling scenarios characterized by small sample sizes, high noise levels, and nonlinearity. Consequently, SVM has gradually emerged as a representative nonlinear estimation tool for the real-time soft sensing of DXN emission concentrations. Utilizing the small-sample dataset collected in Literature [200], Literature [202] developed a soft sensing model for DXN emission concentrations based on an adaptive selective ensemble least squares-SVM framework. Experimental results demonstrated promising generalization performance. However, this approach remains at the laboratory data validation stage; its adaptability and robustness under actual MSWI operating conditions require further verification.

In summary, SVM provides a viable approach for nonlinear modeling with limited samples in the real-time soft sensing of DXN emission concentrations. However, challenges persist in several areas, including input feature selection, model scale adaptation, robustness to operational condition variations, and integration with MSWI mechanistic knowledge. Further research is therefore essential to enhance its reliability and generalizability for real-time deployment.

1.4.1.3.3 Shallow Neural Networks

Shallow neural networks typically refer to feedforward neural networks containing one or two hidden layers, constituting a common class of algorithms for constructing nonlinear mapping models in soft sensing. Their core principle lies in leveraging nested nonlinear activation functions to build complex functional relationships between inputs and outputs, thereby overcoming the limitations of traditional regression models in capturing nonlinear couplings. Based on the datasets for Modular, Waterwall, and RDF furnace types compiled in Literature [201], Literature [200] constructed neural network models with single, double, and multiple hidden layers, respectively, to enhance the descriptive capability for the nonlinear characteristics of DXN. Subsequent research further introduced intelligent optimization algorithms to optimize network parameters [203], aiming to mitigate the propensity of shallow neural networks to converge to local optima and to improve model accuracy and generalization capability.

To address the challenges of limited sample size, high variability, and strong noise in DXN data, Literature [27] utilized four years of offline DXN assay data from an MSWI power plant. By employing correlation analysis and principal component analysis (PCA), the study screened 13 key features from an initial set of 23 process variables, upon which a shallow neural network model was constructed. Sensitivity analysis was further applied to identify three critical factors influencing DXN. Nonetheless, the aforementioned methods exhibit shortcomings regarding the comprehensiveness of feature selection, the understanding of inter-variable coupling, and model stability.

In summary, shallow neural networks provide an early, effective modeling framework for the nonlinear soft sensing of DXN emission concentrations, yet their limitations are also pronounced. As data for DXN modeling are typically characterized by limited sample size, high noise levels, and strong variability, shallow networks are prone to issues during training, such as overfitting, unstable generalization performance, sensitivity to initial weights, and convergence to local optima. Furthermore, due to a lack of constraints derived from MSWI process mechanisms, their generalization capability becomes further constrained under cross-operating-condition, non-steady-state, and complex operational scenarios exhibiting the memory effect. These shortcomings ultimately limit the reliability and interpretability of such models in practical engineering applications.

1.4.1.3.4 Deep Neural Networks (DNNs)

DNNs, by virtue of their multi-layered nonlinear architecture enabling hierarchical feature extraction, possess a superior capacity for function approximation and representation of complex features. Theoretically, the formation process of DXN involves intricate mechanisms spanning distinct zones, temperature fields, and phases, making DNNs a potentially advantageous tool for real-time soft sensing modeling. However, the existing body of research applying DNNs to DXN modeling is limited, with the notable work in [204] being a primary example. Based on long-term operational data obtained via online indirect detection at an MSWI power plant, the study conducted a comparative evaluation of multiple modeling strategies. The results indicated that a DNN utilizing sequential time-series input achieved the best performance in terms of model accuracy, thereby providing initial validation of DNN feasibility for this task. A key unresolved challenge, however, lies in acquiring training data that is both highly reliable and cost-effective, a limitation that currently hinders the practical industrial deployment of such models.

In conclusion, while DNNs are theoretically well-suited to the inherent complexity of DXN formation mechanisms, they face at least two major constraints. On one hand, due to the high cost of obtaining reliable DXN detection truth values, the modeling data are typically characterized by limited sample size, highly skewed distribution, high noise levels, and strong uncertainty. Given their large number of parameters and high degree of freedom, DNNs are consequently highly susceptible to overfitting and unstable generalization performance when trained on insufficient data. On the other hand, DNNs suffer from inherent drawbacks such as structural complexity and weak interpretability, which conflict with the requirements for model transparency and traceability within the MSWI process.

1.4.1.3.5 Genetic Programming (GP)

GP, an evolutionary-based symbolic regression method, automatically generates interpretable mathematical expressions through simulated biological evolution to delineate nonlinear mapping relationships between inputs and outputs. Unlike the "black box" nature of neural networks, GP directly yields explicit functions based on a predefined symbol set, offering advantages such as strong interpretability and independence from a priori model structure. This makes it particularly suitable for complex industrial processes characterized by high nonlinearity and incomplete mechanistic understanding. In the context of DXN soft sensing modeling, utilizing DXN emission concentration data for Modular, Waterwall, and RDF furnace types compiled in Literature [200], the literature [201] employed GP to construct a nonlinear soft sensing model, expressed as follows:

$$
\begin{cases}
\mathrm{DXN} = \mathrm{CO} \times \mathrm{Log}\left[\left(\mathrm{CO} - 59.775/\mathrm{HCL}\right) \times \mathrm{Log}\left(4.013 \times \mathrm{Temp2}\right)\right] + 276.157 \\
\qquad \times\ \mathrm{CO} \times \mathrm{ESP} \\
\mathrm{DXN} = \left(\mathrm{Templ} \times \mathrm{CO} \times \mathrm{HCL}\right)/\left(1.5868 \times \mathrm{PM} \times \mathrm{Temp2}\right) \\
\mathrm{DXN} = \left\{\left[\left(\mathrm{CO} \times \mathrm{HCL}\right)/\left(210.4297 - \mathrm{Templ}\right)\right] \times \mathrm{Log}\left(\mathrm{CO}\right)\right\}/1000 + 0.09456
\end{cases}
\tag{1.18}
$$

where Temp2 denotes the flue gas temperature above the combustion chamber; ESP is a binary variable that equals 1 when an ESP is employed and 0 otherwise; Templ represents the flue gas temperature at the stack outlet; and PM signifies the particulate matter concentration at the stack outlet.

Studies have demonstrated that GP outperforms traditional multivariate regression models in terms of model accuracy, enabling a more effective capture of the nonlinear characteristics inherent in DXN emissions. However, given the generally limited volume of DXN data, GP models still face challenges such as insufficient generalization capability, sensitivity to noise, and expression instability due to an expansive search space. These limitations indicate that, although GP has enhanced the generalization capability of nonlinear soft sensing modeling, its engineering application requires further optimization through strategies incorporating data augmentation and mechanistic constraints.

1.4.1.3.6 Tree-Structured Algorithms

Tree-structured algorithms are characterized by strong interpretability, clear architecture, and moderate nonlinear modeling capability. Unlike black box models such as neural networks or SVM, tree models represent the relationship between inputs and outputs through a hierarchical partitioning of the input space into a tree-like structure, inherently offering interpretability and transparent model architecture. Furthermore, ensemble-based algorithms like RF, gradient boosting decision trees

(GBDT), and deep forest regression (DFR) exhibit robust generalization performance in scenarios with limited samples and high noise levels. Consequently, they have emerged as effective modeling tools for addressing the characteristics of DXN data, which include small sample sizes, sparse distributions, and complex nonlinearity. Currently, real-time soft sensing methods for DXN emission concentrations based on tree-structured algorithms primarily encompass approaches utilizing parallel/serial forest algorithms, DFR algorithms, and hybrid enhanced forest algorithms.

1.4.1.3.6.1 Parallel/Serial Forest-Based Algorithms To enhance the generalization capability of parallel/serial forest-based algorithms in soft sensing modeling for DXN emission concentrations, related research has progressed along multiple avenues, including sample expansion, ensemble structure optimization, and model selection mechanisms. Addressing the issue of limited sample size in DXN modeling data, Tang et al. proposed a soft sensing method for DXN emission concentrations based on boundary extension, sample interpolation, and selection [205]. Their approach augments the modeling dataset through domain extension and hybrid interpolation to generate a large number of virtual samples, followed by a virtual sample selection strategy to filter high-quality samples for training the soft sensing model. This methodology effectively mitigates the problem of insufficient model generalization performance caused by small sample sizes. Regarding the enhancement of model accuracy and robustness, an ensemble learning framework that integrates both parallel and serial forest algorithms was constructed [206]. This methodology involves resampling the original data to construct multiple base learners using CART trees within a RF sub-module. Subsequently, a GBDT sub-module employs residual learning to iteratively improve the model's generalization performance. This process ultimately yields an integrated model with a hybrid serial/parallel architecture. Experimental results demonstrate that this framework significantly outperforms single-tree models in terms of both accuracy and stability. Further advancing this line of research, to address the issues of high model complexity and insufficient generalization performance exhibited by traditional forest algorithms under complex operating conditions, a selective ensemble model based on Bayesian inference was proposed [207]. This method first employs bagging to generate multiple data subsets, from which corresponding binary tree sub-models are constructed. Bayesian inference is then utilized to compute posterior information, enabling a weighted integration of the optimal sub-models.

1.4.1.3.6.2 DFR-Based Algorithms Deep forest (DF) achieves hierarchical feature extraction and representation enhancement through a multi-layer cascade of forests, endowing it with stronger nonlinear fitting capability than traditional tree models, while retaining high interpretability and relatively low training requirements. For the soft sensing of DXN emission concentrations, Tang et al. [208] proposed a DF regression (DFR) algorithm based on cross-layer fully-connected structures. By introducing cross-layer information flow between model layers, this approach enables the sharing and complementary integration of features from different levels, thereby significantly enhancing feature representation capacity and demonstrating the advantage of maintaining good fitting performance even under small-sample conditions. Building on this, a Simplified DFR method with residual fitting (SDFR-ref) was developed to reduce model complexity [209]. This method first performs feature reduction on high-dimensional process variables through mutual information and significance testing, and subsequently employs a residual fitting strategy with a step factor within the forest layers to improve the accuracy and stability of hierarchical learning. To achieve higher accuracy and stronger adaptability to varying operating conditions, an improved deep forest regression (ImDFR) was further proposed model that increases the diversity of forest sub-models [210], which resulted in additional gains in the accuracy of the DXN soft sensing model.

1.4.1.3.6.3 Hybrid Enhanced Forest-Based Algorithms Building upon traditional tree-structured forests and DF algorithms, some studies have further integrated mechanisms such as fuzzy inference systems and broad learning systems to enhance the modeling capability of tree-structured algorithms when dealing with data characterized by small sample sizes, strong noise, and time-varying properties, thereby forming the category of "hybrid enhanced forest algorithms." To address the significant uncertainties present in the modeling data, Takagi-Sugeno (T-S) fuzzy regression trees (TSFRT) and their ensemble variants were proposed [211]. This approach employs T-S fuzzy rules within local regions after node splitting to achieve nonlinear mapping, enabling smoother capture of continuous relationships between variables compared to traditional tree structures, and has achieved superior fitting accuracy in the task of DXN emission concentration soft sensing modeling. Furthermore, targeting the time-varying nature of the MSWI process, Literature [212] proposed a fuzzy-tree broad learning system (FTBLS) comprising offline training and online update/detection phases. In the offline phase, weighted k-means is used to construct a typical sample pool to reduce training costs. The online phase employs recursive Principal component analysis (PCA) to identify the time-varying characteristics of the MSWI process, and upon detection of abnormal operating conditions, utilizes fast incremental learning to update the model. Building on this foundation, Literature [34] proposed a Hybrid Broad Forest Regression algorithm that integrates broad learning, ensemble learning, and latent feature extraction. The algorithm first employs a hybrid forest group (e.g., comprising both RF and complete RF) to generate high-dimensional feature mappings. Subsequently, latent feature extraction is performed based on contribution rate and information measurement criteria to achieve key information retention, redundancy suppression, and model complexity reduction. The extracted latent features are then used to train a feature enhancement layer to augment feature representation capacity. Finally, efficient training is realized through incremental learning and Moore-Penrose pseudoinverse solutions. Experimental results demonstrate its commendable modeling accuracy and robustness. Evidently, by combining the structural advantages of tree-structured algorithms with the flexible reasoning capability of fuzzy systems and the efficient feature mapping capability of broad learning, the aforementioned methods exhibit superior expressiveness and adaptability in the task of real-time soft sensing of DXN emission concentrations.

In summary, the distinctive advantages of tree-structured algorithms in the task of real-time soft sensing for DXN emission concentrations are manifested in their clear model architecture, strong interpretability, ability to handle nonlinear relationships, and capacity to adapt to limited sample sizes and high-dimensional features. Traditional DT algorithms possess inherent strengths in feature identification and model interpretability. Ensemble-based forest algorithms, integrating multiple DTs, further enhance model stability and robustness. DFR algorithms achieve greater feature representation capacity through a multi-layer cascade structure, thereby maintaining sound generalization performance even in data-scarce scenarios. Hybrid enhanced forest algorithms, by incorporating mechanisms such as fuzzy inference and broad learning, enable flexible expression, rapid incremental learning, and improved adaptability to time-varying operating conditions for tree-structured models. However, the performance of these algorithms requires further improvement in areas such as enhancing dynamic modeling capabilities, incorporating mechanistic constraints, and achieving effective online incremental learning. Advancing these aspects is essential for constructing more stable, reliable, and industrially viable real-time soft sensing models for DXN.

1.4.1.3.7 Other Methods

Beyond the DXN emission concentration soft sensing models based on regression analysis, SVM, neural networks, and tree-structured algorithms discussed previously, some studies have explored hybrid modeling strategies that fuse simulated mechanistic data with process data. These strategies aim to mitigate the modeling challenges arising from the scarcity of DXN samples and the complexity of the MSWI process. For instance, the literature [213] proposed a hybrid modeling framework for constructing DXN emission concentration models. The framework first constructs a white-box surrogate model based on a simulated mechanistic dataset using a linear regression decision tree (LRDT). It then introduces semi-supervised transfer learning and a constrained incremental random weight neural network to achieve the fusion of a virtual data-driven mechanistic model with a real historical data-driven LRDT model, thereby enhancing the adaptability of the soft sensing model across multiple operating conditions. However, several critical questions warrant further investigation. These include how to effectively compensate for the simplification errors inherent in the assumptions of simulation-based mechanistic models, how to overcome the potential underfitting of data-driven models caused by the scarcity of real DXN samples, and how to address issues related to the rationality of model transfer, the credibility of virtual samples, and the weight allocation for multi-source data.

1.4.1.3.8 Summary

In summary, while real-time soft sensing methods demonstrate significant potential for DXN emission concentration detection, the "small-sample" issue remains a fundamental bottleneck constraining their accuracy. Due to the long cycles, high costs, and low sampling frequency associated with offline DXN emission concentration detection, most studies can only obtain a limited number of modeling samples that are temporally sparse and exhibit significant noise. This scarcity hinders soft sensing models from adequately learning and mapping the complex mechanisms underlying DXN formation and transformation, leading to widespread issues such as overfitting, parameter instability, and sensitivity to noise. Although recent efforts have attempted to mitigate data scarcity through techniques like VSG and hybrid data modeling, these approaches remain largely validated only on localized operating conditions or specific datasets. The authenticity, mechanistic consistency, and cross-condition applicability of the generated samples have yet to be fully substantiated.

1.4.2 Current Research Status on the Detection of DXN Generation Concentrations

DXN formation predominantly occurs within the furnace of an MSWI process, encompassing stages such as high-temperature pyrolysis, incomplete combustion, de novo synthesis, and gas/solid-phase catalytic reactions within the flue gas cooling zone. Compared to the detection of DXN emission concentrations at the stack outlet of an MSWI process, modeling the generation concentration relies more heavily on a comprehensive characterization of various factors. These include the compositional changes of materials and flue gas inside the furnace, disturbances in flame combustion states, and the overall operating conditions of the incinerator. Recognizing that a single type of process data is often insufficient to fully characterize DXN formation mechanisms, research has increasingly shifted toward constructing detection models for DXN generation concentrations using multi-modal data fusion mechanisms. Within this context, Literature [214] proposed a multi-modal data-driven soft sensing modeling strategy for DXN generation concentrations. The strategy first performs temporal scale matching and alignment of multi-source heterogeneous data based on empirical knowledge. Concurrently, it introduces a transfer learning-based CNN to extract combustion state features from furnace images. Finally, an adaptive DFR algorithm with cross-layer fully-connected structures is employed to construct separate soft sensing models for flame image features and process variables, thereby achieving intelligent detection of DXN generation concentrations.

In summary, existing research has sufficiently demonstrated the unique advantages of multi-modal data in characterizing the DXN formation and has validated its potential for improving the modeling accuracy and feature representation capability of generation concentration detection. However, due to challenges in the acquisition and synchronization of multi-modal data, significant temporal scale discrepancies among different data sources, and the extreme scarcity and uneven quality of labeled samples, current research on DXN generation concentration detection remains largely confined to the exploration and validation stage for specific MSWI processes.

1.4.3 Current Research Status on the Detection of DXN Whole-Lifecycle Concentration

As informed by the mechanisms of DXN formation, it persists throughout the whole MSWI process, exhibiting complex coupled characteristics spanning multiple zones, temperature fields, and phases. Relying solely on formation or emission concentrations measured at points such as the boiler or stack outlet is insufficient for a comprehensive assessment of its generation and transformation patterns. The literature [215] developed a multi-point real-time soft sensing modeling framework for DXN concentration. This framework addresses the distributional heterogeneity of

DXN data across different process stages using auxiliary modeling techniques and employs various enhanced tree-structured algorithms to achieve DXN concentration modeling for multiple stages of the MSWI process. Building upon this, Tang et al. [216] constructed a systematic framework for DXN whole lifecycle concentration detection. Their approach first involved developing a numerical simulation model for the DXN whole lifecycle by integrating software tools such as FLIC, Aspen Plus, and Aspen Adsorption. Subsequently, experimental design and implementation were conducted to obtain virtual simulation data under various operating conditions. Finally, a Takagi-Sugeno (T-S) fuzzy forest regression algorithm was adopted to construct an interpretable detection model for DXN whole-lifecycle concentration.

In summary, this line of research has pioneered an end-to-end modeling strategy that integrates numerical simulation-derived mechanistic data, experimental design, and interpretable model development. It represents a significant advancement toward constructing a comprehensive DXN concentration detection framework for the whole MSWI process. To further strengthen this foundation, future efforts should focus on more effectively incorporating real historical operational data. Such integration is essential for enabling more refined emission control and robust risk early-warning systems.

1.4.4 Current Research Status on Early Warning of DXN Emission Concentrations

The core challenge in early warning of DXN emissions lies in the extreme scarcity of high-quality samples available for training predictive models. Existing research predominantly focuses on VSG techniques, aiming to enhance the robustness and generalization capability of early-warning models by augmenting the number of samples covering various risk categories. Following this paradigm, Literature [217] employed GANs to expand a real-world DXN dataset. By stratifying DXN emissions into multiple risk levels, the study used GANs to generate virtual samples corresponding to each risk category for training a DXN emission risk early-warning model. Building on this, the literature [218] enhanced the VSG process by integrating an active learning mechanism. Their approach first uses GANs conditioned on DXN risk levels to generate candidate virtual samples. An active learning mechanism is then applied to screen and evaluate these candidates. Finally, an early-warning model for DXN emissions is constructed based on a hybrid training set comprising both real and selected virtual samples, effectively alleviating modeling difficulties under small-sample conditions. Furthermore, considering the characteristics of DXN data (e.g., high dimensionality, strong noise, and substantial uncertainty), the literature [219] proposed a VSG framework combining FNNs with GANs for DXN emission concentration risk early-warning. This framework first employs a feature selection strategy to mitigate high-dimensional noise and redundant features. A generative adversarial fuzzy neural network (GAFNN) is then utilized to generate a virtual sample set capable of characterizing uncertainty. After filtering low-quality samples via a multi-constraint selection mechanism, a DF model is employed to build the warning classifier. Results indicate that this method offers superior risk identification performance and generalization capability. To further improve the interpretability of virtual samples, the literature [220] proposed a virtual sample enhancement strategy based on explainable space mapping and adversarial generation mechanisms to support the construction of a DXN emission risk early-warning model. The strategy first maps high-dimensional features into a two-dimensional, interpretable latent space using a Variational Autoencoder (VAE). The distribution of latent variables within this space is then enriched through multi-angle rotation operations. Subsequently, GANs are applied to this enriched space to generate candidate virtual samples. Finally, qualified virtual samples are mixed with real samples to construct the early-warning model.

In summary, current research on early-warning of DXN emission concentrations primarily employs virtual sample augmentation as the core strategy to overcome the bottleneck of small-sample modeling, having made preliminary progress in aspects such as the distributional consistency, representativeness, and interpretability of the generated virtual samples. Nevertheless, existing methods still require further in-depth exploration regarding the assessment of virtual sample authenticity, validation of their adaptability across varying operating conditions, and their deployment for online early-warning in practical industrial scenarios.

1.4.5 Discussion and Analysis

1.4.5.1 Challenges in Offline Direct Detection Methods

Although offline direct detection methods offer the highest measurement accuracy, they exhibit significant limitations in practical engineering applications. Chromatography, capable of precise separation and quantification of DXN congeners, is internationally recognized as the standard reference method. However, its workflow is highly intricate, which encompasses sample collection, extraction, cleaning, and analysis. Moreover, it has a typical detection cycle ranging from days to weeks. Furthermore, it relies on expensive instrumentation and highly skilled personnel, making it unsuitable for meeting the real-time and continuous monitoring requirements of actual MSWI processes. Laser mass spectrometry offers advantages in selectivity, sensitivity, and response speed, enabling rapid detection of multiple congeners. Nevertheless, it requires prior knowledge of the spectral signatures of target congeners and imposes extremely stringent demands on experimental conditions, which hinders its widespread adoption. Biological and immunoassay methods are more advantageous in terms of detection cycle and cost, allowing for rapid, high-throughput sample screening. A fundamental limitation of both approaches, however, is that they can only provide the TEQ of a DXN sample, failing to discriminate the concentrations of individual congeners.

From the perspective of MSWI process control, the most critical issue with offline direct detection methods is the inherent temporal latency introduced by laboratory-based offline analysis. Whether chromatography, laser mass spectrometry, or biological/immunoassays, none can provide minute- or even hour-level real-time data. This inability hinders their application in supporting critical functions such as abnormal condition identification, emission peak early-warning, real-time feedback control, and dynamic operational optimization of the MSWI process. Consequently, while these methods hold an irreplaceable benchmark status in terms of measurement accuracy, they offer limited utility for process control, online detection, and real-time decision-making. In practice, they are primarily employed as calibration standards and data sources for online indirect detection equipment and real-time soft sensing models.

1.4.5.2 Limitations of Online Indirect Detection Methods

Online indirect detection methods consist of two core components: an online detection system for indicators/associated compounds and a mapping model that relates these compounds to DXN concentrations. Although this technical approach offers a clear theoretical advantage in terms of real-time capability, its practical engineering application in actual MSWI processes faces multiple limitations.

From the perspective of the online detection system for indicators/associated compounds, the existing challenges are manifested in several aspects: (1) Current online detection equipment for these compounds typically involves multiple devices for flue gas sampling, pretreatment, and analysis, which are generally characterized by structural complexity and high sensitivity to operational environments; (2) The flue gas from MSWI processes exhibits properties such as high temperature, high humidity, high particulate loading, and high corrosiveness. Consequently, online detection of indicators/associated compounds places exceptionally stringent demands on the stability of sampling probes, pretreatment systems, and analytical instruments; (3) During long-term operation, online detection devices are highly susceptible to issues like contamination, clogging, signal drift, and incur high maintenance costs. Coupled with the substantial initial capital investment, these factors make it difficult to achieve stable, long-term deployment in most MSWI plants; (4) The composition of MSW varies drastically over time and across different regions. Differences in chlorine-containing components, metal catalyst content, and flue gas treatment equipment can significantly alter the formation patterns of indicators, leading to considerable variability in the response characteristics of the same indicator across different plants or even at different times within the same plant. Collectively, these factors fundamentally constrain the applicability of online indirect detection methods in the practical MSWI process.

From the perspective of the mapping model that relates indicators/associated compounds to DXN, the limitations are evident in the following aspects: (1) The mapping relationship between an indicator/associated compound and DXN is inherently dependent on specific operating conditions and is contextually limited. Since DXN concentrations are influenced by a combination of factors, including furnace temperature, flue gas oxygen content, and the memory effect, its correlation with a single indicator typically fluctuates significantly with changes in operating conditions. The modeling data available from industrial sites are extremely limited, with most studies relying on datasets comprising merely dozens to a few hundred data points. Such limited data are insufficient to encompass the diverse scenarios of different furnace types, load fluctuations, tail gas treatment strategies, and seasonal variations in MSW composition. Consequently, mapping models often perform well under specific, calibrated conditions but suffer from rapid accuracy degradation and insufficient generalization capability when applied across different operating scenarios; (2) The memory effect of DXN introduces a lagged relationship between emission concentrations and instantaneous indicator measurements. This lag may span hours, days, weeks, or even months. Conventional static mapping models struggle to capture this asynchronous characteristic, thereby further compromising the reliability of their predictions. (3) Most existing studies rely on a single indicator to construct the mapping relationship, neglecting the multivariable coupling features arising from multiple reaction pathways and multi-stage formation mechanisms. This oversight renders the model overly sensitive to fluctuations in a particular indicator, thereby limiting both its stability and robustness.

In summary, despite its advantage in real-time detection capability, the online indirect detection method faces significant challenges in the engineering deployment and maintenance of the indicator/associated compound detection system. Furthermore, the mapping models between these compounds and DXN exhibit inherent shortcomings in multi-condition adaptability, cross-scenario generalization capability, and the characterization of time-lag effects. Collectively, these factors severely restrict its widespread adoption in industrial environments.

1.4.5.3 Challenges in Real-Time Soft Sensing and Early-Warning Methods

Although existing research has made preliminary progress in areas such as real-time soft sensing and early-warning for DXN, significant shortcomings persist under the constraints of sample scarcity, variable operating conditions, and unobservable mechanistic processes. These shortcomings are particularly evident in the efficiency of real-virtual data fusion, knowledge transfer, semi-supervised learning, and the development of high-confidence, whole-lifecycle interval detection methods.

1.4.5.3.1 Lack of Integrated Whole-Lifecycle Detection and Early-Warning Models Based on Real-Virtual Data Fusion

To address the critical bottleneck of severely insufficient DXN modeling sample data, some research has begun to employ multi-software coupling strategies to generate virtual simulation-based mechanistic data spanning various operating conditions and process stages. While such virtual data exhibit significant advantages in terms of operational coverage and stage completeness, they can, to some extent, compensate for the deficiency of real modeling data. The existing research predominantly focuses on DXN emission concentrations at the stack outlet and lacks measurements from other stages. In addition, their credibility remains to be further validated. Moreover, current research predominantly remains at the stage of using virtual simulation data merely as supplementary samples to augment real historical data, failing to develop mature learning frameworks that effectively fuse real and virtual data. This gap results in the constructed whole-lifecycle detection and early-warning models having inherent limitations in accurately representing actual MSWI processes. Specifically, numerical simulation models heavily rely on simplified mechanisms, empirical parameters, and idealized boundary conditions. This inevitably leads to a distributional bias between the virtual simulation data and real operational conditions, making direct use of such data for modeling prone to introducing systematic errors. Furthermore, existing studies often simply aggregate virtual data as additional training samples, lacking sophisticated fusion frameworks to address critical issues such as credibility assessment and dynamic weight allocation for the virtual data.

Consequently, the stability, reliability, and interpretability of the fused models are severely compromised under complex dynamic operating conditions.

In summary, the core challenge in constructing DXN's whole-lifecycle detection and early-warning models currently lies in developing a unified fusion framework capable of effectively integrating virtual simulation data with real operational data. Such a framework is essential to fully harness the complementary strengths inherent in these two distinct data types, thus enabling the effective development of robust and reliable whole-lifecycle models.

1.4.5.3.2 Lack of Whole-Lifecycle Detection and Early-Warning Models Based on Knowledge Transfer and Semi-supervised Learning

In actual industrial processes, while labeled samples with DXN ground-truth values are extremely scarce, the DCS continuously records a vast amount of process data encompassing multiple stages of the MSWI process. These data, which dynamically evolve, include parameters such as furnace temperature, primary/secondary air flow rates, and flue gas composition. These unlabeled data streams constitute a crucial foundation for constructing DXN's whole-lifecycle detection and early-warning models. However, existing research almost exclusively relies on the limited labeled samples of DXN emission concentrations, failing to establish effective learning frameworks that can systematically leverage this abundance of unlabeled data. Consequently, current models exhibit insufficient learning of the transfer relationships and dynamic coupling characteristics between different stages of the MSWI process, rendering them inadequate to meet the demands of whole-lifecycle detection and early-warning for DXN. Furthermore, significant variations exist among different MSWI plants regarding furnace configuration, feedstock composition, and flue gas cleaning systems. Models lacking knowledge transfer and domain adaptation capabilities are therefore typically effective only within a single plant under specific operating conditions. Although knowledge transfer and semi-supervised learning have been widely and successfully applied in various industrial sectors such as petroleum, chemical, and metallurgy, demonstrating strong adaptability, their application in the field of DXN's whole-lifecycle management remains immature. The primary reasons for this are threefold: (1) The cross-stage coupled complexity of DXN formation mechanisms makes it difficult to define the key features for transfer; (2) The unlabeled process data are characterized by high noise levels and strong dependency on operating conditions, making pseudo-labels generated through direct semi-supervised learning prone to error accumulation and risk; (3) Current research lacks standardized data structures capable of bridging different furnace types and process stages, which impedes the unified construction of cross-domain models.

In conclusion, current research on DXN's whole-lifecycle detection and early-warning has yet to establish a knowledge transfer and semi-supervised learning framework that can effectively leverage massive amounts of unlabeled process data while adapting to multiple operating conditions and varying furnace configurations across different plants. This deficiency prevents existing real-time soft sensing models and early-warning methods from overcoming the bottleneck imposed by data scarcity and hinders the realization of genuine cross-stage, cross-furnace-type online real-time detection and risk early-warning for DXN throughout its whole lifecycle.

1.4.5.3.3 Lack of High-Confidence Whole-Lifecycle Interval Detection Models Aligned with Field Expert Cognition

Despite the substantial volume of research on real-time soft sensing and early-warning for DXN concentrations, existing methods still fall short of meeting the practical requirements of actual MSWI plants for high confidence, interpretability, and regulatory compliance in detection and warning outcomes. Currently, the developed models predominantly provide single-point estimates or warning signals, lacking high-confidence interval ranges that align with the cognitive mechanisms of field experts. Furthermore, existing research outcomes struggle to explain the rationale behind these point estimates or warnings from a mechanistic perspective and are even less capable of tracing the key process stages or variables responsible for anomalous readings or elevated risk levels. This lack of transparency makes it difficult to gain the trust and acceptance of domain experts in practical engineering applications. More critically, under conditions characterized by scarce DXN formation and emission concentration data, the infeasibility of directly measuring in-furnace DXN concentrations, incomplete mechanistic observability, and frequent operational fluctuations, reliance on single-point detection or warning values often masks underlying risk-prone variability. This limitation undermines their support for high-stakes scenarios such as abnormal condition early-warning and emission reduction control.

In summary, the current framework for DXN whole-lifecycle detection and early-warning still lacks a high-confidence interval detection mechanism for DXN concentrations that is both aligned with the cognitive mechanisms of field experts and offers explainable mechanistic chains. Future efforts should therefore prioritize the development of an integrated DXN whole-lifecycle interval detection system that synergistically combines uncertainty modeling, interpretable soft sensing, and cross-stage causal inference. This advancement is essential to ensure that predictive outcomes possess both high accuracy and high credibility, thereby genuinely meeting the cognitive expectations of domain experts and the requirements for sound engineering decision-making.

1.5 RESEARCH STATUS OF DXN EMISSION OPTIMIZATION CONTROL

This chapter systematically reviews recent advances in the optimization control of DXN emission concentrations from the dual perspectives of process routes and process control. It aims to provide a reference for constructing efficient and controllable emission reduction strategies.

1.5.1 Current Status from the Chemical/Environmental Engineering Perspective of Optimization Control

From the chemical/environmental engineering perspective of process control, current optimization of DXN emission concentrations primarily relies on two pathways: combustion condition optimization and cleaning treatment technologies. Regarding combustion optimization, Literature [221] demonstrated the dual inhibitory effects of thiourea on both DXN and NO_x emissions under industrial fluidized-bed furnace conditions. The literature [222] constructed a DXN degradation kinetic model based on electron beam irradiation, showcasing the potential of novel flue gas treatment technologies. Through a comparison of RDF incineration mode and MSW-coal co-combustion, the literature [223] indicated that RDF offers higher energy recovery efficiency and is more conducive to pollution emission control. Literature [224] identified key parameters for the simultaneous suppression of DXN and NO_x in solid waste gasification-combustion experiments, providing a reference for source control within the furnace. In terms of cleaning treatment, the literature [225] noted that while ACI can efficiently adsorb DXN from flue gas, it essentially constitutes a "transfer" rather than an "elimination" of pollutants, making it difficult to fundamentally mitigate emission risks. The literature [226] emphasized that the selection of cleaning treatment technologies should be based on a comprehensive lifecycle assessment of MSWI, considering emission reduction effectiveness, economic viability, and equipment compatibility. In general, most grate-fired MSWI plants in developing countries, such as China, currently adopt the "good combustion practices plus cleaning treatment" strategy summarized in [227] as their primary technological pathway. However, significant gaps remain compared to developed nations in areas such as feedstock pretreatment, combustion regulation, and operational experience accumulation. These gaps lead to substantial volatility in actual DXN emissions and the persistent issue of the NIMBY syndrome.

In summary, the chemical and environmental disciplines have accumulated substantial knowledge regarding DXN formation mechanisms and a wealth of experience in process improvements for its control. However, these studies predominantly focus on newly constructed plants or rely on the framework of existing facilities. For the latter scenario, implementing such findings often entails significant structural adjustments to the existing infrastructure, combustion systems, or cleaning units, which are associated with high costs and prolonged retrofitting periods. Moreover, most of these methodologies are predicated on fixed operating conditions or steady-state assumptions, rendering them ill-equipped to handle the dynamic characteristics commonly present in actual MSWI systems, such as load fluctuations, variations in fuel composition, and equipment aging. Therefore, to achieve stable compliance, predictable regulation, and real-time optimization of DXN emissions, reliance solely on traditional process improvement pathways is inadequate. It is imperative to introduce methodologies from control science to compensate for the essential dynamic regulatory capabilities required for practical engineering implementation. These methodologies include index modeling, optimization decision-making, and intelligent control technologies.

1.5.2 Current Status of Optimization Decision-Making and Intelligent Control from the Control Science Perspective

From the perspective of control science, the optimization control of DXN emission concentrations in the MSWI process constitutes a typical complex industrial process operation optimization problem characterized by multivariable interactions, strong coupling, and nonlinearity. Studies have shown that DXN emission concentrations are closely correlated with multiple types of process variables. These include manipulated variables (MVs) such as air and fuel distribution; controlled variables (CVs) such as furnace temperature and flue gas oxygen content; auxiliary variables (AVs) such as grate temperature; and environmental indicators (EIs) such as the concentrations of CO, HCl, NO_x, and particulate matter. Since DXN cannot be detected online, in actual industrial practice, domain experts typically rely on empirical adjustments based on variations in readily measurable pollutants like CO to achieve indirect control of DXN. This approach, however, is notably stochastic and subjective, making it difficult to ensure stable compliance with DXN emission limits over extended periods of operation.

In this context, an operational optimization framework that integrates a performance index model with intelligent optimization algorithms has gradually emerged as a mainstream approach in the process industries. Within this framework, the performance index model is responsible for constructing real-time calculable process performance indicators, thereby providing continuously available feedback signals for emission reduction optimization. The intelligent optimization algorithms, building upon this foundation, search within complex constraint and high-dimensional decision spaces for optimal operating parameter combinations that concurrently address emission control, safety, and energy efficiency. Although dedicated operational optimization research specifically for DXN emissions is still lacking, a well-established set of optimization control methodologies with clear transfer potential has been developed for conventional pollutants such as NO_x and SO_2. For instance, some studies have introduced evolutionary state assessment mechanisms [228, 229] to achieve adaptive tuning of algorithm parameters, significantly enhancing convergence efficiency and global optimization capability. Other research has employed operating condition identification [230] coupled with incremental update strategies [231] to address model mismatch issues caused by dynamic multi-condition variations. Additionally, multi-time-scale optimization frameworks have been constructed [232, 233] to accommodate differences in the response speeds of various variables, thereby improving the stability and effectiveness of operational optimization. In addition, Tang et al. developed a module hardware-in-loop simulation platform in terms of the end-edge-cloud mode that can verify these intelligent optimization control algorithms [234].

Existing research has sufficiently demonstrated that intelligent optimization methods possess excellent adaptability and powerful search capabilities in addressing industrial operational optimization problems characterized by nonlinearity, multi-objectives, and multiple constraints. This provides a transferable methodological foundation for the optimization control of DXN emission concentrations. However, compared to conventional pollutants, DXN exhibits more complex formation mechanisms, stronger hysteresis (memory effect), and even fewer actual measurement values available. To achieve systematic, interpretable, and deployable optimization control for DXN, it remains imperative to deeply integrate soft sensing models, whole-lifecycle mechanistic understanding, multi-condition robust optimization strategies, and closed-loop control technologies. The ultimate goal is to construct a truly industrially viable optimization control system for DXN emission concentrations.

1.5.3 Discussion and Analysis

Although existing research has achieved certain progress in process regulation and operational optimization, the current research framework nevertheless falls short of the objective, from a control science perspective, of establishing a comprehensive whole-lifecycle optimization control system for DXN emission concentrations. Significant gaps remain in several critical areas, including generation source control during the waste combustion and heat exchange stages, dynamic regulation of the memory effect, synergistic optimization of formation and emission concentrations, and multi-objective collaborative control of trace pollutants alongside conventional pollutants. A systematic and readily deployable control mechanism encompassing these aspects has yet to be developed. Therefore, this section analyzes and discusses the shortcomings of current control research from the four dimensions, aiming to clarify the key breakthrough directions for future optimization control of DXN emission concentrations.

1.5.3.1 Lack of Optimization Control for Minimizing DXN Generation Concentrations during the Solid Waste Combustion and Waste Heat Exchange Stages

The solid waste combustion and waste heat exchange stages constitute the fundamental source of DXN generation at the boiler outlet. Key MVs (e.g., air and material distribution strategies, primary/secondary air flow rates and their ratios, and grate pushing speed), process states (e.g., temperature field distribution, local oxygen content, combustion uniformity), and system constraints (e.g., flue gas residence time distribution) collectively determine the generation intensity of DXN precursors and the subsequent de novo synthesis reactions. These factors represent the most significant sources influencing DXN generation concentrations. However, existing research on DXN optimization control almost exclusively focuses its objectives on the emission side at the stack outlet, emphasizing the adjustment of relevant MVs and the optimization of sorbent consumption in the flue gas cleaning stage to achieve "emission concentration minimization." This approach lacks dedicated optimization control strategies targeting the DXN formation process itself during the combustion and heat exchange stages. Evidently, this cleaning treatment concentration-centered control paradigm results in a deficiency of proactive suppression mechanisms for the DXN formation stage. Consequently, upstream disturbances (e.g., load fluctuations and variations in MSW composition) cannot be promptly addressed or mitigated. These effects, when transmitted to the flue gas cleaning system, amplify DXN re-synthesis behavior and significantly increase the likelihood of triggering the memory effect, thereby making long-term stable compliance at the emission outlet difficult to achieve. Due to the absence of efficient index models capable of characterizing DXN generation concentrations, current optimization control is forced to rely on feedback from DXN concentrations measured at the stack outlet. This necessitates lagged regulation at the final emission stage, whose response speed and disturbance rejection performance are insufficient to meet industrial requirements.

Therefore, DXN optimization control urgently requires a strategic shift from "cleaning treatment feedback re-optimization" to "source reduction pre-control." The critical path forward is to construct models capable of characterizing DXN generation concentrations during the combustion and waste heat exchange stages and to integrate them into the whole-lifecycle optimization control framework. This approach is evidently pivotal for proactively inhibiting DXN generation, reducing system sensitivity, and enhancing the stability of emission reduction across the whole lifecycle.

1.5.3.2 Incomplete Dynamic Optimization Control Mechanisms for the Uncertain Memory Effect

The memory effect is a critical factor leading to delayed emission peaks, post-transitional emission rebounds, and unstable DXN emissions. Its origin lies in the long-term adsorption and accumulation of DXN on equipment surfaces, ash deposits, and filtration media, followed by transient release behaviors triggered under non-steady-state conditions such as temperature disturbances, flow fluctuations, or soot-blowing operations. However, from the perspective of control science, existing optimization control frameworks have neither explicitly modeled this dynamic behavior, which spans multiple time scales and exhibits strong time-delay characteristics, nor have they developed a corresponding control system tailored to it. The current research gap manifests primarily in two aspects: (1) The lack of state characterization for the "accumulation-release" dynamic mechanism and the absence of corresponding MVs to control it. Existing studies have not constructed intermediate state estimation models, such as those reflecting internal DXN inventory, surface coverage, or changes in adsorbed states within ash deposits. This omission prevents the DCS system from proactively identifying accumulation risks to implement "feedforward" control, forcing it to resort to passive adjustments only after anomalous DXN emissions occur. Consequently, it becomes difficult to promptly suppress the peak emissions caused by delayed release; (2) Current ash deposit control strategies predominantly rely on fixed-frequency rapping with predetermined cycles. They lack dynamic compensation mechanisms for non-steady-state operations, such as temperature jumps or load disturbances. This deficiency hinders the effective blocking of conditions that trigger the memory effect and fails to maintain adequate constraints on the memory effect behavior during operational transitions.

In summary, establishing a dynamic optimization control mechanism to address the uncertain memory effect has become a critical and urgently needed focal point within the DXN optimization control framework. Future research must focus on developing dynamic state models capable of characterizing both the accumulation quantity and release rate of DXN within incineration equipment. These models must then be integrated into existing frameworks for dynamic soft sensing, optimized emission reduction decision-making, and multivariable intelligent control. This integration is essential for achieving the early identification and proactive intervention of delayed release risks, thereby enhancing the stability and predictability of whole-lifecycle emission regulation for DXN concentrations.

1.5.3.3 Lack of Intelligent Optimization Control Considering the Whole Lifecycle of DXN

MVs such as "air and material distribution" during the combustion and waste heat exchange stages, sorbent consumption in the flue gas cleaning stage, and the overall operational cost of the incineration process are strongly coupled with and mutually constrained by DXN formation and emission concentrations. This interdependence manifests concretely as follows: higher generation concentrations during the combustion and heat exchange stages increase the pressure on controlling DXN re-synthesis in the cooling zones of the latter stage. Simultaneously, the cleaning treatment units are forced to operate under conditions requiring persistently high sorbent consumption. This not only challenges the maintenance of low emissions but also significantly escalates costs associated with sorbent usage, catalyst maintenance, and corresponding energy consumption. The prevailing research framework generally lacks an intelligent optimization control mode that considers the whole lifecycle of DXN, primarily manifested in two aspects: (1) The absence of a comprehensive optimization model capable of simultaneously capturing the relationships among DXN generation concentration, emission concentration, and economic costs. This gap makes it difficult to strike a balance between emission compliance, safety margins, and operational costs; (2) The lack of a dynamic collaborative optimization mechanism for multi-objective trade-offs. Key MVs (e.g., feeding rate, grate speed, primary/secondary air flow rates and their ratios) are often controlled in an isolated, stage-specific, and objective-specific manner. This fragmented approach fails to achieve coordinated regulation of cost and emissions across the whole-lifecycle concentration profile of DXN.

In summary, from the perspective of control science, achieving optimal control for the reduction and efficiency enhancement of DXN emission concentrations necessitates the construction of its whole-lifecycle concentration model. Furthermore, it requires the introduction of multi-objective optimization, operational constraint coordination, and real-time optimization strategies. This integrated approach enables the system to achieve a dynamic balance among emission compliance, safe operation, and economic benefits.

1.5.3.4 Lack of Research on Multiscale, Multi-objective Optimization Considering Both Trace and Conventional Pollutants

From the perspective of ecological and environmental protection, the MSWI process necessitates the synergistic optimization and control of both trace and conventional pollutants. These emission indicators are highly homologous and coupled in their formation mechanisms, exhibit significant differences in detection time scales, and often present conflicting trade-offs when economic performance metrics of the incineration plant are considered. For instance, increasing the oxygen content in the furnace can help suppress the generation of DXN precursors and products of incomplete combustion, but it may concurrently lead to elevated NO_x emission concentrations. Similarly, injecting sufficient sorbent in the flue gas cleaning stage aids in the efficient removal of DXN, yet it increases fly ash generation and the associated subsequent treatment costs, as well as immediate operational expenses. Consequently, the optimization control of DXN emission concentrations cannot solely aim at minimizing a single indicator; it requires systematic synergistic regulation among multiple pollutant classes. However, current research lacks emission reduction and efficiency enhancement optimization strategies that integrate DXN, conventional pollutants, and material consumption into a unified objective framework. The primary shortcomings are manifested in three aspects: (1) The absence of a comprehensive performance index model capable of characterizing the coupling relationships and conflict characteristics among multiple pollutants. This gap makes it difficult for control strategies to achieve reasonable trade-offs among multiple constraints; (2) Existing optimization algorithms primarily target single objectives or multi-objectives in a weighted-sum form, which inadequately represent the true Pareto front relationships among various pollutants, including DXN. This approach may lead to optimization results that deviate from the global optimum; (3) The lack of a dynamic multi-objective optimization mechanism tailored to actual operating conditions, which fails to address the multi-pollutant synergistic regulation problem under disturbances such as boiler load fluctuations and variations in MSW composition.

In summary, future optimization control of DXN emission concentrations must transition from "single-pollutant optimization" to "multi-pollutant synergistic optimization." This requires the development of a performance index system capable of characterizing the mechanistic coupling relationships among multiple pollutants. Furthermore, it necessitates the introduction of advanced mechanisms such as dynamic multi-objective evolutionary algorithms, dynamic Pareto front optimization, and multi-time-scale coordination strategies. The ultimate goal is to achieve a harmonious balance among emission safety, operational economy, and overall environmental synergy.

1.6 BOOK'S STRUCTURE

Based on the former instruction, the book's structure is shown in Figure 1.21.

Figure 1.21 shows that the book consists of three principal parts:

- The first part relates to the DXN characteristics analysis of a typical MSWI process (Chapters 2–4).
- The second part address the DXN with numerical simulation modeling (Chapters 5 and 6).
- The third part describes the DXN with AI-driven detection modeling and risk warning (Chapters 7–11).

1.6.1 Part I

Chapter 2: This chapter outlines the process flow of a typical MSWI power plant utilizing a grate furnace, focusing on intelligent optimization control for pollution reduction. It provides an overview of the application of AI algorithms from both academic research and industrial application perspectives, detailing the entire process of AI modeling, AI maintenance, AI decision-making, AI optimization, and AI control. Additionally, the chapter analyzes the challenges associated with modeling the entire lifecycle of DXN within the MSWI process. Several key challenges are identified, including the lack of comprehensive testing and analysis of DXN throughout the entire process at industrial sites, the absence of a simulation model that accurately represents the actual industrial processes, and the need for a detection and warning mechanism with explainable AI.

Chapter 3: A MSWI power plant in Beijing is taken as an example, the whole-lifecycle DXN process is divided into three key parts: generation, adsorption, and emission. During the complete maintenance operating period of this MSWI power

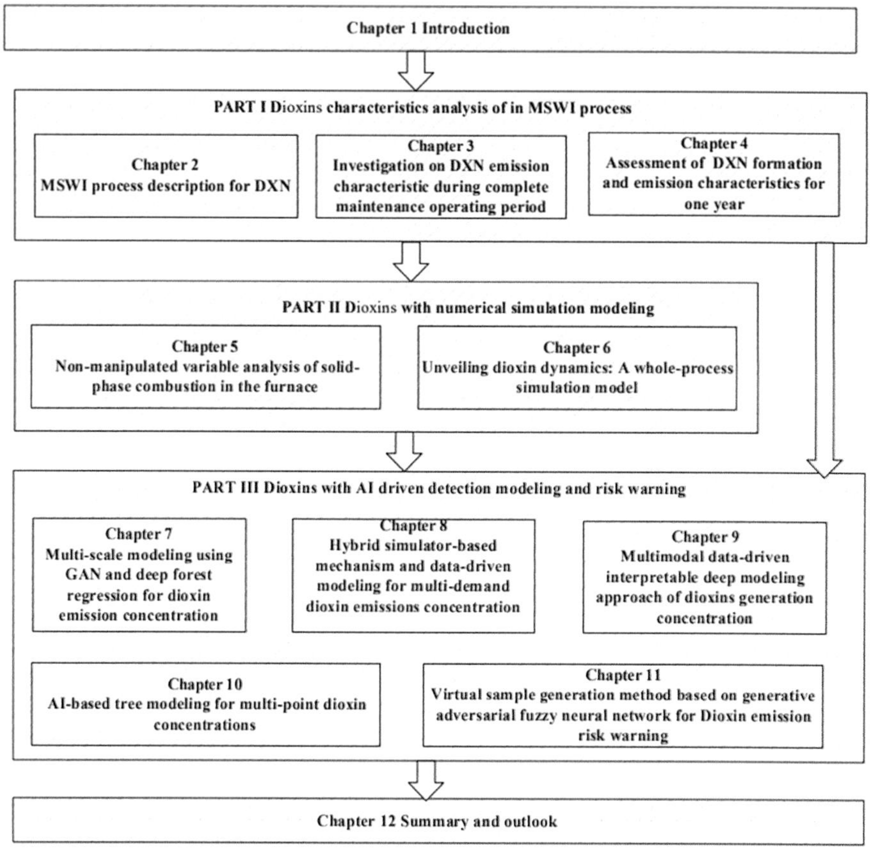

FIGURE 1.21 Book's structure.

plant, flue gas and fly ash sampling are conducted separately at different locations. This comprehensive sampling cycle encompasses various operational phases, including shutdown, cooling, maintenance, heating, startup, and normal operations. The shutdown process (SDP) begins from normal operation, followed by the shutdown and subsequent cooling phases. In contrast, the startup process (SUP) starts with heating, followed by the startup phase, and then transitions to normal operation. Detection and analysis were carried out during both the SDP and SUP stages in terms of the generation, adsorption, and emission concentration of DXN.

Chapter 4: Research on the impact of operating conditions, environmental changes, and operational duration on the generation and emissions of PCDD/Fs remains unresolved. In this chapter, emissions of PCDD/Fs and 17 congeners were tracked and investigated continuously over the course of one year at a MSWI power plant in Beijing. Consistent with Chapter 3, the whole cycle DXN process is divided into three parts: generation, adsorption, and emission, with separate sampling conducted for flue gas and fly ash at different locations. The results indicated the concentration of PCDD/Fs at the boiler outlet, stack inlet, and bag filter across both normal and abnormal operational conditions, ranges from 2.11E−02 to 41.86 ng I-TEQ/Nm^3, 7.00E−04 to 6.76 ng I-TEQ/Nm^3, and 1.12 to 2.90E+03 ng I-TEQ/Nm^3, respectively.

1.6.2 Part II

Chapter 5: Based on actual data from an MSWI power plant in Beijing, a single-factor analysis of non-MVs in solid-phase combustion within the furnace was conducted using biorthogonal numerical simulation experiments. Initially, the solid-phase combustion analysis of the MSWI process was performed, identifying the primary non-MVs. Subsequently, a numerical model was established under benchmark working conditions using FLIC software. Through the biorthogonal experiment, several sets of numerical model inputs were designed to generate mechanistic data across multiple operating conditions. Finally, a multi-condition numerical simulation experiment was conducted to study solid-phase combustion under various scenarios and analyze the influence of non-MVs. The simulation results indicated that the maximum solid temperature reached 1,360 K under benchmark operating conditions, while it varied between 1,120 and 1,470 K across five different conditions.

Chapter 6: The progression of DXN generation, adsorption, and emission concentrations within the MSWI process remains unclear, particularly when considering the specific components of MSW. This chapter proposes a CNSM for the whole-lifecycle of DXN in terms of actual physical size and operating parameters for an MSWI power plant. Based on the existing knowledge of MSW combustion and DXN mechanisms, a model is developed by utilizing FLIC and Aspen simulation software. The model encompasses six key stages to a whole-lifecycle DXN simulation, i.e., precipitation and formation, high-temperature pyrolysis, high-temperature gas-phase synthesis, low-temperature catalytic synthesis, adsorption on AC, and emission to the atmosphere. Simulated experiments demonstrate the model's effectiveness in representing the evolution of DXN concentrations throughout the process.

1.6.3 Part III

Chapter 7: DXN emission concentration soft sensor modeling must tackle problems such as missing data, data scale mismatch, and small samples. This chapter proposes a multi-scale missing data modeling using an improved Generative Adversarial Network (GAN) and an improved DFR. First, the short-time-scale samples and features are divided according to the missing data. Second, the missing data are filled by GAN. Third, a certain range of the short-time-scale input data is averaged to match the long-time-scale output data. Finally, each layer of the DFR model using matching data fully connects the parallel and cascade forest algorithms to improve diversity and accuracy. The effectiveness and rationality of the proposed method are verified on the real DXN dataset.

Chapter 8: Prevailing research on DXN emission concentrations predominantly relies on data-driven models, often overlooking the potential advantages of incorporating combustion mechanism knowledge. This chapter proposes a hybrid modeling strategy that integrates a simulator-based mechanism model with an enhanced regression DT-based data model. Utilizing virtual mechanism data obtained through numerical simulations that couple FLIC and Aspen Plus, we constructed a white-box surrogate model employing a multiple-input multiple-output LRDT. To establish a relationship with DXN emission concentrations, we implemented a semi-supervised transfer learning mapping model. This model was then combined with a novel ensemble LRDT model based on real historical data, utilizing a constrained incremental random weight neural network. The efficacy of this approach was validated through an industrial case study conducted in Beijing.

Chapter 9: The harsh environment of the MSWI process hinders the use of conventional equipment for detecting DXN generation concentrations. The available data types related to DXN generation concentrations encompass two modalities: numerical data and image data. This chapter presents an advanced tree-based interpretable deep modeling approach that employs a multi-modal data-driven strategy. Initially, the time scale of the multi-modal data is adjusted to align with the sampling period of DXN, leveraging mechanistic knowledge. Subsequently, we propose an adaptive DFR algorithm based on cross-layer full connection (ADFR-clfc) for modeling numerical process data and recorded operational data. Additionally, a CNN feature extraction method, utilizing transfer learning in conjunction with ADFR-clfc, is employed to model the image data. Finally, the DXN generation concentration is derived by taking the arithmetic average of the outputs from the aforementioned models. The proposed method is validated using approximately one year of data collected from an MSWI plant in Beijing.

Chapter 10: Most research primarily focuses on developing models for DXN emissions at the stack location, often neglecting the online measurement of DXN generation and absorption concentrations. This chapter emphasizes modeling multiple points regarding generation, adsorption, and emission concentrations. It employs machine learning techniques based on advanced tree-based data-driven deep and broad learning algorithms. State-of-the-art tree-based models, including the adaptive DFR algorithm based on cross-layer full connection, the tree broad learning system, fuzzy forest regression, and auxiliary modeling technologies, are utilized to address diverse data characteristics. These characteristics encompass high-dimensional small samples, low-dimensional ultra-small size samples, and medium-dimensional small samples across various phases related to DXN. The proposed tree-based model for DXN is validated using nearly one year of authentic data from an MSWI plant in Beijing.

Chapter 11: Key performance indicators, such as production quality and pollutant emission concentrations, are challenging to measure online due to limited detection technologies and high economic costs. The modeling samples often exhibit high dimensionality, significant uncertainty, and small sample sizes, which do not meet the requirements of traditional machine learning algorithms. This chapter proposes a VSG method based on a GAFNN. First, an adaptive feature selection algorithm utilizing RF techniques is employed to reduce the input features of the original real samples. Next, candidate virtual samples are generated using GAFNN to address the issues of uncertainty and small sample sizes. These virtual samples are then filtered through a multi-constrained selection mechanism to enhance their quality. Finally, a DF classification model is constructed based on a mixed dataset that includes both the original real samples and the selected virtual samples. The effectiveness of the proposed method is validated using benchmark datasets as well as real DXN emission data.

REFERENCES

1. Gómez-Sanabria A, Kiesewetter G, Klimont Z, et al. Potential for future reductions of global GHG and air pollutants from circular waste management systems. *Nature Communications*, 2022, 13(1): 106.
2. Ali K, Kausar N, Amir M. Impact of pollution prevention strategies on environment sustainability: role of environmental management accounting and environmental proactivity. *Environmental Science and Pollution Research*, 2023, 30(38): 88891–88904.
3. Tang J, Xia H, Yu W, Qiao J F. Research status and prospects of intelligent optimization control for municipal solid waste incineration process. *Acta Automatica Sinica*, 2023, 49(10): 2019–2059.
4. Qiao J F, Guo Z H, Tang J. Dioxin emission concentration measurement approaches for municipal solid wastes incineration process: a survey. *Acta Automatica Sinica*, 2020, 46(6): 1063–1089.
5. Naveenkumar R, Iyyappan J, Pravin R, et al. A strategic review on sustainable approaches in municipal solid waste management and energy recovery: role of artificial intelligence, economic stability and life cycle assessment. *Bioresource Technology*, 2023, 379: 129044.
6. Kammen D M, Sunter D A. City integrated renewable energy for urban sustainability. *Science*, 2016, 352(6288): 922–928.
7. Korai M S, Mahar R B, Uqaili M A. The feasibility of municipal solid waste for energy generation and its existing management practices in Pakistan. *Renewable and Sustainable Energy Reviews*, 2017, 72: 338–353.
8. Ding H X, Qiao J F, Huang W M, et al. Cooperative event-triggered fuzzy-neural multivariable control with multi-task learning for municipal solid waste incineration process. *IEEE Transactions on Industrial Informatics*, 2024, 20(1): 765–774.
9. Qiao J F, Sun J, Meng X. Event-triggered adaptive model predictive control of oxygen content for municipal solid waste incineration process. *IEEE Transactions on Automation Science and Engineering*, 2024, 21(1): 463–474.
10. Hunsinger H, Jay K, Vehlow J. Formation and destruction of PCDD/F inside a grate furnace. *Chemosphere*, 2002, 46: 1263–1272.
11. Hershkowitz A, Salerni E. Municipal solid waste incineration in Japan. *Environmental Impact Assessment Review*, 1989, 9(3), 257–278.
12. Onishi K. Fuzzy control of municipal refuse incineration plant. *Automatic Measurement Control Society*, 1991, 27(3): 326–332.
13. Schuler F, Rampp F, Martin J, Wolfrum J. TACCOS-A thermography-assisted combustion control system for waste incinerators. *Combustion and Flame*, 1994, 99(2): 431–439.
14. Zipser S, Gommlich A, Matthes J, Keller H B. Combustion plant monitoring and control using infrared and video cameras. *IFAC Proceedings Volumes*, 2006, 39(7): 249–254.
15. Silpa K, Lisa C Y, Perinaz B T, Frank V W. *What a Waste 2.0: A Global Snapshot of Solid Waste Management to 2050*. Washington, DC: World Bank, 2018.
16. Gu L, He K, Liu H W. Study on waste heat value characteristics and its impact on waste incineration power generation project in China. *Chinese Society for Environmental Sciences. CSES* 2022, 2022: 439–445.

17. Yamada T, Asari M, Miura T, et al. Municipal solid waste composition and food loss reduction in Kyoto City. *Journal of Material Cycles and Waste Management*, 2017, 19(4): 1351–1360.
18. Sun J, Meng X, Qiao J F. Data-driven optimal control for municipal solid waste incineration process. *IEEE Transactions on Industrial Informatics*, 2023, 19(12): 11444–11454.
19. Ding H X, Tang J, Qiao J F. Data-driven modeling and self-organizing control of municipal solid waste incineration process. *Acta Automatica Sinica*, 2023, 49(3): 550–566.
20. Li X, Zhang C, Li Y, Zhi Q. The status of municipal solid waste incineration (MSWI) in china and its clean development. *Energy Procedia*, 2016, 104: 498–503.
21. Phillips K, Longhurst P J, Wagland S T. Assessing the perception and reality of arguments against thermal waste treatment plants in terms of property prices. *Waste Management* 2014, 34(1): 219–225.
22. Bai J, Sun X, Zhang C, et al. Mechanism and kinetics study on the ozonolysis reaction of 2, 3, 7, 8-TCDD in the atmosphere. *Journal of Environmental Sciences*, 2014, 26(1): 181–188.
23. Sofian K, Fatin S. Dioxins and furans: a review from chemical and environmental perspectives. *Trends in Enviromental analytical Chemistry*, 2018, 17: 1–13.
24. Peng Y, Lu S, Li X, et al. Formation, measurement, and control of dioxins from the incineration of municipal solid wastes: recent advances and perspectives. *Energy & Fuels*, 2020, 34(11): 13247–13267.
25. Li H Y, Zhang S T, Zhao X H. Detection methods of dioxins emitted from municipal solid waste incinerator. *Journal of Fuel Chemistry and Technology*, 2005, 33(3): 379–384.
26. Sun N, Sun S H, Wang M Q. Research progress of dioxins' detection using chemical-luciferase gene expression (CALUX). *Asian Journal of Ecotoxicology*, 2014, 9(3): 391–397.
27. Bunsan S, Chen W Y, Chen H W, et al. Modeling the dioxin emission of a municipal solid waste incinerator using neural networks. *Chemosphere*, 2013, 92: 258–264.
28. Urano K, Kato M, Nagayanagi Y, et al. Convenient dioxin measuring method using an efficient sampling train, an efficient HPLC system and a highly sensitive HRGC/LRMS with a PTV injector. *Chemosphere*, 2001, 43(4): 425–431.
29. Hung P C, Chang S H, Buekens A, Chang M B. Continuous sampling of MSWI dioxins. *Chemosphere*, 2016, 145: 119–124.
30. Watanabe N, Kawamoto K, Asada S, et al. Surrogate study for dioxins from municipal waste incinerator in startup condition: applicability as a dioxin control indicator and an organohalogen emission warning. *Journal of Material Cycles and Waste Management*, 2010, 12(3): 254–263.
31. Jiang Y M, Ma X Y, Li X. Towards virtual sample generation with various data conditions: a comprehensive review. *Information Fusion*, 2025, 117: 102874.
32. Wang K, Li J, Tsung F. Distribution inference from early-stage stationary data streams by transfer learning. *IISE Transactions*, 2022, 54(3): 303–320.
33. Wen J M, Su A, Wang X L, et al. Virtual sample generation for small sample learning: a survey, recent developments and future prospects. *Neurocomputing*, 2025, 615: 128934.
34. Xia H, Tang J, Cui C L, Qiao J F. Soft sensing method of dioxin emission in municipal solid waste incineration process based on broad hybrid forest regression. *Acta Automatica Sinica*, 2023, 49(02): 343–365.
35. Raudys Š. Trainable fusion rules. *II. small sample-size effects. Neural Networks*, 2006, 19(10): 1517–1527.

36. Tang J, Jia M Y, Liu Z, et al. Modeling high dimensional frequency spectral data based on virtual sample generation technique. In: *Proceedings of 2015 IEEE International Conference on Information and Automation*, Lijiang, China, 8–10 August 2015. IEEE, 2015. 1090–1095
37. Raudys S J, Jain A K. Small sample size effects in statistical pattern recognition: recommendations for practitioners. *IEEE Transactions on Pattern Analysis and Machine Intelligence*, 1991, 13(3): 252–264.
38. Duin R P W. Small sample size generalization. In: *Proceedings of 9th Scandinavian Conference on Image Analysis*, Uppsala, Sweden, June 6–9, 1995. 1995, 1–6.
39. Zhong K, Han M, Han B. Data-driven based fault prognosis for industrial systems: a concise overview. *IEEE/CAA Journal of Automatica Sinica*, 2019, 7(2): 330–345.
40. Ji C, Sun W. A review on data-driven process monitoring methods: characterization and mining of industrial data. *Processes*, 2022, 10(2): 335.
41. Xu Y J, Liu X, Cao Xin, et al. Artificial intelligence: a powerful paradigm for scientific research. *The Innovation*, 2021, 2(4), 100179.
42. Ali S, Abuhmed T, El-Sappagh S, et al. Explainable Artificial Intelligence (XAI): what we know and what is left to attain Trustworthy Artificial Intelligence. *Information Fusion*, 2023, 99: 101805.
43. Sollich P, Krogh A. Learning with ensembles: how over-fitting call be useful. *In Advances in Neural Information Processing Systems*, 1996, 9: 190–196.
44. Rokach L. Ensemble-based classifiers, *Artificial Intelligence Review*, 2010, 33: 1–39.
45. Wozniak M, Grana M, Corchado E. A survey of multiple classifier systems as hybrid systems, *Information Fusion*, 2014, 16: 3–17.
46. Ho T K. The random subspace method for constructing decision forests, *IEEE Transactions on Pattern Analysis and Machine Intelligence*, 1998, 20(8): 832–844.
47. Zhu Z, Zhang Y. Flood disaster risk assessment based on random forest algorithm, *Neural Computing and Applications*, 2022, 34: 3443–3455.
48. Zheng Y, Xu Z, Wang X. The fusion of deep learning and fuzzy systems: a state-of-the-art survey. *IEEE Transactions on Fuzzy Systems*, 2022, 30(8): 2783–2799.
49. Zadeh L A. Fuzzy sets. *Information and Control*, 1965, 8(3): 338–353.
50. Zadeh LA. Fuzzy logic. *Computer*, 1988, 21(4): 83–93.
51. Ojha V, Abraham A, Snášel V. Heuristic design of fuzzy inference systems: a review of three decades of research. *Engineering Applications of Artificial Intelligence*, 2019, 85: 845–864.
52. Niyogi P, Girosi F, Poggio T. Incorporating prior information in machine learning by creating virtual examples. *Proceedings of the IEEE*, 1998, 86(11): 2196–2209.
53. Yang J, Yu X, Xie Z Q, et al. A novel virtual sample generation method based on Gaussian distribution. *Knowledge-Based Systems*, 2011, 24(6): 740–748.
54. Cho S, Jang M, Chang S. Virtual sample generation using a population of networks. *Neural Processing Letters*, 1997, 5(2): 21–27
55. Li D C, Chen L S, Lin Y S. Using functional virtual population as assistance to learn scheduling knowledge in dynamic manufacturing environments. *International Journal of Production Research*, 2003, 41(17): 4011–4024.
56. Huang C F, Moraga C. A diffusion-neural network for learning from small samples. *International Journal of Approximate Reasoning*, 2004, 35(2): 137–161.
57. Li D C, Wu C S, Tsai T I. Using mega-trend-diffusion and artificial samples in small data set learning for early flexible manufacturing system scheduling knowledge. *Computers and Operations Research*, 2007, 34(4): 966–982.

58. Li D C, Wu C S, Tsai T I, et al. Using mega-fuzzification and data trend estimation in small data set learning for early FMS scheduling knowledge. *Computers & Operations Research*, 2006, 33(6): 1857–1869.
59. Li D C, Fang Y H, Lai Y Y, et al. Utilization of virtual samples to facilitate cancer identification for DNA microarray data in the early stages of an investigation. *Information Sciences: An International Journal*, 2009, 179(16): 2740–2753.
60. Lin Y S, Li D C. The generalized-trend-diffusion modeling algorithm for small data sets in the early stages of manufacturing systems. *European Journal of Operational Research*, 2010, 207(1): 121–130.
61. Li D C, Wen I H. A genetic algorithm-based virtual sample generation technique to improve small data set learning. *Neurocomputing*, 2014, 143(16): 222–230.
62. Olie K. Chlorodibenzo-p-dioxins and chlorodibenzofurans and trace components of fly ash and flue gas of some municipal incinerators in The Netherlands. *Chemosphere*, 1977, 6(8): 103–108.
63. Choudhry G G, Hutzinger O. Mechanistic aspects of the thermal formation of halogenated organic compounds including polychlorinated dibenzo-p-dioxins. *Toxicological & Environmental Chemistry*, 1999, 5: 67–93.
64. Ritter E R, Bozzelli J W. Pathways to chlorinated dibenzodioxins and dibenzofurans from partial oxidation of chlorinated aromatics by oh radical: thermodynamic and kinetic insights. *Combustion Science and Technology*, 1994, 101(1–6): 153–1569.
65. Ballschmiter K, Zoller W, Buchert H, Class Th. Correlation between substitution pattern and reaction pathway in the formation of polychlorodibenzofurans. *Fresenius' Zeitschrift für Analytische Chemie*, 1985, 322(6): 587–594.
66. Stanmore B R. The formation of dioxins in combustion systems. *Combustion and Flame*, 2004, 136(3): 398–427.
67. Shaub W M, Tsang W. Dioxin Formation in Incinerators. *Environmental Science & Technology*, 1983, 17(12): 721–730.
68. Wikström E, Ryan S, Touati A, Tabor D, Gullett B K. Origin of carbon in polychlorinated dioxins and furans formed during sooting combustion. *Environmental Science and Technology*, 2004, 38(13): 3778–3784.
69. Weber R, Hagenmaier H. Mechanism of the formation of polychlorinated dibenzo-p-dioxins and dibenzofurans from chlorophenols in gas phase reactions. *Chemosphere*, 1999, 38(3): 529–549.
70. Xu F, Yu W, Gao R, et al. Dioxin formations from the radical/radical cross-condensation of phenoxy radicals with 2-chlorophenoxy radicals and 2,4,6-trichlorophenoxy radicals. *Environmental Science & Technology*, 2010, 44(17): 6745–6751.
71. Oh J E, Gullett B, Ryan S, Touati A. Mechanistic relationships among Pcdds/Fs, Pcns, Pahs, Ciphs, And Cibzs in municipal waste incineration. *Environmental Science & Technology*, 2007, 41(13): 4705–4710.
72. Hung P C, Chang S H, Chang M B. Removal of chlorinated aromatic organic compounds from mwi with catalytic filtration. *Aerosol and Air Quality Research*, 2014, 14(4): 1215–1222.
73. Wilhelm J, Stieglitz L, Dinjus E, Will R. Mechanistic studies on the role of pahs and related compounds in PCDD/F formation on model fly ashes. *Chemosphere*, 2001, 42(5): 797–802.
74. Milligan M S, Altwicker E. The relationship between de novo synthesis of polychlorinated dibenzo-p-dioxins and dibenzofurans and low-temperature carbon gasification in fly ash. *Environmental Science & Technology*, 1993, 27(8): 1595–1601.

75. Lenoir D, Wehrmeier A, Sidhu S S, Taylor P H. Formation and inhibition of chloroaromatic micropollutants formed in incineration processes. *Chemosphere*, 2001, 43(1): 107–114.
76. Zhu L, Bozzelli J W. Kinetics and thermochemistry for the gas-phase keto–enol tautomerism of phenol ↔ 2,4-cyclohexadienone. *The Journal of Physical Chemistry A*, 2003, 107(19): 3696–3703.
77. Evans C S, Dellinger B. Mechanisms of dioxin formation from the high-temperature oxidation of 2-chlorophenol. *Environmental Science & Technology*, 2005, 39(1): 122–127.
78. Truong H, Lomnicki S, Dellinger B. Mechanisms of molecular product and persistent radical formation from the pyrolysis of hydroquinone. *Chemosphere*, 2008, 71(1): 107–113.
79. Tao Z, Li Z. A Kinetics study on reactions of C_6H_5O with C_6H_5O and O_3 at 298 K. *International Journal of Chemical Kinetics*, 1999, 31(1): 65–72.
80. Janoschek R, Fabian W M F. Thermodynamic properties of chlorinated phenols, cyclo-c5 compounds, and derived radicals from G3MP2B3 calculations. *Journal of Molecular Structure*, 2003, 661–662: 635–645.
81. Libit L, Hoffmann R. Detailed orbital theory of substituent effects. Charge transfer, polarization, and the methyl group. *Journal of the American Chemical Society*, 2002, 96(5): 1370–1383.
82. Armstrong D R, Cameron C, Nonhebel D C, Perkins P G. Oxidative coupling of phenols. Part 8. A theoretical study of the coupling of phenoxyl radicals. *Journal of the Chemical Society*, 1983, 5: 575–579.
83. Sidhu S, Edwards P. Role of phenoxy radicals in PCDD/F formation. *International Journal of Chemical Kinetics*, 2002, 34(9): 531–541.
84. Altarawneh M, Mackie J C, Kennedy E M, Machie J C. Mechanisms for PCDF and PCB formation from fires. *EMBO Journal*, 2007, 29(29): 2788–2801.
85. Sommeling P M, Mulder P, Louw R. Formation of PCDFs during chlorination and oxidation of chlorobenzene in chlorine/oxygen mixtures around 340°C. *Chemosphere*, 1994, 29(9–11): 2015–2018.
86. Liu J, Qi Z, Li X, et al. Effect of oxygen content on the thermal desorption of polychlorinated biphenyl-contaminated soil. *Environmental Science and Pollution Research*, 2015, 22(16): 12289–12297.
87. Buser H R. Formation of polychlorinated dibenzofurans (PCDFs) and dibenzo-p-dioxins (PCDDs) from the pyrolysis of chlorobenzenes. *Chemosphere*, 1979, 8(6): 415–424.
88. Bani H E H, Hammad M, Matar A, et al. Numerical analysis of the incineration of polychlorinated biphenyl wastes in rotary kilns. *Journal of Environmental Chemical Engineering*, 2016, 4(1): 624–632.
89. Hou S, Altarawneh M, Kennedy E M, Mackie J C, Weber R, Dlugogorski B Z. Formation of polychlorinated dibenzo-p-dioxins and dibenzofurans (PCDD/F) from oxidation of 4,4′-Dichlorobiphenyl (4,4′-DCB). *Proceedings of the Combustion Institute*, 2019, 37(1): 1075–1082.
90. Vejerano E P, Holder A L, Marr L C. Emissions of polycyclic aromatic hydrocarbons, polychlorinated dibenzo-p-dioxins, and dibenzofurans from incineration of nanomaterials. *Environmental Science & Technology*, 2013, 47(9): 4866–4874.
91. Wang W, Simonich S, Giri B, et al. Atmospheric concentrations and air–soil gas exchange of polycyclic aromatic hydrocarbons (PAHs) in remote, rural village and urban areas of Beijing–Tianjin Region, North China. *Science of the Total Environment*, 2011, 409(15): 2942–2950.

92. Zhang R, Zhuang T, Zhang Q, et al. Mechanistic studies on the dibenzofuran and dibenzo-p-dioxin formation reactions from anthracene. *Science of the Total Environment*, 2019, 662: 41–47.
93. Lomnicki S, Dellinger B. Formation of PCDD/F from the pyrolysis of 2-chlorophenol on the surface of dispersed copper oxide particles. *Proceedings of the Combustion Institute*, 2002, 29(2): 2463–2468.
94. Cains P W, Mccausland L J, Fernandes A R, et al. Polychlorinated dibenzo-p-dioxins and dibenzofurans formation in incineration: effects of fly ash and carbon source. *Environmental Science & Technology*, 1997, 31(3): 776–785.
95. Tuppurainen K, Halonen I, Ruokojärvi P, et al. Formation of PCDDs and PCDFs in municipal waste incineration and its inhibition mechanisms: a review. *Chemosphere*, 1998, 36(7): 1493–1511.
96. Mosallanejad S, Dlugogorski B Z, Kennedy E M, et al. Formation of PCDD/Fs in oxidation of 2-chlorophenol on neat silica surface. *Environmental Science & Technology*, 2016, 50(3): 1412–1418.
97. Dickson L C, Karasek F W. Mechanism of formation of polychlorinated dibenzo-p-dioxins produced on municipal incinerator flyash from reactions of chlorinated phenols. *Journal of Chromatography A*, 1987, 389: 127–137.
98. Qian Y, Zheng M, Liu W, et al. Influence of metal oxides on PCDD/Fs formation from pentachlorophenol. *Chemosphere*, 2005, 60(7): 951–958.
99. Ballschmiter K, Braunmiller I, Niemczyk R, et al. Reaction pathways for the formation of polychloro-dibenzodioxins (PCDD) and—dibenzofurans (PCDF) in combustion processes: ii. Chlorobenzenes and chlorophenols as precursors in the formation of polychloro-dibenzodioxins and—dibenzofurans in flame chemistry. *Chemosphere*, 1988, 17(5): 995–1005.
100. Hell K, Stieglitz L, Altwicker E R, et al. Reactions of 2,4,6-trichlorophenol on model fly ash: oxidation to CO and CO_2, condensation to PCDD/F and conversion into related compounds. *Chemosphere*, 2001, 42(5): 697–702.
101. Karasek F W, Dickson L C. Model studies of polychlorinated dibenzo-p-dioxin formation during municipal refuse incineration. *Science*, 1987, 237(4816): 754–756.
102. Nganai S, Lomnicki S, Dellinger B. Ferric oxide mediated formation of PCDD/Fs from 2-monochlorophenol. *Environmental Science & Technology*, 2009, 43(2): 368–373.
103. Lomnicki S, Truong H, Vejerano E, et al. Copper oxide-based model of persistent free radical formation on combustion-derived particulate matter. *Environmental Science & Technology*, 2008, 42(13): 4982–4988.
104. Stieglitz L, Vogg H. On formation conditions of PCDD/PCDF in fly ash from municipal waste incinerators. *Chemosphere*, 1987, 16(8): 1917–1922.
105. Faengmark I, Stroemberg B, Berge N, et al. Influence of postcombustion temperature profiles on the formation of PCDDs, PCDFs, PCBzs, and PCBs in a pilot incinerator. *Environmental Science & Technology*, 1994, 28(4): 624–629.
106. Gullett B K, Lemieux P M, Dunn J E. Role of combustion and sorbent parameters in prevention of polychlorinated dibenzo-p-dioxin and polychlorinated dibenzofuran formation during waste combustion. *Environmental Science & Technology*, 1994, 28(1): 107–118.
107. Zhao B, Hu X D, Lu J. Analysis and discussion on formation and control of dioxins generated from municipal solid waste incineration process. *Journal of the Air & Waste Management Association*, 2022, 72: 1063–1082.
108. Zhang M, Buekens A. De novo synthesis of dioxins: a review. *International Journal of Environment and Pollution*, 2016, 60(1–4): 63–110.

109. Mckay G. Dioxin characterisation, formation and minimisation during municipal solid waste (MSW) incineration: review. *Chemical Engineering Journal*, 2002, 86(3): 343–368.
110. Chang M B, Huang T F. The effects of temperature and oxygen content on the PCDD/PCDFs formation in MSW fly ash. *Chemosphere*, 2000, 40(2): 159–164.
111. Schwarz G, Stieglitz L. Formation of organohalogen compounds in fly ash by metal-catalyzed oxidation of residual carbon. *Chemosphere*, 1992, 25(3): 277–282.
112. Zhang M, Yang J, Buekens A, et al. PCDD/F catalysis by metal chlorides and oxides. *Chemosphere*, 2016, 159: 536–544.
113. Zhang M, Buekens A, Li X. Brominated flame retardants and the formation of dioxins and furans in fires and combustion. *Journal of Hazardous Materials*, 2016, 304: 26–39.
114. Weber P. Der Chlortransfer bei der bildung von polychlorierten organischen spurenstoffen (PCDD/PCDF) in industriellen thermischen prozessen (Müllverbrennung). Dissertation, Heideelberg, Universität Heidelberg, 2000.
115. Chen T. Study on formation mechanisms and control technologies of dioxins during municipal solid waste incineration. Dissertation, Zhejiang University, 2006.
116. Olie K, Addink R, Schoonenboom M. Metals as catalysts during the formation and decomposition of chlorinated dioxins and furans in incineration processes. *Journal of the Air & Waste Management Association (1995)*, 1998, 48(2): 101–105.
117. Takaoka M, Yamamoto T, Shiono A, et al. The effect of copper speciation on the formation of chlorinated aromatics on real municipal solid waste incinerator fly ash. *Chemosphere*, 2005, 59(10): 1497–1505.
118. Luijk R, Akkerman D M, Slot P, et al. Mechanism of formation of polychlorinated dibenzo-p-dioxins and dibenzofurans in the catalyzed combustion of carbon. *Environmental Science & Technology*, 1994, 28(2): 312–321.
119. Pekárek V, Punčochář M, Bureš M, et al. Effects of sulfur dioxide, hydrogen peroxide and sulfuric acid on the de novo synthesis of PCDD/F and PCB under model laboratory conditions. *Chemosphere*, 2007, 66(10): 1947–1954.
120. Chin Y T, Lin C, Chang C G P, et al. PCDD/F formation catalyzed by the metal chlorides and chlorinated aromatic compounds in fly ash. *Aerosol and Air Quality Research*, 2012, 12(2): 228–236.
121. Fujimori T, Takaoka M, Takeda N. Influence of Cu, Fe, Pb, and Zn Chlorides and Oxides on formation of chlorinated aromatic compounds in MSWI fly ash. *Environmental Science & Technology*, 2011, 45(4): 1746.
122. Liao J, Buekens A, Olie K, et al. Iron and copper catalysis of PCDD/F formation. *Environmental Science and Pollution Research*, 2016, 23(3): 2415–2425.
123. Yang J, Yan M, Li X, et al. Formation of dioxins on NiO and $NiCl_2$ at different oxygen concentrations. *Chemosphere*, 2015, 133: 97–102.
124. Öberg T, Öhrström T, Bergström J. Metal catalyzed formation of chlorinated aromatic compounds: a study of the correlation pattern in incinerator fly ash. *Chemosphere*, 2007, 67(9): S185–S190.
125. Brouwers H J H, Rj E V. Chemical reaction of fly ash. In: *Proceedings of the 11th International Congress on the Chemistry of Cement (ICCC)*, Durban, South Africa, 11–16 May 2003. 2003.
126. Vogg H, Stieglitz L. Thermal behavior of PCDD/PCCDF in fly ash from muncipal incinertors. *Chemosphere*, 1986, 15:1373–1378.
127. Addink R, Olie K. Role of oxygen in formation of polychlorinated dibenzo-p-dioxins/dibenzofurans from carbon on fly ash. *Environmental Science & Technology*, 1995, 29(6): 1586–1590.

128. Wikström E, Ryan S, Touati A, et al. Importance of chlorine speciation on de novo formation of polychlorinated dibenzo-p-dioxins and polychlorinated dibenzofurans. *Environmental Science & Technology*, 2003, 37(6): 1108–1113.
129. Everaert K, Baeyens J, Degrève J. Entrained-phase adsorption of PCDD/F from incinerator flue gases. *Environmental Science & Technology*, 2003, 37(6): 1219–1224.
130. Tejima H, Nakagawa I, Shinoda T A, et al. PCDDs/PCDFs reduction by good combustion technology and fabric filter with/without activated carbon injection. *Chemosphere*, 1996, 32(1): 169–175.
131. Chi K H, Chang S H, Huang C H, et al. Partitioning and removal of dioxin-like congeners in flue gases treated with activated carbon adsorption. *Chemosphere*, 2006, 64(9): 1489–1498.
132. Guo Y, Li Y, Zhu T, et al. Modeling of dioxin adsorption on activated carbon. *Chemical Engineering Journal*, 2016, 283: 1210–1215.
133. Zhou X J, Li X D, Ni M J, et al. Removal efficiencies for 136 tetra- through octa-chlorinated dibenzo-p-dioxins and dibenzofuran congeners with activated carbons. *Environmental Science and Pollution Research*, 2015, 22(22): 17691–17696.
134. Li W, Lin X, Yu M, et al. Experimental study on PCDD/Fs adsorption onto nano-graphite. *Aerosol and Air Quality Research*, 2016, 16(12): 3281–3289.
135. Zhou X, Li X, Xu S, et al. Comparison of adsorption behavior of PCDD/Fs on carbon nanotubes and activated carbons in a bench-scale dioxin generating system. *Environmental Science and Pollution Research*, 2015, 22(14): 10463–10470.
136. Chang Y M, Hung C Y, Chen J H, et al. Minimum feeding rate of activated carbon to control dioxin emissions from a large-scale municipal solid waste incinerator. *Journal of Hazardous Materials*, 2009, 161(2): 1436–1443.
137. Li X, Li Z, Luo L A. Adsorption isotherm and kinetics of dibenzofuran on granular activated carbons. *Chinese Journal of Chemical Engineering*, 2005, 5: 131–134.
138. Li X, Li Z, Luo L A. Adsorption kinetics of dibenzofuran in activated carbon packed bed. *Chinese Journal of Chemical Engineering*, 2008, 16(2): 203–208.
139. Lin W Y, Wang L C, Wang Y F, et al. Removal characteristics of PCDD/Fs by the dual bag filter system of a fly ash treatment plant. *Journal of Hazardous Materials*, 2008, 153(3): 1015–1022.
140. Chen M W, Lin T C, Wang L C, et al. The PCDD/F removal efficiency of a medical waste incinerator dual-bag filter system. *Aerosol and Air Quality Research*, 2014, 14: 1223–1231.
141. Trivedi J, Majumdar D. Memory effect driven emissions of persistent organic pollutants from industrial thermal processes, their implications and management: a review. *Journal of Environmental Management*, 2013, 119: 111–120.
142. Zimmermann R, Blumenstock M, Heger H J, et al. Emission of nonchlorinated and chlorinated aromatics in the flue gas of incineration plants during and after transient disturbances of combustion conditions: delayed emission effects. *Environmental Science & Technology*, 2001, 35(6): 1019–1030.
143. Wikström E, Ryan S, Touati A, et al. In situ formed soot deposit as a carbon source for polychlorinated dibenzo-p-dioxins and dibenzofurans. *Environmental Science & Technology*, 2004, 38(7): 2097–2101.
144. Weber R, Sakurai T, Ueno S, et al. Correlation of PCDD/PCDF and CO values in a MSW incinerator—indication of memory effects in the high temperature/cooling section. *Chemosphere*, 2002, 49(2): 127–134.
145. Chang M B, Lin J J. Memory effect on the dioxin emissions from municipal waste incinerator in taiwan. *Chemosphere*, 2001, 45(8): 1151–1157.

146. Wevers M, De F R. Dioxin emission reduction of a municipal waste incinerator by injection of activated carbon-the abatement of memory effects. *Organohalogen Compounds* 1998, 36: 343–346.
147. Li H W, Wang L C, Chen C C, et al. Influence of memory effect caused by aged bag filters on the stack PCDD/F emissions. *Journal of Hazardous Materials*, 2011, 185(2): 1148–1155.
148. Lin X, Wang X, Ying Y, et al. Formation pathways, gas-solid partitioning, and reaction kinetics of PCDD/Fs associated with baghouse filters operated at high temperatures: a case study. *Science of the Total Environment*, 2023, 857: 159551.
149. Wu J, Ying Y, Ma Y, et al. Influence of different loads on PCDD/F removal by SCR during municipal solid waste incineration. *Chemosphere*, 2023, 338: 139516.
150. Sidhu S S, Maqsud L, Dellinger B, et al. The homogeneous, gas-phase formation of chlorinated and brominated dibenzo-p-dioxin from 2, 4, 6-trichloro-and 2, 4, 6-tribromophenols. *Combustion and Flame*, 1995, 100(1–2): 11–20.
151. Huang H, Buskens A. Comparison of dioxin formation levels in laboratory gas-phase flow reactors with those calculated using the SHAUB-Tsang mechanism. *Chemosphere*, 1999, 38(7): 1595–1602.
152. Babushok V I, Tsang W. Gas-phase mechanism for dioxin formation. *Chemosphere*, 2003, 51(10): 1023–1029.
153. Khachatryan L, Asatryan R, Dellinger B. Development of expanded and core kinetic models for the gas phase formation of dioxins from chlorinated phenols. *Chemosphere*, 2003, 52(4): 695–708.
154. Gerasimov G Y. *Combustion, Explosion, and Shock Waves*. Springer Science and Business Media LLC, 2001, 37(2): 148–152.
155. Altwicker E R. Relative rates of formation of polychlorinated dioxins and furans from precursor and de novo reactions. *Chemosphere*, 1996, 33(10): 1897–1904.
156. Stanmore B R. Modeling the formation of PCDD/F in solid waste incinerators. *Chemosphere*, 2002, 47(6): 565–573.
157. Prisciandaro M, Piemonte V, di Celso G M, et al. Thermodynamic features of dioxins' adsorption. *Journal of Hazardous Materials*, 2017, 324: 645–652.
158. Chen, H. L., Zheng, D. Q., Zhang, J., et al. Simulation study on dioxins in municipal waste incineration flue gas treated by irradiation. *Environmental Science & Technology* 2013, 36(S1): 236–240.
159. Sun R, Ismail T M, Ren X, et al. Numerical simulation of gas concentration and dioxin formation for msw combustion in a fixed bed. *Journal of Environmental Management*, 2015, 157: 111–117.
160. Tungalag A, Lee B, Yadav M, et al. Yield prediction of MSW gasification including minor species through ASPEN Plus simulation. *Energy*, 2020, 198: 117296.
161. Xia H, Tang J, Aljerf L, Chen J K. Unveiling dioxin dynamics: a whole-process simulation study of municipal solid waste incineration. *Science of the Total Environment*, 2024, 954: 176241.
162. Weickhardt C, Zimmermann R, Boesl U, Schlag E W. Laser mass spectrometry of dibenzodioxin, dibenzofuran and two isomers of dichlorodibenzodioxins: selective ionization. *Rapid Communications in Mass Spectrometry*, 2010, 7(3): 183–185.
163. Tsuruga S, Suzuki T, Takatsudo Y, et al. On-line monitoring system of P5CDF homologues in waste incineration plants using VUV-SPI-IT-TOFMS. *Environmental Science & Technology* 2007, 41(10): 3684–3688.
164. Morii S, Suzuki T, Seki K, Okada M, Okuno S, Tanaka M. Dioxins and their precursors on-line monitoring method for incineration plants. *Proceedings of the Symposium on Environmental Engineering* 2004, 2004(14): 231–234.

165. Heger H J, Zimmermann R, Dorfner R, et al. On-line emission analysis of polycyclic aromatic hydrocarbons down to PPTV concentration levels in the flue gas of an incineration pilot plant with a mobile resonance enhanced multiphoton ionization time-of-flight mass spectrometer. *Analytical Chemistry* 1999, 71(1): 46–57.
166. Zimmermann R, Heger H J, Blumenstock M, et al. On-line measurement of chlorobenzene in waste incineration flue gas as a surrogate for the emission of polychlorinated dibenzo-p-dioxins/furans (I-TEQ) using mobile resonance laser ionization time-of-flight mass spectrometry. *Rapid Communications in Mass Spectrometry* 1999, 13(5): 307–314.
167. Hafner K, Zimmermann R, Rohwer E R, et al. A capillary-based supersonic jet inlet system for resonanceenhanced laser ionization mass spectrometry: principle and first on-line process analytical applications. *Analytical Chemistry* 2001, 73(17), 4171–4180.
168. Allen M G. Diode laser absorption sensors for gas dynamic and combustion flows. *Measurement Science and Technology*, 1998, 9(4): 545–575.
169. Li Z Y, Wei J. Laser mass spectrometry: principle and applications in environmental monitoring. *Chinese Journal of Quantum Electronics.* 2001, 18(1): 1–8.
170. Wu W Z, Schwirzer S M G, Schramm K W, et al. Rapid bioassay as indicator of potentially harmful effects for dioxin-like compounds in sample of Ya-Er Lake, China: requirements for clean-up and comparison to chemical analysis. *Fresenius Environmental Bulletin*, 1996, 5(7–8): 374–379.
171. Harrison R O, Carlson R E. An immunoassay for TEQ screening of dioxin/furan samples: current status of assay and applications development. *Chemosphere*, 1997, 34(5–7): 915–928.
172. Harrison R O, Eduljee G H. Immunochemical analysis for dioxins – progress and prospects. *Science of the Total Environment*, 1999, 239(1–3): 1–18.
173. Li W, Xu Y, Wu W Z. Combined toxicities of dioxins indicated by EROD in rat hepatoma cells in vitro. *Current Zoology*, 2001, 47(1): 64–70.
174. Rappe C, Bergqvist P A, Buser H R, et al. Analysis of polychlorinated dibenzo-p-dioxins and dibenzofurans. In: *Environmental Specimen Banking and Monitoring as Related to Banking*, Lewis R A, Stein N, Lewis C W (eds.). Springer Netherlands, 1984, 323–330.
175. Zhang Z R, Xu S Q, Zhou Y K. Detection of dioxin-type chemicals by luciferase reporter gene. *Chinese Journal of Analytical Chemistry*, 2001, 29(7): 825–827.
176. Overmeire I V, Clark G C, Brown D J, et al. Trace contamination with dioxin-like chemicals: evaluation of bioassay-based TEQ determination for hazard assessment and regulatory responses. *Environmental Science & Policy*, 2001, 4(6): 345–357.
177. Shang F J. Study on dioxin correlation model and on-line monitoring master dissertation. Dissertation, Zhejiang University, China, 2015.
178. Öberg T, Bergström J. Hexachlorobenzene as an indicator of dioxin production from combustion. *Chemosphere*, 1985, 14(8): 1081–1086.
179. Pandelova M, Lenoir D, Schramm K W. Correlation between PCDD/F, PCB and PCBz in coal/waste combustion. *Influence of various inhibitors. Chemosphere*, 2006, 62(7): 1196–1205.
180. Kaune A, Lenoir D, Nikolai U, Kettrup A. Estimating concentrations of polychlorinated dibenzo-p-dioxins and dibenzofurans in the stack gas of a hazardous waste incinerator from concentrations of chlorinated benzenes and biphenyls. *Chemosphere*, 1994, 29(9–11): 2083–2096.
181. Yoneda K, Ikeguchi T, Yagi Y, et al. A research on dioxin generation from the industrial waste incineration. *Chemosphere* 2002, 46(9–10): 1309–1319.

182. Manninen H, Perkio A, Vartiainen T, Ruuskanen J. Formation of PCDD/PCDF – effect of fuel and fly ash composition on the formation of PCDD/PCDF in the cocombustion of refuse-derived and packaging-derived fuels multivariate analysis. *Environmental Science and Pollution Research International*, 1996, 3(3): 129–134.
183. Kuzuhara S, Sato H, Tsubouchi N, et al. Effect of nitrogencontaining compounds on polychlorinated dibenzo-p-dioxin/dibenzofuran formation through de novo synthesis. *Environmental Science and Pollution Research International* 2005, 39(3): 795–799.
184. Gullett B K, Oudejans L, Tabor D, et al. Near-real-time combustion monitoring for PCDD/PCDF indicators by GC-REMPI-TOFMS. *Environmental Science & Technology*, 2012, 46(2): 923–928.
185. Zimmermann R, Lenoir D, Kettrup A, Nagel H, Boesl U. On-line emission control of combustion processes by laser-induced resonance-enhanced multi-photon ionization/ mass spectrometry. *Symposium on Combustion*, 1996, 26(2): 2859–2868.
186. Blumenstock M, Zimmermann R, Schramm K W, Kettrup A. Identification of surrogate compounds for the emission of PCDD/F (I-TEQ value) and evaluation of their on-line real-time detectability in flue gases of waste incineration plants by REMPI–TOFMS mass spectrometry. *Chemosphere*, 2001, 42(5): 507–518.
187. Liu W, Jiang J, Hou K, et al. Online monitoring of trace chlorinated benzenes in flue gas of municipal solid waste incinerator by windowless VUV lamp single photon ionization TOFMS coupled with automatic enrichment system. *Talanta*, 2016, 161: 693–699.
188. Cao X, Stevens W R, Tang S, et al. Atline measurement of 1, 2, 4-trichlorobenzene for polychlorinated dibenzo-p-dioxin and dibenzofuran International Toxic Equivalent Quantity prediction in the stack gas. *Environmental Pollution* 2019, 244: 202–208.
189. Dyke P H, Foan C, Fiedler H. PCB and PAH releases from power stations and waste incineration processes in the UK. *Chemosphere* 2003, 50(4): 469–480.
190. Li X D, Yin X F, Lu S Y, et al. The correlation between PAHs and dioxins formation during coal and municipal solid waste coincineration process. *Journal of Engineering Thermophysics*, 2006, 27(4): 691–694.
191. Tian F L. Source apportionment of persistent toxic substances in typical environmental media by receptor modeling: polycyclic aromatic hydrocarbons, dioxins and polychlorinated biphenyls. Dissertation, Dalian University of Technology, China, 2009.
192. Yin X, Li X, Lu S. On-line real-time monitoring of trace organic pollutant formation in the simulated flue gas. *Proceedings of the CSEE*, 2007, 27(17): 29–33.
193. Tuppurainen K, Aatamila M, Ruokojärvi P, et al. Effect of liquid inhibitors on PCDD/F formation. prediction of particle-phase PCDD/F concentrations using PLS modelling with gas-phase chlorophenol concentrations as independent variables. *Chemosphere*, 1999, 38 10: 2205–2217.
194. Yamada M, Hashimoto Y, Suga M, et al. Real-time monitoring of chlorobenzenes in flue gas, *Organohalogen Compounds* 2001, 54: 380–383.
195. Yamada M, Hashimoto Y, Suga M, et al. An online system for monitoring dioxin precursor in flue gas. *Organohalogen Compounds* 2000, 45: 149–152.
196. Öberg T, Bergström J G. Emission and chlorination pattern of PCDD/PCDF predicted from indicator parameters. *Chemosphere*, 1987, 16: 1221–1230.
197. Nakui H, Koyama H, Takakura A, Watanabe N. Online measurements of low-volatile organic chlorine for dioxin monitoring at municipal waste incinerators. *Chemosphere*, 2011, 85(2): 151–155.
198. Hasberg W, May H, Dorn I. Description of the residence-time behaviour and burnout of PCDD, PCDF and other higher chlorinated aromatic hydrocarbons in industrial waste incineration plants. *Chemosphere*, 1989, 19(1–6): 565–571.

199. Tillman D A. *Incineration of Municipal and Hazardous Solid Wastes*. Elsevier, 2012.
200. Chang N B, Huang S H. Statistical modelling for the prediction and control of PCDDs and PCDFs emissions from municipal solid waste incinerators. *Waste Management & Research*, 1995, 13: 379–400.
201. Chang N B, Chen W C. Prediction of PCDDs/PCDFs emissions from municipal incinerators by genetic programming and neural network modeling. *Waste Management & Research*, 2000, 18: 41–351.
202. Tang J, Qiao J F. Soft sensor for dioxin emission concentration in waste incineration process based on selective ensemble kernel learning algorithm. *Journal of Chemical Industry and Engineering*, 2019, 70(2): 696–706.
203. Tang J, Qiao J F, Li W T. Simplified stochastic configuration network-based optimized soft measuring model by using evolutionary computing framework with its application to dioxin emission concentration estimation. *International Journal of System Control and Information Processing*, 2018, 4(2): 332–365.
204. Wen C J, Lin X Q, Ying Y X, et al. Dioxin emission prediction from a full-scale municipal solid waste incinerator: deep learning model in time-series input, *Waste Management*, 2023, 170: 93–102.
205. Tang J, Xia H, Aljerf L, et al. Prediction of dioxin emission from municipal solid waste incineration based on expansion, interpolation, and selection for small samples. *Journal of Environmental Chemical Engineering* 2022, 10: 108314.
206. Xia H, Tang J, Qiao J, Yan A, Guo Z. Soft measuring method of dioxin emission concentration for MSWI process based on RF and GBDT. In: *2020 Chinese Control and Decision Conference (CCDC)*, Hefei, China, 22–24 August 2020. 2020, 2173–2178.
207. Xu C, Tang J, Xia H, Yu W, Qiao J F. Dioxin emission concentration prediction using the selective ensemble algorithm based on bayesian inference and binary tree. *IEEE Transactions on Instrumentation and Measurement*, 2023, 72: 2528514.
208. Tang J, Xia H, Zhang J. et al. Deep forest regression based on cross-layer full connection. *Neural Computing and Applications*, 2021, 33: 9307–9328.
209. Xia H, Tang J, Aljerf Lai, Cui C L, Gao B Y, Ukaogo P O. Dioxin emission modeling using feature selection and simplified DFR with residual error fitting for the grate-based MSWI process. *Waste Management*, 2023, 168: 256–271.
210. Xia H, Tang J, Aljerf L. Dioxin emission prediction based on improved deep forest regression for municipal solid waste incineration process. *Chemosphere*, 2022, 294: 133716.
211. Xia H, Tang J, Yu W, et al. Takagi–Sugeno fuzzy regression trees with application to complex industrial modeling. *IEEE Transactions on Fuzzy Systems*, 2023, 31(7): 2210–2224.
212. Xia H, Tang J, Yu W, Qiao J F. Online measurement of dioxin emission in solid waste incineration using fuzzy broad learning. *IEEE Transactions on Industrial Informatics*, 2024, 20(1): 358–368.
213. Xia H, Tang J, Yu W, Qiao J F. Hybrid simulator-based mechanism and data-driven for multidemand dioxin emissions intelligent prediction in the MSWI process. *IEEE Transactions on Industrial Electronics*, 2024, 71(10): 13221–13231.
214. Xia H, Tang J, Pan X T, et al. Multimodal data-driven interpretable deep modeling approach of dioxins generation for municipal solid waste incineration processes. *IEEE Transactions on Instrumentation and Measurement*, 2024, 73: 2534618.
215. Xia H, Tang J, Aljerf L, Wang T Z, Gao B Y, Alajlani M. AI-based tree modeling for multi-point dioxin concentrations in municipal solid waste incineration. *Journal of Hazardous Materials*, 2024, 480: 135834.

216. Tang J, Cui W W, Chen J K, et al. Full life cycle prediction model construction for dioxins in municipal solid waste incineration process: method of coupling numerical simulation and fuzzy forest regression. *Chemical Industry and Engineering Progress*, 2025, 44(8): 4628–4647.
217. Cui C L. Risk early-warning modeling for dioxin emissions in MSWI processes based on generative adversarial networks. Master's thesis, Beijing University of Technology, 2024.
218. Tang J, Cui C L, Xia H, Qiao J F. A risk early-warning model for dioxin emissions in MSWI processes based on an active learning mechanism GAN. *Journal of Beijing University of Technology*, 2023, 49(5): 507–522.
219. Cui C L, Tang J, Xia H, et al. Virtual sample generation method based on generative adversarial fuzzy neural network. *Neural Computing and Applications*, 2023, 35: 6979–7001.
220. Cui C L, Tang J, Qiao J F, Xia H. A novel data-driven sample augmentation method using interpretable space and adversarial mechanism for dioxin risk warning. *Engineering Applications of Artificial Intelligence*, 2025, 159: 111690.
221. Lin X, Yan M, Dai A, Zhan M, Fu J, Li X, Chen T, Lu S, Buekens A, Yan J. Simultaneous suppression of PCDD/F and NOx during municipal solid waste incineration. *Chemosphere*, 2015, 126: 60–66.
222. Gerasimov G. Modeling study of polychlorinated dibenzo-p-dioxins and dibenzofurans behavior in flue gases under electron beam irradiation. *Chemosphere*, 2016, 158: 100–106.
223. Havukainen J, Zhan M, Dong J, Liikanen M, Deviatkin L, Li X, Horttanainen M. Environmental impact assessment of municipal solid waste management incorporating mechanical treatment of waste and incineration in Hangzhou, China. *Journal of Cleaner Production*, 2017, 141: 453–461.
224. Zhang R, Luo Y, Yin R. Experimental study on dioxin formation in an MSW gasificationcombustion process: an attempt for the simultaneous control of dioxins and nitrogen oxides. *Waste Management*, 2018, 82: 292–301.
225. Chen Z, Lin X, Lu S, Li X, Yan J. Suppressing formation pathway of PCDD/Fs by S-N-containing compound in full-scale municipal solid waste incinerators. *Chemical Engineering Journal*, 2019, 359: 1391–1399.
226. Wen Z, Di J, Liu S, Han J, Lee J C K. Evaluation of flue-gas treatment technologies for municipal waste incineration: a case study in Changzhou, China. *Journal of Cleaner Production*, 2018, 18: 812–920.
227. Mukherjee A, Debnath B, Ghosh S K. A review on technologies of removal of dioxins and furans from incinerator flue gas. *Procedia Environmental Sciences*, 2016, 35: 528–540.
228. Cui Y Y, Meng X, Qiao J F. Intelligent optimization setting method for air volume in municipal solid waste incineration process. *Control and Decision*, 2023, 38(02): 318–326.
229. Wang T Z, Tang J, Aljerf L, et al. Emission reduction optimization of multiple flue gas pollutants in municipal solid waste incineration power plant, *Fuel*, 2025, 381: 133382.
230. Cui Y Y, Meng X, Qiao J F. Multi-condition operational optimization with adaptive knowledge transfer for municipal solid waste incineration process. *Expert Systems with Applications*, 2024, 238: 121783.
231. Qiao J Q, Cui Y Y, Meng X. Dynamic multi-objective operation optimization of municipal solid waste incineration process based on transfer learning. *IEEE Transactions on Automation Science and Engineering*, 2025, 22: 9338–9352.

232. Huang W M, Meng X, Qiao J F. Dynamic collaborative optimization method for municipal solid waste incineration process. *Scientia Sinica Informations*, 2025, 55(05): 1200–1220.
233. Huang W, Meng X, Qiao J. Cooperative dynamic multiobjective optimization with multi-time-scale for MSWI process, *IEEE Transactions on Industrial Electronics*, 2025, 72(12): 14093–14105.
234. Tang J, Wang T Z, Tian H, Yu W. AI-and security-empowered end-edge-cloud modular platform in complex process industries: a case study on municipal solid waste incineration. *Sensors*, 2025, 25(22): 6973.

Part I

Dioxins Characteristics Analysis of in MSWI Process

2 MSWI Process Description for Dioxins

2.1 DESCRIPTION OF GRATE FURNACE TYPE MSWI PROCESS

The MSWI process involves six subsystems: solid waste fermentation, solid waste combustion, waste heat exchange, steam power generation, flue gas cleaning, and flue gas emission, as illustrated in Figure 2.1.

Figure 2.1 illustrates the process flow of MSWI.

Municipal solid waste (MSW) is transported to the plant by dedicated compactor collection vehicles, weighed at a weighbridge, and then unloaded into the MSW deposit pool from the unloading platform. Within the pool, the MSW is thoroughly crushed, mixed, and stacked using a crane grab, which facilitates natural fermentation and dehydration by microorganisms [1]. This fermentation process, which typically lasts five to seven days, increases the calorific value of the solid portion by approximately 30%, thereby enhancing its efficiency and stability during combustion. The grab lifts the fermented MSW and transfers it to the hopper, where it slides down the chute and is pushed into the incinerator by the feeder. Before entering the combustion stage, the MSW is dried through the heat radiation from the furnace wall and the blowing and baking of preheated primary air. During combustion, additional air is introduced to provide the necessary oxygen; in some cases, other media may also assist in the combustion process. After high-temperature combustion, the combustible components are fully burned, generating heat, while the non-combustible ash is pushed out of the furnace by the burnout grate. As the high-temperature flue gas from MSW combustion passes through various boiler heating surfaces, it undergoes absorption and cooling. Toxic substances and heavy metals are meticulously treated through processes such as denitrification, desulfurization, dust removal, and ash collection. These processes transform harmful emissions into non-toxic gases that comply with stringent environmental standards, allowing them to be safely discharged into the atmosphere through a chimney with the assistance of an induced draft fan [2]. Meanwhile, deionized water in the waste heat boiler absorbs the heat generated by incineration, converting it into high-temperature steam. This steam expands to generate power, driving the turbine and generator to produce electricity [3].

The primary goal of the MSWI process is the harmless treatment of MSW, with power generation or heat production serving as secondary objectives. Typically, the power output of the steam turbine generator in an MSWI plant is closely aligned with the operational state of the incinerator, and the external power grid does not restrict the dispatching of generator power. Consequently, the main objective of the automatic control system in the MSWI process is to achieve stable combustion of MSW,

DOI: 10.1201/9781003634195-3

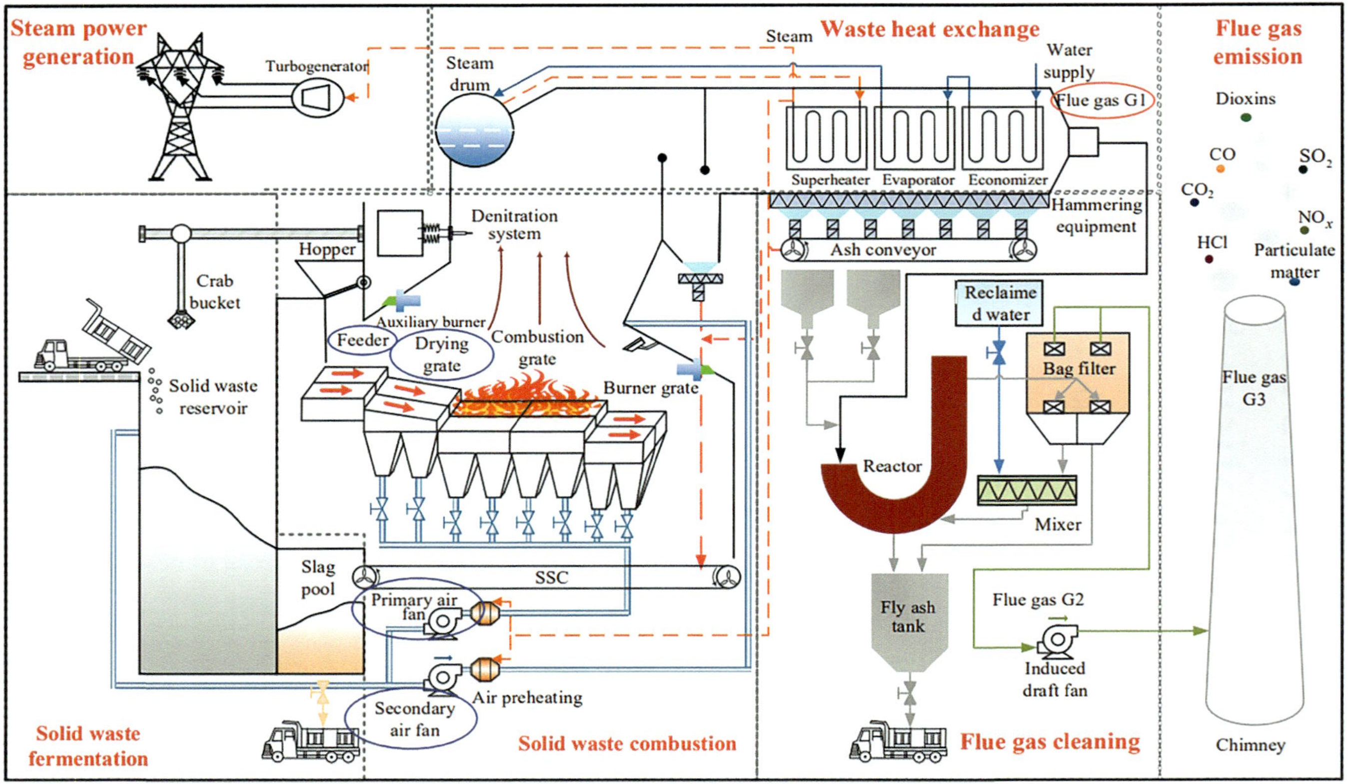

FIGURE 2.1 MSWI process flow diagram.

ensure consistent boiler steam production, minimize the loss on ignition rate of the slag, and reduce pollutant emissions as much as possible.

The detailed descriptions of each subsystem are as follows.

2.1.1 Solid Waste Fermentation Subsystem

The solid waste fermentation subsystem is responsible for receiving and storing MSW. Typically, MSW is transported by vehicles to the plant, weighed, recorded by a weighing system, and then dumped into the deposit pool via the unloading platform and port. Larger-sized MSW items require coarse crushing before entering the pool.

The primary purpose of storing MSW in the deposit pool is for dewatering and fermentation. A grab is used to stir the MSW, ensuring even distribution of its components while also removing some mud and sand. The MSW deposit pool is generally designed to hold enough waste for five to seven days of incineration.

This subsystem includes the MSW deposit pool, grab, crusher, feed hopper, and fault removal and monitoring devices. The MSW deposit pool serves as a space for storage, mixing, and the removal of MSW items. A typical large MSWI power plant features one deposit pool that supplies three to four incinerators, each equipped with multiple feed hoppers, usually serviced by one to two cranes and grabs positioned above the pit.

Operators monitor screens or visually check the speed at which MSW slides into the furnace from the feed hopper to regulate the feeding frequency. If large-sized MSW obstructs the feed port, the fault removal device in the feed hopper can push these items back into the deposit pool. Additionally, operators can use the grab to lift oversized items and transfer them to the crusher located above the deposit pool for crushing.

2.1.2 Solid Waste Combustion Subsystem

The solid waste combustion subsystem is the core of an MSWI power plant, determining the overall process flow and equipment configuration [4]. This subsystem typically includes an incinerator, feeder, combustion air supply equipment, auxiliary fuel supply and combustion equipment, reagent supply equipment, as well as slag discharge and treatment devices [5]. Key components within the furnace are the grate and the combustion chamber [6]. The grate is typically a mechanically movable surface where MSW is flipped and burned. Above the grate, the combustion chamber provides several seconds of residence time for combustion gases. Primary air is introduced from below the grate to mix with the MSW layer, while secondary air from above enhances the mixing of the combustion gases.

The combustion process of MSW involves several key steps: (1) evaporation of moisture on the solid surface; (2) evaporation of moisture within the solid; (3) ignition and combustion of volatile components in the solid; (4) surface combustion of solid carbon; and (5) completion of combustion. These stages can be grouped into two main processes: drying (steps 1 and 2) and combustion (steps 3–5).

Combustion is further categorized into primary and secondary stages, with the primary stage initiating combustion and the secondary stage completing it. MSW combustion primarily relies on decomposition combustion, as primary air alone is insufficient for complete combustion [7]. Its role is to facilitate the burning of volatile components and the decomposition of macromolecular elements. During primary combustion, the concentration of CO_2 may sometimes be reduced, which is significantly influenced by temperature.

In secondary combustion, flammable gases and particulate carbon generated during primary combustion undergo homogeneous gaseous combustion [8]. The efficiency of this process, as indicated by CO concentration, is crucial not only for energy recovery but also for minimizing the formation of harmful dioxins, thereby ensuring environmental safety. Therefore, the incineration process in a grate furnace must be designed based on these combustion mechanisms and characteristics.

Specific process parameters in the furnace include: (1) Combustion temperature: maintained above 850°C, with 900°C being optimal; (2) Flue gas residence time: exceeding 2 seconds; (3) CO concentration: kept below 100 mg/m^3 (1-hour average); (4) Stable combustion: efforts should be made to avoid instantaneous CO concentrations exceeding 100×10^{-6} g/m^3; (4) Daily management: thermometers, continuous CO analyzers, and continuous O_2 analyzers should be established to monitor and control combustion process parameters in real-time.

2.1.3 Waste Heat Exchange Subsystem

After cooling, the high-temperature flue gas from the incinerator can be treated through waste heat recovery and water spray cooling [9]. Waste heat boilers recover energy in three main ways: by producing steam for power generation, facilitating combined heat and power (CHP) systems, and providing hot water.

Waste heat boilers in MSWI plants can be categorized as either duct-type or integrated. Duct-type boilers are similar to conventional boilers; they allow fully combusted MSW in the incinerator to enable flue gas heat exchange, thereby reducing its temperature and generating steam or hot water [10]. In contrast, integrated boilers combine waste heat recovery and incineration processes, often utilizing water-cooled walls as combustion chambers [11].

2.1.4 Steam Power Generation Subsystem

In the power generation process, waste heat boilers transfer heat from the flue gas to water, converting it into superheated steam at precise pressure and temperature levels, which is critical for efficient energy conversion. This steam drives a turbo generator, effectively transforming thermal energy into electrical energy [12].

Practically, significant heat loss occurs during this conversion process, influenced by factors such as the waste calorific value, boiler thermal efficiency, and turbo generator efficiency [13]. Implementing CHP systems can enhance overall efficiency, as the turbine and generator efficiency during steam power generation typically ranges from 62% to 67%, while direct heating efficiency tends to be comparatively higher.

2.1.5 Flue Gas Cleaning Subsystem

Flue gas from the furnace is the primary source of pollution in the MSWI process [14], containing a diverse range of particulate and gaseous pollutants. To achieve compliant emission standards, a flue gas cleaning system is essential for removing these pollutants.

Particulate pollutants are typically removed using several methods, including gravity settling, centrifugal separation, electrostatic precipitation, and bag filtration. Gaseous pollutants such as SO_2, NO_2, HCl, and organic gases are treated through absorption, adsorption, and redox technologies [15, 16].

The main equipment involved in this flue gas cleaning process includes: (1) Settling chambers: which remove larger particles by gravity; (2) Cyclone dust collectors: which utilize centrifugal force to capture finer particles; (3) Electrostatic precipitators: which apply electric charges to facilitate the removal of particulates; (4) Scrubbers: which neutralize gaseous pollutants through chemical reactions; and (5) Bag filters: which trap particulate matter and DXN within fabric bags.

2.1.6 Flue Gas Emission Subsystem

After dust removal, flue gas is discharged into the atmosphere by an induced draft fan and is monitored by a continuous emission monitoring system (CEMS). CEMS continuously tracks flue gas emissions, detecting concentrations of key pollutants such as particulate matter (PM), CO, SO_2, NO_x, HCl, and O_2, as well as monitoring the flow rate, temperature, and pressure of the flue gas. CO_2 is also monitored primarily for greenhouse gas reporting purposes. The system displays, prints, and transmits data to the distributed control system (DCS) and regulatory authorities.

CEMS consists of several components, including a gas pollutant content analysis system, dust concentration monitoring, flow detection systems, and data collection and analysis systems. The gas pollutant content analysis system is composed of sampling probes, pipelines, pre-treatment systems, and analyzers. Common analysis methods include dilution, extraction, and direct measurement, with the extraction method being the most prevalent in MSWI power plants.

2.2 MSWI PROCESS DESCRIPTION IN TERMS OF DXN WHOLE LIFECYCLE CONCENTRATION

From the perspective of the whole lifecycle of DXN, encompassing its generation, decomposition, combustion, re-synthesis, adsorption, and emission, a schematic diagram of the typical MSWI process based on grate furnaces is illustrated in Figure 2.2.

As illustrated in Figure 2.2, the whole lifecycle mechanisms of DXN within the MSWI process can be categorized into six core zones: the solid-phase combustion zone, the gas-phase combustion zone, the high-temperature heat exchange zone, the low-temperature heat exchange zone, the flue gas cleaning zone, and the flue gas emission zone. The characteristics of each zone pertinent to DXN concentration are detailed as follows.

(1) Solid-Phase Combustion Zone: After being fed into the incinerator via the feeder, the combustible components of MSW pass sequentially through the drying, combustion stage 1, combustion stage 2, and burnout grates, where

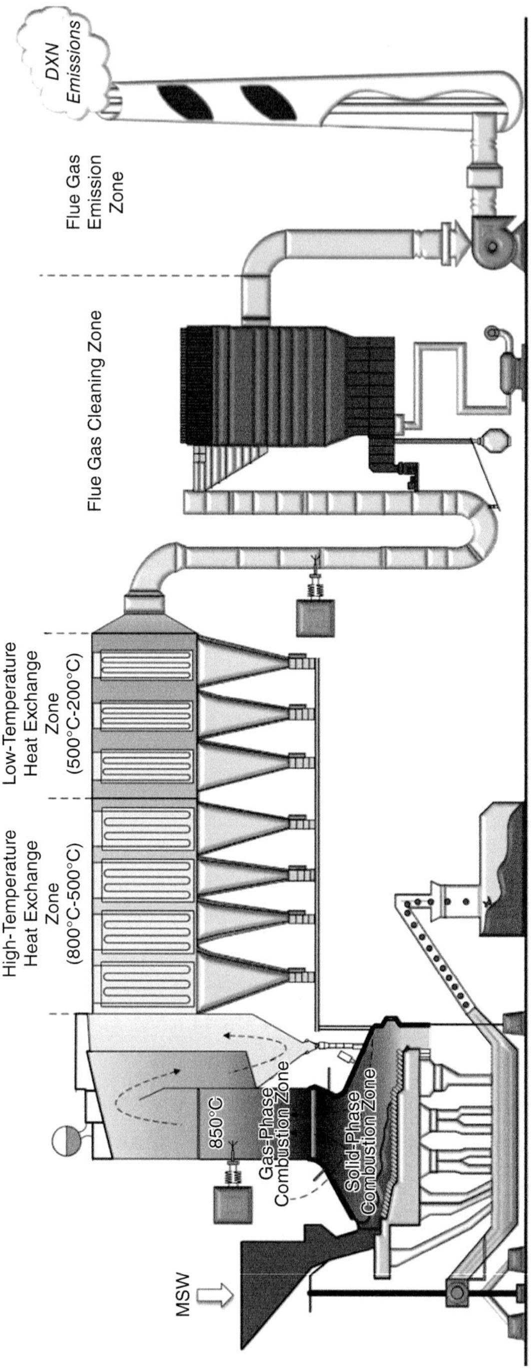

FIGURE 2.2 MSWI process in terms of DXN whole lifecycle concentration.

the combustion process is ultimately completed. In addition to the DXN inherently present in the raw MSW itself [17], under local oxygen-deficient conditions within this zone, incomplete combustion promotes the formation of DXN and releases key precursors. Consequently, this zone represents a critical stage for DXN generation; specifically, DXN is predominantly formed during the solid-phase combustion process;

(2) Gas-Phase Combustion Zone: In this zone, DXN and its precursors contained in volatiles are decomposed into carbon dioxide and HCl under high-temperature conditions. To ensure the complete destruction of DXN within the furnace, this zone must meet stringent operational criteria: the flue gas temperature must be maintained above 850°C, the residence time of the high-temperature gas must exceed 2 seconds, and sufficiently high flue gas turbulence must be guaranteed;

(3) High-Temperature Heat Exchange Zone: The temperature in this zone ranges from approximately 800°C down to 500°C, and this zone corresponds to the heat exchange section of the waste heat recovery boiler. The high-temperature flue gas releases its thermal energy through heat exchange equipment, which is utilized for steam generation or heating other media. During this process, high-temperature gas-phase synthesis reactions of DXN can occur [18]. Among these, the oxidative coupling of chlorophenols constitutes one of the primary pathways for DXN re-synthesis [19, 20];

(4) Low-Temperature Heat Exchange Zone: The temperature in this zone ranges between 500°C and 200°C. It is primarily where low-temperature heterogeneous catalytic reactions [21], including the surface-catalyzed reactions of precursors [22] and *de novo* synthesis [23], occur. To effectively inhibit the re-synthesis of DXN, it is essential to strictly control the residence time of flue gas within this temperature range through measures such as flue gas quenching;

(5) Flue Gas Cleaning Zone: The cooled flue gas sequentially passes through a deacidification reactor and a baghouse filter. Lime and activated carbon are injected upstream in this zone to ensure effective removal of acidic gases, heavy metals, and DXN. It is noteworthy that a not yet fully elucidated "memory effect" exists in this region. In some processes, a Selective Catalytic Reduction (SCR) unit is also included for denitrification, which can also influence DXN emission concentrations.

(6) Flue Gas Emission Zone: In this final zone, gases containing trace amounts of HCl, SO_2, NO_x, DXN, and other substances are drawn by the induced draft fan into the stack for eventual atmospheric release. Notably, a "memory effect" is also observed within the stack itself. Since no further treatment occurs at this stage, the DXN concentration after flue gas purification is generally considered equivalent to that in the emission zone.

Given that DXN exhibits complex, multi-zone, and cross-mechanism coupled characteristics during the MSWI process, and that intricate nonlinear coupling and dynamic evolutionary relationships exist between these zones, existing control measures struggle to achieve unified, source-controllable, and whole-process synergistic management. Therefore, it is crucial to construct an integrated "theory-characterization-monitoring-regulation" pathway that spans the whole lifecycle of DXN.

2.3 INTELLIGENT OPTIMIZATION CONTROL DESCRIPTION OF MSWI PROCESS

The MSWI process is a complex system involving multiple industrial components that convert municipal solid waste into electricity, non-toxic flue gas, and non-combustible ash. To optimize operational efficiency, combustion quality, material consumption, and pollutant emission concentrations, it is crucial to coordinate and control various subsystems with a focus on emission reduction and stable operation, facilitating the whole process optimization of this intricate industrial process [24].

The MSWI process can be summarized into four main stages: solid waste fermentation, solid waste combustion, waste heat exchange, and flue gas cleaning. In the solid waste fermentation stage, MSW undergoes biological fermentation and dehydration in storage tanks for three to seven days. After this period, the fermented waste is fed into hoppers by mechanical grabs for incineration. The primary variable at this stage is the MSW calorific value, which is crucial for optimizing the overall MSWI process. During the solid waste combustion stage, the fermented MSW is introduced into the incinerator, where it is dried, burned, and fully combusted using push feeders and grates, resulting in the formation of ash and high-temperature flue gas. Key manipulated variables (MVs) at this stage include the MSW feed rate, grate speed, and furnace air intake, while key controlled variables (CVs) encompass furnace temperature, flue gas oxygen content, steam flow, and the combustion line. In the waste heat exchange stage, high-temperature flue gas transfers heat through water-cooled walls and equipment, generating high-temperature, high-pressure steam that drives the turbo generator. Some of the steam produced is also utilized for heating purposes. Controlling the cooling rate is essential, with the main MV being the quantity of boiler feed water and the primary CV being the boiler steam flow. Finally, in the flue gas cleaning stage, the flue gas undergoes various purification processes, including selective non-catalytic reduction (SNCR) for denitrification, semi-dry desulfurization, activated carbon adsorption, and bag filtration. Flue gas that meets emission standards is then discharged into the atmosphere through a chimney. Key MVs in this stage are the amounts of urea, activated carbon, and lime consumed, while the main environmental indicator (EI) is the pollutant emission concentration.

Based on the above analysis, the key to controlling the combustion process lies in ensuring stable combustion through effective operations related to "air distribution and material distribution." The actuators for key MVs include damper blades for primary and secondary air, grate equipment responsible for drying, burning, and burnout speed, as well as their associated hydraulic power units.

The key CVs in this process include: (1) Furnace temperature (FT): This variable is essential for maintaining optimal combustion conditions and maximizing energy production; (2) Flue gas oxygen content (FGOC): Monitoring FGOC ensures that combustion occurs efficiently, minimizing unburned materials and emissions; (3) Steam flow (SF): This variable reflects the amount of steam generated, which is crucial for driving the turbo generator and for heating applications; (4) Combustion line: This encompasses various parameters that indicate the overall effectiveness of the combustion process. A detailed description of these CVs follows.

The combustion process of MSW involves various physical and chemical reactions, heat and mass transfer, and multiphase fluid flow, all occurring under conditions of high temperature, high pressure, and complex multiphase interactions. This complexity makes the stable control of FT particularly critical. FT is typically measured using thermocouples, which are essential for characterizing combustion stability and have a direct correlation with pollutant emissions. Precise control of this parameter is necessary for efficient operation. The "3T + E" principle dictates that the FT should be maintained between 850°C and 950°C. This FT range ensures the complete combustion of MSW, enhances the degree of burnout, achieves the desired volume and mass reduction rates, and minimizes pollutant concentrations in the emitted flue gas.

FGOC is closely linked to incineration efficiency and pollutant emissions. In practical operations, sufficient FGOC is essential for ensuring complete combustion, as well as for effective desulfurization and denitrification within the incinerator. However, an excess of FGOC can lead to increased flue gas losses, higher fan power consumption, and larger exhaust gas treatment requirements, resulting in the loss of heat and dust, and heightened emissions of fuel-derived NO_x pollutants. Conversely, a low excess air coefficient can result in incomplete combustion, leading to increased heat loss, reduced incineration efficiency, and the production of harmful gases such as dioxins, CO, and SO_2. Therefore, it is crucial to maintain FGOC within an optimal range to improve incineration efficiency, minimize pollutant emissions, and enhance overall economic benefits.

During the MSWI process, waste is burned on the grate with oxygen supplied to both the solid waste layer and the gas phase through the grate and furnace walls. The resulting flue gas, which contains fly ash particles, transfers heat to the steam system, subsequently converting this heat into thermal and/or electrical energy. From an economic perspective, maximizing waste treatment and energy output are mutually supportive; more material processed translates to greater energy generation. However, considerations regarding equipment lifespan impose constraints on various process variables, particularly those that affect corrosion, such as FT and boiler SF. This creates a direct conflict between economic objectives, which necessitate high FT and SF, and the operational longevity of the equipment, which can be compromised under these conditions. Therefore, stabilizing SF is essential for maintaining both operational stability and equipment durability [25].

Taking the key CVs of the MSWI process described above as an example, the multivariable intelligent optimization control problem can be articulated as follows:

$$
\left\{Y_{\text{Tem}}^{*}, Y_{\text{Oxy}}^{*}, Y_{\text{Ste}}^{*}\right\} = F_{\text{arg}}\left(\min\left\{\left\| R - R^{*}\right\|\right\}\right)
$$

$$
\text{s.t.}\begin{cases}
\text{Equality constraints}: \begin{cases}
\hat{R} = f_R(R, \left\{Y_{\text{Tem}}, Y_{\text{Oxy}}, Y_{\text{Ste}}\right\}, D) \\
\left\{\hat{Y}_{\text{Tem}}, \hat{Y}_{\text{Oxy}}, \hat{Y}_{\text{Ste}}\right\} = f_Y(\left\{Y_{\text{Tem}}, Y_{\text{Oxy}}, Y_{\text{Ste}}\right\}, \\
\qquad \left\{U_{\text{Tem}}, U_{\text{Oxy}}, U_{\text{Ste}}\right\}, D) \\
\left\{\hat{U}_{\text{Tem}}, \hat{U}_{\text{Oxy}}, \hat{U}_{\text{Ste}}\right\} = f_U((Y_{\text{Tem}}^{*} - Y_{\text{Tem}}), (Y_{\text{Oxy}}^{*} - Y_{\text{Oxy}}), \\
\qquad (Y_{\text{Ste}}^{*} - Y_{\text{Ste}}), D, \theta_{\text{Tem}}, \theta_{\text{Oxy}}, \theta_{\text{Ste}})
\end{cases} \\
\text{Inequality constraints}: \begin{cases}
R^{\min} \le R \le R^{\max} \\
Y_{\text{Tem}}^{\min} \le Y_{\text{Tem}} \le Y_{\text{Tem}}^{\max} \\
Y_{\text{Oxy}}^{\min} \le Y_{\text{Oxy}} \le Y_{\text{Oxy}}^{\max} \\
Y_{\text{Ste}}^{\min} \le Y_{\text{Ste}} \le Y_{\text{Ste}}^{\max} \\
U_{\text{Tem}}^{\min} \le U_{\text{Tem}} \le U_{\text{Tem}}^{\max} \\
U_{\text{Oxy}}^{\min} \le U_{\text{Oxy}} \le U_{\text{Oxy}}^{\max} \\
U_{\text{Ste}}^{\min} \le U_{\text{Ste}} \le U_{\text{Ste}}^{\max}
\end{cases}
\end{cases}
\tag{2.1}
$$

where $\left\{Y_{\text{Tem}}, Y_{\text{Oxy}}, Y_{\text{Ste}}\right\}$, $\left\{Y_{\text{Tem}}^{*}, Y_{\text{Oxy}}^{*}, Y_{\text{Ste}}^{*}\right\}$, and $\left\{\hat{Y}_{\text{Tem}}, \hat{Y}_{\text{Oxy}}, \hat{Y}_{\text{Ste}}\right\}$ are the actual, desired, and dynamically modeled estimated values of the FT, FGOC, and SF, respectively; R, R^{*}, and $\hat{R}$ are the actual, desired, and dynamically modeled estimated values of the EIs such as NOx and DXN; U is the MV of the basic control loop; U_{Tem} and $\hat{U}_{\text{Tem}}$ are the actual output value and the dynamically modeled estimated value of the FT controller; U_{Oxy} and $\hat{U}_{\text{Oxy}}$ are the actual output value and the dynamically modeled estimated value of the FLOC controller; U_{Ste} and $\hat{U}_{\text{Ste}}$ are the actual output value and the dynamically modeled estimated value of the SF controller; $f_R(\cdot)$, $f_Y(\cdot)$, and $f_U(\cdot)$ are the unknown nonlinear dynamic functions of the EIs, controlled objects, and controllers, respectively; D is external disturbances including MSW composition fluctuations, operating condition drifts, equipment wear, and maintenance; $\theta_{\text{Tem}}, \theta_{\text{Oxy}}$ and θ_{Ste} are the parameters of the FT, FGOC, and SF controllers, respectively; superscripts min and max denote the minimum and maximum values of the variables.

Therefore, the various CVs-controlled processes within an MSWI power plant are interdependent and must be coordinated to maintain stable operation within specified parameters. To comply with environmental regulations and optimize operational efficiency, the control objectives are as follows: (1) Maintain an FT above 850°C with minimal fluctuations; (2) Ensure consistent thermal output, reflected in uniform SF; (3) Stabilize the FGOC in the flue gas at the boiler outlet. Achieving these targets is crucial for effective operation. However, due to the uncertain composition and calorific value of MSW, the combustion process often operates under dynamic and variable conditions. Therefore, effective intelligent optimization control is essential to meet operational goals and guide the design of MSWI process control systems.

2.4 AI APPLICATION IN POLLUTION REDUCTION OF MSWI PROCESS

As a typical example of a complex process industry, the MSWI system can be analyzed from the perspective of control disciplines. Figure 2.3 illustrates the logical implementation diagram and corresponding AI algorithm positions within the AI-enhanced MSWI pollution reduction process. This diagram encompasses both the field-controlled objects relevant to industrial operations and the virtual controlled objects intended for academic research.

Figure 2.3 illustrates that the controlled objects in the MSWI process are crucial for both AI algorithm research and practical applications. This includes both real

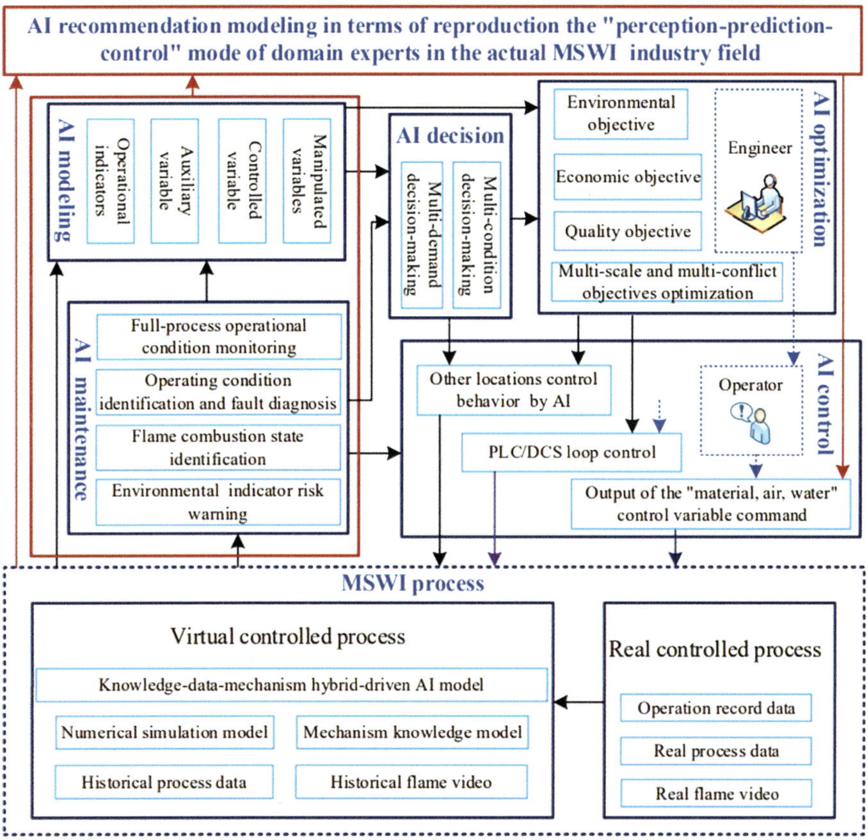

FIGURE 2.3 Logic diagram of the AI-based MSWI pollution reduction process.

MSWI-controlled objects and virtual MSWI-controlled objects. The real controlled objects are used to verify AI algorithms through actual process data and flame videos. In contrast, the virtual controlled objects necessitate the development of mechanism knowledge models, numerical simulation models, and hybrid intelligent models that integrate knowledge, data, and mechanisms based on historical process data and flame videos derived from the real objects. AI operations and maintenance encompass tasks such as flame combustion status recognition, EI risk warnings, operation status monitoring, condition recognition, and fault diagnosis. AI modeling involves creating models for EIs, CVs, MVs, and auxiliary variables. AI decision-making addresses multiple demands and operating conditions, leveraging insights from AI operations and AI modeling to optimize multi-scale and multi-objective tasks related to environmental protection, economic efficiency, and product quality. AI control utilizes AI operations and optimization to achieve closed-loop control of MSWI-controlled objects, employing strategies such as PLC/DCS basic loop control and advanced loop control across various operational levels. Additionally, the AI recommendation algorithm aims to simulate the embodied intelligence of domain experts, facilitating an automated "perception-cognition-decision-execution" mode.

2.4.1 AI Modeling for MSWI Process

AI modeling can be categorized into at least four groups: (1) Controlled object models for verifying various control algorithms in terms of academic research; (2) EI prediction and soft sensing detection models for easy-to-measure and difficult-to-measure pollution emission concentration; (3) Soft sensing detection models for product quality indicators; (4) Prediction models for CVs and MVs. The prospects for each category are described as follows.

The core challenge lies in the "data-mechanism-generalization" problem. The data faces several issues, including being "more conventional and less abnormal," "more single-modal and less cross-modal," and "more unlabeled and less precisely labeled." The scarcity of abnormal case samples hinders the model's ability to fully learn the operating conditions. Additionally, the MSWI process itself is highly coupled and nonlinear, while most AI models function as black boxes, complicating their integration into industrial mechanisms. This often results in "excellent fitting but incorrect physical interpretations." Furthermore, the dynamic drift of operating conditions can lead to a significant drop in the accuracy of laboratory models when applied to real-world scenarios.

2.4.2 AI Maintenance for MSWI Process

AI maintenance includes combustion state recognition, local and global operating states monitoring, and potential anomalies detection in terms of the whole process across different operational levels. The changes in the AI maintenance model construction phases include a lack of abnormal samples, unclear classifications of anomaly types, and difficulties in explaining and understanding the uncertain changes or occurrences of anomalies.

Focus areas for the AI maintenance model include dynamic adaptation, human-machine collaboration, and system compatibility. Data drift caused by equipment aging and changes in raw materials can degrade the accuracy of AI maintenance models, making real-time monitoring challenging. While operation and maintenance personnel possess a strong understanding of the equipment, they often lack expertise in AI, and AI engineers may not always be familiar with industrial operations. This creates barriers to effective human-machine collaboration. The interface between AI systems and MSWI industrial systems is often closed and employs diverse protocols, complicating the direct integration of AI maintenance model results into automated operation and maintenance processes.

2.4.3 AI Decision for MSWI Process

An AI decision involves determining the appropriate actions to take for multiple requirements in the MSWI process under various operating conditions and scenarios. For example, when faced with an abnormal EI, the AI decision system must decide which MVs to adjust or what setpoint value to assign to the CVs.

The challenges in AI decision-making for the MSWI process include explainability, multi-objective balance, and real-time performance. AI decisions are often treated as black-box solutions, making it difficult to explain the underlying logic in terms relevant to the MSWI process. For example, when suggesting an adjustment in grate speed, the system may fail to clarify its impact on pollutant emission concentrations. This lack of transparency can lead domain experts to resist adopting these recommendations due to a lack of trust. Additionally, multiple objectives in terms of pollution emission reduction, operating efficiency improvement, and energy consumption reduction often conflict with one another, complicating the AI's ability to dynamically adjust the weights of these objectives. As a result, the outputs may be "theoretically optimal but practically unfeasible."

2.4.4 AI Optimization for MSWI Process

The environmental, product, and economic indicators of MSWI exhibit characteristics such as multiple conflicts, various constraints, dynamic time-varying conditions, and multiple spatiotemporal scales. AI optimization focuses on a multi-objective optimization task decomposition mechanism based on MSWI process characteristics, a multi-operational indicator fusion mechanism, a collaborative optimization algorithm, and optimization planning algorithms under multiple operating conditions and time scales.

The central issue lies in the disconnection between constraint adaptation, global optima, and practical implementation. Industrial constraints in the MSWI process restrict the optimization space, causing AI to become trapped in local optima. This results in conflicts between optimization goals and safety requirements. Moreover, the optimization objectives change dynamically with varying operating conditions. Consequently, it becomes challenging for the model to promptly adjust the switching weights in response to these changing conditions, leading to mismatched output parameters. The ideal parameters suggested by AI often exceed the equipment's control accuracy and fail to account for response delays, making them difficult to implement effectively in real MSWI process operations.

2.4.5 AI Control for MSWI Process

The MSWI process is highly nonlinear, characterized by significant delays and time-varying dynamics. Regional and seasonal variations in MSW components and calorific value, coupled with differing levels of experience and maintenance skills among domain experts, result in multiple stable and non-stationary operating conditions. Consequently, AI control research focuses on several advanced strategies, including multivariable PID controllers, neural network fuzzy controllers, multimodal feature-driven combustion line controllers, multivariable model predictive controllers, and hierarchical self-organizing controllers, among others.

The core challenges in AI control include dynamic response, safety robustness, and multi-loop coupling. Effective AI control requires precise modeling of dynamic parameters that can change due to equipment aging and variations in MSW. Such variability increases the likelihood of control overshoot or oscillations. Additionally, multiple control loops within the system are interdependent; controlling one loop can inadvertently cause other parameters to exceed acceptable limits. Implementing multi-loop collaborative control is complex and presents significant challenges, particularly under the real-time constraints inherent in the MSWI process.

2.5 DIFFICULTIES IN THE DXN WHOLE LIFECYCLE MODELING IN ACTUAL MSWI PROCESS

2.5.1 Lack of DXN Whole Process Testing and Analysis in Industrial Sites

Currently, most existing studies on the detection of DXN concentrations in the MSWI process focus primarily on emission concentration at the stack outlet location, neglecting the generation concentrations at the boiler outlet location. This means that research conducted at industrial sites has not thoroughly examined the impacts of real operating conditions, environmental changes, and operational time on the generation, adsorption, and emission concentrations of DXN. As a result, it is challenging to support comprehensive modeling of the whole lifecycle of DXN.

Furthermore, these existing studies do not consider the generation and emission concentrations of DXN during the start-up and shutdown phases of the MSWI process. Therefore, it becomes difficult to understand the changing characteristics of DXN concentrations in these special operational conditions. It is essential to detect and analyze DXN concentrations at various locations throughout the complete operating cycle of the MSWI process. This will provide robust data to enhance our understanding of the DXN lifecycle mechanism and aid in the construction of a DXN detection and warning model.

2.5.2 Lack of DXN Whole Process Simulation Model for Actual Industrial Process

Understanding the mechanisms of DXN generation, adsorption, and emission is essential for developing effective DXN soft sensing models. While existing basic theories of DXN explain the macroscopic evolution of DXN generation, adsorption, and emission processes, there remains a significant gap between theoretical predictions and actual measured whole lifecycle DXN concentrations. Additionally, many physical and chemical reactions involving DXN occur rapidly, and phenomena such as the "memory effect" complicate accurate descriptions.

Experimental studies indicate that the concentration distribution of DXN throughout the whole MSWI process is influenced by various factors, including the physicochemical properties of MSW, the specific characteristics, and the combustion operation control strategies employed in the MSWI process. Given this complexity, it is crucial to closely align our understanding of the whole lifecycle DXN mechanism with the actual MSWI process. Therefore, designing a comprehensive numerical simulation model that captures the whole process of DXN generation, adsorption, and emission in real MSWI operations is necessary. This model will help elucidate the actual characteristics of the DXN mechanism and contribute to improved monitoring and modeling strategies.

2.5.3 Lack of DXN Detection and Warning Mechanism for Explainable AI

Existing research indicates that the evolution mechanism of DXN involves multiple stages, including MSW pre-treatment, MSW combustion, waste heat recovery, flue gas purification, and flue gas emissions within the MSWI process. However, accurately describing the actual changes in DXN during these stages presents significant challenges, making it difficult to effectively analyze the correlation between MSWI process variables and DXN lifecycle concentrations.

This complexity suggests potential correlations among process variables, combustion status, and DXN concentrations across various stages of the MSWI process. MVs, CVs, auxiliary variables, and flame image features can all be utilized to characterize DXN concentrations throughout its entire lifecycle. Collectively, this results in a rich dataset comprising over a hundred dimensions of input features. Therefore, it is essential to construct an interpretable multimodal data-driven DXN detection and warning model that retains the integrity of these structured and non-structured data. Such a model will enhance our ability to control DXN concentrations effectively and support timely decision-making and optimization in the MSWI process.

2.6 SUMMARY

This chapter outlines the typical process flow based on grate furnaces and provides an overview of intelligent optimization control in the MSWI process aimed at achieving a reduction in pollution emissions. It details the requirements for AI algorithms related to modeling, operation, decision-making, optimization, and control within the MSWI framework. Additionally, it summarizes the challenges associated with modeling the entire lifecycle of DXN in the actual MSWI process.

REFERENCES

1. Xin M, Sun Y, Li W, et al. Exposure levels of PCDD/Fs from co-landfill scenario with MSW and stabilized fly ash during hydrolysis-acidogenesis. *Journal of Cleaner Production*, 2023, 394: 136343.
2. Liu J, Xie G, Wang Z, et al. Manufacture of alkali-activated cementitious materials using municipal solid waste incineration (MSWI) ash: immobilization of heavy metals in MSWI fly ash by MSWI bottom ash. *Construction and Building Materials*, 2023, 392: 131848.
3. Fan C, Wang B, Ai H, et al. A comparative study on solidification/stabilization characteristics of coal fly ash-based geopolymer and Portland cement on heavy metals in MSWI fly ash. *Journal of Cleaner Production*, 2021, 319: 128790.
4. Magnanelli E, Tranås O L, Carlsson P, et al. Dynamic modeling of municipal solid waste incineration. *Energy*, 2020, 209: 118426.
5. Liu J, Goel A, Kua H W, et al. Evaluating the urban metabolism sustainability of municipal solid waste management system: an extended exergy accounting and indexing perspective. *Applied Energy*, 2021, 300: 117254.
6. Qi W, Geng C, Zhu F, et al. Complementary vitrification of municipal solid waste incineration fly ash from grate furnaces and fluidised bed incinerators via a co-reduction process. *Waste Management*, 2024, 184: 92–100.
7. Hu F, Li P, Cheng P, et al. A pilot-scale experimental study on MILD combustion of sawdust and residual char solid waste blend using low-temperature preheating air. *Fuel*, 2023, 342: 127768.
8. Jiang P, Parvez A M, Meng Y, et al. Novel two-stage fluidized bed-plasma gasification integrated with SOFC and chemical looping combustion for the high efficiency power generation from MSW: a thermodynamic investigation. *Energy Conversion and Management*, 2021, 236: 114066.
9. Hu D, Wang Y, Li J, et al. Energy saving control of waste heat utilization subsystem for fuel cell vehicle. *IEEE Transactions on Transportation Electrification*, 2024, 10(2): 3192–3205.
10. Zoghi M, Habibi H, Choubari A Y, Ehyaei M A. Exergoeconomic and environmental analyses of a novel multi-generation system including five subsystems for efficient waste heat recovery of a regenerative gas turbine cycle with hybridization of solar power tower and biomass gasifier. *Energy Conversion and Management*, 2021, 228: 113702.
11. Patel P, Vaezi M, Billal M M, Kumar A. Development of data-intensive techno-economic models for the assessment of a biomass, waste heat, and MSW integrated waste-to-electricity facility. *Resources, Conservation & Recycling Advances*, 2023, 20: 200188.
12. Malik S, Verma S, Gupta A, et al. Performability evaluation, validation and optimization for the steam generation system of a coal-fired thermal power plant. *MethodsX*, 2022, 9: 101852.
13. Wang B, Gupta R, Bei L, et al. A review on gasification of municipal solid waste (MSW): syngas production, tar formation, mineral transformation and industrial challenges. *International Journal of Hydrogen Energy*, 2023, 48(69): 26676–26706.
14. Lu M, Ge W, Xia Y, et al. Upcycling MSWI fly ash into green binders via flue gas-enhanced wet carbonation. *Journal of Cleaner Production*, 2024, 440: 141013.

15. Cai P T, Chen T, Chen B, et al. The impact of pollutant emissions from co-incineration of industrial waste in municipal solid waste incinerators. *Fuel*, 2023, 352: 129027.
16. Sun W, Shao Y, Zhao L, Wang Q. Co-removal of CO_2 and particulate matter from industrial flue gas by connecting an ammonia scrubber and a granular bed filter. *Journal of Cleaner Production*, 2020, 257: 120511.
17. Wei J, Li H, Liu J. Fate of dioxins in a municipal solid waste incinerator with state-of-the-art air pollution control devices in China. *Environmental Pollution*, 2021, 289: 117798.
18. Zhong R, Cai J, Yan F, et al. Process tracing and partitioning behaviors of PCDD/Fs in the post-combustion zone from a full-scale municipal solid waste incinerator in southern China. *Environmental Technology & Innovation*, 2021, 23: 101789.
19. Altarawneh M, Dlugogorski B Z, Kennedy E M, Mackie J C. Mechanisms for formation, chlorination, dechlorination and destruction of polychlorinated dibenzo-p-dioxins and dibenzofurans (PCDD/Fs). *Progress in Energy and Combustion Science*, 2009, 35(3): 245–274.
20. Peng Y Q, Chen J H, Lu S Y, et al. Chlorophenols in municipal solid waste incineration: A review. *Chemical Engineering Journal*, 2016, 292: 398–414.
21. Lomnicki S, Dellinger B. A detailed mechanism of the surface-mediated formation of PCDD/F from the oxidation of 2-chlorophenol on a CuO/Silica surface. *The Journal of Physical Chemistry A*, 2003, 107(22): 4387–4395.
22. Zhang S, Chen Z L, Lin X Q, et al. Kinetics and fusion characteristics of municipal solid waste incineration fly ash during thermal treatment. *Fuel*, 2020, 279: 118410.
23. Zhang M, Buekens A. De Novo synthesis of dioxins: a review. *International Journal of Environment and Pollution*, 2017, 60(1–4): 63–110.
24. Chai T Y. Challenges of optimal control for plant-wide production processes in terms of control and optimization theories. *Acta Automatica Sinica*, 2009, 35(6): 641–649.
25. Leskens M, Van't Veen P P, Van Kessel L B M, et al. Improved economic operation of MSWC plants with a new model based PID control strategy. *IFAC Proceedings Volumes*, 2010, 43(5): 655–660.

3 Investigation on Dioxins Emission Characteristic During Complete Maintenance Operating Period

3.1 PROBLEM DESCRIPTION

In developing country such as China, waste-to-energy initiatives are strongly encouraged and supported due to their benefits in terms of the harmlessness, reduction, and recycling of municipal solid waste (MSW) [1–3]. The emission characteristics of DXN [4–6], also known as polychlorinated dibenzo-*p*-dioxins and dibenzofurans (PCDD/F), have garnered significant public and governmental attention as unintentional persistent organic pollutants (POPs) resulting from MSWI plants [7–9]. Key reactions contributing to the formation of DXN [4, 10] include homogeneous synthesis (500°C–800°C) [11, 12], precursor synthesis (200°C–400°C) [4, 13], and de novo synthesis [14, 15]. The highly toxic nature of DXN has triggered prominent social issues, such as the "not in my backyard (NIMBY)" conflict related to MSWI facilities and concerns over health risks. Therefore, investigating the actual emission characteristics and mass concentrations of DXN is crucial for addressing these challenges. Such investigations can provide essential support for upgrading air pollution control devices (APCDs) and assist in developing soft sensing models for DXN, enabling effective online detection.

The MSWI plant is a significant source of DXN since its discovery in flue gas [16]. Consequently, extensive research has been conducted on the emission characteristics of DXN during normal operations [17]. In Germany, the emission characteristics of DXN in both the gas phase and particulate matter within the combustion chamber and boiler system were studied [18]. Their findings indicated that the highest concentrations of DXN are observed at the downstream end of the boiler, approximately at 270°C. Following this, it is discovered that advanced APCDs can effectively reduce DXN emission concentration at the stack to as low as 0.05 ng international toxic equivalents quantity (I-TEQ)/Nm^3 under most operating conditions [19]. Recommendations for emission reductions during normal operations by examining the formation characteristics of PCDD/F above the grate and in the flue of the incinerator were provided [20]. In Japan, the relation of "3T"

DOI: 10.1201/9781003634195-4

principle (temperature, time, and turbulence) to DXN emissions from both intermittent and continuous furnace types is investigated. Their results indicated that increased turbulence in the combustion process is beneficial for controlling DXN emissions [21]. In Italy, it is demonstrated that advanced APCDs can significantly reduce the mass concentration of PCDD/F by comparing measurements taken before and after APCD enhancements [22], aligning with the findings in [19]. In the Czech Republic, studies on DXN emission levels in stack gas have reported that emissions primarily ranged from 0.32 to 1.0 ng TEQ/Nm^3 [23]. In the Taiwan province of China, it was revealed that the mass concentration of DXN exhibited a unimodal trend, with the peak concentration occurring at the exit of the electrostatic precipitator (EP).

Furthermore, it was estimated that the mean equivalent concentration of DXN is 0.0511 ng I-TEQ/Nm^3 based on 23 exhaust gas samples collected between 2003 and 2006, with total DXN equivalent concentrations ranging from 0.0171 to 0.0904 ng I-TEQ/Nm^3 [24]. In Zhejiang province of China, it was observed that the mass flow of DXN exhibited a significant resynthesis phenomenon [25], where the mass concentration increased nearly 60-fold after APCDs, rising from 1.34 to 79.2 ng/Nm^3 in a typical circulating fluidized bed (CFB) MSWI [26].

The removal efficiencies of DXN following treatment with a semi-dry scrubber, activated carbon (AC), and bag filter (BF) were found to be 82.2% and 91.8%, respectively, in the mechanical grate furnace [27]. Moreover, the concentration of DXN emissions in the exhaust gas was measured at 0.013 ng I-TEQ/Nm^3, which is significantly below the legally permitted limit. It was confirmed that changes in operating conditions have a substantial impact on DXN emissions [28], with stable conditions being essential for maintaining lower concentrations than the emission limit. The studies conducted on DXN emissions under normal operating conditions indicate that most MSWI plants can comply with the regional permitted limits [29].

The unstable conditions of the MSWI process, particularly during startup and shutdown phases, are primarily responsible for increased DXN emissions, leading to the formation of a "memory effect" [30]. As a result, DXN emissions during these stages have become a focal point of analysis. Studies have investigated DXN emissions in small MSWI power plants with batch-type operations during startup and shutdown phases [31], revealing that DXN concentrations during unsteady operations are significantly higher than those during stable operations. Furthermore, there was a strong correlation between DXN emissions and the products of incomplete combustion, such as carbon monoxide (CO) concentrations. Subsequently, the emission characteristics of DXN in continuously operating MSWI plants were analyzed [32], with findings indicating that DXN emissions during the startup stage represent the largest proportion of total annual DXN emissions.

In Germany, it was reported that the DXN emission concentration during a cold startup procedure is approximately half that of normal operational levels [33]. This report highlights the improvements associated with cold startups and the necessity for purging during shutdowns. Based on optimization suggestions [34], increasing the primary air volume can help reduce the carbon monoxide (CO) produced by incomplete combustion, while injecting inert material upstream of the cloth bag can further decrease DXN emissions. Moreover, investigations have shown that the concentration of DXNs is less affected during the shutdown stage compared to the significant increases observed during the startup stage [35], consistent with previous findings [32].

In Italy, a study on DXN emissions from mechanical grate furnaces burning MSW and medical waste (comprising 95% and 5%, respectively) revealed that under transient operating conditions, the total release of DXN increased by 50% compared to steady-state operation [22]. In the Taiwan province of China, it was found that DXN concentrations exceeded the emission limit by more than 40 times, with DXN accounting for 60% of annual emissions from the intermittent incinerator during the startup process [36]. Both startup and shutdown phases were associated with increased DXN concentrations during continuous MSWI operation, with startup being the dominant factor, consistent with previous findings [35]. During the startup stage, the mass distribution of DXN accounted for 14.6% of the total emissions, including flue gas and fly ash (FA) [24]. Furthermore, it was observed that more than 90% of the annual DXN emissions from the MSWI plant occurred during the cold startup period when the number of operations exceeded five times [37]. In the Czech Republic, it was indicated that the DXN concentration in exhaust gas before and after shutdowns is significantly higher than during other operational periods [23].

Consequently, the startup and shutdown stages are identified as the primary sources of DXN emissions from MSWI plants [23, 38, 39]. However, there is a scarcity of reports focusing on the emission characteristics of DXN during the complete maintenance operating period (CMOP), which includes normal operation, shutdown, cooling, maintenance, heating, startup, and return to normal operation. This lack of data does not effectively support accurate annual emission assessments of DXNs from MSWI plants [40] and fails to provide relevant information for the DXN emission reduction strategies [16].

In this chapter, the emission characteristics of DXN in an MSWI plant in Beijing during the CMOP were studied. The investigation was organized into two operational stages: the shutdown process (SDP) stage, which encompasses the transition from normal operation to shut down and cooling, and the startup process (SUP) stage, which includes heating, startup, and return to normal operation. During the CMOP, flue gas sampling was conducted at the boiler outlet (BO) and stack inlet (SI), while FA was collected at the BF. The DXN removal efficiency of APCDs was also investigated. Additionally, the contribution of DXNs to annual emissions from the MSWI plant was identified and evaluated throughout this project period. To the best of our knowledge, this study provides the first dataset containing the mass concentrations of 17 different DXN congeners in exhaust gas and FA obtained from real MSWI plants during the annual scheduled CMOP.

The explanations of abbreviations used in this chapter are provided in Table 3.1.

TABLE 3.1
Explanation of Abbreviations in Chapter 3

Abbreviation	Description
DXN	Dioxins
PCDD/F	Polychlorinated dibenzo-*p*-dioxin/dibenzofuran
MSWI	Municipal waste incineration processes
CMOP	Complete maintenance operating period
SDP	Shutdown process
SUP	Startup process
TEQ	Toxic equivalent quantity
FB-FA	Flue gas and BF fly ash
APCD	Air pollution control device
MSW	Municipal solid waste
I-TEQ	International toxic equivalents quantity
CFB	Circulating fluidized bed
AC	Activated carbon
BF	Bag filter
CO	Carbon monoxide
FA	Fly ash
BO	Boiler outlet
SI	Stack inlet
MEE	Ministry of Ecology and Environment
HRGC/HRMS	High-Resolution Gas Chromatography High-Resolution Mass Spectrometry
USEPA	U.S. Environmental Protection Agency
LCV	Lower calorific value

3.2 MATERIALS AND METHODS

3.2.1 Basic Information – General Operation About the Selected MSWI Plant

The mechanical-grate-type MSWI plant commenced operations in 2009 and is equipped with two lines that collectively have a daily processing capacity of 800 tons. Each line features a stepped (inclined and horizontal) forward-pushing mechanical grate, primary and secondary air systems, and a single-drum horizontal boiler with a rated evaporation capacity of 74 tons per hour for energy recovery, with flue gas flowing horizontally. The boiler system includes a heavy hammer ash cleaner designed to enhance the cleaning efficiency of the superheater, evaporator, and economizer. The facility is also outfitted with APCDs that incorporate selective non-catalytic reduction (SNCR) technology for denitrification [41], semi-dry deacidification using lime [42], and AC [43] injection, along with BF [44]. These measures are implemented to reduce PCDD/F emissions, ensuring compliance with the legal standard of 0.1 ng I-TEQ/Nm^3, as illustrated in Figure 3.1.

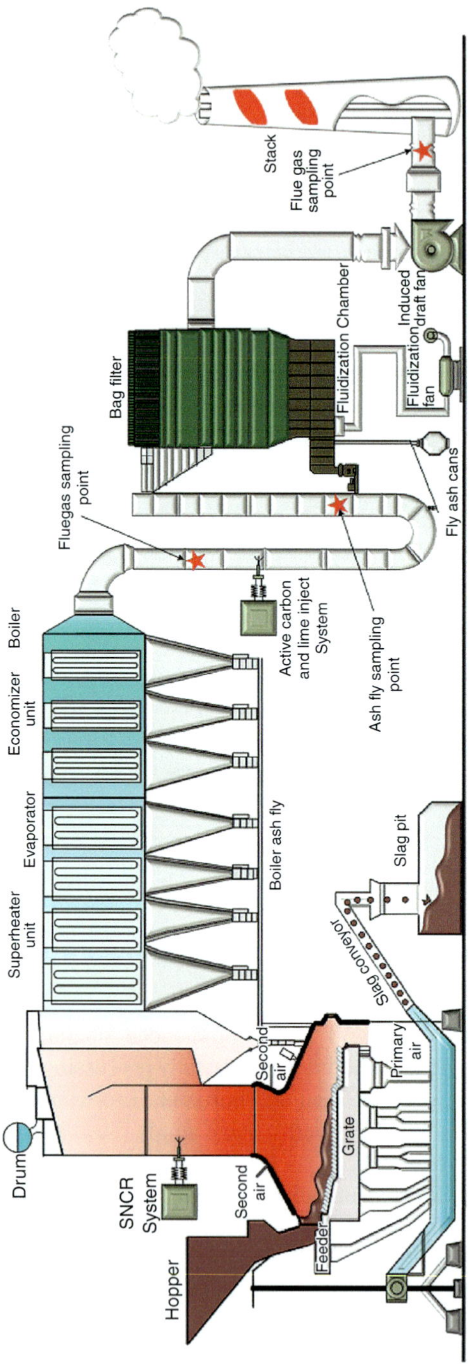

FIGURE 3.1 Process flow of the studied MSWI plant in Beijing with sampling locations.

In accordance with the latest regulations issued by the Ministry of Ecology and Environment (MEE) in China regarding the application and management of automatic monitoring data for MSWI plants, these facilities are required to accurately document the operational conditions and any abnormalities in automatic monitoring for each incinerator. This is aimed at enhancing environmental protection and regulating flue gas emissions.

The general operating conditions outlined in the maintenance plan for MSWI power plants include six phases: shutdown, cooling, maintenance, heating, startup, and normal operation, collectively referred to as the "SDP" (Shutdown) – "Maintenance" – "SUP" (Startup) cycle. The specific marking phases are as follows:

(1) Normal Operation and Shutdown: The furnace temperature must not fall below 850°C.
(2) Cooling: The furnace temperature should decrease from 850°C to below 400°C under normal circumstances.
(3) Maintenance: The oxygen content must not be lower than 2% of the standard state (which is 21%).
(4) Heating: The furnace temperature should be raised from below 400°C under normal conditions. Additionally, the duration of this phase should not exceed 168 hours following the repair or transformation of the refractory material in the furnace.
(5) Startup: This phase involves feeding the MSW into the incinerator until stable operating conditions are achieved, maintaining the furnace temperature above 850°C, at which point the heating phase is considered complete.

3.2.2 DXN Sampling Schemes at Different Operational Stages

The DXN sampling schemes at various operational stages are presented as follows.

(1) In the SDP stage, the sampling is divided into three phases:
   1. Normal Operation Phase (P1): The flue gas sampling points are located at the BO and the SI, while the FA sampling point is positioned beneath the BF. During this phase, the MSWI plant operates normally with the furnace temperature exceeding 850°C, and the APCDs are functioning properly.
   2. Shutdown Phase (P2): The flue gas and FA sampling points remain the same as in P1. At this stage, feeding into the hopper is halted, the furnace temperature is maintained above 850°C, and the APCDs continue to operate normally.
   3. Cooling Phase (P3): The locations of the sampling points are identical to those in P1. The MSWI plant enters the cooling phase after the MSW has been completely burned. The furnace temperature decreases according to the cooling curve, and the APCDs are operating normally.

(2) In the Maintenance Stage: Following the cooling phase, maintenance work is conducted based on the condition of the grate furnace and its associated devices; this aspect is not detailed here.
(3) In the SUP stage: The flue gas and FA sampling points are the same as those in the SDP stage. This stage is divided into three phases as follows:
    1. Heating Phase (P4): The MSWI plant enters the heating phase after ignition, where the furnace temperature is raised using the startup burner according to the heating curve.
    2. Startup Phase (P5): When the furnace temperature reaches 850°C, MSW is introduced into the system. The furnace is then maintained at 850°C without the use of the burner.
    3. Normal Operation Phase (P6): Normal operation begins once the startup phase stabilizes for approximately 4 hours.

3.2.3 Flue Gas and Fly Ash Sample Collection

As illustrated in Figure 3.1, flue gas and FA sampling operations are conducted at the BO, SI, and BF. The flue gas sampling device comprises a sampling pipe with an inner diameter of 7 mm for the sampling nozzle, a quartz filter cartridge, a gas phase adsorption unit equipped with a condensing unit, a flow measurement control device, and sensors for temperature, pressure, and flow, as shown in Figure 3.2.

According to the National Environmental Protection Standard (HJ 77.2-2008) of China, the inner diameter of the sampling ports is 100 mm at both the BO and the SI, with flue cross-sectional areas of 4.930 and 4.725 m^2, respectively. The sampling method employed is a single-hole multi-point isokinetic tracking sampling technique. Each sampling session lasts over 2 hours, and the volume of flue gas for each specimen is approximately 3–4 Nm^3, as depicted in Figure 3.3. For FA sampling during the flue gas sampling period, an equal time interval method is applied, with samples collected approximately every hour from the bottom of the BF using the sampler in the current flue gas, as shown in Figure 3.4.

3.2.4 Sample Analysis Process

The solid phase (quartz filter cartridge) and gas phase (XAD-2 resin and condensate water) from the aforementioned flue gas samples were mixed and analyzed to produce an experimental report in the laboratory. The experimental analysis consists of the following components:

(1) Sample Extraction: Initially, the pickled quartz filter cartridge and XAD-2 resin were dried, followed by Soxhlet extraction for 16–24 hours. Next, the condensate water collected during flue gas sampling was separated using liquid–liquid extraction. Finally, the extracted samples were concentrated and combined for purification.

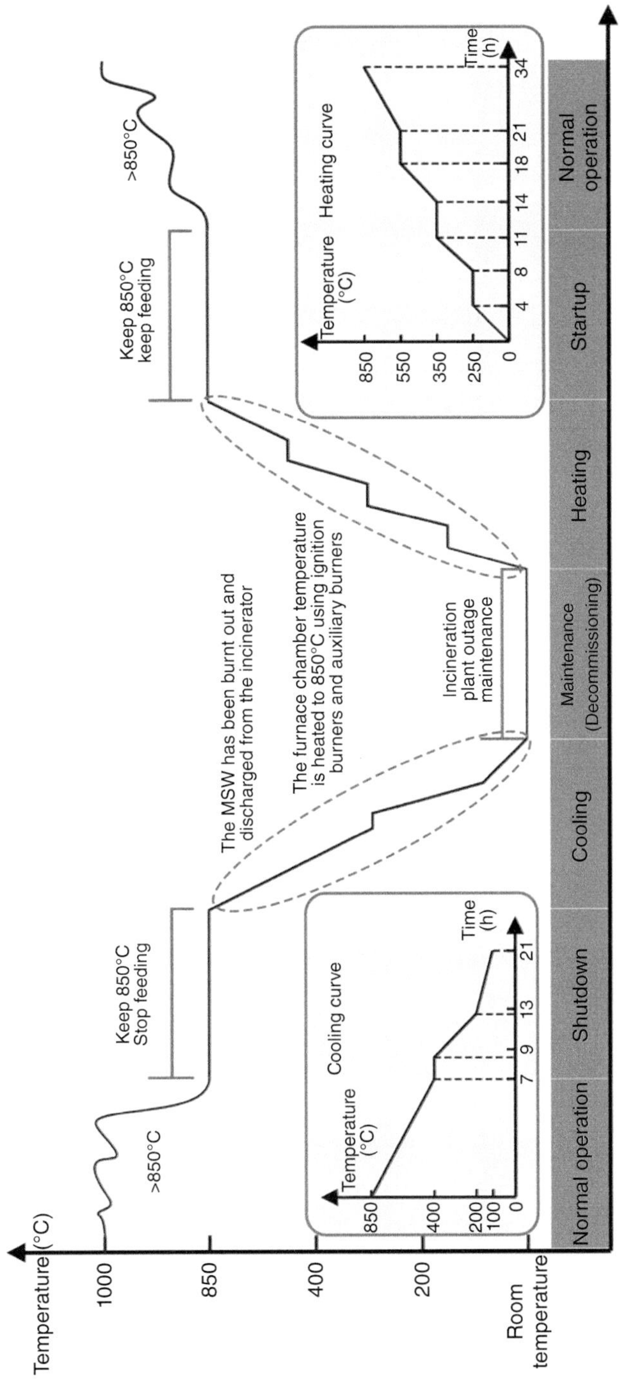

FIGURE 3.2 Temperature curve of the MSWI plants during the CMOP procedures.

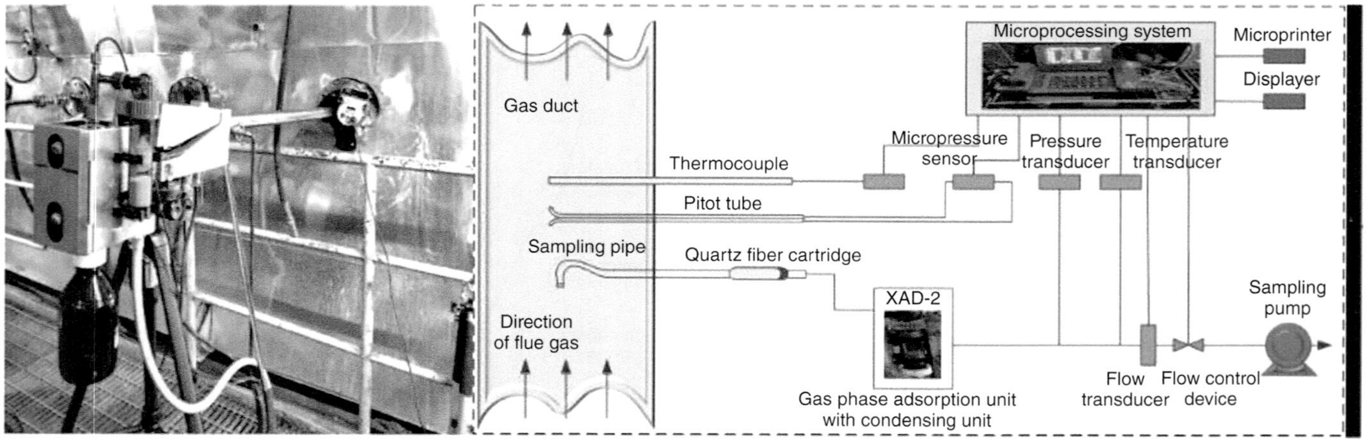

FIGURE 3.3 Sampling inside the flue gas pipe.

FIGURE 3.4 Sampling fly ash.

(2) Sample Purification: The samples were purified using an activated carbon-silica gel (SG) column. The eluate was then evaporated under a stream of high-purity nitrogen to remove excess solvent, resulting in a concentrated volume of 2 mL of aqueous pure DXN.
(3) Instrument Analysis: Following EPA Method 1668A, high-resolution gas chromatography/high-resolution mass spectrometry (HRGC/HRMS) (Thermo Scientific Trace 1310) was employed for analysis due to its high sensitivity. A chromatographic column (60 m × 0.25 mm I.D., 0.25 μm film thickness) was utilized to separate PCDD/Fs and determine their concentrations. The thermal program for the gas chromatograph maintained the column temperature at 150°C for 3 minutes, then increased to 230°C at a rate of 20°C/min and held for 18 minutes. The temperature was subsequently raised to 235°C at a rate of 5°C/min and maintained for 10 minutes before being elevated to 320°C at a rate of 4°C/min for 3 minutes.

According to the MSWI standard for pollution control in China (GB 18485-2014), the TEQs of PCDD/F were calculated using the I-TEQ factors published by the U.S. Environmental Protection Agency (USEPA) in 1989. The resulting PCDD/F concentrations were all normalized to dry air at 101.325 kPa and 237 K (with oxygen at 11%).

3.3 RESULTS AND DISCUSSION

3.3.1 Operational Conditions and Pollutants Emissions

The profiles of furnace temperature and conventional pollutants emitted at the SI during the shutdown and startup stages are illustrated in Figure 3.5. The concentrations of all readily measurable gases with the CEMS system, including NO_x, SO_2, HCl, and CO, were recalculated based on an oxygen content of 11%.

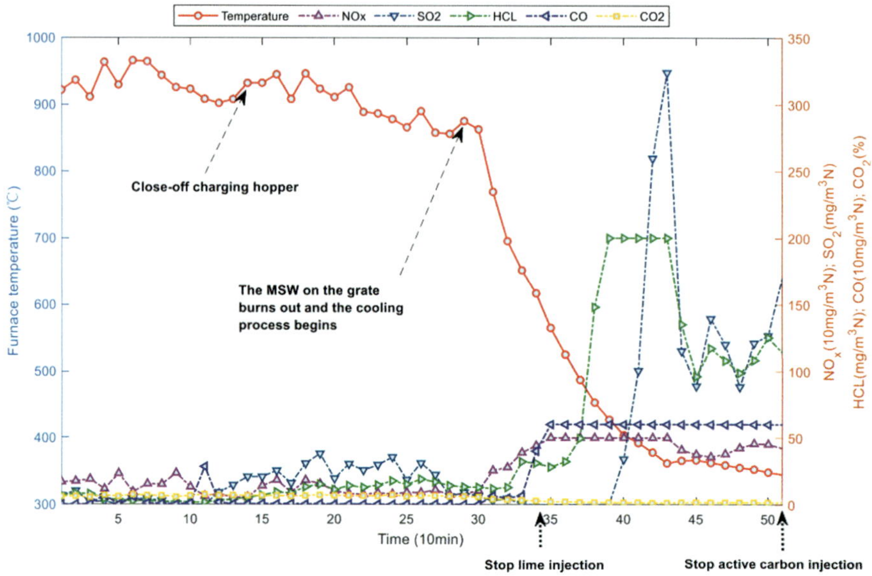

(a) SDP stage

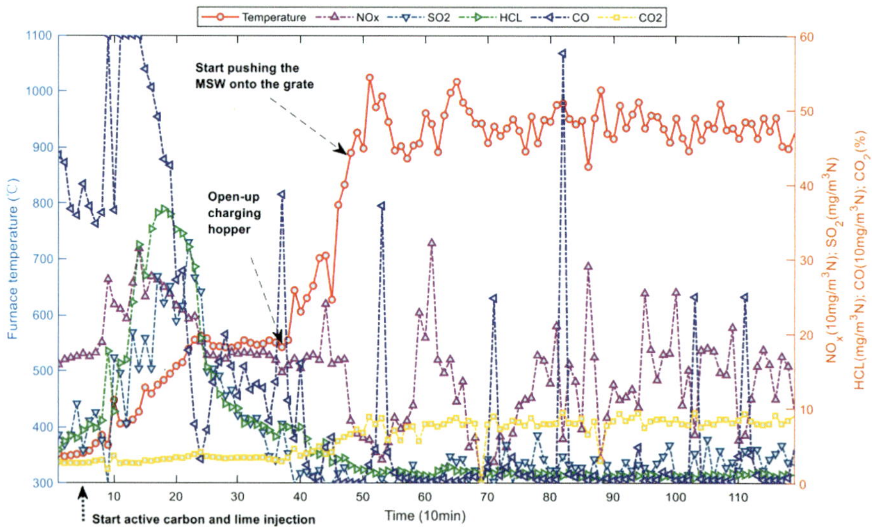

(b) SUP stage

FIGURE 3.5 Operating conditions during the (a) SDP and (b) SUP stage.

During the SDP stage, both the furnace temperature and conventional pollutant emissions remained relatively stable under normal operating conditions. After 140 minutes of normal operation, the feeding hopper was closed, marking the beginning of the shutdown stage. An auxiliary burner was used to maintain the furnace temperature above 850°C until the MSW was completely burned. Following this, the furnace temperature began to decline, while the concentrations of SO_2, HCl, and CO experienced slight increases. The shutdown phase lasted approximately 130 minutes before transitioning into the cooling phase. Notably, lime and AC injections were halted at 340 and 510 minutes, respectively. The cooling phase extended for about 240 minutes, which facilitated a decrease in furnace temperature to around 350°C, aided by the burner. Subsequently, the burner was turned off to allow the furnace to cool naturally to room temperature. During this cooling process, the emission concentrations of conventional pollutants significantly increased. The peak concentrations observed were 500 mg/m^3·N for NO_x, 320 mg/m^3·N for SO_2, 200 mg/m^3·N for HCl, and 600 mg/m^3·N for CO, occurring shortly after the cessation of lime injection.

In the SUP stage, the furnace temperature reached 350°C after 9 hours and was maintained at that level for 2 hours. Subsequently, the temperature increased to 550°C at a rate of 67°C per hour. During this period, lime and AC were sprayed. However, the emission concentrations of conventional pollutants remained extremely high, with peak levels of NO_x, SO_2, HCl, and CO measuring 300, 30, 40, and 600 mg/m^3·N, respectively. During the 2-hour constant-temperature phase at 550°C, pollutant emissions exhibited a gradual downward trend. A satisfactory blocking effect was achieved by reducing air intake, which was accomplished by opening the hopper baffle. This allowed MSW to enter the filling hopper, in line with the optimization recommendations in [34]. Once the constant temperature of 550°C was reached, the furnace temperature rose to 850°C at a rate of 87°C per hour, during which emissions of conventional pollutants continued to decrease. When the temperature reached 850°C, MSW could be pushed into the furnace. The furnace temperature stabilized between 900°C and 1,000°C, at which point the emissions of conventional pollutants returned to normal levels alongside regular feeding operations.

The SDP and SUP stages have a significant impact on conventional pollutant emissions. Notably high concentrations were observed even with the early implementation of APCDs for decontamination during the SUP stage. Additionally, low-temperature combustion conditions tend to produce large amounts of CO. The CO concentration serves as an indicator of the incinerator's combustion efficiency, reflecting the presence of products of incomplete combustion (PICs). A sharp increase in CO levels can notably influence the emission concentrations of PCDD/Fs [21]. Furthermore, during the SDP and SUP stages, low HCl concentrations were generated due to the use of light diesel burners, which resulted in substantial HCl production. This phenomenon is primarily attributed to chlorine deposition on the tube walls of the superheater, evaporator, and economizer [45].

3.3.2 DXN in the Flue Gas

The results and discussion regarding DXN in flue gas during the SDP and SUP stages are presented below.

3.3.2.1 Results and Discussion for SDP Stage

The profiles of DXN concentrations in the flue gas at the BO and SI, along with furnace temperature, boiler inlet temperature, and CO concentration from the continuous emission monitoring system, are presented in Figure 3.6 and Table 3.2.

During the normal operation stage (the 12 hours preceding the preparation for shutdown), the DXN concentrations at the BO measured 0.41 and 0.19 ng I-TEQ/m^3, significantly lower than the process design value of 3 ng I-TEQ/m^3. In contrast, the

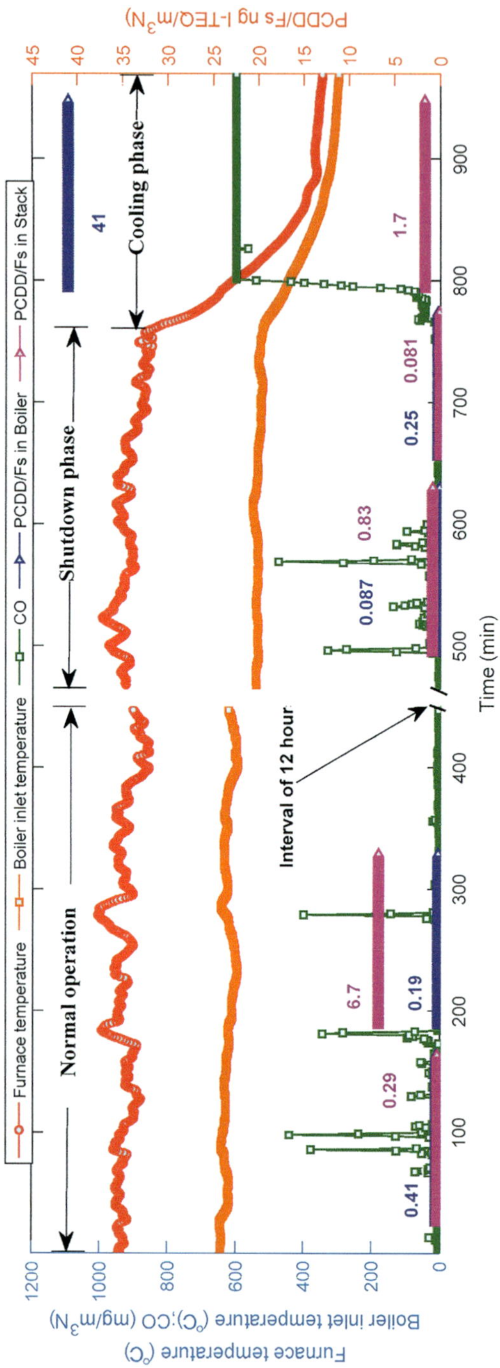

FIGURE 3.6 PCDD/F concentration of the BO and the SI during the SDP stage.

TABLE 3.2
Measurement Results of Flue Gas at the BO and the SI During the SDP Stage

	Boiler					Stack				
	P1		P2		P3	P1		P2		P3
	S1	S2	S3	S4	S5	S1	S2	S3	S4	S5
Oxygen (%)	5.5		8		17.6	8.15		12.5		18.5
Particulates (mg/m^3)	4.08×10^3		1.19×10^3		623	1.1		1.1		1.6
HCl (mg/m^3)	1.23		0.31		0.61	ND		ND		ND
Copper (mg/m^3)	0.125		6.87×10^{-2}		/	3.48×10^{-3}		1.50×10^{-3}		/
DXN concentration (ng/m^3)	0.64	0.29	0.13	0.28	14.23	0.37	8.99	0.79	0.06	0.42
DXN TEQ (ng I-TEQ/ m^3 N)	0.41	0.19	0.09	0.25	41.86	0.29	6.76	0.83	0.08	1.69

"S1, S2, S3, and S4" represent the first, second, third and fourth sampling, respectively.
"ND" represents not detected.
"/" represents no sampling.

DXN concentrations at the SI were recorded at 0.29 and 6.7 ng I-TEQ/m^3, respectively. The observed inversion in DXN concentrations can primarily be attributed to operational conditions. Based on the trend in CO levels, it is evident that a substantial amount of PICs [21] was generated in the prior combustion phase, leading to an increase in DXN concentration in the subsequent measurement period. This finding aligns with the investigation conducted in [35].

Following the scheduled shutdown, the DXN concentrations were measured at 0.087, 0.25, and 41 ng I-TEQ/m^3 at the BO, and 0.83, 0.081, and 1.7 ng I-TEQ/m^3 at the SI. The results indicate a significant increase in DXN emission concentrations, surpassing the design value by approximately 13–17 times at the SI, particularly during the cooling stage. These findings are consistent with those reported in [31, 32].

The observed increases may be attributed to the decline in thermal radiation within the furnace due to the absence of MSW, which weakens the decomposition of DXNs in the flue gas. Additionally, the MSWI process may facilitate the synthesis of DXNs as oxygen levels rise. Furthermore, there is a notable proportional relationship between DXN concentrations and CO levels during the shutdown stage. It is also worth noting that the DXN levels increased more rapidly before the closure of the charging hopper, indicating a substantial rise in DXN concentrations both before and after the shutdown, which supports the conclusions of Šyc et al. [23].

3.3.2.2 Results and Discussion for SUP Stage

The continuous sampling period lasted approximately 13 hours while the feeding hopper was open (with the furnace temperature reaching 550°C). The DXN emission curves are illustrated in Figure 3.7 and Table 3.3.

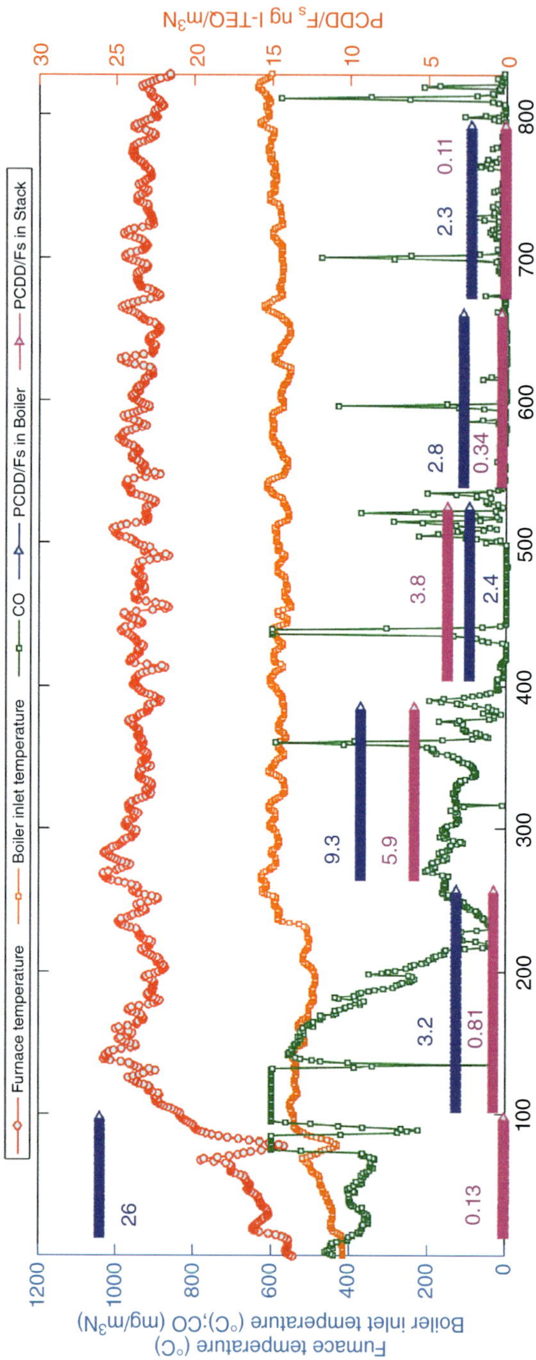

FIGURE 3.7 DXN concentration of the BO and the SI during the SUP stage.

TABLE 3.3
Measurement Results of Flue Gas, BO and SI During the SUP Stage

	Boiler						**Stack**					
	P1		**P2**		**P3**		**P1**		**P2**		**P3**	
	S1	**S2**	**S3**	**S4**	**S5**	**S6**	**S1**	**S2**	**S3**	**S4**	**S5**	**S6**
Oxygen (%)	15.2	7.4	8.1	7.1	6.5	6.9	14.8	10.4	7.9	8.2	8.0	9.3
Particulates (mg/m^3)	524		4225		4570		8.07		1.1		1.2	
HCl (mg/m^3)	0.1*		0.6*		0.39		37.9		5.3*		5.2	
Copper (mg/m^3)	5.95×10^{-2}		4.41×10^{-2}		3.07×10^{-2}		ND		ND		ND	
DXNs concentration (ng/m^3)	15	4.4	12	3.3	4.2	3.2	0.082	0.87	7.7	4.8	0.44	0.13
DXNs TEQ (ng I-TEQ/m^3 N)	26	3.2	9.3	2.4	2.8	2.3	0.13	0.81	5.9	3.8	0.34	0.11

"S1, S2, S3, and S4" represent the first, second, third and fourth sampling.
"ND" represents not detected.
"*" represents less than the detection limit is measured by 1/2 of the detection limit.

During the heating stage, the DXN emission concentration at the BO reached a maximum level approximately nine times higher than the design value. This finding aligns with the results in [32], although the concentration observed in this study is significantly lower than theirs. This discrepancy may be attributed to improvements in equipment performance and the accumulation of expertise in the MSWI process. The DXN emissions exhibited a bimodal profile, with concentration values ranging from 2.3 to 26 ng I-TEQ/m^3 at the BO. The peaks occurred during the heating phase and the initial stage of normal operation, measuring 26 and 9.3 ng I-TEQ/m^3, respectively. This phenomenon can be explained by the low heat load in the furnace during the heating process, which hampers the complete combustion of MSW.

In contrast, the DXN emissions at the SI followed a unimodal profile, with a peak concentration of 5.9 ng I-TEQ/m^3 observed in the early stage of normal operation. The typical concentration range at the SI was 0.11–5.9 ng I-TEQ/m^3. As the normal operation time progressed after the startup phase, the DXN emission concentration gradually decreased, approaching 0.1 ng I-TEQ/m^3 at the SI. A certain correlation was noted between CO and DXN concentrations. However, elevated DXN levels do not necessarily coincide with high CO concentrations. Conversely, when DXN exposure is high, CO concentrations tend to increase.

Overall, the SDP and SUP stages are the primary periods responsible for the surge in DXN emissions in the MSWI power plant [23, 38, 39]. Concentration peaks of DXNs were observed immediately before the shutdown phase and after the startup

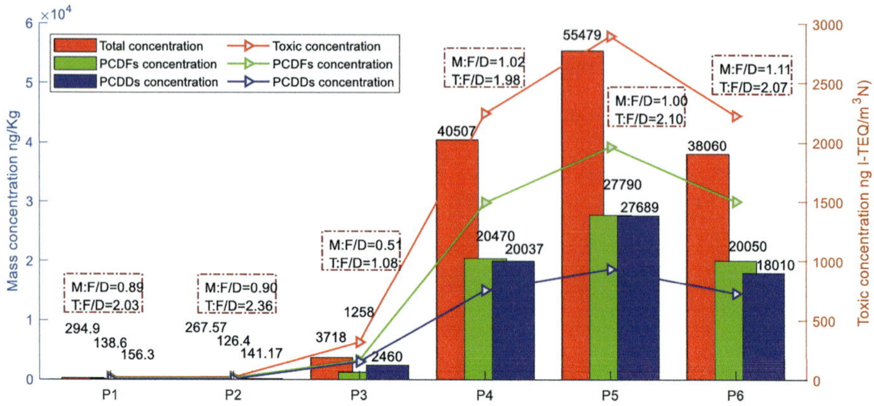

FIGURE 3.8 DXN concentrations of fly ash during the CMOP procedures.

phase at the SI, consistent with the findings in [23]. Additionally, DXN concentration peaks were noted at the BO during both the cooling and heating phases. Furthermore, the fluctuations in DXN concentrations are strongly correlated with CO levels, reinforcing the conclusions drawn in [32].

3.3.3 DXN in the Fly Ash

The sampling of FA at phases P1, P2, P3, P4, and P5 encompasses the entire SDP and SUP stages, with the growth of DXN illustrated in Figure 3.8.

3.3.3.1 Results and Discussion of SDP Stage

During the SDP stage, significant levels of DXN were observed exclusively in the cooling phase, with mass concentrations reaching 3,718 ng/kg and toxic concentrations at 320 ng I-TEQ/kg – both of which are an order of magnitude higher than normal levels. This phenomenon is primarily attributed to the de novo synthesis reaction occurring within the boiler system, which operates optimally in the temperature range of 200°C–450°C [37]. During the cooling stage, the boiler inlet temperature was approximately 600°C, as illustrated in Figure 3.5, facilitating the rapid production of substantial amounts of DXN [38]. This finding aligns with the results in [20], where the DXN TEQ concentration ratio was approximately equal to 1. In contrast, the concentration remained at normal levels during the combustion of MSW in the furnace. These results indicate that a sufficient supply of MSW is crucial for establishing the necessary temperature window for the decomposition of DXNs.

3.3.3.2 Results and Discussion of SUP Stage

During the SUP stage, the DXN concentration in FA exceeds normal levels by three orders of magnitude and surpasses the peak value recorded during the cooling phase by one order of magnitude. The transformation of DXN concentrations in FA exhibits a single peak profile, with a maximum value reached during the normal operation phase after the grate charging layer was fully paved (mass concentration: 55,479 ng/kg; toxic concentration: 2,900 ng I-TEQ/kg). This indicates that DXNs were primarily generated from incomplete combustion within the furnace [20], as evidenced by a DXN ratio greater than 1 during the SUP stage, which is also reflected in the CO concentration profile shown in Figure 3.7.

In general, DXN concentrations in FA during the SUP stage are significantly higher than those observed during the SDP stage. This phenomenon can be explained by several factors: (1) In terms of incomplete combustion, the SUP stage exhibits more substantial inefficiencies compared to the SDP stage; (2) The HCl concentration is higher during the SUP stage than in the SDP stage; (3) The particulate concentration is also greater during the SUP stage compared to the SDP stage.

3.3.4 DXN Congener Profiles in the Flue Gas

The results and discussion regarding DXN congener profiles during the SDP and SUP stages are presented below.

3.3.4.1 Results and Discussion of SDP Stage

The TEQ concentration distribution of 17 DXN congeners and the ratio of PCDFs/PCDDs concentration in flue gas at the BO and SI are shown in Figure 3.9.

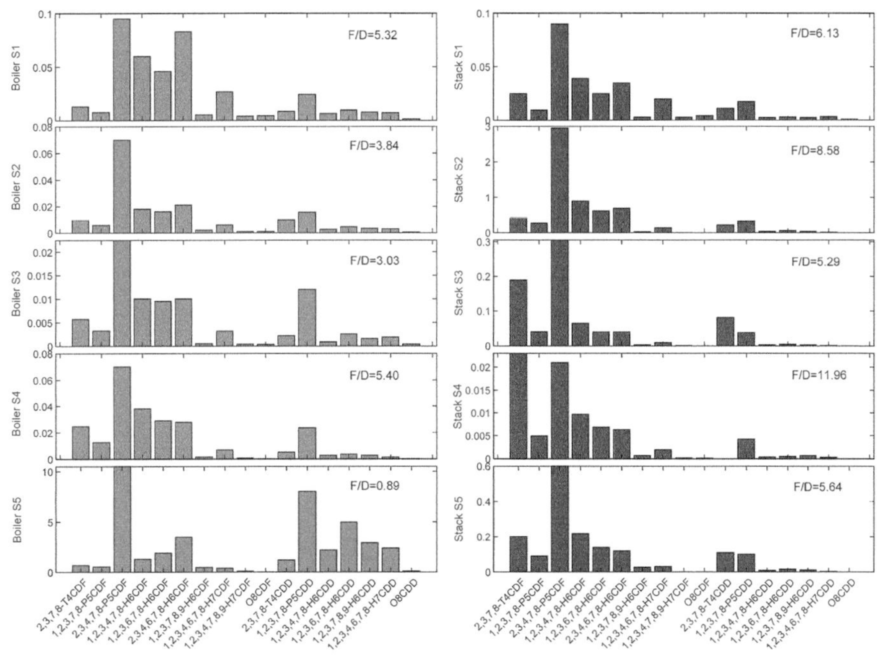

FIGURE 3.9 DXNs congener TEQ concentration of flue gas during the SDP stage.

During the normal operation phase (boiler S1–S2 and stack S1–S2), the major congener concentrations were predominantly composed of 5–6 chlorinated PCDFs, accounting for 71.1% and 68.9% of the total concentration in the BO and 67.3% and 80.5% in the SI, respectively. Among these, 2,3,4,7,8-P5CDF was the most prevalent, contributing 23.2% and 36.8% to the total concentration in the BO, and 30.1% and 43.7% in the SI.

In the shutdown phase (boiler S3–S4 and stack S3–S4), the concentration of 5–6 chlorinated PCDFs decreased compared to levels observed during normal operation, comprising 63.5% and 70.7% of the total concentration in the BO and 59.4% and 60.2% in the SI. Concurrently, the TEQ concentrations of 2,3,7,8-T4CDF and 1,2,3,7,8-P5CDD showed a notable increase, with the concentration of 2,3,7,8-T4CDF (28.3%) surpassing that of 2,3,4,7,8-P5CDF (25.8%) in S4.

During the cooling phase (boiler S5 and stack S5), the proportion of PCDD congeners (52.8%) was higher than that of PCDFs (47.2%) in the BO, with 5–7 chlorinated PCDDs being the dominant group, constituting 49.6%. However, the concentration distribution of PCDD congeners at the SI remained consistent with the aforementioned results.

Consequently, the concentration of PCDFs is significantly higher than that of PCDDs in both the BO and SI under most conditions. This indicates that PCDFs are the primary components of DXNs generated and discharged from the MSWI power plant. Among these, 2,3,4,7,8-P5CDF is the main contributor to DXN concentrations, aligning with previous findings [37]. Therefore, mapping the overall DXN emission concentration based on the concentration of a single substance, specifically 2,3,4,7,8-P5CDF, is warranted.

Considering that the ratio of PCDF/PCDD for S1–S4 is substantially greater than 1, it can be concluded that the DXN emission concentration was primarily influenced by unstable combustion conditions within the furnace, as noted in [20]. During the SDP procedures, the TEQ concentration distribution in the furnace when combusting MSW differs from that when not combusting MSW. Moreover, the PCDF/PCDD ratio was less than 1 in the DXN emissions at S5, which can be attributed to the de novo synthesis reaction occurring due to FA deposited in the boiler (as no MSW was being burned at this time). This suggests that the de novo synthesis reaction produces more PCDD than PCDF. Additionally, the overall distribution of DXNs remains relatively unchanged, with the most significant difference observed in PCDD concentrations after flue gas purification by APCDs. This finding confirms that AC and bag filters have superior adsorption capacities for PCDDs during the SDP stage.

3.3.4.2 Results and Discussion of SUP Stage

Figure 3.10 shows the TEQ concentration distribution of 17 DXN congeners and the ratio of PCDF/PCDD concentration in flue gas at BO and SI during the SUP stage.

During the heating phase (boiler S1 and stack S1), the major congener concentrations in the BO were as follows: 2,3,4,7,8-P_5CDF (41.2%), 1,2,3,4,7,8-H_6CDF (8.2%), 1,2,3,6,7,8-H_6CDF (8.2%), 2,3,4,6,7,8-H_6CDF (13%), and 1,2,3,7,8-P_5CDD (9.2%). The proportions of other congeners were relatively high, except for highly chlorinated PCDFs (such as 1,2,3,7,8,9-H_6CDF, 1,2,3,4,7,8,9-H_7CDF, and O_8CDF) and highly chlorinated PCDDs (notably O_8CDD) in the SI. In both the BO and SI,

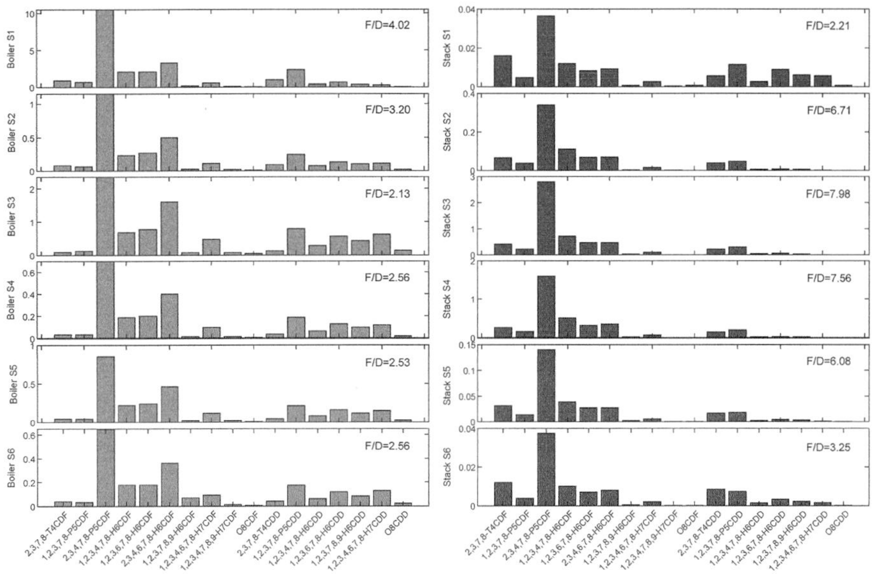

FIGURE 3.10 DXNs congener TEQ concentration of flue gas during the SUP stage.

2,3,4,7,8-P_5CDF exhibited the highest concentration. The decrease in the PCDF/PCDD ratio before and after flue gas purification indicates that PCDFs are adsorbed more effectively than PCDDs.

During the startup phase (boiler S2–S3 and stack S2–S3), the concentration distribution profiles of DXN congeners were nearly identical between the BO and SI. In this phase, 5–6 chlorinated PCDFs dominated the BO (69.2% and 60.4% in S2 and S3, respectively), while 5–7 chlorinated PCDDs accounted for 20.4% and 29% in S2 and S3, respectively. The concentration at the SI was predominantly focused on 5–6 chlorinated PCDFs (76.9% and 80.1% in S2 and S3). Although the overall concentration of DXNs decreased before and after flue gas cleaning, the concentration of 2,3,4,7,8-P_5CDF increased in S3, indicating a notable "memory effect" in the APCDs.

In the normal operation phase (boiler S4–S6 and stack S4–S6), the distribution profiles of DXN congeners were nearly equivalent in the BO. The primary concentrations were centered on the ten congeners of PCDF, while the proportion of the seven congeners of PCDD gradually increased (11.7% in S4, 14.1% in S5, and 23.5% in S6) at the SI. The "memory effect" of the APCD was observed again; results from the S4 experiment indicated that the concentration of 2,3,4,7,8-P_5CDF nearly doubled after cleaning, like the results from S3. However, subsequent experiments (S5 and S6) demonstrated that the memory effect had dissipated, and flue gas cleaning returned to normal levels. Therefore, it can be concluded that furnace startup operations can trigger the "memory effect" in APCDs.

In both the BO and SI, the concentration of the seven PCDD congeners is significantly lower than that of the ten PCDF congeners, with 2,3,4,7,8-P5CDF exhibiting the highest proportion among all homologs [36]. Consistent with the SUP stage, 2,3,4,7,8-P5CDF remains the principal component of DXNs. The PCDF/PCDD ratios consistently exceed 1 at all times (ranging from 2.13 to 7.98), which differs from the trends observed during the SDP stage. This suggests that DXN emission concentrations are primarily attributed to unstable combustion conditions within the furnace, as concluded in [20]. Similar to the SDP stage, the removal efficiency of the APCD equipment tends to favor the seven PCDD components. Furthermore, achieving stable combustion that complies with emission standards requires at least 12 hours of operation to ensure that conventional indicators of the MSWI process reach normal levels.

3.3.5 DXN Removal Efficiency

The APCDs include AC injection, lime injection, and BF devices between the BO and SI. The removal efficiencies of APCDs for DXN, PCDF, and PCDD are shown in Figure 3.11.

3.3.5.1 Results and Discussion of SDP Stage

The highest removal efficiency of DXNs via APCDs during the normal operation phase is 28% (27% for PCDFs and 36% for PCDDs). However, in subsequent measurements, the removal efficiencies for DXNs, PCDFs, and PCDDs were negative

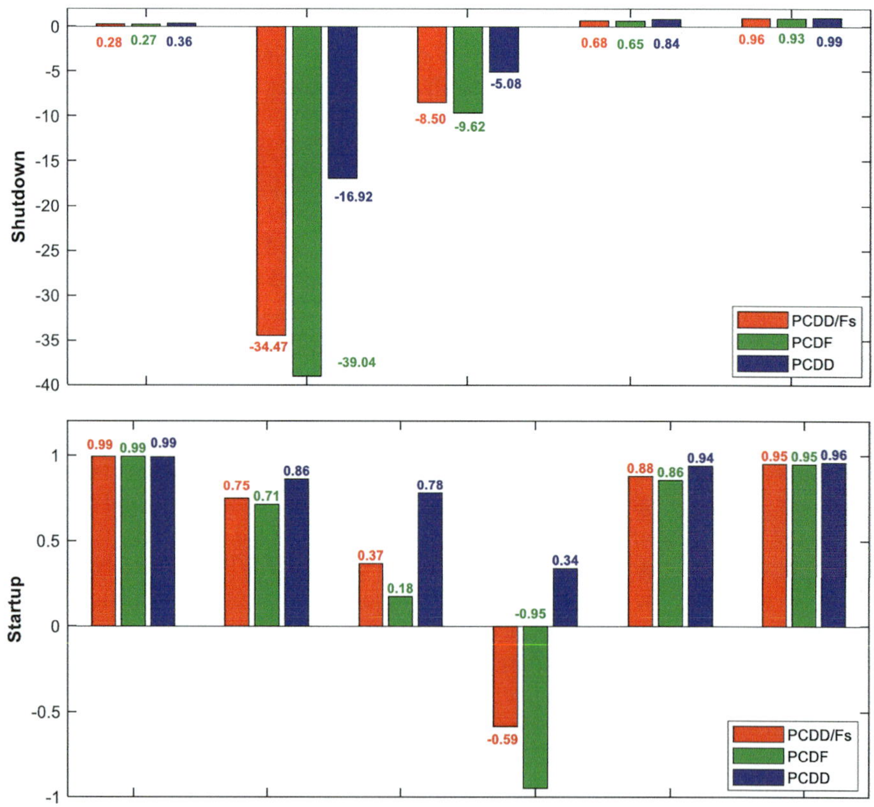

FIGURE 3.11 DXN and congener removal efficiencies during the SDP and SUP stages.

(−34.47% and −8.5%). Consequently, the concentrations of DXNs, PCDFs, and PCDDs at the SI were equal to or higher than those at the BO. This phenomenon can be attributed to the "memory effect" of the APCDs [27], which may lead to increased final DXN emissions even during the normal operation of MSWI power plants [36].

During the shutdown and cooling phases, the removal efficiencies of the APCDs gradually improved, reaching 68% during shutdown and 96% during cooling. This indicates that APCDs effectively remove gaseous DXNs during these phases, aligning with the design intentions of the MSWI process.

3.3.5.2 Results and Discussion SUP Stage

In the SUP stage, the removal efficiencies of APCDs exhibited a cosine curve profile. Initially, the removal efficiency was at its highest, with a maximum value of 99% for DXNs observed during the heating phase. However, this efficiency gradually decreased to a minimum of −59% (−95% for PCDFs and 34% for PCDDs) at the onset of the normal operation phase. This result suggests that the concentration of PCDFs exceeded the upper limit of the currently AC adsorption capacity, leading to a gradual increase and eventual return to normal levels. Ultimately, the last recorded removal efficiency reached 95% (−95% for PCDFs and 96% for PCDDs).

Overall, the removal efficiency of APCDs during the SUP stage is significantly higher than that in the SDP stage, with PCDD removal efficiencies surpassing those of PCDFs. This finding is consistent with the results in [38], indicating that AC has a stronger capacity for PCDD removal compared to PCDFs. Therefore, additional effective removers and inhibitors are necessary for PCDF cleaning to reduce DXN emissions. The negative adsorption efficiency can be attributed to the "memory effect" of DXNs. Experimental results suggest that the "memory effect" manifests when the DXN concentration in boiler flue gas is low, aligning with the conclusions in [46].

3.3.6 DXN Emission Efficiency

In modern MSWI plants, the DXN emission concentration in the flue gas at the SI is typically less than or equal to 0.05 ng I-TEQ/m^3 during normal operations [47]. This studied MSWI plant operates for over 8,000 hours per year, whereas the CMOP occurs only once annually, corresponding to the shutdown and startup phases that take place once each year. Tables 3.2 and 3.3 present the DXN emissions measured at the SI during the SDP and SUP stages. The calculation of DXN emissions from the flue gas based on actual operational data is detailed as follows

$$\text{PCDD/F of shutdown} \approx \left(\frac{1}{5}\sum_{i=1}^{5} Si\right)\left(\text{ng I-TEQ/m}^3\,\text{N}\right) \times \text{Flue gas}\left(\text{Nm}^3/\text{h}\right) \times \text{Shutdown process}\left(\text{h}\right) \tag{3.1}$$

$$\text{PCDD/F of normal operation} \approx 0.05\left(\text{ng I-TEQ/m}^3\,\text{N}\right) \times \text{Flue gas}\left(\text{Nm}^3/\text{h}\right) \times \text{Operation process}\left(\text{h}\right) \tag{3.2}$$

$$\text{PCDD/F of startup} \approx \left(\frac{1}{6}\sum_{i=1}^{6} Si\right)\left(\text{ng I-TEQ/m}^3\text{ N}\right) \times \text{Flue gas}\left(\text{Nm}^3/\text{h}\right) \times \text{Startup process}\left(\text{h}\right) \quad (3.3)$$

As a result, the annual DXN emission can be calculated as follows:

$$\begin{aligned}\text{PCDD/F} &\approx \text{PCDD/F of shutdown} + \text{PCDD/F of normal operation} \\ &\quad + \text{PCDD/F of startup} \\ &\approx \frac{1}{5}\times 9.601\times 117{,}000\times 27 + 0.05\times 135{,}000\times 8{,}000 + \frac{1}{6}\times 11.09\times 115{,}000 \\ &\quad \times 36 \\ &\approx 6{,}065{,}912 + 54{,}000{,}000 + 7{,}652{,}100 \\ &\approx 67{,}718{,}012\left(\text{ng I} - \text{TEQ}\right) \end{aligned} \quad (3.4)$$

The total amount of DXN emitted from the MSWI plant is approximately 67,718,012 ng I-TEQ per year, with CMOP accounting for 9.0% and 11.3%. Consequently, the annual DXN emissions represent over 20% of the total emissions. This finding aligns with the results of in [32, 36], although the proportion of emissions is significantly lower than previously reported. Furthermore, the percentage of CMOP emissions is inconsistent with earlier studies [32, 36]. These discrepancies can be attributed to the measurement of DXN emission concentrations taken after opening the feed hopper, at which point the furnace temperature has reached 550°C. This timing may overlook the contributions of DXN emissions arising from memory effects and de novo synthesis during the heating phase, specifically between 0°C and 500°C [32]. Therefore, it is likely that the actual proportion of DXN emissions during the SUP stage is greater than indicated.

3.3.7 Effects of MSW Composition, Combustion Properties, and Fuel

Table 3.4 presents the composition of MSW during the SDP and SUP stages, which includes volatile matter, ash, fixed carbon, moisture, sulfur, carbon, hydrogen, oxygen, nitrogen, and chlorine, as well as the calorific values, which encompass both high calorific value and low calorific value.

The air/MSW ratio is a crucial factor in determining DXN emissions and can be calculated as follows:

$$\text{AFR}_{\text{Stoich}} = \frac{m_{\text{Air}}}{m_{\text{MSW}}} = \frac{m_{\text{Oxy}}/23\%}{12x+y+16z} = \frac{\left(16^{*}2^{*}\dfrac{4x+y-2z}{4}\right)/23\%}{12x+y+16z} \quad (3.5)$$

where $\text{AFR}_{\text{Stoich}}$ is the stoichiometry air-fuel ratio, m_{Air} denotes the quality of air, m_{MSW} indicates the quality of the MSW, and m_{Oxy} represents the quality of oxygen.

TABLE 3.4
MSW Composition and Calorific Value in the SDP and SUP Stage

Composition	SDP Stage	SUP Stage	Unit
Volatile on dry basis	65.89	65.31	%
Ash on dry basis	20.24	19.85	%
Fixed carbon on dry basis	13.87	14.84	%
Moisture	40.93	45.94	%
Sulfur on dry basis	0.1	0.24	%
Carbon on dry basis	51.32	51.04	%
Hydrogen on dry basis	6.66	6.64	%
Oxygen on dry basis	21.51	21.76	%
Nitrogen on dry basis	0.17	0.47	%
Chlorine on dry basis	0.29	0.63	%
Calorific value	SDP stage	SUP stage	Unit
High calorific value on dry basis	19.11	19.46	MJ/kg
Low calorific value as received basis	9.54	8.72	MJ/kg

In addition, the formula $12x + y + 16z$ is extracted from the stoichiometry of the following reaction:

$$C_xH_yO_z + \left(\frac{4x + y - 2z}{4}\right)O_2 \rightarrow xCO_2 + \frac{y}{2}H_2O \tag{3.6}$$

According to Eq. (3.5), the air/MSW ratios for the SDP and SUP stages are 4.3006 and 3.9265, respectively. For the sake of convenience in analysis and discussion, the air/MSW ratio is treated as equivalent to the excess air coefficient, which can be denoted as follows:

$$\alpha_{EAC} = \frac{21\%}{21\% - c_{O_2}} \tag{3.7}$$

where α_{EAC} represents the excess air coefficient, and c_{O_2} expresses the oxygen concentration. The results of the excess air coefficient at the SDP and SUP stages are shown in Figure 3.12.

Figure 3.13 presents the combustion properties (such as excess air rate), incinerator setting parameters (including furnace temperature, steam flow, and oxygen levels), and control parameters (including primary air, secondary air, urea, lime, and AC) during the SDP and SUP stages.

An ignition burner is utilized for P2 in the SDP stage and for P4 in the SUP stage, with No. 0 diesel oil as the fuel. The composition and usage statistics of the fuel are presented in Table 3.5.

Detailed discussion and analysis are as follows.

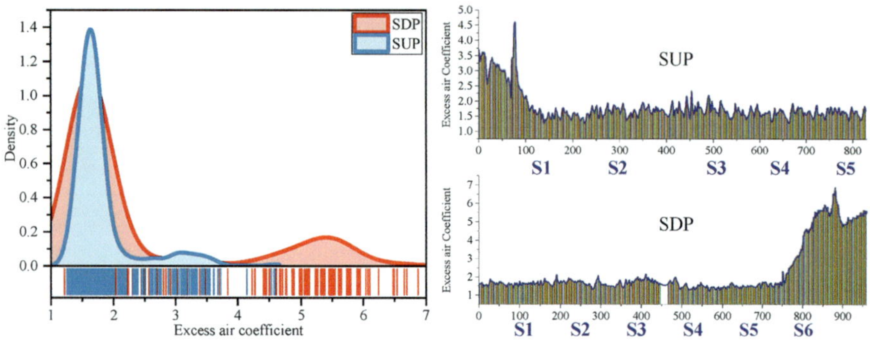

FIGURE 3.12 Excess air coefficient during the SDP and SUP stages.

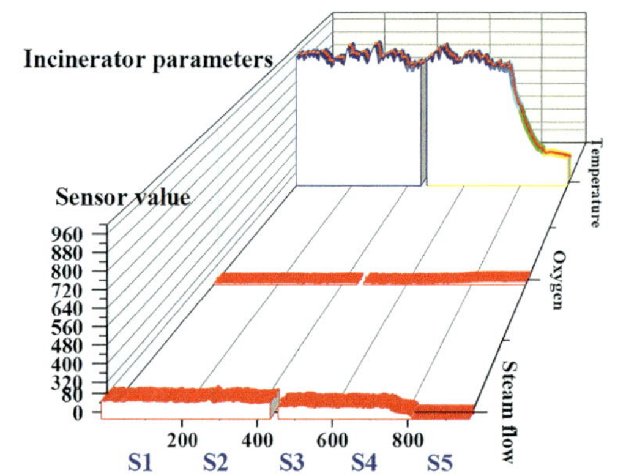

a) Incinerator parameters in SDP

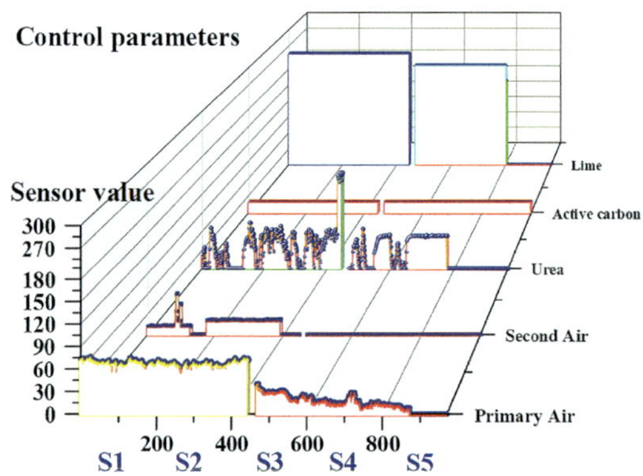

b) Control parameters in SDP

FIGURE 3.13 Incinerator and control parameters during the SDP and SUP stages. (a) Incinerator parameters in SDP. (b) Control parameters in SDP.

(Continued)

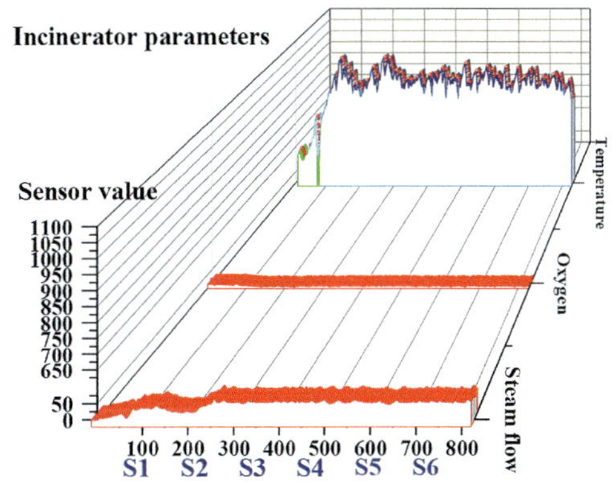

c) Incinerator parameters in SUP

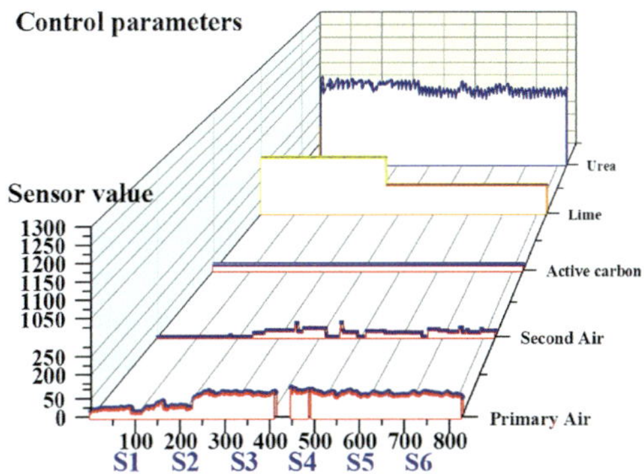

d) Control parameters in SUP

FIGURE 3.13 (Continued) (c) Incinerator parameters in SUP. (d) Control parameters in SUP.

TABLE 3.5
Fuel Composition and Usage in the SDP and SUP Stage

Composition	SDP Stage	SUP Stage	Unit
Sulfur	2.4	2.4	mg/kg
Ash	0.002	0.002	%
Total calorific	45.73	45.73	MJ/kg
Net calorific	42.73	42.73	MJ/kg
Consumption	6500	18160	kg

The composition of MSW fed into the furnace is similar in both the SDP and SUP stages. However, the levels of sulfur, nitrogen, and chlorine are lower in the SDP stage (S = 0.10%, N = 0.17%, Cl = 0.29%) compared to the SUP stage (S = 0.24%, N = 0.47%, Cl = 0.63%). The lower calorific values (LCV) for the two stages are 9.54

MJ/kg for SDP and 8.72 MJ/kg for SUP, both exceeding the 6.688 MJ/kg specified in the process design, indicating that the MSW is easy to burn. Under normal operating conditions, with furnace temperatures exceeding 850°C, the mean values of DXN emissions are 0.3 ng I-TEQ/m^3 (BO) and 3.495 ng I-TEQ/m^3 (SI) in the SDP stage, while in the SUP stage, the mean values are 4.2 ng I-TEQ/m^3 (BO) and 2.5375 ng I-TEQ/m^3 (SI). This suggests that an increase in the proportion of chlorine in MSW may lead to higher DXN emissions.

The excess air coefficients during the SDP and SUP stages are primarily distributed in the range of 1–2, with a concentration around 1.6. The overall trend is inversely related: the SDP stage shows an increase to 7, while the SUP stage experiences a decrease to 1.6. The production and emission of DXN reach their maximum levels during the rapid fluctuations of the excess air coefficient in both the SDP and SUP stages, indicating that significant changes in the excess air coefficient greatly affect the concentration of DXN. However, there is no significant correlation between the concentration of DXN and the excess air coefficient during the stabilization period (i.e., when the excess air coefficient remains stable within the interval [1, 2]).

The operational parameters of the MSWI power plant during the SDP and SUP stages include both controlled variables and manipulate variables. The controlled variables selected are furnace temperature, steam flow, and oxygen content at the BO. The manipulate variables consist of primary air flow, secondary air flow, urea feed, lime feed, and AC feed. In the SDP stage, the operational characteristics indicate that the furnace temperature and steam flow decrease, while the oxygen consumption for combustion gradually reduces, leading to an increase in the oxygen content of the flue gas as manipulate variables decline. During this period, DXN generation and emission exhibit an overall upward trend. Conversely, in the SUP stage, the behavior of manipulate variables opposes that observed in the SDP stage. The concentration of DXN generated during the heat exchange stage peaks when the furnace temperature rises to 850°C. However, the concentration of DXN formation diminishes once the temperature reaches this threshold. Additionally, both primary and secondary air flows display abnormal patterns, with control outputs unexpectedly dropping close to zero during P2 sampling. This anomaly corresponds to a maximum DXN emission value of 5.9 ng I-TEQ/m^3 observed in the P2 sampling stage, suggesting a potential link to this operational issue. These findings highlight the critical role of manipulate variables in influencing incinerator operations, as well as their impact on DXN generation and emissions.

The consumption of No. 0 diesel oil in the SUP stage is significantly higher than that in the SDP stage, approximately 2.8 times greater. Since the sulfur content of No. 0 diesel is much higher than that of MSW, the substantial use of No. 0 diesel to elevate furnace temperatures during both the SDP and SUP stages leads to a rapid increase in SO_2 concentration (as shown in Figure 3.2). Typically, the chlorine content in No. 0 diesel oil is overlooked; therefore, this chapter does not account for the impact of No. 0 diesel combustion on DXN generation and emissions.

3.3.8 Summary of Results and Discussion

This study presents the emission characteristics and relevant data regarding DXN in a grate-based MSWI power plant during the CMOP. It includes two flue gas sampling points, one FA sampling point, the distribution of 17 homologs at each sampling point, the removal efficiency of APCDs, the overall emission efficiency of the MSWI power plant, and the effects of MSW composition and combustion properties. Grate-based MSWI technology is the most widely implemented globally, and the findings from this investigation provide valuable insights for future research, which can be summarized as follows:

(1) Numerical Simulation and Technology Modification: The DXN emission results during CMOP indicate that efforts to reduce DXN emissions should focus more on the shutdown and startup periods.
(2) DXN Modeling and Optimized Emission: Experimental results show that even when the furnace temperature exceeds 850°C, DXN concentrations remain abnormally high before shutdown and after startup. Therefore, it is essential to consider diagnostic technologies for incineration conditions and control strategies tailored to different emission levels to achieve greater modeling accuracy and lower DXN emissions.
(3) Assessment of the Surrounding Environment of the MSWI Plant: The associated data and findings from the shutdown and startup periods can effectively support the assessment of DXN concentrations in the atmosphere and soil around the MSWI plant. Additionally, this information can aid in estimating the contribution of the MSWI plant to overall DXN emissions and the DXN congener profile in the region.

3.4 CHAPTER SUMMARY

The DXN emission characteristics during the CMOP of the mechanical grate-based MSWI power plant, which is the most widely used in China, were investigated. The conclusions are as follows:

(1) Elevated Conventional Pollutant Emissions: During the CMOP, the emissions of conventional pollutants (NO_x, SO_2, HCl, and CO) are significantly higher than under normal operating conditions, often approaching the upper limit of the emission standard, particularly during the cooling and heating phases.
(2) Peak DXN Emissions: DXN emissions reached their highest levels at the BO location during the "cooling" and "heating" phases, with concentrations of 41 and 26 ng I-TEQ/m^3, respectively. At the SI location, DXN concentrations were found to be 34 and 118 times higher than the normal level (0.05 ng I-TEQ/m^3) during the startup and shutdown periods. Additionally, DXN emissions remained elevated for approximately 12 hours before shutdown

and after startup. A notable correlation was observed between DXN and CO concentrations: higher CO levels corresponded with increased DXN concentrations. Among the 17 homologs of DXN, 5–6 chlorinated PCDFs accounted for the largest proportion.

(3) Fly Ash DXN Concentrations: The DXN concentration in FA produced during the SUP stage was 55,279 ng I-TEQ/kg, which is 15 times greater than the concentration during the SDP stage (3,718 ng I-TEQ/kg). This suggests a strong "memory effect" in locations such as boilers and APCD systems.

(4) APCD Removal Efficiency: The removal efficiency of APCDs, including AC, lime, and bag filters, was significantly better in the SUP stage compared to the SDP stage. This indicates that a considerable amount of DXNs is generated by memory effects within the APCD equipment during the SDP stage. Furthermore, the removal efficiency for PCDD by APCDs was consistently higher than that for PCDFs under all operating conditions.

(5) Annual DXN Emissions: Based on the experimental study (excluding unsampled time periods), DXN emission concentrations accounted for 20% of the annual emissions during the CMOP. Thus, reducing the frequency of shutdown and startup operations in the MSWI plant can substantially lower its annual DXN emissions.

The comprehensive analysis indicates that DXN emissions in FA have surged due to the CMOP of the MSWI power plant; however, these parameters are manageable and treatable. Future investigations will focus on DXN concentrations in flue gas and the development of online measurement methods for DXN in the MSWI process.

REFERENCES

1. Kammen D M, Sunter D A. City-integrated renewable energy for urban sustainability. *Science*, 2016, 352(6288): 922–928.
2. Ma W C, Wenga T, Frandsen F J, et al. The fate of chlorine during MSW incineration: vaporization, transformation, deposition, corrosion and remedies. *Progress in Energy and Combustion Science*, 2020, 76: 100789.
3. Munir M T, Mohaddespour A, Nasr A T, Carter S. Municipal solid waste-to-energy processing for a circular economy in New Zealand. *Renewable and Sustainable Energy Reviews*, 2021, 145: 111080.
4. Altarawneh M, Dlugogorski B Z, Kennedy E M, Mackie J C. Mechanisms for formation, chlorination, dechlorination and destruction of polychlorinated dibenzo-p-dioxins and dibenzofurans (PCDD/Fs). *Progress in Energy and Combustion Science*, 2009 35: 245–274.
5. Olie K, Vermeulen P L, Hutzinger O. Chlorodibenzo-p-dioxins and chlorodibenzofurans are trace components of fly ash and flue gas of some municipal incinerators in the Netherlands. *Chemosphere*, 1977, 6: 455–459.
6. Karasek F, Dickson L. Model studies of polychlorinated dibenzo-p-dioxin formation during municipal refuse incineration. *Science*, 1987, 237: 754–756.

7. Jin R, Zheng M H, Lammel G, et al. Chlorinated and brominated polycyclic aromatic hydrocarbons: sources, formation mechanisms, and occurrence in the environment. *Progress in Energy and Combustion Science*, 2020, 76: 100803.
8. Qiao J F, Guo Z H, Tang J. Dioxin emission concentration measurement approaches for municipal solid wastes incineration process: a survey. *Acta Automatica Sinica*, 2020, 46: 1063–1089.
9. Lei R R, Xu Z C, Xing Y, et al. Global status of dioxin emission and China's role in reducing the emission. *Journal of Hazardous Materials*, 2021, 418: 126265.
10. Peng Y Q, Lu S Y, Li X D, et al. Formation, measurement, and control of dioxins from the incineration of municipal solid wastes: recent advances and perspectives. *Energy & Fuels*, 2020, 34: 13247–13267.
11. Kim K S, Hong K H, Ko Y H, Kim M G. Emission characteristics of PCDD/Fs, PCBs, chlorobenzenes, chlorophenols, and PAHs from polyvinylchloride combustion at various temperatures. *Journal of the Air & Waste Management Association*, 2004, 54: 555–562.
12. Altwicker E R. Some laboratory experimental designs for obtaining dynamic property data on dioxins. *The Science of the Total Environment*, 1991, 104: 47–72.
13. Peng Y Q, Chen J H, Lu S Y, et al. Chlorophenols in municipal solid waste incineration: a review. *Chemical Engineering Journal*, 2016, 292: 398–414.
14. Stieglitz L. Selected topics on the de novo synthesis of PCDD/PCDF on fly ash. *Environmental Engineering Science*, 1998, 15: 5–18.
15. Zhang M M, Alfons G B. De novo synthesis of dioxins: a review. *International Journal of Environment and Pollution*, 2016, 60: 63–110.
16. Wei J X, Li H, Liu J G. Fate of dioxins in a municipal solid waste incinerator with state-of-the-art air pollution control devices in China. *Environmental Pollution*, 2021, 289: 117798.
17. Ruth L A. Energy from municipal solid waste: a comparison with coal combustion technology. *Progress in Energy and Combustion Science*, 1998, 24: 545–564.
18. Düwel U, Nottrodt A, Ballschmiter K. Simultaneous sampling of PCDD/PCDF inside the combustion chamber and on four boiler levels of a waste incineration plant. *Chemosphere*, 1990, 20: 1839–1846.
19. Gass H C, Jager E, Menke D, Luder K. Long term study for minimization of the PCDD/PCDF – emissions of a municipal solid waste incinerator in Germany. *Organohalogen Compounds*, 1998, 36: 175–178.
20. Hunsinger H, Klaus J, Juergen V. Formation and destruction of PCDD/F inside a grate furnace. *Chemosphere*, 2002, 46: 1263–1272.
21. Tejima H, Shibakawa S, Osumi K, Kawashima M. Dioxin emission behavior in MSW incinerator designed after Japanese guidelines for controlling dioxin. *Chemosphere* 1998, 37: 2309–2314.
22. Mario G, Stefano C, Michele G, et al. Environmental release and mass flux partitioning of PCDD/Fs during normal and transient operation of full scale waste to energy plants. *Chemosphere*, 2007, 67: S118–S124.
23. Šyc M, Fiserova E, Karban J, et al. The effect of transient operations on the levels and congener profiles of PCBz, PCPh and PCDD/F in raw flue gases of MSWI plant. *Chemosphere*, 2015, 118: 261–267.
24. Chen C K, Lin C, Lin Y C, et al. Polychlorinated dibenzo-p-dioxins/dibenzofuran mass distribution in both start-up and normal condition in the whole municipal solid waste incinerator. *Journal of Hazardous Materials*, 2008, 160: 37–44.

25. Zhan M X, Wang T J, Yang J L, et al. The behaviors and relationships of PCDD/Fs and chlorobenzenes in the whole process of one municipal solid waste incinerator. *Aerosol and Air Quality Research*, 2018, 18: 3134–3146.
26. Trinh M M, Kuo C H, Chang M B. Characterization of PCDD/Fs and dl-PCBs emission from combustion of PCB-containing oil in a fluidized-bed incinerator. *Chemosphere*, 2019, 225: 35–42.
27. Ren M H, Lv Z Y, Xu L, et al. Partitioning and removal behaviors of PCDD/Fs, PCBs and PCNs in a modern municipal solid waste incineration system. *The Science of the Total Environment*, 2020, 735: 139134.
28. Huang Y Q, Lu J W, Xie Y S, et al. Process tracing of PCDD/Fs from economizer to APCDs during solid waste incineration: re-formation and transformation mechanisms. *Waste Management*, 2021, 120: 839–847.
29. Karademir A. Health risk assessment of PCDD/F emission from a hazardous and medical waste incinerator in Turkey. *Environment International*, 2014, 30: 1027–1038.
30. Trivedi J, Deepanjan M. Memory effect driven emissions of persistent organic pollutants from industrial thermal processes, their implications and management: a review. *Journal of Environmental Management*, 2013, 119: 111–120.
31. Tejima H, Karatsu Y, Kawashima M, et al. Reduction of dioxin emission on start up and shut down at batch-operational MSW incineration plant. *Chemosphere*, 1993, 27: 263–269.
32. Tejima H, Nishigaki M, Fujita Y, et al. Characteristics of dioxin emissions at startup and shutdown of MSW incinerators. *Chemosphere*, 2007, 66: 1123–1130.
33. Gass H C, Lüder K, Wilken M. PCDD/F-emissions during cold start-up and shut-down of a municipal waste incinerator. *Organohalogen Compounds*, 2002, 56: 193–196.
34. Gass H C, Wilken M, Lüder K. Optimization of the start-up procedures in a municipal waste incinerator– impact on the emissions of dioxins and related compounds. *Organohalogen Compounds*, 2003, 63: 25–28.
35. Neuer-Etscheidt K, Nordsieck H, Liu Y. PCDD/F and other micropollutants in MSWI crude gas and ashes during plant start-up and shut-down processes. *Environmental Science & Technology*, 2006, 40: 342–349.
36. Wang L C, Hsi H C, Chang J E, Yang X Y. Influence of start-up on PCDD/F emission of incinerators. *Chemosphere*, 2006, 67: 1346–1353.
37. Guo C S, Wang M S, Lin S L, et al. Emissions of PCDD/Fs and PCBs during the Cold Start-up of Municipal Solid Waste Incinerators. *Aerosol and Air Quality Research*, 2014, 14: 1593–1604.
38. Li M, Wang C, Cen K, et al. PCDD/F emissions during startup and shutdown of a hazardous waste incinerator. *Chemosphere*, 2017, 181: 645–654.
39. Guo C J, Wang M S, Lin S L, Mi H H. Emissions of PCDD/F and PCBs during the cold start-up of municipal solid waste incinerators. *Aerosol and Air Quality Research*, 2014, 14: 1593–1604.
40. VoPham T, Bertrand K A, Fisher J A, et al. Emissions of dioxins and dioxin-like compounds and incidence of hepatocellular carcinoma in the United States. *Environmental Research*, 2022, 204: 112386.
41. Glarborg P, Miller J A, Ruscic B, Klippenstein S J. Modeling nitrogen chemistry in combustion. *Progress in Energy and Combustion Science*, 2018, 67: 31–68.
42. Anthony E J, Bulewicz E M, Jia L. Reactivation of limestone sorbents in FBC for SO2 capture. *Progress in Energy and Combustion Science*, 2007, 33: 171–210.
43. Caneghem J V, Brems A, Lievens P, et al. Fluidized bed waste incinerators: design, operational and environmental issues. *Progress in Energy and Combustion Science*, 2012, 38: 551–582.

44. Shanthakumar S, Singh D N, Phadke R C. Flue gas conditioning for reducing suspended particulate matter from thermal power stations. *Progress in Energy and Combustion Science*, 2008, 34: 685–695.
45. Lin X Q, Huang Q X, Chen T, et al. PCDD/F and PCBz emissions during start-up and normal operation of a hazardous waste incinerator in China. *Aerosol and Air Quality Research*, 2014, 14: 1142–1151.
46. Giugliano M, Cernuschi S, Grosso M, et al. PCDD/F mass balance in the flue gas cleaning units of an MSW incineration plant. *Chemosphere*, 2002, 46: 1321–1328.
47. Zhong R G, Wang C, Zhang Z T, et al. PCDD/F levels and phase distributions in a full-scale municipal solid waste incinerator with co-incinerating sewage sludge. *Waste Management*, 2020, 106: 110–119.

4 Assessment of Dioxins Formation and Emission Characteristics for One Year

4.1 PROBLEM DESCRIPTION

As unintentionally produced persistent organic pollutants [1], polychlorinated dibenzo-*p*-DXNs and furans are byproducts of various human activities [2], including waste incineration [3], metal production [4], and heat generation [5]. In developing countries such as China, the technology for MSWI has advanced significantly, surpassing that of other industries during the 12th and 13th Five-Year Development Plans [6]. The MSWI technology is beneficial for alleviating the waste problems in densely populated cities and converting waste into energy. However, the emission of DXNs from stack gas and fly ash cannot be overlooked due to their high toxicity and complex degradation processes [7]. Moreover, MSWI is a significant source of DXN emissions among all human activities [2, 8, 9].

Over the past few decades, extensive studies have focused on the emission characteristics and control of DXNs in MSWI processes [10–15]. These studies can be categorized into three stages: the early years (1990–2000), the middle decade (2000–2010), and the recent decade (2010–present). Details are as follows.

In the early years, it was confirmed that the concentration of DXNs is inversely proportional to the flue gas temperature during the boiler heat exchange process; specifically, the concentration of DXNs increases as the flue gas temperature decreases [16]. The "3T" principle (temperature, time, and turbulence) was proposed to control the production of DXNs from the furnace [17]. Additionally, advanced air pollutant control devices (APCDs) such as selective non-catalytic reduction (SNCR), activated carbon injection (ACI), bag filters (BF), alkaline scrubbers, and electrostatic precipitators (ESP) have been employed to manage DXN emissions from the stack [18]. As a result, the DXNs TEQ concentration can be maintained below 0.05 ng I-TEQ/Nm^3 under most normal operating conditions. Subsequently, Oh et al. found that DXNs emissions were highly correlated with the age of the MSWI plant, CO emission concentration, and flue gas dust emission concentration [19]. These findings provide strong support for the continued development and study of MSWI in the coming decade.

In the middle decade, MSWI process technology saw further improvements, leading most studies to focus on APCDs and DXN emissions. Some notable findings include the following: Giugliano et al. demonstrated an increase in DXN

DOI: 10.1201/9781003634195-5

concentrations in flue gas as it passed through the superheater and economizer [20], followed by a gradual reduction after the BF and SCR. Chang et al. indicated that the removal efficiency (RE) of DXNs could be negative when the operating temperature of the APCD falls within the 200°C–350°C range [21], which is the temperature window for de novo synthesis. Kai et al. reported that ACI and BF could remove 98.5% of vapor and solid-phase DXNs [22], with the removal of solid-phase DXNs primarily dependent on the BF. Additionally, Chi et al. further emphasized that using ACI alone for the removal of solid-phase DXNs could yield negative results (−51.7%) in certain cases [23]. The efficiency of APCDs is highly sensitive to operating temperature [24]. Consequently, advanced APCDs can significantly reduce the mass concentration of DXN emissions (to much less than 0.1 ng I-TEQ/Nm^3) in the stack when incinerators and boilers operate under normal conditions [25–29]. Additionally, some researchers have focused on the correlation between the furnace, boiler, and DXNs. Hunsinger et al. confirmed that DXN production is a primary cause of incomplete combustion in the furnace [30], exhibiting a strong correlation with CO concentration (consistent with [19]). Everaert et al. found that ESPs could lower the flue gas temperature at the boiler outlet [31], thereby aiding in the reduction of DXNs through de novo synthesis. Wang et al. also reported that the peak mass concentration of DXNs occurred at the exit of the ESP [32]. Furthermore, Choi et al. investigated 43 intermittent incinerators in Korea and revealed that frequent startup and shutdown of incinerators could lead to significant DXN emissions [33]. During this period, the emission concentrations of DXNs from MSWI plants were below 1.0 ng I-TEQ/Nm^3 (the legal limit), accounting for approximately 85% of such emissions in China [34].

In the recent decade, emission limits for DXNs from MSWI plants have been progressively strengthened worldwide [35, 36]. Current reports indicate that DXN emissions are significantly lower than those recorded in previous decades. Nzihou et al. published findings showing that 124 MSWP plants (comprising 85 large and 39 smaller facilities) operated well below the EU standard of 0.1 ng I-TEQ/Nm^3 [37]. Similarly, Zhang et al. and Liu et al. demonstrated that DXN emissions from large MSWI plants (with a capacity of more than 500 tons/day) could meet international standards in China (i.e., 0.1 ng I-TEQ/Nm^3) [38, 39]. Subsequently, China revised its national standard for DXN emissions from MSWIs to 0.1 ng I-TEQ/Nm^3 in 2014 (GB18485-2014 [40]) to protect human health and promote the sustainable development of the ecological environment. As a result, the investigation of DXN emissions from MSWI plants has garnered increased attention with the tightening of emission limits [41]. Hung et al. reported that DXN concentrations at the stack ranged from 0.42E−01 to 0.65E−01 ng I-TEQ/Nm^3 during continuous sampling over 122 hours (5 days) using ACI and BF [36]. Li et al. examined six MSWI plants (five in Southern China and one in Northern China) and found that only one plant exceeded the emission limit at 0.12 ng I-TEQ/Nm^3 [42]. Additionally, Zhu et al. studied six MSWI plants (two using moving grates and four using fluidized beds) in Northern China and found that their DXN content was below the legal limits, ranging from 0.42E−01 to 0.65E−01 ng I-TEQ/Nm^3 [43]. Modern MSWI plants equipped with advanced APCDs [44–46] have demonstrated sufficient capacity to achieve standard DXN emissions under specific combustion conditions [11].

However, several issues remain unaddressed. The first aspect is the variability in combustion conditions. For example, Ma et al. and Zhong et al. demonstrated that DXN concentrations increase with higher oxygen levels (both primary and secondary air) [47, 48]. The second aspect relates to changes in the operating environment [49], which include waste classification [50] and seasonal variations [51]. The third aspect concerns the accumulation of operational time. For example, continuous operation for 6,000 hours can lead to significant fly ash deposition in the boiler and APCDs [52, 53]. Therefore, accurately and comprehensively estimating the emission characteristics of DXNs in China presents a challenging task, particularly when considering the three aforementioned factors [9].

To reinforce and interpret the formation and emission characteristics of DXNs, this study conducted a one-year investigation at an MSWI plant equipped with SNCR, semi-dry deacidification (SDD), ACI, and BF. The main objectives are outlined as follows: (1) to assess the RE of APCDs over the entire operational cycle (one year); (2) to clarify the actual emissions of DXNs by considering the complete operational cycle; (3) to analyze the relationship between the operational processes of the MSWI plant and DXN emission concentration in China; and (4) to provide data support for artificial intelligence technology, which is crucial for early warning systems regarding DXN emissions in MSWI process. To the best of our knowledge, this is the first one-year follow-up study on a large grate MSWI power plant that includes DXN measurements from both flue gas and fly ash. This research is significant in revealing the actual emissions of MSWI, the RE of APCDs, and other information closely related to DXN emissions.

4.2 MATERIALS AND METHODS

4.2.1 Basic Information and General Operation about the Selected MSWI Plant

This study investigated a grate-type MSWI power plant located in Beijing, China. The facility features two mechanical grate incinerators, each with a capacity of 800 tons/day, operating in parallel. Each incinerator is equipped with a single-drum horizontal boiler and APCDs. The boiler system includes three superheaters, one evaporator, and three economizers, which facilitate the exchange of high-temperature flue gas (ranging from 850°C to 200°C) and generate electric energy. The APCDs incorporate SNCR, SDD, ACI, and BF to ensure that the flue gas emissions from the stack comply with legal pollutant standards (i.e., NO_x, SO_2, HCl, and DXNs) as specified in GB 18485-2014. The profile of the MSWI plant is illustrated in Figure 4.1.

Most MSWI plants primarily rely on manual operation to achieve optimal operating conditions, such as the "3T + E" parameters (combustion temperature, residence time, turbulence, and exhaust oxygen concentration), which help reduce DXN formation in the incinerator and boiler [54]. However, operating conditions can change frequently due to the operational habits of engineers and seasonal fluctuations in MSW composition. Additionally, the large-scale MSWI plant studied typically undergoes shutdown maintenance once a year, most commonly around February. This results in regular operation times for the incinerator exceeding 8,000 hours.

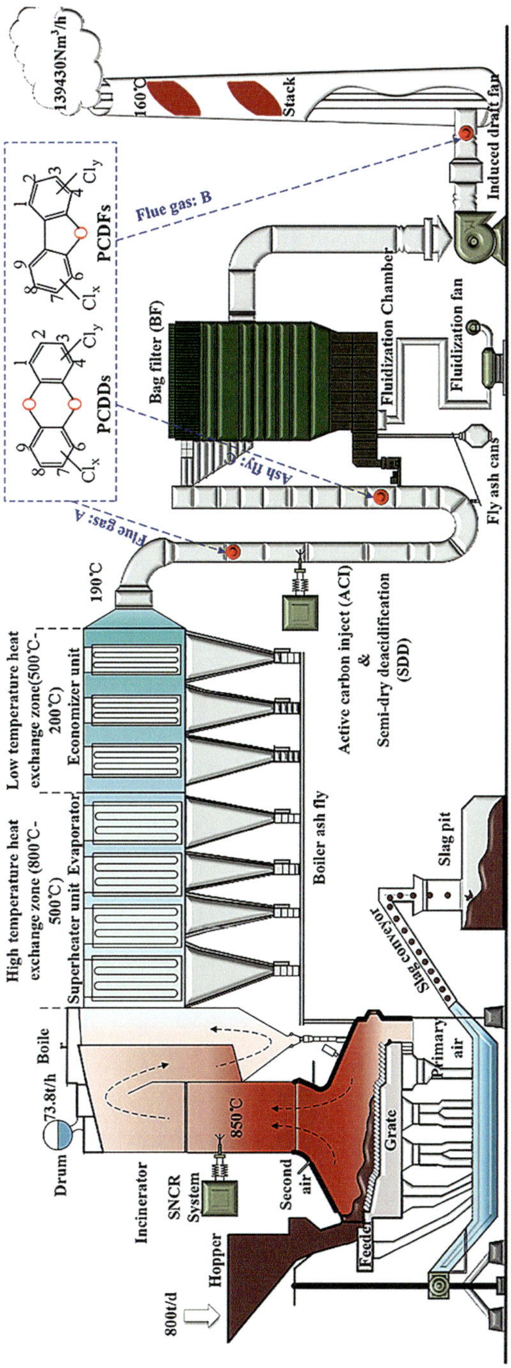

FIGURE 4.1 Full-scale profile of the MSWI plant.

Consequently, the risk of APCDs releasing DXNs increases gradually due to the particle deposition associated with the memory effect [55]. Table 4.1 presents the fluctuations in operational parameters and conventional pollutants over the course of a year.

The annual operating conditions and emissions of conventional pollutants comply with the general standards (GB 18485–2014). However, the combustion conditions exhibit notable instability during the sampling period, as indicated by the statistical variance results. This fluctuation in conditions is a primary factor contributing to the inconsistent emissions of pollutants, particularly NOx and CO [26]. Unstable combustion can also create precursors and temperature windows for various reactions that lead to DXN formation, including high-temperature gas-phase synthesis (500°C–700°C) in the high-temperature heat exchange zone [47], de novo synthesis (250°C–450°C) in the low-temperature heat exchange zone [56], and precursor

TABLE 4.1
Average Concentrations of Regular Pollutants and Operating Conditions within a Year

Parameters	Run-Time				Rated Operation
Operation Parameters[a]	Min	Max	Mean	Variance	
Temp. of incinerator (°C)	9.00E+02	1.11E+03	9.96E+02	1.53E+03	850°C[b]
Inlet temp. of superheater unit (°C)	6.54E+02	8.52E+02	7.60E+02	0.99E+03	N.A.[c]
Exit temp. of economizer (°C)	1.57E+02	1.93E+02	1.82E+02	0.04E+03	190[d]
Steam flow rate (tons/h)	50.73	8.44E+01	74.43	3.73E+01	73.8[d]
Exit O_2 of economizer (%)	4.04	1.02E+01	7.04	0.10E+01	6–9[d]
Inlet temp. of BF (°C)	6.09E+01	1.71E+02	1.55E+02	5.87E+02	N.A.[c]
Exit temp. of BF (°C)	7.04E+01	1.71E+02	1.59E+02	2.16E+02	N.A.
Inlet temp. of stack (°C)	1.19E+02	1.76E+02	1.66E+02	6.16E+01	N.A.
Inlet O_2 of stack (%)	4.51	2.24E+01	8.17	0.69E+01	N.A.
Flow rate at stack (kNm^3/h)	72.37	1.57E+02	128.54	9.59E+01	139.43[d]
Conventional pollutants[e]					
NO_x (mg/Nm^3)	77.08	184.38	141.02	545.25	250[b]
SO_2 (mg/Nm^3)	0	26.27	5.36	33.05	100[b]
HCl (mg/Nm^3)	1.63	13.39	4.95	2.29	60[b]
CO (mg/Nm^3)	1.02	61.48	16.62	177.31	100[b]

[a] Five-minute mean.
[b] National emission standard of China: GB 18485-2014 (mean 1 hour).
[c] Not available.
[d] Technological design value.
[e] One-hour mean.

synthesis (400°C–750°C) between the high- and low-temperature zones [57]. Therefore, measuring the generation, adsorption, and emission characteristics of DXNs under diverse conventional operating conditions is essential for gaining a more realistic understanding of emissions and emission factors from MSWI plants. This knowledge, in turn, provides robust support for implementing online detection of DXN emissions using artificial intelligence technology [54].

4.2.2 Sampling and Analysis

A total of 47 samples were collected from the MSWI power plant between November 2020 and January 2022, in accordance with the national environmental protection standard (HJ 77.2-2008). Each sample included flue gas at the exit of the economizer, flue gas at the stack inlet, and fly ash at the BF inlet. The sampling frequency and distribution information are illustrated in Figure 4.2.

In this study, the Laoying (ZR-3720) dioxin intelligent sampling instrument was used for flue gas sampling. This isokinetic sampling system operates entirely automatically without manual intervention. Isokinetic tracking sampling was employed at Sites A and B (as shown in Figure 4.1), with each sampling session lasting over 2 hours and collecting a volume of approximately 3–4 Nm^3. However, particulate matter (PM) can easily clog the quartz filter cartridge at Site A. The cartridge is replaced when resistance at Site A becomes excessively high. Additionally, sampling at Point B is suspended during the replacement of the quartz filter cartridge at Point A. For fly ash sampling, an equal time interval strategy (approximately 1 hour) was utilized to ensure real-time collection of fly ash samples from the pipeline. The actual sampling working environment and operational program are illustrated in Figure 4.3.

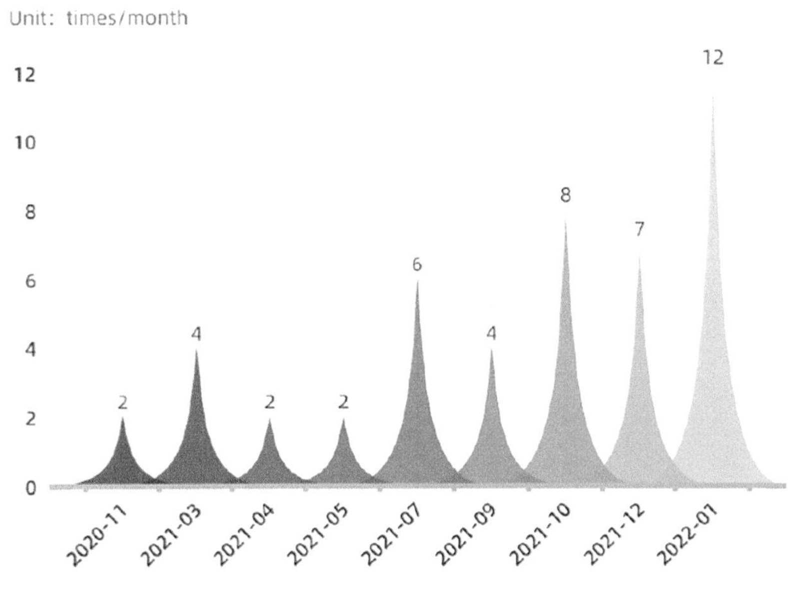

FIGURE 4.2 Sampling date and times.

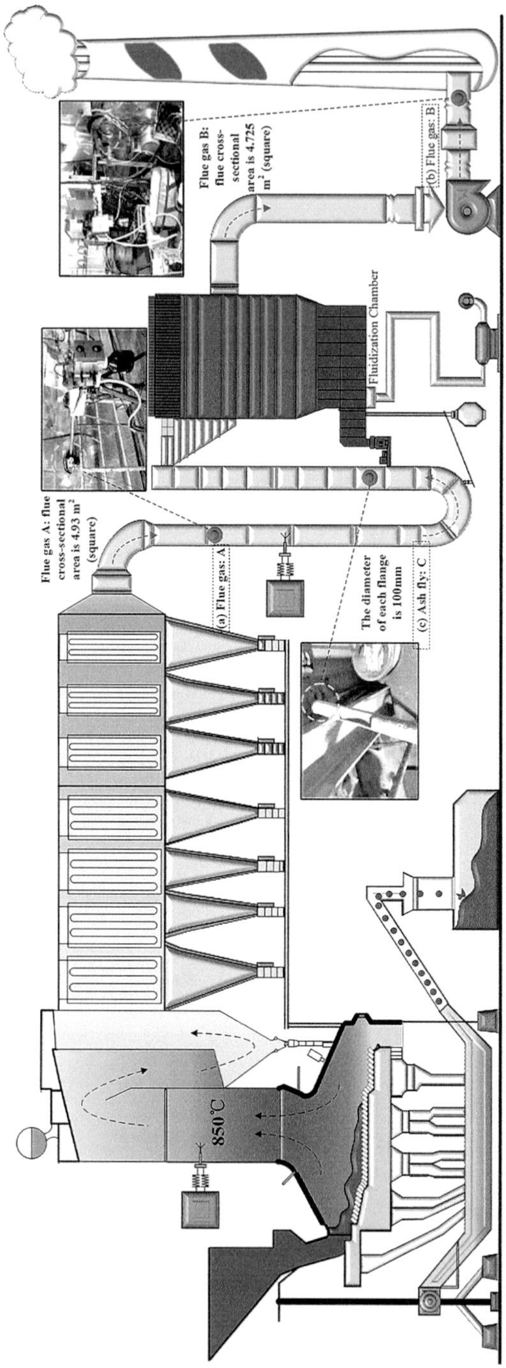

FIGURE 4.3 Sampling the flue gas and ash fly.

Figure 4.3 indicates that the diameter of each flange is 100 mm, with the flue cross-sectional areas at Sites A and B measuring 4.93 and 4.73 m^2, respectively. Given that both cross-sectional areas exceed 4 m^2, 16 traverse points were established for sampling at both sites. The final samples collected included micro-quartz fiber filters (for the particle-phase portion), XAD-2 (for the gas-phase portion), and condensate water (also for the gas-phase portion). In this study, the solid phase (quartz filter cartridge) and gas phase (XAD-2 and condensate water) were combined for laboratory analysis.

The experimental analysis involves three key steps: sample extraction, sample cleanup, and instrumental analysis. For the extraction and cleanup processes, 13C12-labeled compounds are utilized [8, 10, 12, 43, 44].

The sample extraction process is outlined as follows: (1) The pickling quartz filter cartridge and XAD-2 resin were dried, followed by Soxhlet extraction for 16–24 hours; (2) The condensate water collected during flue gas sampling was separated using liquid-liquid extraction; (3) The above samples were then concentrated and combined for cleanup treatment.

The sample cleanup procedure is as follows: (1) The sample was purified using an activated carbon-silica gel (AC-SG) column; (2) The eluate was evaporated using high-purity nitrogen to remove excess solvent, concentrating the sample into 2 mL of aqueous pure DXN for analysis via high-resolution gas chromatography/high-resolution mass spectrometry (HRGC/HRMS) using a Thermo Scientific Trace 1310 system. The recovery rate of the 13C12-labeled compounds is presented in Table 4.1. Any analysis result is discarded if the recovery rate falls outside the range indicated in Table 4.2.

The primary difference in our analysis lies in the optimized temperature program for HRGC/HRMS specifically tailored for DXN components. The program consists of the following steps: (1) The column temperature was initially maintained at 150°C for 3 minutes; (2) It was then gradually increased to 230°C at a rate of 20°C/min and held for 18 minutes; (3) The temperature was further raised to 235°C at a rate of 5°C/

TABLE 4.2
Recovery Rate of Extracted Internal Standard

Compound	Range	Compound	Range
2,3,7,8-TCDD	25%–164%	2,3,7,8-TCDF	24%–169%
1,2,3,7,8-P5CDD	25%–181%	1,2,3,7,8-P5CDF	24%–185%
1,2,3,4,7,8-H6CDD	32%–141%	2,3,4,7,8-P5CDF	21%–178%
1,2,3,6,7,8-H6CDD	28%–130%	1,2,3,4,7,8-H6CDF	32%–141%
1,2,3,7,8,9-H6CDDa	—	1,2,3,6,7,8-H6CDF	28%–130%
1,2,3,4,6,7,8-H7CDD	23%–140%	2,3,4,6,7,8-H6CDF	28%–136%
O8CDD	17%–157%	1,2,3,7,8,9-H6CDF	29%–147%
		1,2,3,4,6,7,8-H7CDF	28%–143%
		1,2,3,4,7,8,9-H7CDF	26%–138%
		O8CDF[a]	—

[a] No labeled during extraction and cleanup.

min and maintained for 10 minutes; (4) Finally, the column temperature was escalated to 320°C at a rate of 4°C/min and held for an additional 3 minutes.

The toxic equivalent values (TEQ) of DXNs are calculated based on the pollution control standards in China (GB 18485-2014). The calculation can be expressed as follows:

$$\rho_t\left(\text{ng I-TEQ / Nm}^3\right) = \sum_{i=1}^{17} \rho_s^i \cdot \text{TEF}_i \tag{4.1}$$

where ρ_t is the TEQ concentration, ρ_s denotes the measured mass concentration, and TEF_i represents the toxic equivalence factor (TEF) (the value of TEF was provided in Table 4.3).

The TEQ concentration ρ_t is further normalized, as follows:

$$\rho\left(\text{ng I-TEQ / Nm}^3\right) = \frac{21\% - 11\%}{21\% - \varphi\left(O_2\right)} \times \rho_t\left(\text{ng I-TEQ / Nm}^3\right) \tag{4.2}$$

where ρ indicates the TEQ concentration under dry air at 101.325 kPa and 237 K (at oxygen 11%), and $\varphi\left(O_2\right)$ represents the oxygen content in flue gas.

TABLE 4.3
DXNs Toxic Equivalent Factor (TEF)

Compound	I-TEF[a] (1989, EPA)
PCDDs	
2,3,7,8-T5CDD	1
1,2,3,7,8-P5CDD	0.5
1,2,3,4,7,8-H6CDD	0.1
1,2,3,6,7,8-H6CDD	0.1
1,2,3,7,8,9-H6CDD	0.1
1,2,3,4,6,7,8-H7CDD	0.01
O8CDD	0.001
PCDFs	
2,3,7,8-TCDF	0.1
1,2,3,7,8-P5CDF	0.05
2,3,4,7,8-P5CDF	0.5
1,2,3,4,7,8-H6CDF	0.1
1,2,3,6,7,8-H6CDF	0.1
2,3,4,6,7,8-H6CDF	0.1
1,2,3,7,8,9-H6CDF	0.1
1,2,3,4,6,7,8-H7CDF	0.01
1,2,3,4,7,8,9-H7CDF	0.01
O8CDF	0.001

[a] The version of TEF used in China legal standards and limits is the international toxic equivalent factor (I-TEF) provided by U.S. Environmental Protection Agency.

In the remainder of this chapter, the TEQ concentration of DXNs in the flue gas is the normalized value ρ.

4.2.3 Statistical Analysis

The statistical data obtained from the MSWI plant included DXNs (mass and TEQ concentrations), 2,3,7,8-substituted DXN congeners (mass and TEQ concentrations), and process variables (refer to Table 4.1), following the aforementioned sampling and analysis procedures. For any DXN congener mass concentrations that were below the measured detection limit (MDL), the values were adjusted to one-half of the MDL. This subsection details the statistical analysis methods employed to gain a better understanding of DXN formation and emission characteristics.

The RE of DXNs and their congeners is used to assess the performance of APCDs (primarily dependent on ACI and BF), which can be expressed as follows:

$$\beta^{\mathrm{RE}}(\%)=\frac{\rho^{\mathrm{A}}\left(\mathrm{ng\,I\text{-}TEQ/Nm^3}\right)-\rho^{\mathrm{B}}\left(\mathrm{ng\,I\text{-}TEQ/Nm^3}\right)}{\rho^{\mathrm{A}}\left(\mathrm{ng\,I\text{-}TEQ/Nm^3}\right)}\times 100\% \tag{4.3}$$

where β^{RE} indicates the RE of DXNs, and the value range is negative infinity to 100%; ρ^{A} and ρ^{B} denote the TEQ contents of DXNs in Sites A and B, respectively.

To quantify the annual DXNs emissions of the MSWI power plant, the emission factors (EFs) is calculated as follows,

$$\nu^{\mathrm{EFs}}\left(\mu\mathrm{g/kg}\right)=\frac{\rho^{\mathrm{B}}\left(\mathrm{ng\,I\text{-}TEQ/Nm^3}\right)\times\nu^{\mathrm{Gas}}\left(\mathrm{Nm^3/h}\right)\times 10^{-3}\left(\mu\mathrm{g/ng}\right)}{\left(M^{\mathrm{Daily}}/24\right)\left(\mathrm{kg/h}\right)} \tag{4.4}$$

where $\nu^{\mathrm{EFs}}\left(\mu\mathrm{g\,TEQ/kg}\right)$ is the EFs of DXNs emission from the stack, ν^{Gas} is the mean flow rate of flue gas in the stack during sampling period (approximately 2 hours), and M^{Daily} masks the daily processing capacity of MSW.

The annual emission (AE) is further estimated by the EFs, which is shown as follows:

$$\left[m_{\min}^{\mathrm{AE}},\ m_{\max}^{\mathrm{AE}}\right]\left(\mu\mathrm{g}\right)=\left(\bar{\nu}^{\mathrm{EFs}}\left(\mu\mathrm{g/kg}\right)\pm\sigma^{\mathrm{EFs}}\right)\times M^{\mathrm{Year}}\left(\mathrm{kg}\right) \tag{4.5}$$

where $m_{\min}^{\mathrm{AE}}$ and $m_{\max}^{\mathrm{AE}}$ are the minimum and maximum AEs of DXNs from the stack, respectively; M^{Year} represents the year processing capacity of MSW; $\bar{\nu}^{\mathrm{EFs}}$ and σ^{EFs} indicate the mean and the variance of EFs as follows:

$$\begin{cases}\bar{\nu}^{\mathrm{EFs}}\left(\mu\mathrm{g/kg}\right)=\dfrac{1}{47}\displaystyle\sum_{i=1}^{47}\nu_i^{\mathrm{EFs}}\\[2ex]\sigma^{\mathrm{EFs}}=\dfrac{1}{47}\displaystyle\sum_{i=1}^{47}\left(\nu_i^{\mathrm{EFs}}-\bar{\nu}^{\mathrm{EFs}}\right)^2\end{cases} \tag{4.6}$$

where ν_i^{EFs} represents the ith EFs, and 47 is the number of samples in this chapter.

Furthermore, Pearson correlation coefficient (PCC) analysis is employed to examine the relationship between the emissions of DXNs at Sites A, B, and C and the process data, based on the actual operational data from the MSWI. This can be expressed as follows:

$$r^{\mathrm{PC}} = \frac{\sum_{i=1}^{47}\left(x_i - \bar{x}_i\right)\left(\rho_i - \bar{\rho}_i\right)}{\sqrt{\sum_{i=1}^{47}\left(x_i - \bar{x}_i\right)}\sqrt{\sum_{i=1}^{47}\left(\rho_i - \bar{\rho}_i\right)}} \tag{4.7}$$

where r^{PC} represents the PCC between DXNs TEQ content and process variable x_i, and it contains $r_{\mathrm{A}}^{\mathrm{PC}}$ (site A), $r_{\mathrm{B}}^{\mathrm{PC}}$ (site B), and $r_{\mathrm{C}}^{\mathrm{PC}}$ (site C); x_i is the ith process variable, and it also represents the actual physical quantity (e.g., temperature); $\bar{x}_i$ denotes the mean of the ith process variable; and $\bar{\rho}_i$ indicates the mean of DXNs TEQ content.

4.3 RESULTS AND DISCUSSION

4.3.1 DXNs Concentration in MSWI Plant for One Year

The mass concentrations and TEQ of DXNs at Sites A and B are shown in Figure 4.4, and the statistical data are shown in Table 4.4.

The white blocks in the three-dimensional histogram indicate the absence of detectable values. The shutdown occurred on December 24, 2021, and the restart took place on January 2, 2022. All other times are categorized as periods of normal operation. Additionally, the timeframe from November 2020 to December 14, 2021, is considered the normal operating state, while the period from December 24, 2021, to January 2022 is classified as the abnormal state, in accordance with the studied conclusions of Hung et al. [35].

The DXN concentration at Site A can be considered as the cumulative result of the generation process since it is situated before the APCDs. During normal operations, the actual DXN mass concentrations and TEQ values range from 3.46E−02 to 8.45E−01 ng/Nm^3 and from 2.11E−02 to 4.83E−01 ng I-TEQ/Nm^3, respectively [44]. Consequently, the estimated mass concentration and TEQ range of DXNs generated in the flue gas at the boiler outlet are [1.86E−01, 2.50E−01 ng/Nm^3] and [1.33E−01, 1.53E−01 ng I-TEQ/Nm^3], calculated based on the average values (2.12E−01 ng/Nm^3 and 1.34E−01 ng I-TEQ/Nm^3) and variances (2.59E−02 and 0.96E−02). In this study, the results indicate lower DXN concentrations compared to the recent study of three 750 tons/day grate furnaces conducted in [35], which reported values ranging from 1.0 to 2.2 ng I-TEQ/Nm^3. Figure 4.4(b) also demonstrates that the DXN TEQ concentration for the 14 samples collected from flue gas A during normal operations is below 0.1 ng I-TEQ/Nm^3 (as per GB 18485-2014). This suggests that the probability of DXN concentrations generated by flue gas passing through the incinerator and boiler exceeding the national standard (0.1 ng I-TEQ/Nm^3) is only 56.67%, significantly lower than findings in most other investigations [11, 35, 44, 48]. Additionally, both the startup and shutdown phases of the incinerator are associated with maximum DXN emissions [49]. Specifically, the true mass

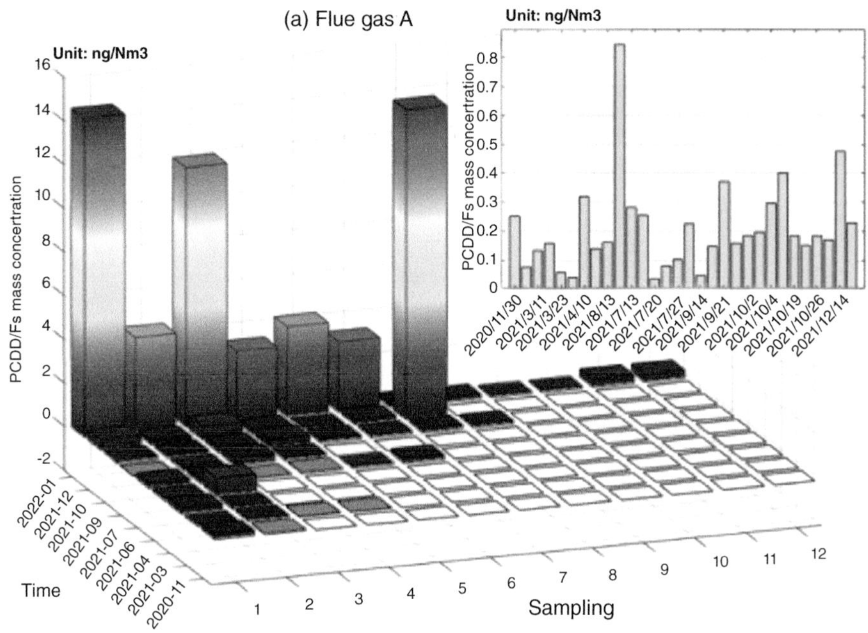

(a)Mass concentration of flue gas A

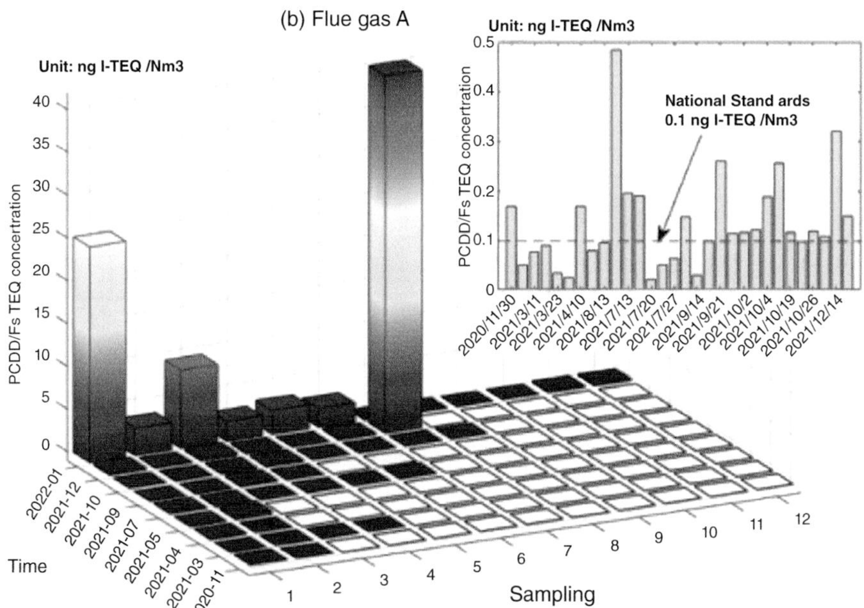

(b)TEQ concentration of flue gas A

FIGURE 4.4 DXNs concentration of flue gas A and B. (a) Mass concentration of flue gas A. (b) TEQ concentration of flue gas A.

(Continued)

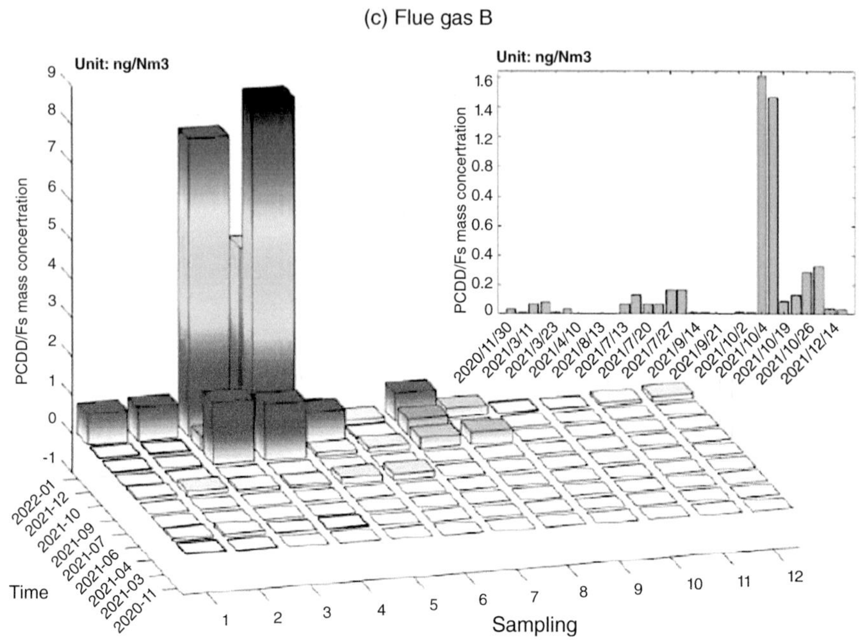

(c)Mass concentration of flue gas B

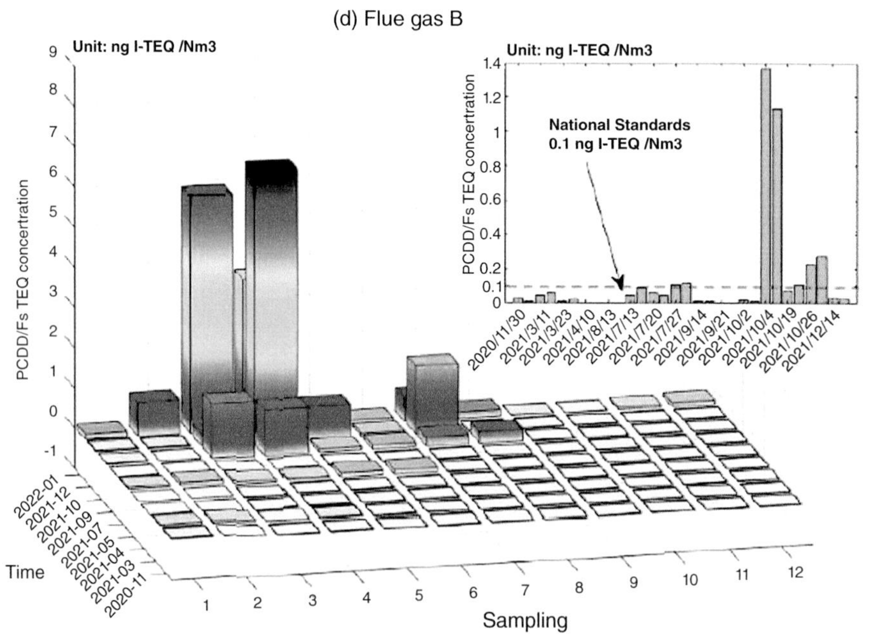

(d)TEQ concentration of flue gas B

FIGURE 4.4 (Continued) (c) Mass concentration of flue gas B. (d) TEQ concentration of flue gas B.

TABLE 4.4
Statistical Data of DXNs Concentration

	Flue Gas A	Flue Gas B	Ash Fly C				
Time	Mass Concentration[a]	TEQ Concentration[b]	Oxygen[c]	Mass Concentration	TEQ Concentration	Oxygen	TEQ Concentration
2020-11-30	0.2503	0.1680	6.1	0.0329	0.0245	7.6	N.M.[d]
	0.0773	0.0509	5.8	0.0107	0.0081	7.8	
2021-03-11	0.1320	0.0767	3.8	0.0647	0.0459	6.9	16.4474
	0.1572	0.0903	3.6	0.0814	0.0550	6.2	
2021-03-28	0.0573	0.0343	4.3	0.0092	0.0075	8.7	1.1161
	0.0408	0.0243	4.2	0.0316	0.0234	7.5	
2021-04-10	0.3189	0.1687	2.1	0.0017	0.0011	5.7	17.1096
	0.1415	0.0804	3.4	0.0035	0.0022	5.2	
2021-05-13	0.1633	0.0955	3.9	0.0010	0.0007	6.6	29.637
	0.8453	0.4830	3.5	0.0021	0.0014	5.9	
2021-07-13	0.2826	0.1949	6.5	0.0659	0.0439	6	15.6492
	0.2550	0.1903	7.6	0.1316	0.0908	6.5	
2021-07-20	0.0346	0.0211	4.6	0.0690	0.0605	9.6	31.797
	0.0790	0.0520	5.8	0.0699	0.0463	5.9	
2021-07-27	0.1022	0.0639	5	0.1630	0.1079	5.9	40.296
	0.2254	0.1463	5.6	0.1636	0.1106	6.2	
2021-09-14	0.0459	0.0287	5	0.0131	0.0098	7.7	15.7153
	0.1478	0.0979	5.9	0.0118	0.0088	7.7	
2021-09-21	0.3727	0.2606	6.7	0.0031	0.0022	6.9	28.0668

(Continued)

TABLE 4.4 (Continued)
Statistical Data of DXNs Concentration

	Flue Gas A	Flue Gas B	Ash Fly C				
Time	Mass Concentration[a]	TEQ Concentration[b]	Oxygen[c]	Mass Concentration	TEQ Concentration	Oxygen	TEQ Concentration
	0.1602	0.1144	7	0.0022	0.0015	5.8	
2021-10-02	0.1832	0.1160	5.2	0.0192	0.0154	8.5	9.3085
	0.1977	0.1206	4.6	0.0116	0.0094	8.6	
2021-10-14	0.2978	0.1885	5.2	1.6089	1.3635	9.2	45.4651
	0.4018	0.2559	5.3	1.4638	1.1347	8.1	
2021-10-19	0.1836	0.1169	5.3	0.0906	0.0775	9.3	23.6983
	0.1507	0.0966	5.4	0.1336	0.1095	8.8	
2021-10-26	0.1862	0.1179	5.2	0.2887	0.2292	8.4	21.6056
	0.1681	0.1071	5.3	0.3239	0.2745	9.2	
2021-12-14	0.4757	0.3214	6.2	0.0362	0.0342	10.4	37.45
	0.2296	0.1481	5.5	0.0326	0.0302	10.2	
2021-12-24	0.6423	0.4117	5.4	0.3650	0.2944	8.6	20.806
	0.2906	0.1887	5.6	8.9889	6.7586	7.7	
	0.1291	0.0873	6.2	0.7923	0.8253	11.4	21.599
	0.2826	0.2523	9.8	0.0601	0.0812	13.6	
	14.2310	41.8559	17.6	0.4224	1.6895	18.5	316.64
2022-01-02	14.6550	25.2672	15.2	0.8160	0.1316	14.8	2124.65
	4.3730	3.2154	7.4	0.8662	0.8492	10.8	

	11.9180	9.2388	8.1	7.7171	5.8909	7.9	2895.7
	3.2745	2.3558	7.1	4.7977	3.7482	8.2	
	4.1540	2.8648	6.5	0.4439	0.3415	8	2226.7
	3.2389	2.2971	6.9	0.1257	0.1075	9.3	
2022-01-11	0.2284	0.1575	6.5	0.7332	0.6060	8.9	47.387
	0.2095	0.1360	5.6	0.2298	0.1715	7.6	
2022-01-18	0.1058	0.0715	6.2	0.0186	0.0135	7.2	N.M. d
	0.1116	0.0734	5.8	0.0099	0.0075	7.8	
2022-01-22	0.3940	0.2680	6.3	0.0755	0.0559	7.5	49.138
	0.3904	0.2638	6.2	0.1335	0.0934	6.7	

[a] The unit is ng/Nm^3.

[b] The unit is ng I-TEQ/Nm^3.

[c] The unit is ng %The oxygen content in the flue gas is used to calculate the TEQ value of DXNs emission (see Eq. (4.2) for the calculation process).

[d] Not measurement.

concentrations and TEQ values during the shutdown phase are 14.23 ng/Nm^3 and 41.86 ng I-TEQ/Nm^3, respectively, while those during the startup phase are 14.66 ng/Nm^3 and 25.27 ng I-TEQ/Nm^3, respectively.

The DXN concentration in flue gas B, which is the location monitored for environmental protection, must have a TEQ concentration below 0.1 ng I-TEQ/Nm^3. During normal operations, the actual mass concentration of DXNs ranges from 3.46E−01 to 8.45E−01 ng/Nm^3, with a mean value of 1.65E−01 ng/Nm^3. Notably, the DXN TEQ values exceed the 0.1 ng I-TEQ/Nm^3 threshold (as per GB18485-2014) in seven samples, including two high-concentration instances of 1.36 and 1.12 ng I-TEQ/Nm^3 (refer to Figure 4.4(d)). The rate of emissions exceeding the standard is alarmingly high at 23.33%, which aligns with earlier findings by Li et al. [15]. Furthermore, the two peaks in concentration observed on October 14, 2021, and December 26, 2021, as shown in Figure 4.4(d), indicate a significant increase in DXNs when the flue gas passes through the APCDs. The cause of the observed inversion can be attributed to the memory effect of the APCDs, a conclusion also noted by Wang et al. [13] and Trivedi et al. [55]. However, the temperature conditions (155.33°C ± 0.59E+03°C) are insufficient to facilitate the various reactions that lead to DXN generation in the operating environment of the APCDs, as indicated by the process data in Table 4.1. The APCD equipment can release substantial quantities of DXNs due to this memory effect after prolonged periods of continuous operation. As illustrated in Figure 4.4(c) and (d), the shutdown and startup phases are significant sources of DXN emissions in the MSWI plant, with 11 sampling detection values nearly exceeding the 0.1 ng I-TEQ/Nm^3 threshold. Specifically, total emissions during the shutdown and startup phases amount to 20.72 ng I-TEQ/Nm^3, which is more than four times the total emissions recorded during other periods (4.88 ng I-TEQ/Nm^3).

Furthermore, the distribution of DXN TEQ concentrations in fly ash at Site C is illustrated in Figure 4.5 (detailed statistical information can be found in Table 4.4). It is important to note that fly ash samples were not collected or monitored for DXN concentrations on November 30, 2020, and January 18, 2022.

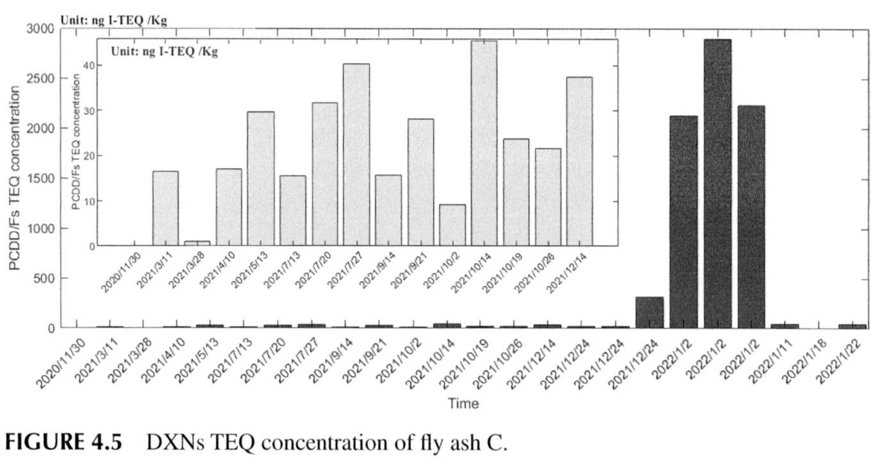

FIGURE 4.5 DXNs TEQ concentration of fly ash C.

Sampling position C is situated between the ACI and BF to facilitate the synchronous collection of fly ash as well as flue gases A and B. As illustrated in Figure 4.5, the TEQ concentration of DXNs in fly ash is significantly higher than that in flue gases A and B [53]. During normal operating periods, the DXN TEQ values range from 1.12 to 45.47 ng I-TEQ/kg, with an average of 23.81 ng I-TEQ/kg. In contrast, during abnormal operating conditions, the DXN TEQ ranges from 20.81 to 2,895.70 ng I-TEQ/kg, with a mean of 962.83 ng I-TEQ/kg. This indicates that DXN concentrations increase by more than 40 times under unstable operating conditions (shutdown and startup) [49].

In general, operational conditions significantly influence the generation and emissions of DXNs. Unstable operating conditions, such as shutdowns and startups, result in increased generation and emissions of DXNs in both flue gas and fly ash. Regarding flue gas, the impact of the shutdown phase (41.86 ng I-TEQ/Nm^3) is more pronounced than that of the startup phase (25.27 ng I-TEQ/Nm^3). Conversely, for fly ash, the startup phase has a more substantial impact compared to the shutdown phase. Therefore, it is essential to develop optimized operational strategies for controlling DXNs that are tailored to these varying operating conditions.

4.3.2 Congener Distribution of DXNs

The mass concentrations and TEQ of 17 DXN congeners from 47 samples are presented in Figure 4.6. Each sampling point (A, B, and C) is categorized into two conditions: normal and abnormal, to illustrate the distribution of the congeners. In Figure 4.6, the profile is interpreted as follows. The height of each bar represents the mean value of all samples, while the upper and lower error limits are calculated by adding and subtracting the variance from the mean value. Each circle depicts the actual value of the congeners, and the dotted line indicates the maximum value observed across all samples.

Flue gas A is illustrated in Figure 4.6(a)–(d), revealing that the concentration of PCDF congeners is significantly higher than that of PCDD congeners across the overall distribution. The mass concentration ratio of PCDF to PCDD under normal conditions is 3.65 ± 2.17 (greater than 1), indicating that the generation of DXNs is primarily concentrated during waste heat exchange. Notably, the contribution of 4–6 chlorinated PCDFs accounts for 74.70% by mass and 74.76% by TEQ, while 4–5 chlorinated PCDDs represent 17.46% by mass and 17.48% by TEQ, marking them as the predominant components of DXN content [51]. Among all 17 congeners, 2,3,4,7,8-P_5CDF exhibited the highest concentrations, accounting for 40.81% by mass and 40.92% by TEQ [38, 51]. It recorded the maximum mass concentration of 4.05E−01 ng/Nm^3 and the highest TEQ of 2.314E−01 ng I-TEQ/Nm^3. Under abnormal conditions, the mass concentration ratio of PCDFs to PCDDs is 3.37 ± 1.36, which is lower than the ratio observed during normal operating conditions. This finding suggests that the primary generation mechanisms of DXNs involve incomplete combustion and waste heat exchange. The major contributors to DXN concentrations include 2,3,4,7,8-P_5CDF (31.31% by mass and 30.84% by TEQ), 2,3,4,6,7,8-H_6CDF (13.31% by mass and 11.54% by TEQ), 1,2,3,7,8-P_5CDD (11.16% by mass and 13.85% by TEQ), and 1,2,3,6,7,8-H_6CDD (6.20% by mass and 7.65% by TEQ).

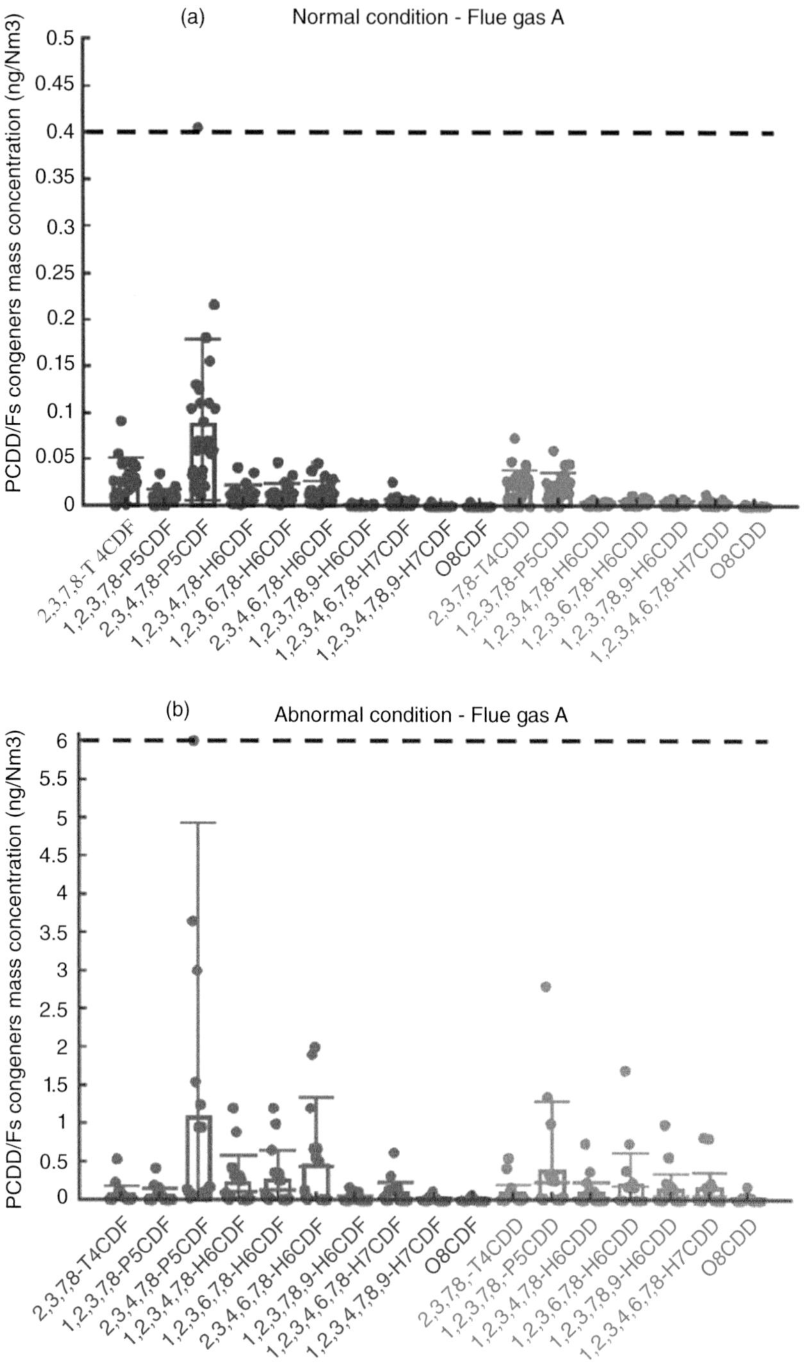

FIGURE 4.6 DXNs congeners distribution profile. (a and c) Normal condition – flue gas A. (b and d) Abnormal condition – flue gas A. (e and g) Normal condition – flue gas B. (f and h) Abnormal condition – flue gas B. (i) Normal condition – ash fly C. (j) Abnormal condition – ash fly C.

(Continued)

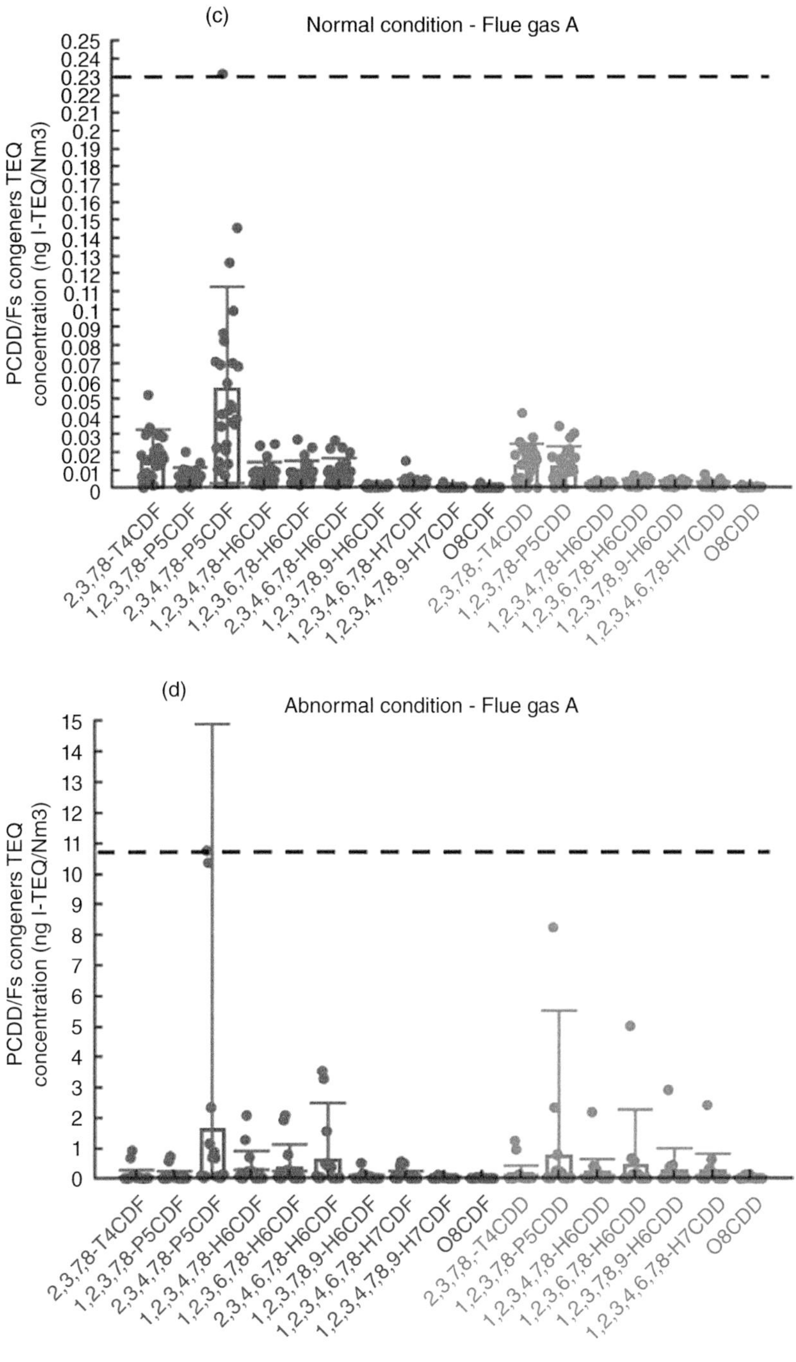

FIGURE 4.6 (Continued)

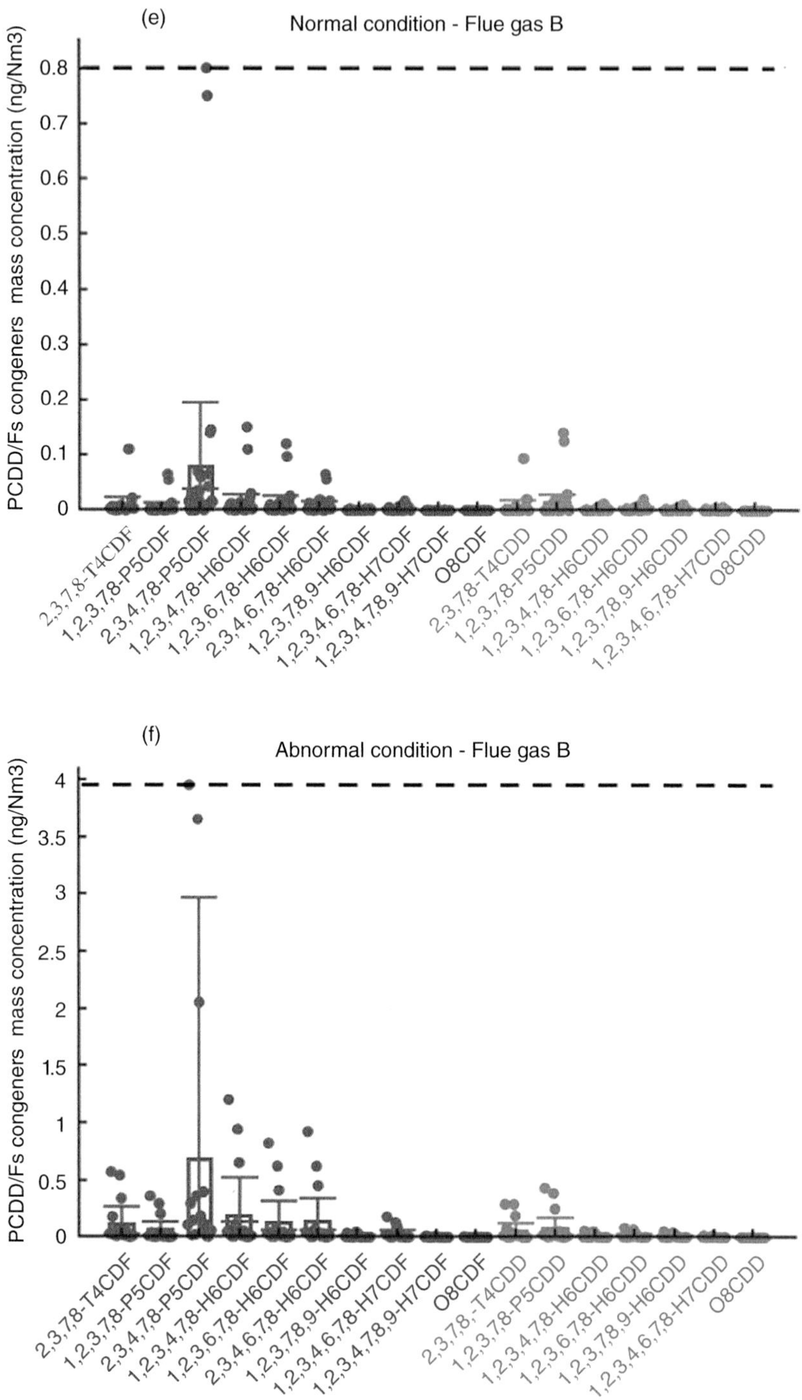

FIGURE 4.6 (Continued)

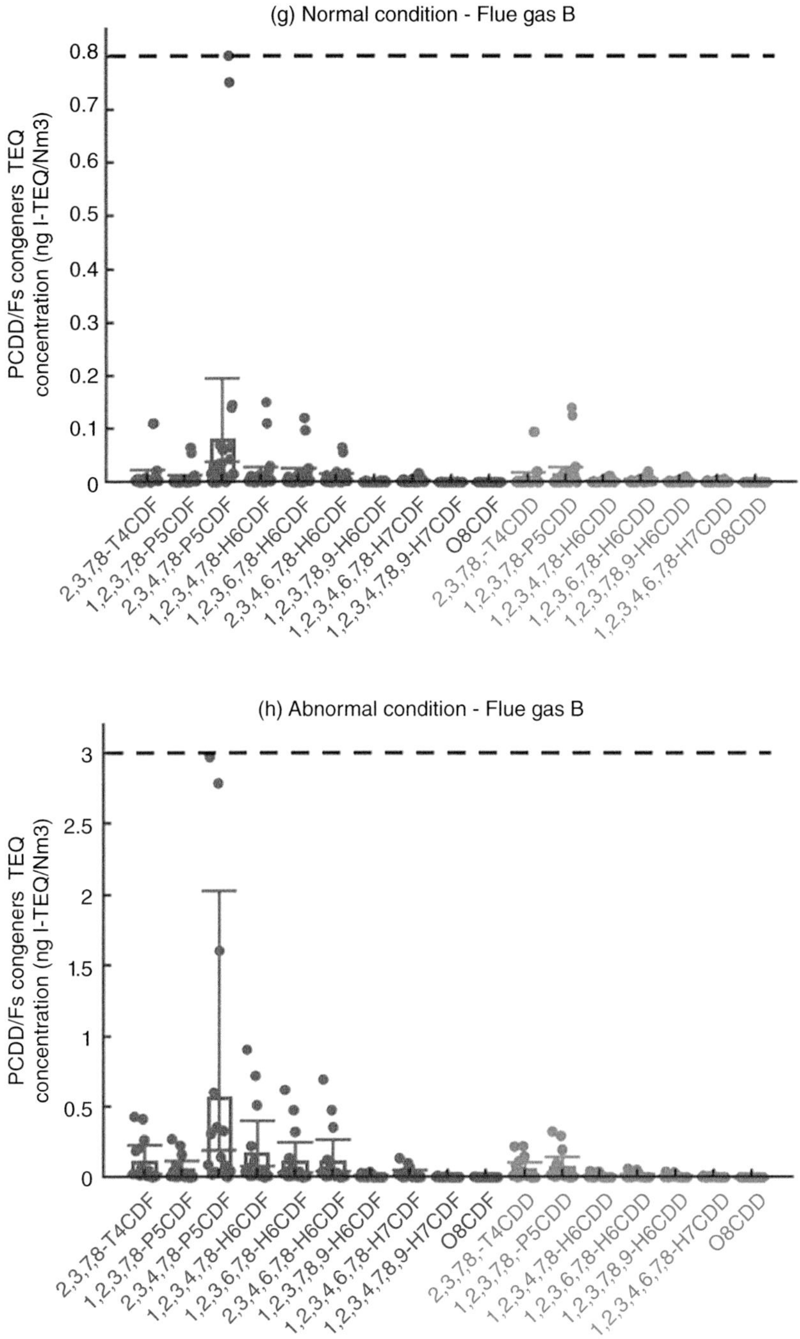

FIGURE 4.6 (Continued)

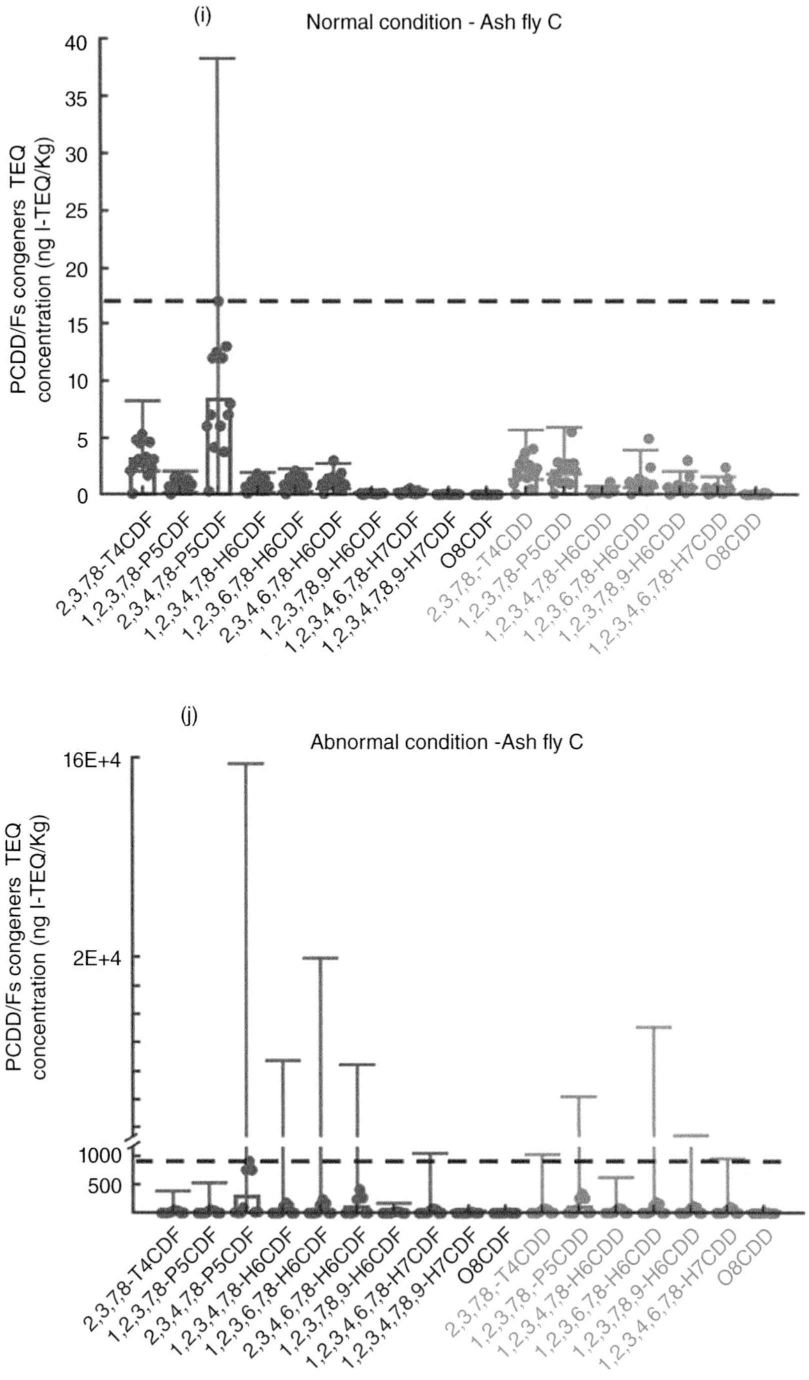

FIGURE 4.6 (Continued)

After purifying the flue gas using APCDs, the concentrations of 17 congeners in flue gas B are presented in Figure 4.6(e)–(h). The scatter diagram reveals two samples with abnormally high concentrations (notably on October 14, 2021, as depicted in Figure 4.4(d)). Consequently, the concentration levels of these congeners exceed national standards, particularly for 2,3,4,7,8-P_5CDF. Following ACI adsorption and BF dedusting under normal conditions, the congeners with the highest concentrations are 2,3,4,7,8-P_5CDF (47.58% by mass and 77.73% by TEQ) and 1,2,3,7,8-P_5CDD (8.54% by mass and 8.46% by TEQ). The distribution of the 17 congeners under abnormal conditions is markedly different from that under normal conditions. Specifically, DXN concentrations primarily originate from PCDFs, which constitute 88.46% by mass and 88.11% by TEQ [15, 43]. The major contributors include 2,3,7,8-T_4CDF (7.62% by mass and 8.11% by TEQ), 2,3,4,7,8-P_5CDF (44.25% by mass and 43.53% by TEQ), 1,2,3,4,7,8-H_6CDF (12.57% by mass and 12.54% by TEQ), 1,2,3,6,7,8-H_6CDF (8.34% by mass and 8.27% by TEQ), and 2,3,4,6,7,8-H_6CDF (8.88% by mass and 8.72% by TEQ). These results indicate that the design of the ACI and BF systems is more effective in purifying PCDDs compared to PCDFs [51].

Fly ash C is depicted in Figure 4.6(i) and (j), showing that the mass concentration ratio of PCDFs to PCDDs is lower than that of flue gases A and B, with values of 2.46 $\pm$ 0.60 under normal conditions and 1.89 $\pm$ 0.19 under abnormal conditions. Specifically, the percentages of PCDFs and PCDDs are 68.19% and 31.81% for normal conditions, and 66.84% and 33.16% for abnormal conditions, respectively. Consistent with the distribution observed in flue gas, the contribution of 2,3,4,7,8-P_5CDF to fly ash is 36.51% under normal conditions and 33.3% under abnormal conditions. This congener exhibits the largest fluctuation with varying operational conditions, with an upper limit of approximately 40 ng I-TEQ/kg under normal conditions and around 16,000 ng I-TEQ/kg during abnormal conditions [38]. Under normal conditions, the primary concentrations are found in 4–5 chlorinated PCDFs (53.64%) and 4–5 chlorinated PCDDs (18.57%). In contrast, during abnormal conditions, the major contributors include 5 chlorinated PCDFs (35.21%) and 4–5 chlorinated PCDDs (26.32%).

In summary, the impact of operating conditions on the distribution of congeners in flue gas and fly ash differs significantly. For the generation of DXNs in flue gas A, operating conditions affect both PCDFs and PCDDs. In contrast, for the emissions of DXNs in flue gas B, only PCDFs are more influenced by these conditions. Additionally, both PCDFs and PCDDs play a role in the variations observed under different operating conditions.

4.3.3 Removal Efficiency of APCDs

Using the TEQ concentration of DXNs at positions A and B as examples, the distribution of risk equivalent (RE) values under normal and abnormal conditions is illustrated in Figure 4.7. The left *Y*-axis displays the actual TEQ concentration of DXNs, serving as the reference axis for the histogram, while the right *Y*-axis represents the RE values calculated using Eq. (4.3).

Under normal conditions (Figure 4.7(a)), the maximum RE reaches 99.71% (consistent with Wei et al. [11]), while the minimum RE is −623.34%. The actual standard

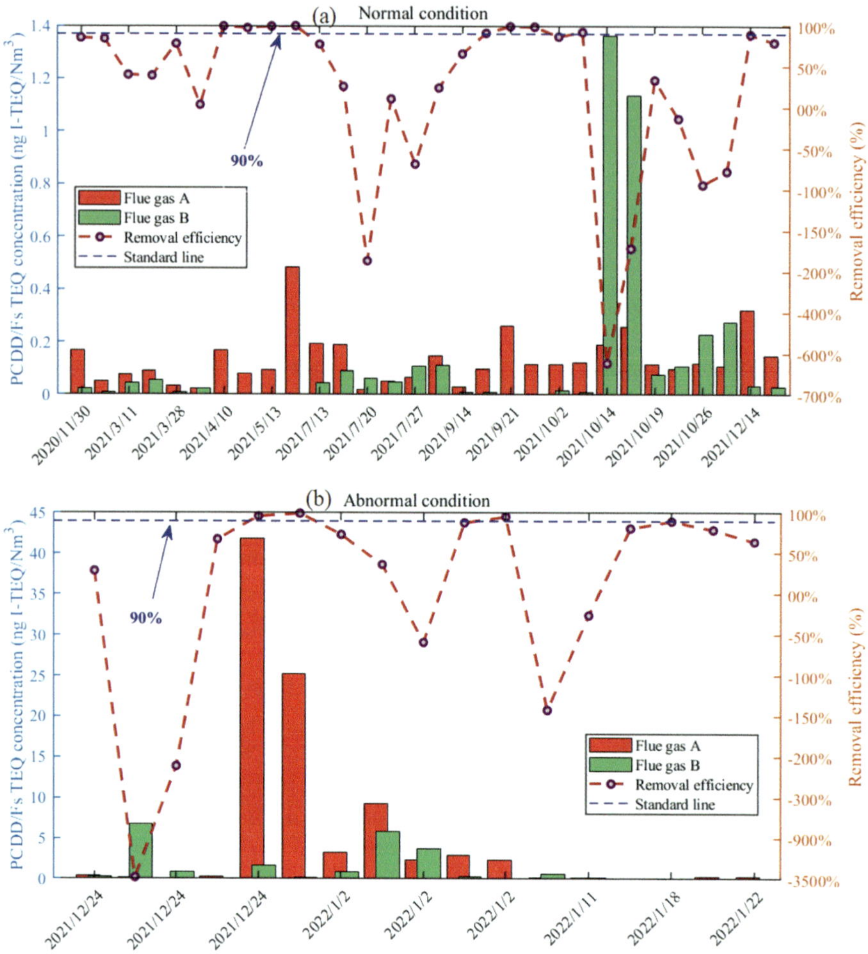

FIGURE 4.7 Removal efficiency of DXNs TEQ concentration. (a) Normal condition. (b) Abnormal condition.

rate is only 26.67% when using a RE of 90% as the benchmark during the annual operational period. Further analysis reveals that the proportion of negative RE values is also 26.67%, indicating that the ACI and BF systems do not consistently achieve stable DXN removal. Notably, the proportions of high-efficiency removal (greater than 90%) and negative-efficiency removal (less than 0%) are equal. Meanwhile, the low-efficiency removal rate, falling between 0% and 90%, accounts for 50% of the samples. Additionally, the average RE calculated from 30 samples is 4.71%, with a variance of 2.46, highlighting significant fluctuations in RE. These variations may primarily result from combustion conditions and the memory effect of APCDs [58].

To examine the impact of operating conditions on the RE of DXNs, it is evident that the fluctuations under abnormal working conditions are more pronounced, with a mean RE of −223.38% and a variance of 75.92, as illustrated in Figure 4.7(b). Furthermore, only 17.35% of the measurements reach the standard line for RE, which is approximately 10% lower than during normal operation. The proportion of RE values falling between 0% and 90% is about 52.94%, reflecting an increase of 2.94% compared to normal conditions. The lowest RE values of −3481.66% and −845.36% occur during shutdown stages, indicating incomplete combustion conditions within the incinerator. Notably, on the 9th day following startup, a second minimum RE value of −284.76% is observed, suggesting that the startup phase has a more significant impact on the overall operation of the incinerator compared to the shutdown phase.

Additionally, the distribution of RE for the 17 DXN congeners under both normal and abnormal conditions is presented in Figure 4.8. Each histogram represents the

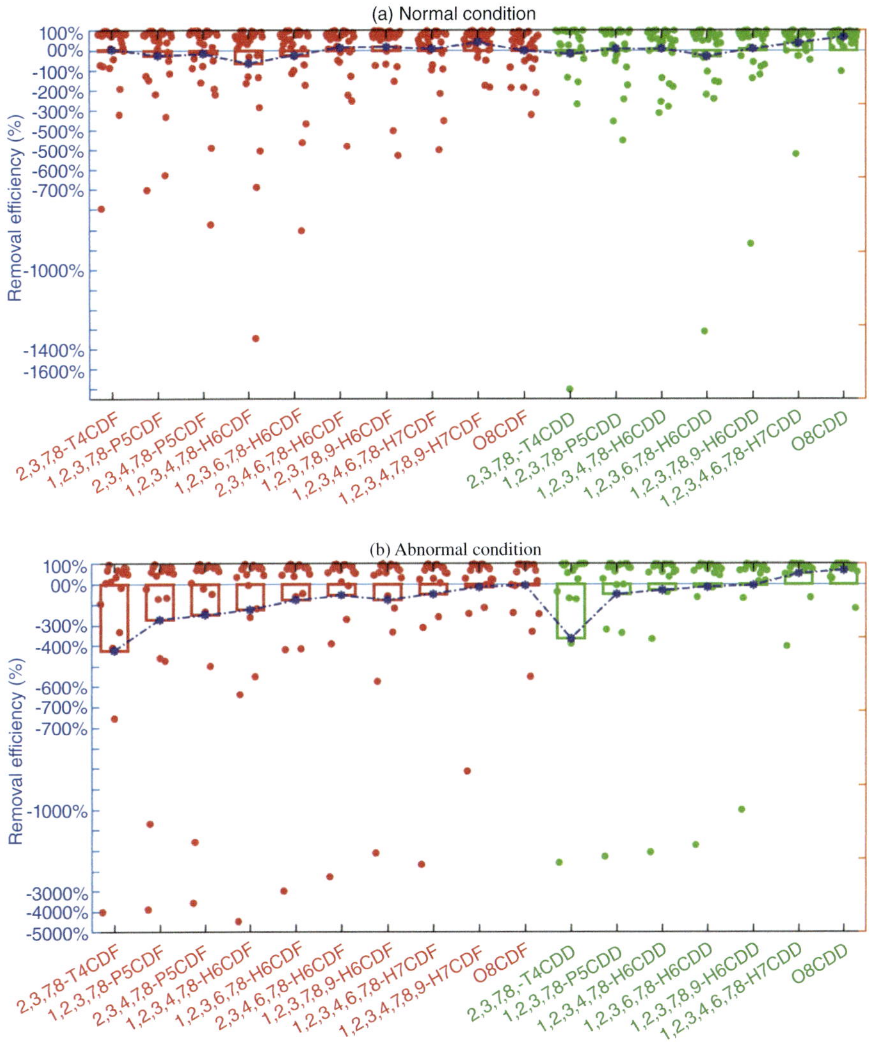

FIGURE 4.8 Removal efficiency of 17 DXNs congeners. (a) Normal condition. (b) Abnormal condition.

average of all data points, with each point corresponding to the actual TEQ concentration of the congeners.

In Figure 4.8(a), the proportion of average RE values greater than 0% is 58.82%, encompassing 5 PCDD congeners and 5 PCDF congeners. Overall, the RE for most congeners is positive in the majority of cases. The performance of the ACI and BF systems in removing PCDDs is significantly higher than that for PCDFs under normal conditions, as indicated by the distribution of scattered points in Figure 4.8(a). Notably, the congeners 1,2,3,4,7,8-H_6CDF (with a minimum RE of −1444.98%), 2,3,7,8-T_4CDD (minimum RE of −6257.62%), and 1,2,3,6,7,8-H_6CDD (minimum RE of −1411.75%) exhibit the greatest sensitivity to fluctuations in operating conditions.

Under abnormal conditions, the average RE of congeners is negative, with a mean range between −482.48% and −3.62%, with the exceptions of 1,2,3,4,6,7,8-H_7CDF (mean of 52.84%) and O_8CDF (mean of 70.56%), as shown in Figure 4.8(b). For both PCDDs and PCDFs, the average RE tends to increase gradually with higher levels of chlorination. In other words, the ACI and BF systems exhibit the lowest RE for low chlorinated congeners (such as 2,3,7,8-T_4CDF and 2,3,7,8-T_4CDD), while the highest RE values are observed in highly chlorinated congeners (such as O_8CDF and O_8CDD).

4.3.4 Emission Factor and Annual Emission of DXNs

To calculate the EFs, it is necessary to obtain the hourly capacity of MSW, the real-time flue gas flow, and the TEQ emissions of DXNs. The MSW treatment capacity is estimated based on a rated capacity of 800 tons/day, while the flue gas flow is determined by averaging the recorded data from the distributed control system (DCS) during the sampling period. The statistical data are presented in Table 4.5.

The EFs for each day are calculated using Eq. (4.4), and the results are further illustrated in Figure 4.9.

From the estimated EFs curve based on 47 samples, it is evident that the EFs during shutdown (mean of 1.33E-02μg I-TEQ / kg, upper limit of 1.58E − 02 μg I-TEQ / kg, and lower limit of 1.07E-02 μg I-TEQ / kg) and startup (mean of 1.42E-02μg I-TEQ / kg, upper limit of 1.63E-02 μg I-TEQ / kg, and lower limit of 1.21E-02 μg I-TEQ / kg) are significantly higher than those observed during normal operation. Using the mean value as the statistical index, the EF under normal conditions is 5.16E-04 μg I-TEQ / kg, while under abnormal conditions, it is 0.58E-02 μg I-TEQ / kg. Thus, the EFs during shutdown and startup exceed the average value during normal operation by factors of 25.70 and 27.55, respectively. Based on the 47 samples collected throughout the year 2021, the estimated mean EFs for that year is 0.18E-02 μg I-TEQ / kg, with a variance of 1.78E–05, resulting in a range of EFs between $\left[0.18\text{E-02}\,\mu g\ \text{I-TEQ / kg}, 0.19\text{E-02}\,\mu g\ \text{I-TEQ / kg}\right]$.

TABLE 4.5
Statistical Data of Flow Rate and DXNs Emission at Stack

Times	MSW Capacity (kg/h)[a]	Flow Rate at Stack (kNm³/h)		DXNs (ng I-TEQ/Nm³)						
		Mean[b]	Standard deviation[c]	1	2	3	4	5	6	Mean[d]
2020/11/30	33,333.3	135.8297	3.3713	0.0245	0.0081	/	/	/	/	0.0163
2021/3/11		134.4120	3.2715	0.0459	0.0550	/	/	/	/	0.0505
2021/3/28		136.3170	3.0659	0.0075	0.0234	/	/	/	/	0.0155
2021/4/10		131.1118	1.7816	0.0011	0.0022	/	/	/	/	0.0017
2021/5/13		127.9461	1.9405	0.0007	0.0014	/	/	/	/	0.0011
2021/7/13		126.1314	6.6637	0.0439	0.0908	/	/	/	/	0.0674
2021/7/20		124.1119	23.0839	0.0605	0.0463	/	/	/	/	0.0534
2021/7/27		125.6647	4.2488	0.1079	0.1106	/	/	/	/	0.1093
2021/9/14		132.5958	2.4528	0.0098	0.0088	/	/	/	/	0.0093
2021/9/21		136.2358	5.6676	0.0022	0.0015	/	/	/	/	0.0019
2021/10/2		130.4500	2.6379	0.0154	0.0094	/	/	/	/	0.0124
2021/10/14		132.6642	3.0738	1.3635	1.1347	/	/	/	/	1.2491
2021/10/19		131.7025	2.7085	0.0775	0.1095	/	/	/	/	0.0935
2021/10/26		130.0859	2.5858	0.2292	0.2745	/	/	/	/	0.2519
2021/12/14		113.0249	13.6343	0.0342	0.0302	/	/	/	/	0.0322
2021/12/24	16,666.7[e]	114.4933	22.3887	0.2944	6.7586	0.8253	0.0812	1.6895	/	1.9298
2022/1/2		128.3943	18.6422	0.1316	0.8492	5.8909	3.7482	0.3415	0.1075	1.8448
2022/1/11	33,333.3	116.7864	16.1062	0.6060	0.1715	/	/	/	/	0.3888
2022/1/18		125.8375	2.6102	0.0135	0.0075	/	/	/	/	0.0105
2022/1/22		125.6036	1.6053	0.0559	0.0934	/	/	/	/	0.0747

[a] Calculate the hourly processing capacity using 800 tons/day.
[b] Mean value of flue gas volume during sampling.
[c] Variance value of flue gas volume during sampling.
[d] Arithmetic mean of DXNs.
[e] MSW capacity is set to half of normal conditions during shutdown and startup.

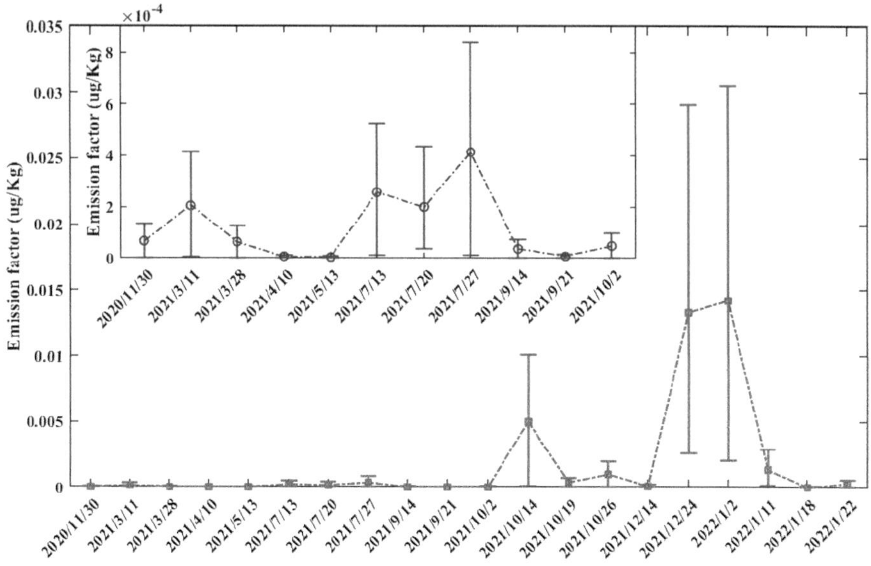

FIGURE 4.9 Emission factors of each sampling period.

Next, the daily treatment capacity of 800 tons and an annual operation time of 8,000 hours are utilized as inputs of AE. The calculation process is outlined as follows:

$$\begin{cases} m_{\min}^{\mathrm{AE}}\left(\mu\mathrm{g\ I\text{-}TEQ}\right) = \left(0.0018\left(\mu\mathrm{g\ I\text{-}TEQ}/\mathrm{kg}\right) + 1.7778\mathrm{E}-05\right) \times \left(800\left(\mathrm{tons}/\mathrm{day}\right)\right. \\ \qquad \times \left(1{,}000/24\right) \times 8{,}000\left(\mathrm{h}\right) \\ \qquad = 480{,}000\left(\mu\mathrm{g}\right) \\ m_{\max}^{\mathrm{AE}}\left(\mu\mathrm{g\ I\text{-}TEQ}\right) = \left(0.0018\left(\mu\mathrm{g\ I\text{-}TEQ}/\mathrm{kg}\right) - 1.7778\mathrm{E}-05\right) \times \left(800\left(\mathrm{tons}/\mathrm{day}\right)\right. \\ \qquad \times \left(1{,}000/24\right) \times 8{,}000\left(\mathrm{h}\right) \\ \qquad = 506{,}667\left(\mu\mathrm{g}\right) \end{cases} \tag{4.8}$$

Consequently, the minimum and maximum AE values are 4.80E + 05 μg I-TEQ and 5.07E + 05μg I-TEQ, respectively. Thus, the total estimated DXN emissions from the MSWI powe plant amount to approximately 4.93E + 05 μg I-TEQ [2].

4.3.5 Correlation Analysis between DXNs and Operation Parameter

The incineration conditions can be characterized by the sensor data collected from the DCS. This study includes 176 process data points collected by the DCS (Table 4.6), with data stored at a frequency of once every second. The statistical data of conventional pollutants and operating conditions within a year are presented in Table 4.7.

TABLE 4.6
Process Variables in Studied MSWI Power Plant

Number	Tag	Unit
1	1st flue gas temperature of combustion chamber	°C
2	2nd flue gas temperature of combustion chamber	°C
3	3rd flue gas temperature of combustion chamber	°C
4	4th flue gas temperature of combustion chamber	°C
5	5th flue gas temperature of combustion chamber	°C
6	6th flue gas temperature of combustion chamber	°C
7	7th flue gas temperature of combustion chamber	°C
8	8th flue gas temperature of combustion chamber	°C
9	9th flue gas temperature of combustion chamber	°C
10	1st flue gas temperature of combustion chamber rear arch	°C
11	2nd flue gas temperature of combustion chamber rear arch	°C
12	10th flue gas temperature of combustion chamber	°C
13	11th flue gas temperature of combustion chamber	°C
14	12th flue gas temperature of combustion chamber	°C
15	Temperature of left drying grate inside	°C
16	Temperature of left drying grate outside	°C
17	Temperature of right drying grate inside	°C
18	Temperature of right drying grate outside	°C
19	Temperature of left inside of furnace wall	°C
20	Temperature of left outside of furnace wall	°C
21	Temperature of right inside of furnace wall	°C
22	Temperature of right outside of furnace wall	°C
23	Temperature of left burning grate inside (stages 1–1)	°C
24	Temperature of left burning grate outside (stages 1–1)	°C
25	Temperature of right burning grate inside (stages 1–1)	°C
26	Temperature of right burning grate outside (stages 1–1)	°C
27	Temperature of left burning grate inside (stages 1–2)	°C
28	Temperature of left burning grate outside (stages 1–2)	°C
29	Temperature of right burning grate inside (stages 1–2)	°C
30	Temperature of right burning grate outside (stages 1–2)	°C
31	Temperature of left burning grate inside (stages 2–1)	°C
32	Temperature of left burning grate outside (stages 2–1)	°C
33	Temperature of right burning grate inside (stages 2–1)	°C
34	Temperature of right burning grate outside (stages 2–1)	°C
35	Temperature of left burning grate inside (stages 2–2)	°C
36	Temperature of left burning grate outside (stages 2–2)	°C
37	Temperature of right burning grate inside (stages 2–2)	°C
38	Temperature of right burning grate outside (stages 2–2)	°C
39	Left temperature of slag outlet	°C
40	Right temperature of slag outlet	°C
41	Flue gas temperature of left first channel inlet	°C

(Continued)

TABLE 4.6 (Continued)
Process Variables in Studied MSWI Power Plant

Number	Tag	Unit
42	Flue gas temperature of right first channel inlet	°C
43	Flue gas temperature of left protective tube inlet	°C
44	Flue gas temperature of right protective tube inlet	°C
45	Flue gas temperature of left three-stage superheater inlet	°C
46	Flue gas temperature of right three-stage superheater inlet	°C
47	Flue gas temperature of left evaporator inlet	°C
48	Flue gas temperature of right evaporator inlet	°C
49	Flue gas temperature of left economizer inlet	°C
50	Flue gas temperature of right economizer inlet	°C
51	Flue gas temperature of economizer outlet	°C
52	Air temperature of primary fan heater outlet	°C
53	Air temperature of burning grate inlet	°C
54	Air temperature of drying grate inlet	°C
55	Air temperature of second fan heater outlet	°C
56	Temperature of cooling air of left furnace wall outlet	°C
57	Temperature of cooling air of right furnace wall outlet	°C
58	Temperature of cooling air of left grate wall outlet	°C
59	Temperature of cooling air of right grate wall outlet	°C
60	Air temperature of drying grate outlet	°C
61	Air temperature of burning grate outlet	°C
62	Biogas flow rate of drying grate	m^3/h
63	Biogas flow rate of burning grate	m^3/h
64	Steam temperature of first-stage superheater outlet	°C
65	Steam temperature of two-stage superheater outlet	°C
66	Steam temperature of three-stage superheater outlet	°C
67	Water temperature of economizer outlet	°C
68	1st flue gas temperature of bag filter inlet	°C
69	2nd flue gas temperature of bag filter inlet	°C
70	3rd flue gas temperature of bag filter inlet	°C
71	Temperature of left bag filter inlet	°C
72	Temperature of right bag filter inlet	°C
73	Temperature of fluidizing fan heater outlet	°C
74	1st flue gas temperature of left bag filter outlet	°C
75	2nd flue gas temperature of left bag filter outlet	°C
76	3rd flue gas temperature of left bag filter outlet	°C
77	1st flue gas temperature of right bag filter outlet	°C
78	2nd flue gas temperature of right bag filter outlet	°C
79	3rd flue gas temperature of right bag filter outlet	°C
80	Flue gas temperature of induced fan outlet	°C
81	Air flow rate of left drying grate (stages 1)	km^3 N/h
82	Air flow rate of right drying grate (stages 1)	km^3 N/h
83	Air flow rate of left drying grate (stages 2)	km^3 N/h
84	Air flow rate of right drying grate (stages 2)	km^3 N/h

(Continued)

TABLE 4.6 (Continued)
Process Variables in Studied MSWI Power Plant

Number	Tag	Unit
85	Air flow rate of left burning grate (stages 1–1)	km^3 N/h
86	Air flow rate of right burning grate (stages 1–1)	km^3 N/h
87	Air flow rate of left burning grate (stages 1–2)	km^3 N/h
88	Air flow rate of right burning grate (stages 1–2)	km^3 N/h
89	Air flow rate of left burning grate (stages 2–1)	km^3 N/h
90	Air flow rate of right burning grate (stages 2–1)	km^3 N/h
91	Air flow rate of left burning grate (stages 2–2)	km^3 N/h
92	Air flow rate of right burning grate (stages 2–2)	km^3 N/h
93	Air flow rate of left burnout grate	km^3 N/h
94	Air flow rate of right burnout grate	km^3 N/h
95	Air flow rate of second fan	km^3 N/h
96	Feedwater flow rate of economizer	tons/h
97	2th feedwater flow rate of economizer	tons/h
98	1th feedwater flow rate of economizer	tons/h
99	Cooling water flow of primary superheater	tons/h
100	Cooling water flow of second superheater	tons/h
101	Desuperheating water flow of primary and secondary superheaters	tons/h
102	Steam flow rate of boiler	tons/h
103	Flue flow rate of stack inlet	km^3 N/h
104	Water flow rate of left FGD mixer	kg/h
105	Water flow rate of right FGD mixer	kg/h
106	Flue flow rate of left FGD reactor inlet	km^3 N/h
107	Flue flow rate of right FGD reactor inlet	km^3 N/h
108	Supply flow rate of urea	L/h
109	Supply flow rate of urea diluent	L/h
110	Supply flow rate of lime	kg/h
111	Supply flow rate of active carbon	kg/h
112	Air pressure of primary fan outlet	kPa
113	Air pressure of second fan outlet	kPa
114	Negative pressure of furnace	Pa
115	Flue gas pressure of economizer outlet	kPa
116	Flue gas pressure of induced fan outlet	Pa
117	Pressure of boiler drum	MPa
118	Steam pressure of three-stage superheater outlet	MPa
119	Flue gas pressure of FGD outlet	kPa
120	Differential pressure of left reactor inlet	Pa
121	Differential pressure of left reactor outlet	Pa
122	Differential pressure of right reactor inlet	Pa
123	Differential pressure of right reactor outlet	Pa
124	Differential pressure of left bag filter	Pa
125	Differential pressure of right bag filter	Pa
126	Inside speed of left feeder	%
127	Outside speed of left feeder	%

(Continued)

TABLE 4.6 (Continued)
Process Variables in Studied MSWI Power Plant

Number	Tag	Unit
128	Inside speed of right feeder	%
129	Outside speed of right feeder	%
130	Inside speed of left drying grate	%
131	Outside speed of left drying grate	%
132	Inside speed of right drying grate	%
133	Outside speed of right drying grate	%
134	Inside speed of left burning grate (stages 1)	%
135	Outside speed of left burning grate (stages 1)	%
136	Inside speed of right burning grate (stages 1)	%
137	Outside speed of right burning grate (stages 1)	%
138	Inside speed of left burning grate (stages 2)	%
139	Outside speed of left burning grate (stages 2)	%
140	Inside speed of right burning grate (stages 2)	%
141	Outside speed of right burning grate (stages 2)	%
142	Inside speed of left burnout grate	%
143	Inside speed of right burnout grate	%
144	Baffle opening of air inlet duct of left drying grate (stages 1)	%
145	Baffle opening of air inlet duct of left drying grate (stages 2)	%
146	Baffle opening of air inlet duct of left burning grate (stages 1–1)	%
147	Baffle opening of air inlet duct of left burning grate (stages 1–2)	%
148	Baffle opening of air inlet duct of left burning grate (stages 2–1)	%
149	Baffle opening of air inlet duct of left burning grate (stages 2–2)	%
150	Baffle opening of air inlet duct of left burnout grate (stages 1)	%
151	Baffle opening of air inlet duct of right drying grate (stages 1)	%
152	Baffle opening of air inlet duct of right drying grate (stages 2)	%
153	Baffle opening of air inlet duct of right burning grate (stages 1–1)	%
154	Baffle opening of air inlet duct of right burn grate (stages 1–2)	%
155	Baffle opening of air inlet duct of right burning grate (stages 2–1)	%
156	Baffle opening of air inlet duct of right burning grate (stages 2–2)	%
157	Baffle opening of air inlet duct of right burnout grate (stages 1)	%
158	Baffle opening of air inlet duct of primary fan	%
159	Baffle opening of air inlet duct of second fan	%
160	Baffle opening of air inlet duct of second fan (bypass)	%
161	Baffle opening of air inlet duct of furnace wall cooling fan	%
162	Baffle opening of air inlet duct of grate wall cooling fan	%
163	Baffle opening of air inlet duct of induced fan	%
164	Speed of induced fan	rpm
165	Frequency of primary fan	Hz
166	Frequency of second fan	Hz
167	Water level of boiler drum	mm
168	1st flue gas oxygen concentration at boiler	%

(Continued)

TABLE 4.6 (Continued)
Process Variables in Studied MSWI Power Plant

Number	Tag	Unit
169	2nd flue gas oxygen concentration at boiler	%
170	3rd flue gas oxygen concentration at stack	%
171	Flue gas dust concentration	mg/m³ N
172	Flue gas NO_x concentration	mg/m³ N
173	Flue gas SO_2 concentration	mg/m³ N
174	Flue gas HCl concentration	mg/m³ N
175	Flue gas CO concentration	mg/m³ N
176	Flue gas CO_2 concentration	%
177	Flue gas HF concentration	mg/m³ N

TABLE 4.7
Statistical Data of Conventional Pollutants and Operating Conditions within a Year

Parameters	Run-Time				Rated
Operation Parameters[a]	Min	Max	Mean	Variance	Operation
Temp. of incinerator (°C)	900.0822	1.1134E+03	996.5906	1.5340E+03	850°C[b]
Inlet temp. of superheater unit (°C)	654.3515	0.8520E+03	760.6708	0.9865E+03	N.A.[c]
Exit temp. of economizer (°C)	157.9453	0.1924E+03	182.1618	0.0391E+03	190[d]
Steam flow rate (tons/h)	50.7345	0.0844E+03	74.4323	0.0373E+03	73.8[d]
Exit O_2 of economizer (%)	4.0471	0.0102E+03	7.0416	0.0010E+03	6–9[d]
Inlet temp. of BF (°C)	60.9000	0.1707E+03	155.3332	0.5870E+03	N.A.[c]
Exit temp. of BF (°C)	70.3600	0.1710E+03	159.3794	0.2155E+03	N.A.
Inlet temp. of stack (°C)	119.1617	0.1763E+03	166.2263	0.0616E+03	N.A.
Inlet O_2 of stack (%)	4.5126	0.0224E+03	8.1751	0.0069E+03	N.A.
Flow rate at stack (kNm³/h)	72.3723	0.1569E+03	128.5433	0.0959E+03	139.430[d]
Conventional pollutants[e]					
NO_x (mg/Nm³)	77.0770	184.3813	141.0154	545.2502	250[b]
SO_2 (mg/Nm³)	0	26.2719	5.3590	33.0487	100[b]
HCl (mg/Nm³)	1.6347	13.3949	4.9533	2.2876	60[b]
CO (mg/Nm³)	1.0223	61.4848	16.6162	177.3057	100[b]

Note:

[a] Five-minute mean.

[b] National emission standard of China: GB 18485-2014 (mean 1 hour).

[c] Not available.

[d] Technological design value.

[e] One-hour mean.

Consequently, the process data are averaged over approximately 2 hours during the sampling period, resulting in a total of 47 samples. Using Eq. (4.7), the correlation between the 176 process variables and the mass concentration of DXN emissions is analyzed, and the results are illustrated in Figure 4.10.

As illustrated in Figure 4.10, it includes the PCC, which are based on an original hypothesis, along with the statistical P values set at a significance level of 0.05 (the alternative hypothesis being that the correlation is not less than 0). The PCC values are categorized as follows: weakly correlated in the range of [−0.1, 0.1], moderately correlated in the ranges of [−0.4, −0.1] and [0.1, 0.4], and strongly correlated in the ranges of [−1, −0.4] and [0.4, 1]. Additionally, the calculations for normal and abnormal conditions are conducted independently during the correlation analysis.

Under normal conditions, the following variables show a strong positive correlation with the mass concentration of DXN emissions: flue gas oxygen concentration (contrary to findings by Oh et al. [19]) with $r^{\mathrm{PC}} = 0.41$, the baffle opening of the air inlet duct for the left burnout grate with $r^{\mathrm{PC}} = 0.43$, the cooling water flow of the primary superheater with $r^{\mathrm{PC}} = 0.40$, and the air flow of the right burning grate (stages 1–2) with $r^{\mathrm{PC}} = 0.82$. Based on the P values, the correlation between these

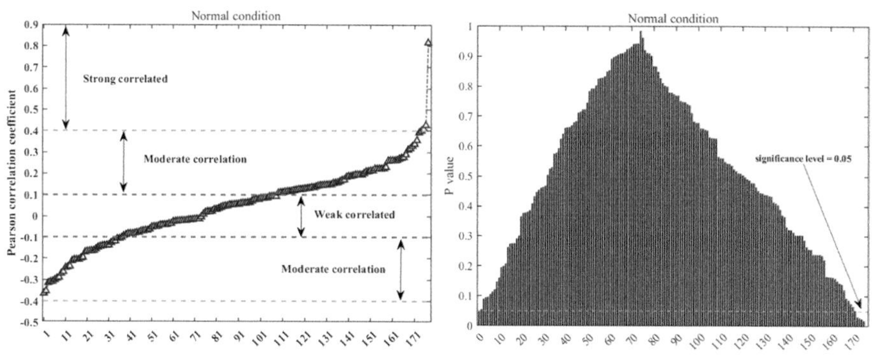

a) Normal condition

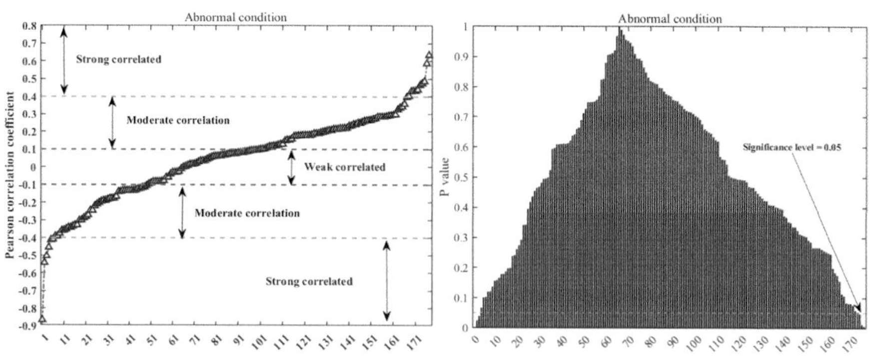

b) Abnormal condition

FIGURE 4.10 Correlation between process variables and DXNs mass concentration. (a) Normal condition. (b) Abnormal condition.

four variables and DXN concentration is statistically significant, leading to the rejection of the original assumption regarding a strong correlation. Additionally, there is a strong negative correlation observed with the fourth flue gas temperature in the combustion chamber, which has a statistical P value of 4.78E–02, indicating that this correlation is not zero. Specifically, 99 process variables (with *P* values greater than 0.05) exhibit moderate correlation, of which 63 are positively correlated, while 36 are negatively correlated. Furthermore, 71 process variables (with *P* values greater than 0.05) show a weak correlation with DXN emissions from the MSWI power plant.

Under abnormal conditions, six process variables demonstrate a strong negative correlation with DXN emissions. These variables include: the temperature of the left drying grate inside with $r^{PC} = -0.50$, the temperature of the right burning grate outside (stages 2–1) with $r^{PC} = -0.41$, the temperature of the cooling air at the right furnace wall outlet with $r^{PC} = -0.54$, the differential pressure of the left BF with $r^{PC} = -0.87$, the baffle opening of the air inlet duct for the left burnout grate (stages 2–2) with $r^{PC} = -0.41$, and the supply flow rate of urea with $r^{PC} = -0.45$. Several variables did not meet the hypothesis criteria, including the differential pressure of the left BF with $p_{\text{value}} = 6.73e-06$, the temperature of the cooling air at the right furnace wall outlet with $p_{\text{value}} = 2.43e-02$, and the temperature of the left drying grate inside with $p_{\text{value}} = 3.87e-02$. Moreover, the speed of right feeder inside $\left(r^{PC} = 0.44\right)$, the 8th flue gas temperature of combustion chamber $\left(r^{PC} = 0.59\right)$, the speed of left feeder inside $\left(r^{PC} = 0.49\right)$, the speed of induced draft fan $\left(r^{PC} = 0.40\right)$, the 2nd flue gas temperature at BF inlet $\left(r^{PC} = 0.43\right)$, the 2nd feedwater flow of economizer $\left(r^{PC} = 0.48\right)$, the 3rd flue gas temperature at right bag outlet $\left(r^{PC} = 0.43\right)$, the 1st flue gas temperature at right BF outlet $\left(r^{PC} = 0.43\right)$, the 2nd air flow of right drying grate $\left(r^{PC} = 0.64\right)$, and the frequency of secondary fan $\left(r^{PC} = 0.47\right)$ have a strong positive correlation to DXNs emissions. In terms of the *P* value, the 2nd air flow of right drying grate $\left(p_{\text{value}} = 0.61\text{E}-02\right)$, the 8th flue gas temperature of combustion chamber $\left(p_{\text{value}} = 0.13\text{E}-01\right)$, and the speed of left feeder inside $\left(p_{\text{value}} = 4.74\text{E}-01\right)$ is not strongly positively correlated with DXNs emissions. Moreover, the proportions of moderately correlated (110 in total) and weakly correlated (50 in total) process variables are similar to those observed under normal conditions. In summary, grate speed, air volume, water flow, gas temperature, and grate temperature all show a strong correlation with DXN emissions from the MSWI power plant.

4.3.6 Atmospheric Emission Assessment and Suggestions

To estimate the impact of environmental changes on DXN emissions, the relationship between the external environment (primarily considering seasonal variations throughout the year) and DXN emissions from the MSWI plant is further analyzed. Figure 4.11 presents the correlation between seasonal changes and DXN emissions in flue gas and fly ash. Each scatter point represents a detected value for a specific season, while the histogram displays the mean value of all detected values for that season.

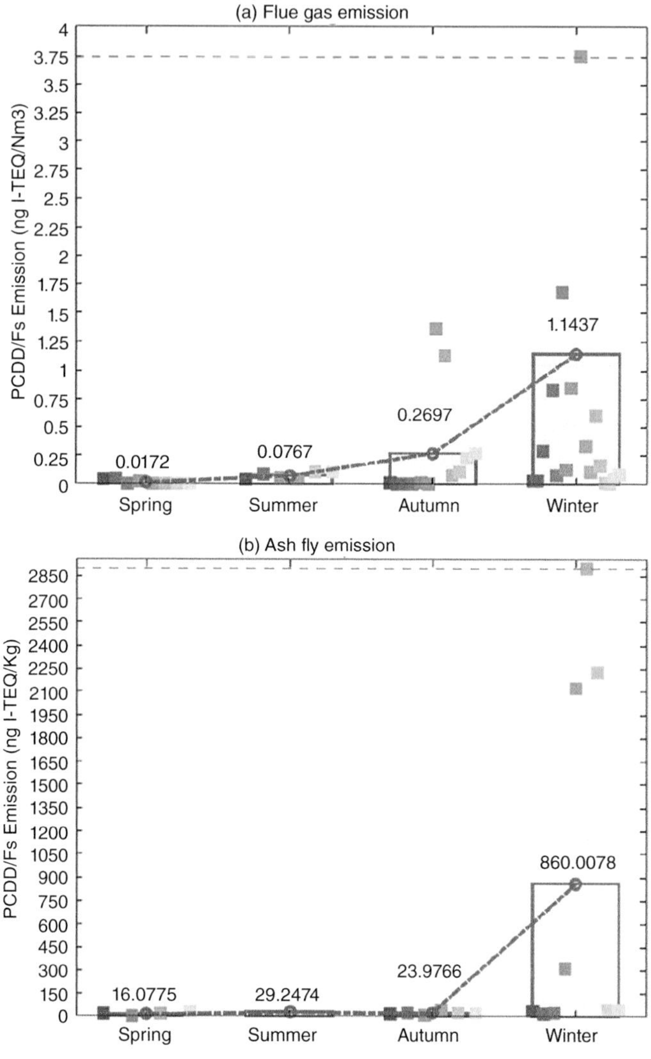

FIGURE 4.11 DXNs TEQ concentration emission in different seasons. (a) Flue gas emission. (b) Ash fly emission.

According to Figure 4.11, DXN emissions increase gradually from spring to winter, with the lowest emissions recorded in spring and the highest in winter, which aligns with the findings of Zhang et al. [51]. Specifically, the DXN emissions for flue gas and fly ash are 1.72E−02 ng I-TEQ/Nm^3 and 16.08 ng I-TEQ/kg, respectively, in spring, while in winter, these values rise to 1.14 ng I-TEQ/Nm^3 and 860.01 ng I-TEQ/kg. The significant increase in emissions during winter can be attributed to two main factors: (1) the annual planned shutdown and overhaul of the MSWI plant occurs in winter (i.e., at the end of the year); and (2) after continuous operation for over 6 months, particles tend to accumulate in the inner surface of the equipment, leading to the occurrence of a memory effect that can release DXNs, particularly from APCDs [55, 58]. Additionally, the lowest emissions recorded in spring can be attributed to maintenance work conducted during winter, which ensures optimal operation of the MSWI power plant. As the operational duration increases, DXN emissions in summer and autumn are significantly higher than those in spring. Specifically, the average emissions of flue gas and fly ash in summer are four times and two times greater, respectively, than those in spring, while in autumn, the average emissions of flue gas and fly ash are 15 times and equal to those of spring, respectively.

In summary, the experimental analysis yields several key findings: (1) startup and shutdown processes are significant sources of DXN emissions; (2) there is a clear upward trend in DXN emissions corresponding to the operational time of the incinerator, with the memory effect of APCDs emerging as a major, non-negligible contributor to these emissions; and (3) the DXN RE of ACI and BF is significantly influenced by the operational conditions of the incineration process.

Several recommendations for reducing DXN emissions from MSWI power plants in the future are proposed. First, auxiliary combustion technology should be implemented to ensure stable combustion and the use of plastic fillers with strong DXN adsorption capacity, aiming for ultra-low emissions during abnormal operating conditions such as shutdowns and startups. Second, regular maintenance of APCDs and the timely replacement or cleaning of malfunctioning equipment can help minimize DXNs released due to the memory effect. Third, the application of artificial intelligence technology can provide intelligent optimization control for MSWI operations, reducing fluctuations in operating conditions that may arise from manual operation.

In terms of integrating artificial intelligence technology, this article presents 47 samples, which include sensor data and DXN emission concentrations. These samples can support the development of AI-driven solutions, such as a data-driven soft sensor model for measuring DXN emission concentrations to aid in operational control, as well as a data-driven indicator model for providing intelligent optimization decisions within the MSWI process. However, the generation and emission mechanisms of DXNs, along with the removal technologies and the memory effect mechanisms of APCDs, require further investigation to effectively support DXN emission reduction efforts.

4.4 SUMMARY

To investigate the impact of operating conditions, environmental changes, and operational time on the generation and emissions of DXNs, this study conducted a comprehensive one-year analysis of DXN emissions from an MSWI power plant with a daily processing capacity of 800 tons, equipped with APCDs including SNCR, SDD, ACI, and BF. A total of 47 flue gas samples (designated as flue gas A and B), 22 fly ash samples, and 47 process data sets (comprising 176 process variables) were collected during the experiment. The substantial sample size allows for a more accurate characterization of DXN emissions in the MSWI process. The analysis encompassed various aspects, including DXN mass concentration and TEQ concentration, distribution of 17 congeners, APCD RE, DXN EFs, and annual emissions. Furthermore, correlation analyses between process variables and DXN emissions, as well as seasonal variations in emissions, were thoroughly examined and discussed. Additionally, several recommendations for reducing DXN emissions from MSWI power plants were provided. Overall, this study addresses a critical gap in understanding the emission characteristics of DXNs from MSWI power plants under long-term operational conditions.

The limitations of this study include the need for further research on the impact of waste composition in the furnace on DXN emissions. Additionally, exploring the relationship between other organic pollutants in flue gas emissions and DXN emissions warrants further investigation. Future research opportunities can be categorized into two main areas. First, there is a need for more detailed studies on the concentration levels of DXNs in the accumulated fly ash within the inner walls of the MSWI. Second, based on the DXN data obtained from this study, there is a demand for research focused on data-driven AI technologies, such as soft sensor detection technology and intelligent optimization control algorithms.

REFERENCES

1. Lin B C, Yang L L, Zheng M H, et al. Synergetic promoting/inhibiting mechanisms of copper/calcium compounds in the formation of persistent organic pollutants and environmentally persistent free radicals from anthracene. *Chemical Engineering Journal*, 2022, 441: 136102.
2. Lei R R, Xu Z C, Xing Y, et al. Global status of dioxin emission and China's role in reducing the emission. *Journal of Hazardous Material*, 2021, 418: 126265.
3. Kammen D M, Sunter, D A. City-integrated renewable energy for urban sustainability. *Science*, 2016, 352(6288): 922–928.
4. Zhou T, Bo X, Qu J B, et al. Characteristics of PCDD/Fs and metals in surface soil around an iron and steel plant in North China Plain. *Chemosphere*, 2019, 216: 413–418.
5. Mutahharah M M, Rozainee M T, Mimi H H. Measurement of PCDD/Fs emissions from a coal–fired power plant in Malaysia and establishment of emission factors. *Atmospheric Pollution Research*, 2014, 5(3): 388–397.
6. Zhuang J, Tang J, Aljerf L. Comprehensive review on mechanism analysis and numerical simulation of municipal solid waste incineration process based on mechanical grate. *Fuel*, 2022, 320: 123826.
7. Zhan M X, Liu Y W, Ye W W, et al. Modification of activated carbon using urea to enhance the adsorption of dioxins. *Environmental Research*, 2022, 204(B): 112035.
8. Zhou H, Meng A H, Long Y Q, et al. A review of dioxin-related substances during municipal solid waste incineration. *Waste Management*, 2015, 36: 106–118.
9. Huang Y, Chen Y, Li Y, et al. Atmospheric emissions of PCDDs and PCDFs in China from 1960 to 2014. *Journal of Hazardous Materials*, 2022, 424(A): 127320.
10. Buekens A, Huang H. Comparative evaluation of techniques for controlling the formation and emission of chlorinated dioxins/furans in municipal waste incineration. *Journal of Hazardous Materials*, 1998, 62(1): 1–33.
11. Wei J X, Li H, Liu J G. Phase distribution of PCDD/Fs in flue gas from municipal solid waste incinerator with ultra-low emission control in China. *Chemosphere*, 2021, 276(9–11): 130166.
12. Li H F, Liu W B, Lu A X, et al. PCDD/Fs emissions from secondary copper production synergistically controlled by fabric filters and desulfurization. *Environmental Pollution*, 2021, 270: 116065.

13. Wang P J, Yan F, Cai J J, et al. Emission levels and phase distributions of PCDD/Fs in a full-scale municipal solid waste incinerator: the impact of wet scrubber system. *Journal of Cleaner Production*, 2022, 337: 130468.
14. Wu G L, Sun Y Z, Xie J H, Lu Y. Research on pollution prevention and control BAT of PCDD/Fs in secondary copper industry. *Ecotoxicology and Environmental Safety*, 2019, 181: 308–311.
15. Li J F, Lv Z W, Du L, et al. Emission characteristic of polychlorinated dibenzo-p-dioxins and polychlorinated dibenzofurans (PCDD/Fs) from medical waste incinerators (MWIs) in China in 2016: a comparison between higher emission levels of MWIs and lower emission levels of MWIs. *Environmental Pollution*, 2017, 221: 437–444.
16. Düwel U, Nottrodt A, Ballschmiter K. Simultaneous sampling of PCDD/PCDF inside the combustion chamber and on four boiler levels of a waste incineration plant. *Chemosphere*, 1990, 20(10–12): 1839–1846.
17. Tejima H, Shibakawa S, Osumi K, Kawashima M. Dioxin emission behavior in MSW incinerator designed after Japanese guidelines for controlling dioxin. *Chemosphere*, 1998, 37(9–12): 2309–2314.
18. Gass H C, Jager E, Menke D, Luder K. Long term study for minimization of the PCDD/PCDF – emissions of a municipal solid waste incinerator in Germany. *Organohalogen Compounds*, 1998, 36: 175–178.
19. Oh E E, Lee K T, Lee J W, Chang Y S. The evaluation of PCDD/Fs from various Korean incinerators. *Chemosphere*, 1999, 38(9): 2097–2108.
20. Giugliano M, Cernuschi S, Grosso M, et al. The flux and mass balance of PCDD/F in a MSW incineration full scale plant. *Chemosphere*, 2001, 43(4–7): 743–750.
21. Chang M B, Lin J J, Chang S H. Characterization of dioxin emissions from two municipal solid waste incinerators in Taiwan. *Atmospheric Environment*, 2002, 36(2): 279–286.
22. Kai H C, Chang M B. Evaluation of PCDD/F congener partition in vapor/solid phases of waste incinerator flue gases. *Environmental Science and Technology*, 2005, 39(20): 8023–8031.
23. Chi K H, Chang S H, Huang C H, et al. Partitioning and removal of dioxin-like congeners in flue gases treated with activated carbon adsorption. *Chemosphere*, 2006, 64(9): 1489–1498.
24. Chi K H, Chang M B, Chang S H. Evaluation of PCDD/F partitioning between vapor and solid phases in MWI flue gases with temperature variation. *Journal of Hazardous Materials*, 2006, 138(3): 620–627.
25. Chi K H, Chang M B, Chang-Chien G P, Lin C. Characteristics of PCDD/F congener distributions in gas/particulate phases and emissions from two municipal solid waste incinerators in Taiwan. *Science of The Total Environment*, 2005, 347(1–3): 148–162.
26. Abad E, Martínez K, Caixach J, Rivera J. Polychlorinated dibenzo-p-dioxins, dibenzofurans and 'dioxin-like' PCBs in flue gas emissions from municipal waste management plants. *Chemosphere*, 2006, 63(4): 570–580.
27. Wang L C, Chang-Chien G P. Characterizing the emissions of polybrominated dibenzo-p-dioxins and dibenzofurans from municipal and industrial waste incinerators. *Environmental Science and Technology*, 2007, 41(4): 1159–1165.
28. Bie R, Li S Y, Wang H. Characterization of PCDD/Fs and heavy metals from MSW incineration plant in Harbin. *Waste Management*, 2007, 27(12): 1860–1869.
29. Mario G, Stefano C, Michele G, et al. Environmental release and mass flux partitioning of PCDD/Fs during normal and transient operation of full scale waste to energy plants. *Chemosphere*, 2007, 67(9): S118–S124.
30. Hunsinger H, Klaus J, Juergen V. Formation and destruction of PCDD/F inside a grate furnace. *Chemosphere*, 2002, 46(9–10): 1263–1272.

31. Everaert K, Baeyens J. The formation and emission of dioxins in large scale thermal processes. *Chemosphere*, 2002, 46(3): 439–448.
32. Wang H C, Hwang J F, Chi K H, Chang C M. Formation and removal of PCDD/Fs in a municipal waste incinerator during different operating periods. *Chemosphere*, 2007, 67(9): S177–S184.
33. Choi K I, Lee S H, Lee D H. Emissions of PCDDs/DFs and dioxin-like PCBs from small waste incinerators in Korea. *Atmospheric Environment*, 2008, 42(5): 940–948.
34. Ni Y W, Zhang H J, Fan S, et al. Emissions of PCDD/Fs from municipal solid waste incinerators in China. *Chemosphere*, 2009, 75: 1153–1158.
35. Chen T, Sun C, Wang T J, et al. Removal of PCDD/Fs and CBzs by different air pollution control devices in MSWIs. *Aerosol and Air Quality Research*, 2020, 20: 2260–2272.
36. Hung P C, Chang S H, Buekens A, Chang M B. Continuous sampling of MSWI dioxins. *Chemosphere*, 2016, 145: 119–124.
37. Nzihou A, Themelis N J, Kemiha M, Benhamou Y. Dioxin emissions from municipal solid waste incinerators (MSWIs) in France. *Waste Management*, 2012, 32: 2273–2277.
38. Zhang G, Hai J, Cheng J. Characterization and mass balance of dioxin from a large-scale municipal solid waste incinerator in China. *Waste Management*, 2012, 32: 1156–1162.
39. Liu W, Tian Z, Li H, et al. Mono- to octa-chlorinated pcdd/fs in stack gas from typical waste incinerators and their implications on emission. *Environmental Science & Technology*, 2013, 47(17): 9774–9780.
40. Ministry of Environmental Protection and State Administration of Quality Supervision. Standard for pollution control on the municipal solid waste incineration. China Environmental Science Press, 2014.
41. Hu B H, Huang Q X, Chi Y, Yan J H. Polychlorinated dibenzo-p-dioxins and dibenzofurans in a three-stage municipal solid waste gasifier. *Journal of Cleaner Production*, 2019, 218: 920–929.
42. Li Y C, Yang Y, Yu G, et al. Emission of unintentionally produced persistent organic pollutants (UPOPs) from municipal waste incinerators in China. *Chemosphere*, 2016, 158: 17–23.
43. Zhu F, Li X F, Lu J W, et al. Chengyang Hong Emission characteristics of PCDD/Fs in stack gas from municipal solid waste incineration plants in Northern China. *Chemosphere*, 2018, 200: 23–29.
44. Wei J X, Li H, Liu J G. Fate of dioxins in a municipal solid waste incinerator with state-of-the-art air pollution control devices in China. *Environmental Pollution*, 2021, 289: 117798.
45. Ren M H, Lv Z Y, Xu L, et al. Partitioning and removal behaviors of PCDD/Fs, PCBs and PCNs in a modern municipal solid waste incineration system. *The Science of the Total Environment*, 2020, 735: 139134.
46. Qian L X, Wang Y F, Liu M L, et al. Performance evaluation of urea injection on the emission reduction of dioxins and furans in a commercial municipal solid waste incinerator. *Process Safety and Environmental Protection*, 2021, 146: 577–585.
47. Ma C, Li B, Chen D M, et al. An investigation of an oxygen-enriched combustion of municipal solid waste on flue gas emission and combustion performance at a 8 MWth waste-to-energy plant. *Waste Management*, 2019, 96: 47–56.
48. Zhong R G, Cai J J, Yan F, et al. Process tracing and partitioning behaviors of PCDD/Fs in the post-combustion zone from a full-scale municipal solid waste incinerator in southern China. *Environmental Technology & Innovation*, 2021, 23: 101789.
49. Chen C K, Lin C, Lin Y C, et al. Polychlorinated dibenzo-p-dioxins/dibenzofuran mass distribution in both start-up and normal condition in the whole municipal solid waste incinerator. *Journal of Hazardous Materials*, 2008, 160(1): 37–44.

50. Zhang L T, Liu G R, Li S, et al. Model framework to quantify the effectiveness of garbage classification in reducing dioxin emissions. *Science of the Total Environment*, 2022, 814: 151941.
51. Zhang C C, Li X X, Zhou Z G. Spatial and temporal variation, source profile of PCDD/Fs in the atmosphere of a municipal waste incinerator in China. *Ecotoxicology and Environmental Safety*, 2019, 184: 109615.
52. Xiong S, Shang F, Chen K, et al. Stable and effective online monitoring and feedback control of PCDD/F during municipal waste incineration. *Molecules*, 2021, 26(14): 4290.
53. Ma Y F, Wang P Y, Lin X Q, et al. Formation and inhibition of Polychlorinated-ρ-dibenzodioxins and dibenzofurans from mechanical grate municipal solid waste incineration systems. *Journal of Hazardous Materials*, 2021, 403: 123812.
54. Xia H, Tang J, Aljerf L. Dioxin emission prediction based on improved deep forest regression for municipal solid waste incineration process. *Chemosphere*, 2022, 294: 133716.
55. Trivedi J, Majumdar D. Memory effect driven emissions of persistent organic pollutants from industrial thermal processes, their implications and management: a review. *Journal of Environmental Management*, 2016, 119: 111–120.
56. Zhang M M, Alfons G B. De novo synthesis of dioxins: a review *International Journal of Environment and Pollution*, 2016, 60(1–4): 63–110.
57. Altarawneh M K, Dlugogorski B Z, Kennedy E M, Mackie J C. Mechanisms for formation, chlorination, dechlorination and destruction of polychlorinated dibenzo-p-dioxins and dibenzofurans (PCDD/Fs) *Progress in Energy and Combustion Science*, 2009, 35: 245–274.
58. Li H W, Wang L C, Chen C C, et al. Influence of memory effect caused by aged bag filters on the stack PCDD/F emissions. *Journal of Hazardous Materials*, 2011, 185(2–3): 1148–1155.

Part II

Dioxins with Numerical Simulation Modeling

5 Non-Manipulated Variable Modeling of Solid-Phase Combustion in the Furnace

5.1 RELATED WORKS

5.1.1 NON-MANIPULATED VARIABLE MODELING DESCRIPTION

The quantity of MSW has surged rapidly due to economic development and rising living standards [1]. According to statistics, the global annual growth rate of MSW has reached 8% [2], with projections suggesting it could reach 9.5 billion tons by 2050 [3]. This trend poses significant challenges for the sustainable development of urban areas and the surrounding environment. Currently, the primary methods for treating MSW include landfilling, composting, and incineration. MSWI is a well-established waste-to-energy technology that offers benefits such as waste reduction, resource recovery, and environmental safety [4, 5]. MSWI has been widely adopted worldwide, particularly in developing countries like China and India.

The MSWI process simulation encompasses both experimental methods and mathematical modeling techniques, with numerical simulation falling under the latter category. This method serves as a flexible tool capable of simulating various actual operating conditions. Shin and Goh modeled MSW combustion on a grate as a one-dimensional fixed bed [6, 7], treating it as a continuous porous medium to establish a mathematical framework. Building on prior research, the team from Sheffield University developed a solid-phase combustion model for grates, known as FLIC [8]. This model effectively simulates the MSWI process and predicts the distribution of velocity, pressure, temperature, and composition. Subsequently, FLIC has been utilized to create models for various incineration plants. Yang and Yu developed a straw-burning model to analyze combustion conditions under an oxygen-enriched atmosphere [9, 10], providing a theoretical foundation for structural design and optimized operational conditions. Additionally, Ryu, Goddard, and Ismail constructed a full-size FLIC model of an MSWI plant to predict gas temperature and composition distribution [11–13], demonstrating good agreement between predicted values and experimental data.

Parameter analysis and optimization have been conducted using the FLIC incineration model. Yang et al. evaluated the impacts of factors such as fuel moisture, fuel size, particle mixing level, volatile matter release rate, primary airflow, and feeding on combustion performance [14–17]. Yan et al. investigated how primary air

DOI: 10.1201/9781003634195-7

temperature influences MSW combustion [18, 19], finding that an increase in temperature enhances the release rate of volatile matter, which subsequently raises the local maximum temperature in the furnace. This provides a theoretical basis for maintaining local high temperatures within a reasonable range. Research by Alima et al. [20] indicated that a low degree of mixing between MSW and air results in insufficient combustion. They proposed increasing the stacking height of MSW as a potential solution. Liang et al. examined the effects of grate speed and oxygen concentration on pollutant emissions [21], noting that NO emissions decrease while CO emissions increase with higher grate speeds. Hu et al. identified the optimal method for controlling NO_x emissions as a combination of operating at 100% load with a 31% excess air coefficient [22]. Lin et al. coupled the FLIC model with computational fluid dynamics (CFD) to investigate the co-disposal of sludges and MSWs [23], determining that an optimal maximum sludge-mixing rate of 13% is suitable for the incinerator. Chen et al. segmented the temperature field within the furnace [24], identifying feeding, grate speed, and primary air as key manipulated variables affecting temperature distribution. Tang et al. utilized FLIC in conjunction with other software to analyze the impacts of grate speed and air ratio across different stages of the MSWI process [25], aiming to determine optimal operating parameters.

The aforementioned research indicates that the FLIC model is widely utilized in MSWI combustion modeling. However, there is a scarcity of studies focusing on the analysis of non-manipulated variable parameters [15, 17]. Given the variations in incinerator structure and industrial parameters, it is essential to create models specific to each actual incinerator. Additionally, the effects of different operating conditions and non-manipulated variable parameters on combustion performance should be examined in the context of robust control aimed at minimizing pollution emissions.

5.1.2 Description of Solid-Phase Combustion in the Furnace

The MSWI process is divided into six stages: solid waste storage and transportation, solid waste combustion, heat recovery, steam generation for electricity production, flue gas cleaning, and flue gas emission. The process details are as follows.

MSWs are collected and transported by vehicles to the MSWI plant, where they undergo fermentation and dehydration to achieve a low calorific value. The MSWs are then hoisted into the feed hopper of the incinerator and pushed into the combustion chamber by a feeder, progressing through drying, combustion stages 1 and 2, and burnout on the grate.

The flue gas generated during combustion is drawn into the waste heat recovery system by an induced draft fan, where it exchanges heat with liquid water in the boiler drum to produce high-temperature steam. The flue gas exiting the boiler subsequently flows into a reactor and a bag filter, where acidic gases, particulates, and adsorbates from activated carbon are removed. Finally, the induced draft fan discharges the treated flue gas from the stack into the atmosphere. The tail gas may contain pollutants such as HCl, SO_2, NO_x, dioxins, and other substances.

Solid-state combustion in the furnace pertains to the combustion of MSW on the grate. In this process, MSWs are modeled as spherical particles that undergo several stages, ultimately transforming into ash. These stages include drying, the release of volatile matter, combustion, and the oxidation of char [26]. Figure 5.1 illustrates the combustion process of MSW particles.

Figure 5.1 illustrates that the drying process is primarily characterized by moisture evaporation during the first stage, which occurs due to radiation heating from the furnace wall and convection heat transfer from the primary air. The second stage involves the release of volatile matter, taking place in the front section of the burning 1 grate. During this stage, the organic components of the MSWs decompose and are emitted as gases, such as C_mH_n and CO. The volatile matter then reacts vigorously with O_2, while the remaining MSWs gradually convert into char in the third stage, which occurs in the rear sections of the burning 1 and 2 grates. This area serves as the main combustion zone. The fourth stage transpires in the burnout grate, where the remaining char continues to react with O_2, resulting in the oxidation of carbon and the production of CO and CO_2. As the char is consumed, the flame extinguishes, leaving only ash behind.

5.1.3 Non-Manipulated Variable Analysis of Influencing the MSWI Process

Manipulated variables refer to the controllable factors within the operational MSWI power plant, including feedstock, grate speed adjustments, and primary air regulation. In contrast, non-manipulated variables cannot be directly controlled by experts through the distributed control system during the operational process. Some key non-manipulated variables that influence solid-phase combustion include the following.

(1) When MSWs are assumed to be spherical particles, larger particles can hinder heat transfer radiation, resulting in lower thermal efficiency [27]. This

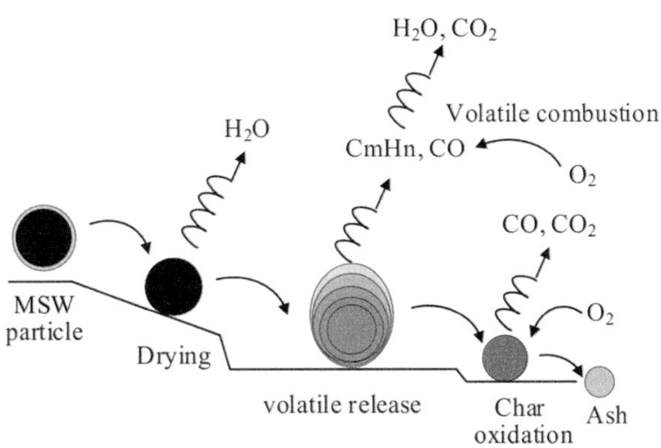

FIGURE 5.1 Combustion process and changes of MSW particles.

leads to the formation of more voids, which increases the amount of air surrounding the MSW particles. As a result, gas flow becomes more efficient, enhancing turbulence and combustion speed.

(2) Particle mixing refers to the vibration of MSW particles on the bed, allowing them to exchange positions with neighboring particles and change their shapes. The degree of particle mixing is influenced by grate movement, which significantly affects the combustion process. Figure 5.2 illustrates the relationship between flue gas oxygen content levels and temperature fluctuations measured at the outlet of the furnace, alongside grate movement [28].

(3) The radiation source within the furnace primarily originates from the furnace wall, with emissivity indicating the wall's ability to radiate energy.

(4) Moisture content predominantly impacts the moisture evaporation process during the initial stage of combustion. MSWs with high water content experience prolonged drying periods, which can weaken the combustion process and result in lower temperatures [29].

(5) MSWs are rich in carbon (C), hydrogen (H), oxygen (O), and nitrogen (N). Theoretically, a higher carbon content correlates with a greater calorific value and improved combustion performance.

In summary, particle size, mixing coefficient, emissivity, moisture content, and the carbon-to-oxygen (C:O) ratio are key factors influencing solid-phase combustion.

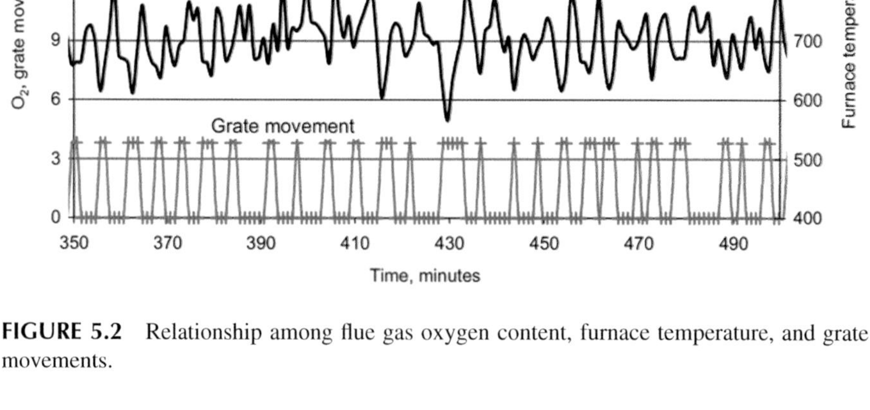

FIGURE 5.2 Relationship among flue gas oxygen content, furnace temperature, and grate movements.

5.2 MODELING STRATEGY

The proposed numerical simulation strategy for non-manipulated variable in terms of solid-phase combustion in the furnace is shown in Figure 5.3.

Figure 5.3 illustrates the strategy proposed in this chapter, which is divided into four modules, each with specific functions:

(1) Numerical Simulation Model Construction for Benchmark Operating Conditions: This module develops a solid-state combustion model within the furnace, comprising four sub-models: water evaporation, volatile matter release, volatile matter combustion, and coke oxidation.
(2) Numerical Simulation Experiment for Biorthogonal Multi-Operating Conditions: This module utilizes a dual orthogonal experimental design to generate virtual simulation data for multiple operating conditions through two orthogonal experiments and simulations.

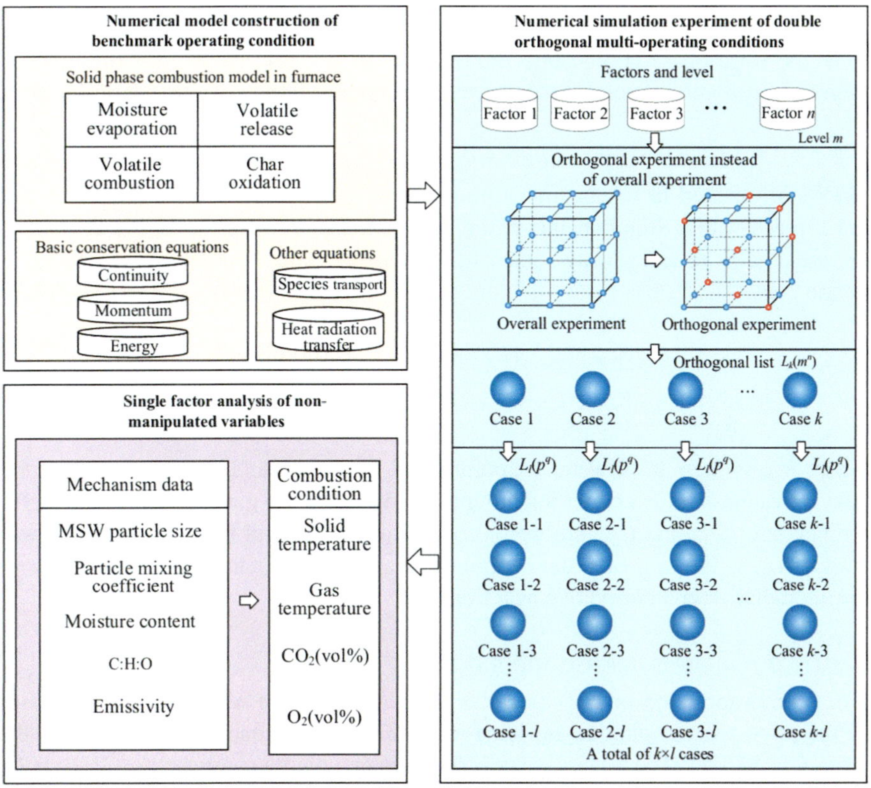

FIGURE 5.3 Numerical simulation strategy for non-manipulated variable in terms of solid-phase combustion in the furnace.

(3) Single-Factor Analysis of Non-Manipulated Variables: This module examines the impact of non-manipulated variables on MSW combustion based on the virtual simulation data, allowing for the identification of a substantial number of combustion conditions.

5.3 MODELING IMPLEMENTATION

5.3.1 Numerical Simulation Model Construction of the Benchmark Operating Condition

The FLIC software was utilized to establish a numerical model representing the benchmark operating conditions of an actual MSWI power plant in Beijing. The model for solid-phase combustion, along with the fundamental conservation equations and other relevant equations, is presented below.

5.3.1.1 Solid-Phase Combustion Model in the Furnace

Figure 5.2 illustrates the numerical model of the MSW combustion process, referred to as the solid-phase combustion model. This model consists of four sub-models: moisture evaporation, volatile release, volatile combustion, and char oxidation [8].

5.3.1.1.1 Moisture Evaporation Sub-model

MSWs are heated to release moisture as a result of radiation from the flame and the furnace walls. Simultaneously, the convection of primary air at the bottom of the grate helps to carry away some of the moisture. The evaporation rate [8, 30] is defined by

$$\begin{aligned} R_{\text{evp}} &= S_a h_s \left(C_{w,s} - C_{w,g} \right), T_s < 100^{\circ}\text{C} \\ R_{\text{evp}} &= Q_{\text{cr}} / H_{\text{evp}}, T_s = 100^{\circ}\text{C} \end{aligned} \tag{5.1}$$

where R_{evp} is the rate of water evaporation; S_a is the particle surface area; h_s is the convective mass transfer coefficient with MSWs when the gas flows; T_s is the MSW temperature; $C_{w,s}$ and $C_{w,g}$ are water content in MSWs and the mixed gas, respectively; H_{evp} is the heat of water evaporation; and Q_{cr} is the heat that MSWs absorb during radiative and convective heat transfer.

5.3.1.1.2 Volatiles Release Sub-model

After the evaporation of moisture, the temperature of the MSW rapidly increases to the point where volatiles are released, while char is generated slowly from the remaining MSWs. The release process of the main volatile components (C_mH_n, H_2O, CO_2, and CO) [8] in MSWs is defined by

$$\text{MSW} \rightarrow \text{Volatile}\left(\text{C}_m\text{H}_n, \text{CO}, \text{CO}_2, \text{H}_2\text{O}\right) + \text{Char} \tag{5.2}$$

The one-step global reaction model [31] is employed due to its simplicity and accuracy. The reaction rate is directly proportional to both the temperature and the residual volatile content, as expressed by the following equation:

$$R_V = \rho_{sb}\left(v_{\infty} - v\right) A_V \exp\left(-\frac{E_V}{RT_s}\right) \quad (5.3)$$

where R_v is the release rate of volatiles; v_{∞} is the final production of the gas released; v is the amount of the gas released at time t; R is the ideal gas constant; A_v is the pre-exponential factor in devolatilization; and E_v is the active state energy of the chemical reaction.

5.3.1.1.3 Volatiles Combustion Sub-model

The gases released in the previous stage combust immediately upon mixing with air. The combustion reactions of C_mH_n and CO are defined by the following equations:

$$C_mH_n + \left(m/2 + n/4\right)O_2 \rightarrow mCO + n/2H_2O \quad (5.4)$$

$$CO + 1/2O_2 \rightarrow CO_2 \quad (5.5)$$

The kinetic rates for the combustion of C_mH_n and CO [32] are expressed as follows:

$$R_{C_mH_n} = 59.8T_gP^{0.3}C^{0.5}_{C_mH_n}C_{O_2}\exp\left(-12{,}200/T_g\right) \quad (5.6)$$

$$R_{CO} = 1.3\times10^{11}C_{CO}C^{0.5}_{H_2O}C^{0.5}_{O_2}\exp\left(-62{,}700/T_g\right) \quad (5.7)$$

where T_g represents the MSW temperature.

The eddy-break-up concept posits that the mixing rate of volatiles with air is proportional to the pressure drop across the MSW bed. Accordingly, the mixing rate, based on the Ergun equation [13], is defined as follows:

$$R_{mix} = C_{mix}\rho_g\left\{150\frac{D_g\left(1-\phi\right)^{2/3}}{d_p^2\phi} + 1.75\frac{V_g\left(1-\phi\right)^{1/3}}{d_p\phi}\right\}\min\left\{\frac{C_{fuel}}{S_{fuel}}, \frac{C_{O_2}}{S_{O_2}}\right\} \quad (5.8)$$

where C_{mix} is the empirical constant; ρ_g is the gas density; C is the mass fraction of the participating mixtures; D_g is the mass diffusion coefficient of the gas; d_p is the MSW particle size; ϕ is the bed void; and S is the coefficient of the combustion reaction.

The combustion rate of volatile matter is determined by taking the minimum of the kinetic rate and the mixing rate, as expressed by the following equation:

$$R = \min\left(R_{kinetics}, R_{mix}\right) \quad (5.9)$$

5.3.1.1.4 Chars Oxidation Sub-model

MSWs gradually form char as volatiles are released. The carbon in the char subsequently reacts with air, producing CO and CO_2 based on the available oxygen (O_2) during combustion [33], as expressed by the following equation,

$$C(s)+\alpha O_2 \rightarrow 2(1-\alpha)CO+(2\alpha-1)CO_2 \tag{5.10}$$

where α represents the stoichiometric coefficient, which ranges from 0.5 to 1 according to Arthur's law [33].

The char's oxidation rate [30] is defined by

$$R_C = \frac{P_{O_2}}{\frac{1}{k_r}+\frac{1}{k_d}} \tag{5.11}$$

where R_C is the oxidation rate of chars; P_{O_2} is the pressure at which O_2 is divided during combustion; k_r is the rate of chemical kinetics; and k_d is the diffusion rate.

5.3.1.2 Basic Conservation Equation

There is an exchange of reaction heat and mass between MSWs and gases within the bed, considering the effects of porous media. Therefore, it is essential to establish conservation equations for continuity, momentum, and energy in the MSW bed.

The mass continuity equations for the gas and solid phases [32] are defined as follows:

$$\frac{\partial(\rho_g\phi)}{\partial t}+\nabla\cdot\left(\rho_g\phi\left(V_g-V_B\right)\right)=S_{sg} \tag{5.12}$$

$$\frac{\partial\rho_{sb}}{\partial t}+\nabla\cdot\left(\rho_{sb}\left(V_s-V_B\right)\right)=S_s \tag{5.13}$$

where V_g and V_s are the gas velocity and MSW particle velocity, respectively; V_B is the speed of moving the boundary; S_{sg} is the rate at which MSWs are converted into gases due to water evaporation, devolatilization, and char oxidation; and S_s is the source term of MSW particle mass.

The accumulation of MSW creates numerous irregular and narrow gaps, resulting in the absence of large-scale turbulence within the MSW bed. The momentum equations for the gas and solid phases [34] are defined as follows:

$$\frac{\partial(\rho_g\phi V_g)}{\partial t}+\nabla\cdot\left(\rho_g\phi\left(V_g-V_B\right)V_g\right)=-\nabla p_g+F(v) \tag{5.14}$$

$$\frac{\partial(\rho_{sb}V_s)}{\partial t}+\nabla\cdot\left(\rho_{sb}\left(V_s-V_B\right)V_s\right)=-\nabla\cdot\sigma-\nabla\cdot\tau+\rho_{sb}g+A \tag{5.15}$$

where p_g is the pressure of the mixed gas in the MSW bed; $F(v)$ is the resistance of MSWs to gas flow in the gap; σ and τ are the normal stress tensor and tangential stress tensor, respectively; and A is the random motion of MSW particles.

The energy equations for the gas and solid phases [8] are defined as follows:

$$\frac{\partial\left(\rho_g \phi H_g\right)}{\partial}+\nabla\cdot\left(\rho_g \phi\left(V_g-V_B\right)H_g\right)=\nabla\cdot\left(\lambda_g \nabla T_g\right)+S_a h'_s\left(T_s-T_g\right)+Q_h \quad (5.16)$$

$$\frac{\partial\left(\rho_{\text{sb}} H_s\right)}{\partial t}+\nabla\cdot\left(\rho_{\text{sb}}\left(V_s-V_B\right)H_s\right)=\nabla\cdot\left(\lambda_s \nabla T_s\right)-S_a h'_s\left(T_s-T_g\right)+\nabla\cdot q_r+Q_{\text{sh}} \quad (5.17)$$

where H_g is gas enthalpy; H_s is solid-phase enthalpy; Q_h is the heat gain or loss of the gas; Q_{sh} is the source term of the MSW particles' heat; q_r is the radiant heat flux; λ_g is the heat dispersion coefficient consisting of turbulence and diffusion; and λ_s indicates the thermal conductivity of the MSW bed.

5.3.1.3 Species Transport and Heat Radiation Transfer Equations

After establishing the conservation equations, the species transport equations [8] used to simulate the interactions within the mixture are presented as follows:

$$\frac{\partial\left(\rho_g Y_{i,g}\phi\right)}{\partial t}+\nabla\cdot\left(\rho_g \phi\left(V_g-V_B\right)Y_{i,g}\right)=\nabla\cdot\left(D_{i,g}\nabla\left(\rho_g Y_{i,g}\right)\right)+S_{Y_{i,g}} \quad (5.18)$$

$$\frac{\partial\left(\rho_{\text{sb}} Y_{i,s}\right)}{\partial t}+\nabla\bullet\left(\rho_{\text{sb}}\left(V_s-V_{\text{B}}\right)Y_{i,s}\right)=\nabla\bullet\left(D_s\nabla\left(\rho_{\text{sb}} Y_{i,s}\right)\right)+S_{Y_{i,s}} \quad (5.19)$$

where $Y_{i,g}$ is the mass fraction of a single component (C_mH_n, CO, CO_2, H_2, and H_2O); $S_{Y_{i,g}}$ is the source term that represents the mass source of volatile gases and chars during evaporation, devolatilization, and combustion; $Y_{i,s}$ is the mass fraction of MSW particle compositions; $S_{Y_{i,s}}$ is the mass source of MSW particle compositions; $D_{i,g}$ is the fluid dispersion coefficient composed of diffusion and turbulence contributions; and D_s is the mixing coefficient of particles within the bed, which depends on the frequency of the grate's motion.

The radiation source in the furnace primarily originates from the walls. Radiant heat transfer facilitates the transport of heat generated by the combustion reactions to the top layer of the bed. This energy is utilized for moisture evaporation, MSW pyrolysis, and other endothermic solid-phase reactions occurring in the upper layer [35]. Typically, this process is described using a four-way radiation model [36].

$$\frac{dI_x^+}{dx}=-\left(k_a+k_s\right)I_x^+ +\frac{1}{4}K_a E_b+\frac{1}{4}k_s\left(I_x^+ +I_x^- +I_z^+ +I_z^-\right) \quad (5.20)$$

$$-\frac{dI_x^-}{dx}=-\left(k_a+k_s\right)I_x^- +\frac{1}{4}K_a E_b+\frac{1}{4}k_s\left(I_x^+ +I_x^- +I_z^+ +I_z^-\right) \quad (5.21)$$

$$\frac{dI_z^+}{dz} = -(k_a + k_s)I_z^+ + \frac{1}{4}K_a E_b + \frac{1}{4}k_s\left(I_x^+ + I_x^- + I_z^+ + I_z^-\right) \tag{5.22}$$

$$-\frac{dI_z^-}{dz} = -(k_a + k_s)I_z^- + \frac{1}{4}K_a E_b + \frac{1}{4}k_s\left(I_x^+ + I_x^- + I_z^+ + I_z^-\right) \tag{5.23}$$

where I_x^+, I_x^-, I_z^+ and I_z^- are the radiation intensities in four directions, respectively; k_a and k_s are the coefficients of absorption and scattering for the radiation process, respectively; E_b is the radiation of the blackbody; k_s is approximately zero; and k_a [16] is defined by

$$k_a = -\frac{1}{d_p}\ln(\phi) \tag{5.24}$$

5.3.2 Numerical Simulation Experiment of Biorthogonal Multi-Operating Conditions

The design process for the orthogonal experiment with double coupling is as follows:

(1) Determine Experimental Factors: Identify the number of experimental factors n and their respective levels m. An orthogonal experiment table $L_k(m^n)$ can be created by substituting the comprehensive experiment with the orthogonal experiment. This table allows for the generation of k experimental cases.
(2) Redefine Experimental Factors: Based on the k cases from the first orthogonal experiment, re-determine their experimental factors p and level numbers q to obtain k orthogonal experimental tables $(L_l(p^q))$.
(3) Obtain Final Experimental Cases: Finally, $k \times l$ experimental cases can be derived from the orthogonal tables.

The orthogonal experimental method of double coupling is utilized to design multiple operating conditions for solid-phase combustion. Tables 5.1 and 5.2 present the orthogonal experimental settings for the first and second stages, respectively.

Table 5.1 indicates that ten parameters have been selected as factors for the biorthogonal experiment. The feeding rate, grate speed, and primary air flow reflect variations in operating conditions. Meanwhile, the moisture content and the mass ratio of carbon to oxygen illustrate fluctuations in the composition of MSWs within a certain range. Additionally, the MSW particle size and mixing coefficient represent the accumulation and movement of particles during grate motion. The parameter ranges are determined based on experience, with five levels set at equal intervals within these ranges.

The factors and levels outlined in Table 5.1 are designated for the first-stage biorthogonal experiment, as shown in Figure 5.3, where $m = 10$ and $n = 5$. An orthogonal design tool is employed to generate an orthogonal list, which automatically matches

TABLE 5.1
Biorthogonal Experiment Settings at the First Stage

	Factor	Level Value
Manipulated variables	Feeding (m/h)	7, 7.5, 8, 8.5, and 9
	Grate speed (t/h)	24.2, 24.7, 25.2, 25.7, and 26.2
	Primary air in drying (Nm^3/min)	268, 274, 280, 286, and 292
	Primary air in burning 1 (Nm^3/min)	477, 488, 499, 510, and 521
	Primary air in burning 2 (Nm^3/min)	211, 216, 221, 226, and 231
	Primary air in burnout (Nm^3/min)	144, 147, 150, 153, and 156
Non-manipulated variables	Particle size (mm)	15, 20, 25, 30, and 35
	Particle mixing coefficient	2E−6, 3E−6, 4E−6, 5E−6, and 6E−6
	Moisture content (%)	48, 49.75, 51.5, 53.25, and 55
	C:O	(58:33), (59:32), (60:31), (61:30), and (62:29)

TABLE 5.2
Biorthogonal Experiment Settings at the Second Stage

Factor	Level Value	Factor	Level Value
Grate speed (m/h)	6.6, 6.7, 6.8, 6.9, and 7	Feeding (tons/h)	24.2, 24.3, 24.4, 24.5, and 24.6
	7.1, 7.2, 7.3, 7.4, and 7.5		24.7, 27.8, 24.9, 25, and 25.1
	7.6, 7.7,7.8, 7.9, and 8		25.2, 25.3, 25.4, 25.5, and 25.6
	8.1, 8.2, 8.3, 8.4, and 8.5		25.7, 25.8, 25.9, 26, and 26.1
	8.6, 8.7, 8.8, 8.9, and 9		26.2, 26.3, 26.4, 26.5, and 26.6
Primary air in drying (Nm^3/min)	255, 256.8, 258.6, 260.4, and 262.2	Primary air in burning 1 (Nm^3/min)	455, 458.2, 461.4, 464.6, and 467.8
	264, 265.8, 267.6, 269.4, and 271.2		471, 474.2, 477.4, 480.6, and 483.8
	273, 274.8, 276.6, 278.4, and 280.2		487, 490.2, 493.4, 496.6, and 499.8
	282, 283.8, 285.6, 287.4, and 289.2		503, 506.2, 509, 513, and 515.8
	291, 292.8, 294.6, 296.4, and 298.2		519, 522.2, 525.4, 528.6, and 531.8
Primary air in burning 2 (Nm^3/min)	203, 204.4, 205.8, 207.2, and 208.6	Primary air in burnout (Nm^3/mi)n	137, 138, 139, 140, and 141
	210, 211.4, 212.8, 214.2, and 215.6		142, 143, 144, 145, and 146
	217, 218.4, 219.8, 221.2, and 222.6		147, 148, 149, 150, and 151
	224, 225.4, 227, 228, and 229.6		152, 153, 154, 155, and 156
	231, 232.4, 233.8, 235.2, and 236.6		157, 158, 159, 160, and 161

the most consistent orthogonal configuration, resulting in $\left(L_{49}\left(5^{10}\right)\right)$; therefore, $k = 49$.

Similarly, the factors and levels presented in Table 5.2 are utilized for the second-stage biorthogonal experiment, where $p = 6$ and $q = 5$. The biorthogonal experiment list $L_{42}\left(5^{6}\right)$ is obtained, and $l = 42$ is designed using the orthogonal design tool.

5.3.3 Single-Factor Analysis of Non-Manipulated Variables

This study investigated the influences of non-manipulated parameters on solid-phase combustion, including MSW particle size, particle mixing coefficient, moisture content, the carbon-to-oxygen ratio, and emissivity. Mechanistic data have been obtained through biorthogonal numerical simulation experiments. Consequently, the analysis of the first four factors can be directly informed by this mechanistic data. Each factor is categorized into five levels, and an operating condition is selected from each group based on the principle of similar operating conditions.

The combustion conditions are characterized by furnace temperature and flue gas concentration. However, the biorthogonal experiment does not account for emissivity. Thus, additional simulations are required for emissivity analysis. Specifically, five emissivity values are assigned for these further simulations. The single-factor analysis conducted in this chapter reflects the influences of non-manipulated variables on solid-phase combustion under complex and variable operating conditions.

5.4 SIMULATION RESULTS AND DISCUSSION

5.4.1 Data Description

Based on the actual design of the MSWI process, the incineration operation is considered stable when the furnace temperature is maintained above 850°C under a given set of operating parameters (e.g., feed rate, grate speed, and primary air flow). This operating condition is referred to as the benchmark operating condition for the studied MSWI power plant in Beijing. Tables 5.3 and 5.4 present the properties of the MSW and the benchmark operating conditions of the MSWI power plant in Beijing.

TABLE 5.3
MSW Properties of the MSWI Power Plant in Beijing

Proximate (As a Received Basis, wt%)		Ultimate (As a Dry Ash-Free Basis, wt%)	
Moisture	48	C	65.2
Volatile matter	33.31	H	8.09
Fixed carbon	8.08	O	24.93
Ash	10.61	N	1.12
		S	0.24

TABLE 5.4
Specifications and Operating Parameters of the MSWI Power Plant in Beijing

Parameter	Value
Capacity (tons/d)	628.8
Type of grate	Reciprocating grate
Grate length × width (m)	11 × 12.9
Grate speed (m/h)	8
Primary airflow (Nm^3/h)	67,500
Air temperature (°C)	200
Air distribution (%)	24.31, 43.35, 19.27, and 13.07

The MSW capacity is 628.8 tons/day, with a grate length of 11 m and a width of 12.9 m. The primary airflow is set at 67,500 Nm^3/h, with an air temperature of 200°C. Primary air enters the MSW bed from four independent sections below the grate, with the flow rates for each section accounting for 24.31%, 43.35%, 19.27%, and 13.07% of the total primary air, respectively.

5.4.2 Simulation Results and Discussion

5.4.2.1 Benchmark Operating Condition

Figure 5.4 illustrates the simulation results for the benchmark operating conditions of the MSWI power plant in Beijing.

Figure 5.4 indicates the following observations:

(1) Radiated energy from the flame and walls is utilized to heat the top layer of the MSW bed within the drying grate, resulting in significant moisture evaporation. Water vapor becomes the predominant gas at the top of the MSW bed, with a temperature of 373 K. Concurrently, primary air passes through the grate and enters the gaps in the MSW, heating the bottom of the bed.

(2) The bed temperature reaches 533 K (the initial temperature) at a depth of 3.3 m, signifying the onset of volatile matter release and the subsequent ignition of the MSWs. The solid temperature rises rapidly, leading to an instantaneous increase in the release rate of volatile matter to 1,046 $kg/m^2 \cdot h$.

(3) The center of the grate serves as the primary combustion area, where oxygen (O_2) is consumed in large quantities. The peak solid temperature reaches 1,360 K, while the highest gas temperature soars to 1,597 K.

(4) The release of volatile substances concludes at the burnout stage, and the combustion rate of char increases rapidly. Following the combustion of the char, the solid temperature gradually decreases to 550 K, while the O_2 content increases significantly, reaching 23% at approximately 9 m.

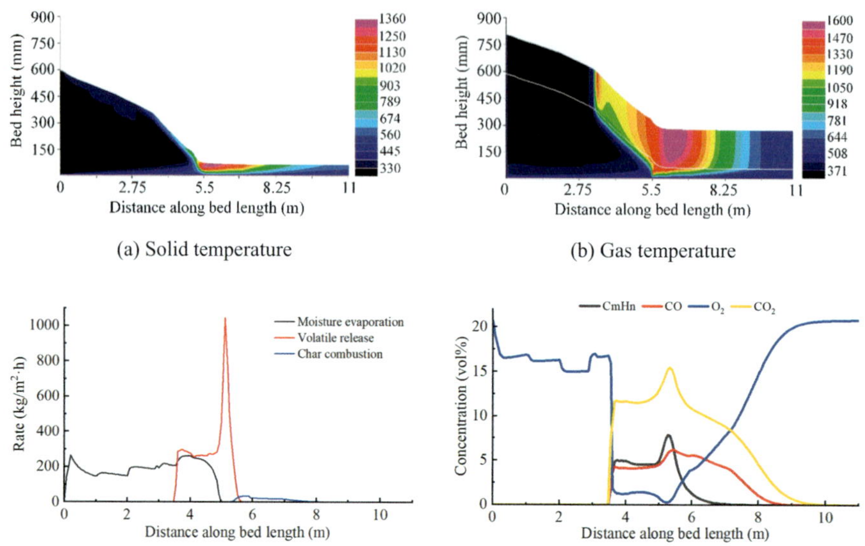

(a) Solid temperature

(b) Gas temperature

(c) Rate

(d) Gas concentration

FIGURE 5.4 Benchmark operating condition. (a) Solid temperature. (b) Gas temperature. (c) Rate. (d) Gas concentration.

5.4.2.2 Biorthogonal Multi-Conditions

The multi-operating conditions designed using the biorthogonal experiment method have been subjected to numerical simulations, as shown in Figures 5.5 and 5.6.

Figures 5.5 and 5.6 reveal some spikes in the raw data, which correspond to abnormal values. After removing these outliers, the maximum solid temperature ranges from 842 to 1,467 K, while the maximum flue gas temperature varies from 1,177 to 1,548 K. The concentrations of C_mH_n, CO, CO_2, and O_2 are found to be within the ranges of 0.603%–1.989%, 1.61%–3.054%, 3.559%–6.911%, and 7.671%–13.13%, respectively. The mechanistic data encompass the incineration scenarios under various operating conditions derived from the biorthogonal experiment method.

Temperature serves as an indicator for assessing the quality of combustion conditions. Five operating conditions, both above and below the reference temperature benchmark, have been selected from the mechanistic data for further analysis. Table 5.5 lists their corresponding parameter settings, while Figures 5.7–5.11 depict the solid temperature and gas concentration.

Figures 5.7–5.11 illustrate the following observations:

(1) In Condition 1, the feed rate is set at 24.3 tons/h, with a grate speed of 7 m/h and primary air flow at 1,069.6 Nm^3/min. These values represent the minimum among the five conditions studied. The slow grate speed facilitates more complete combustion of the MSW, resulting in higher concentrations of C_mH_n, CO, and CO_2 compared to the operating condition benchmark. The solid temperature is also the highest among all conditions, reaching 1,466 K.

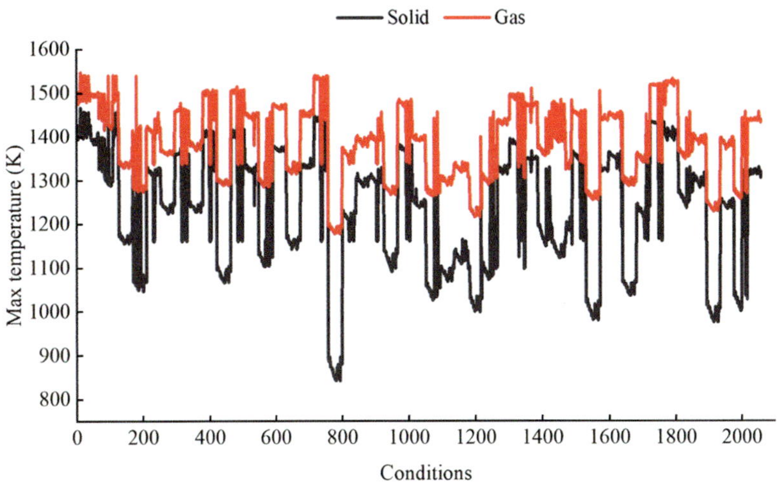

(a) Raw data

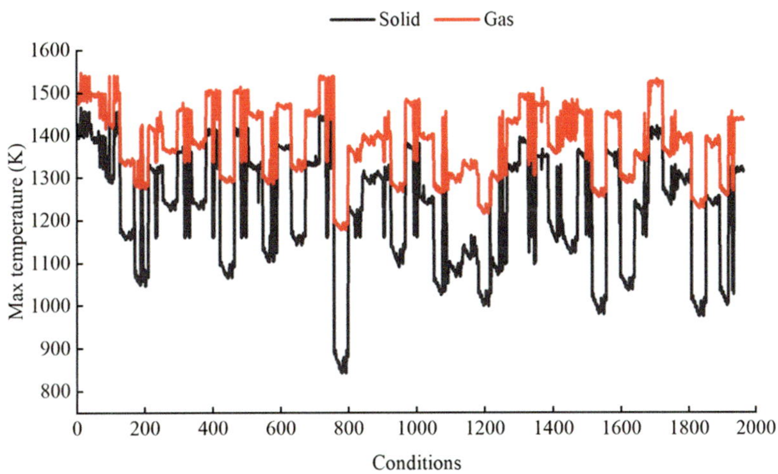

(b) Data without abnormal values

FIGURE 5.5 Maximum temperature under multi-operating conditions. (a) Raw data. (b) Data without abnormal values.

(2) Condition 2 features a grat speed of 7.9 m/h, which is close to the benchmark condition. By reducing the feed rate and increasing the primary air, the combustion of MSWs can still be achieved effectively. Specifically, the reduced feed rate decreases bed thickness, while the appropriate increase in primary air enhances combustion, leading to a solid temperature that exceeds the benchmark condition.

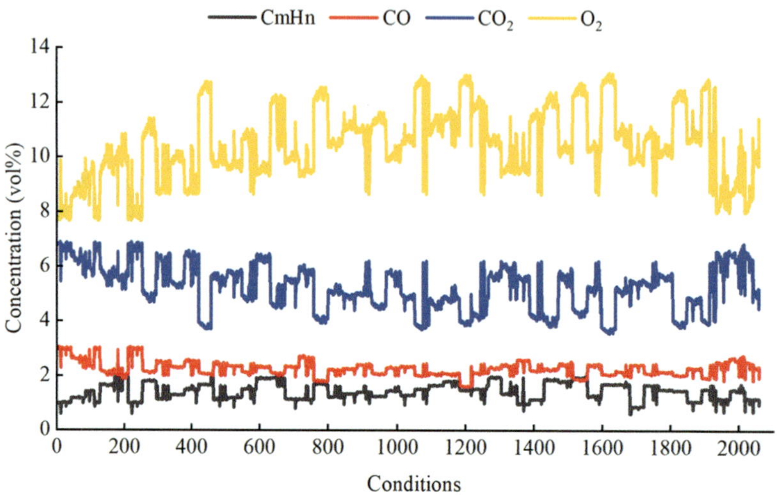

(a) Raw data

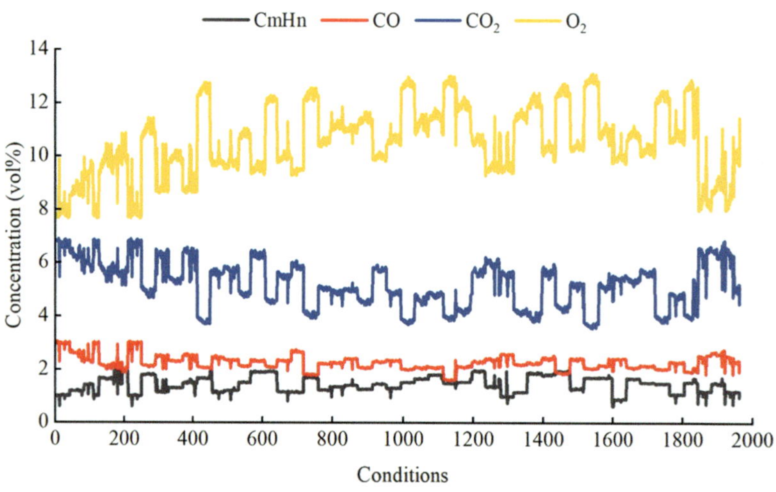

(b) Data without abnormal values

FIGURE 5.6 Gas concentration under multi-operating conditions. (a) Raw data. (b) Data without abnormal values.

(3) In Condition 5, the maximum solid temperature is only 1,120 K. The increased grate speed directly contributes to the incomplete combustion of the MSWs.

(4) Compared to Condition 1, Condition 3 exhibits an increase in both grate speed and primary air, resulting in a 160 K decrease in solid temperature. This observation highlights that the grate speed and primary air volume

TABLE 5.5
Multi-Conditions Parameters

	Feeding (tons/d)	Grate (m/h)	Primary Air (Nm³/min)
Benchmark condition	26.2	8	1,125
Condition 1	24.3	7	1,069.6
Condition 2	25	7.9	1,140.8
Condition 3	24.3	7.4	1,139.8
Condition 4	24.9	7.1	1,126.4
Condition 5	25.2	9	1,144

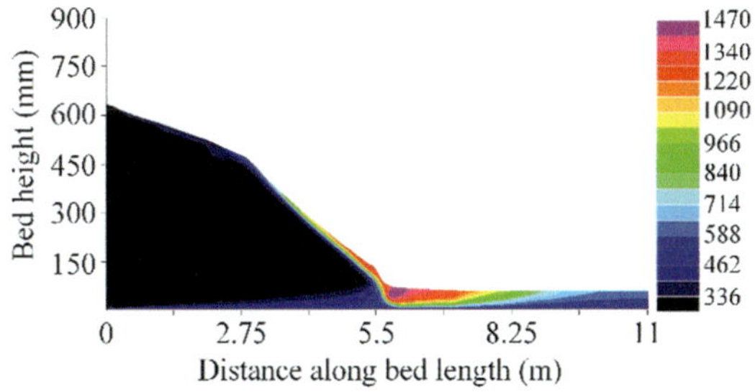

(a) Solid temperature

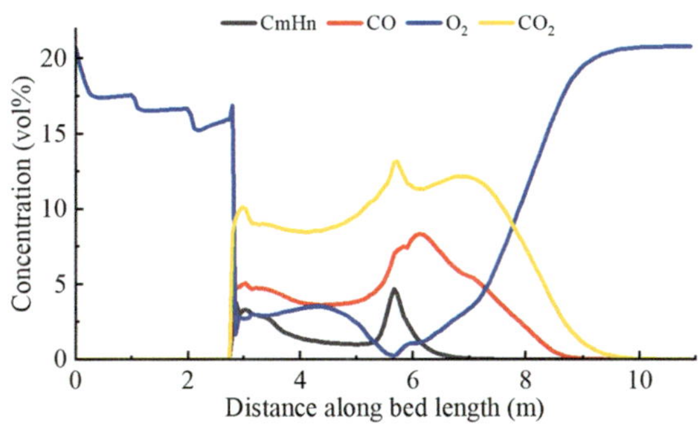

(b) Gas concentration

FIGURE 5.7 Condition 1 for solid-phase combustion stage. (a) Solid temperature. (b) Gas concentration.

cannot be increased simultaneously without compromising combustion efficiency.

(5) When comparing Condition 4–Condition 3, the solid temperature decreases by 60 K, and more unburned C_mH_n is observed in Condition 4. This indicates that increasing the ratio of feed to primary air is not conducive to optimal combustion.

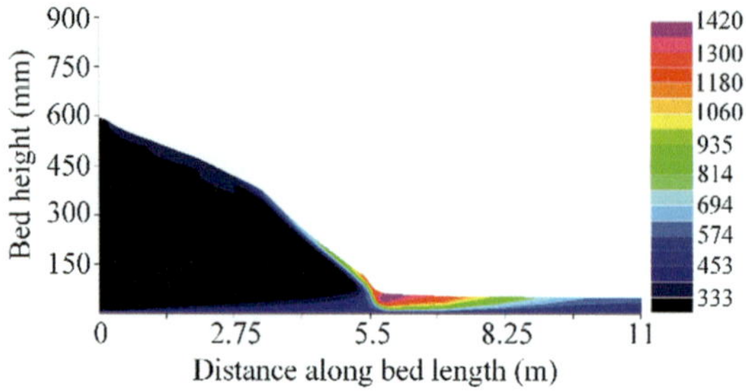

(a) Solid temperature

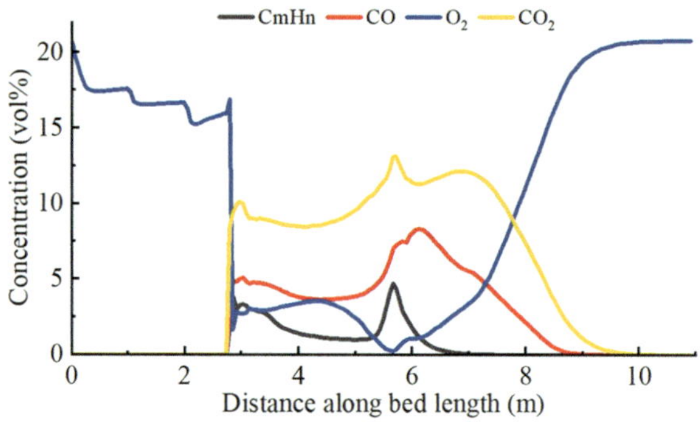

(b) Gas concentration

FIGURE 5.8 Condition 2 for solid-phase combustion stage. (a) Solid temperature. (b) Gas concentration

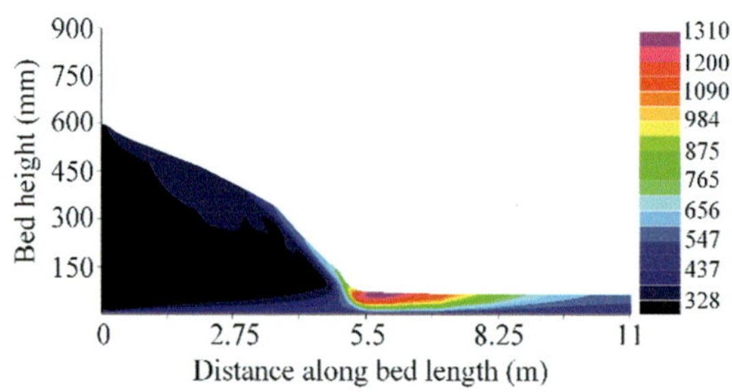

(a) Solid temperature

FIGURE 5.9 Condition 3 for solid-phase combustion stage. (a) Solid temperature.

(Continued)

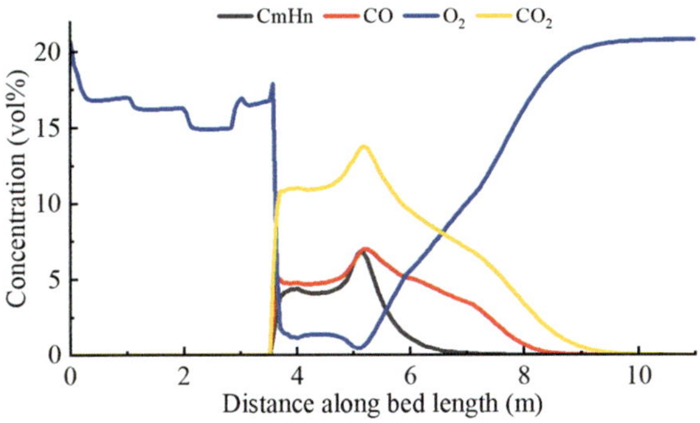

(b) Gas concentration

FIGURE 5.9 (Continued) (b) Gas concentration.

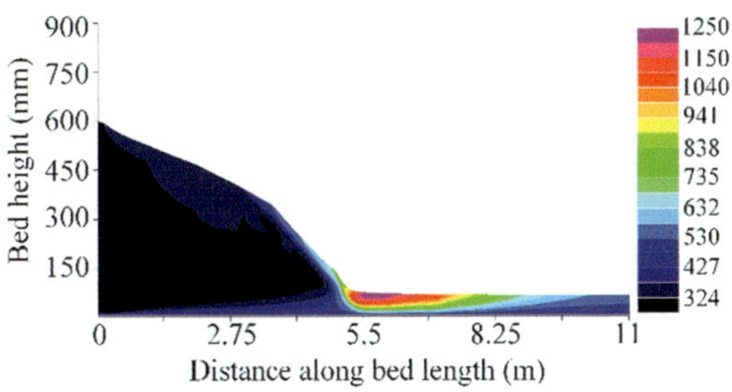

(a) Solid temperature

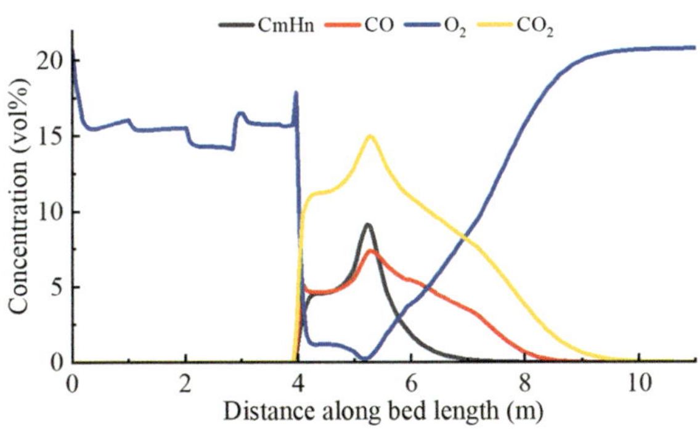

(b) Gas concentration

FIGURE 5.10 Condition 4 for solid-phase combustion stage. (a) Solid temperature. (b) Gas concentration.

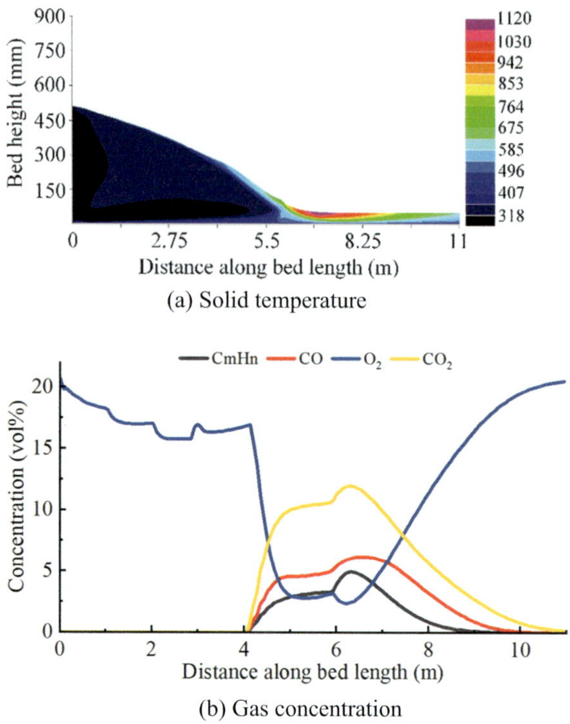

FIGURE 5.11 Condition 5 for solid-phase combustion stage. (a) Solid temperature. (b) Gas concentration.

5.4.2.3 Non-Manipulated Variables in the Single-Factor Analysis

Non-manipulated variables in the simulation model also influence the results. Therefore, a single-factor analysis should be conducted to investigate how these parameters affect solid-phase combustion.

5.4.2.3.1 MSW Particle Size

The simulation assumes that MSWs are represented as spherical particles. Figure 5.12 illustrates the influence of MSW particle size on solid-phase combustion.

As shown in Figure 5.12, both solid temperature and gas temperature increase with larger MSW particle sizes. When the particle size reaches 35 mm, the solid temperature is recorded at 1,417 K, while the gas temperature is 1,604 K. Larger particles facilitate combustion more effectively, which aligns with the findings in Ref. [15]. The accumulation of larger MSW particles creates more gaps, allowing for easier entry of primary air. Additionally, the enhanced contact between MSWs and primary air promotes combustion, resulting in decreased O_2 levels and increased CO_2 production.

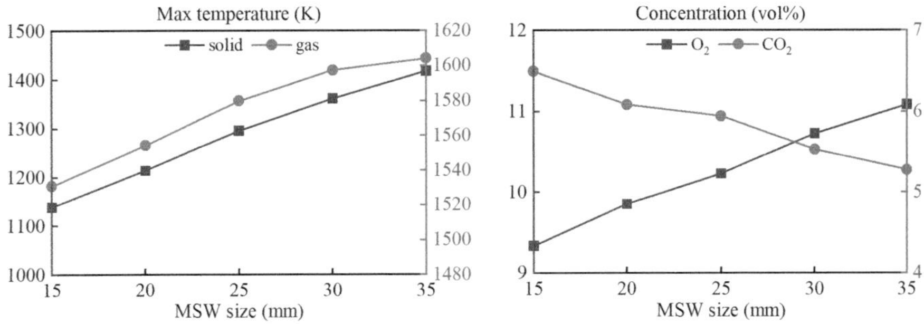

(a) Temperature (b) Gas concentration

FIGURE 5.12 Effect of the MSW particle size on solid-phase combustion. (a) Temperature. (b) Gas concentration.

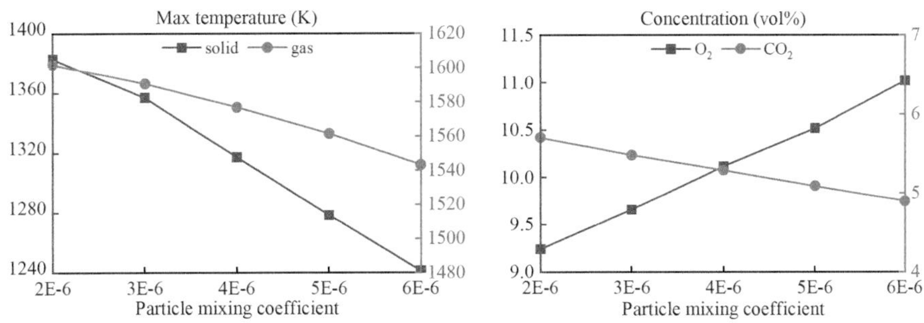

(a) Temperature (b) Gas concentration

FIGURE 5.13 Effect of the particle mixing coefficient on solid-phase combustion. (a) Temperature. (b) Gas concentration.

5.4.2.3.2 Particle Mixing Coefficient

The intensity of particle mixing is influenced by the designed grate-moving frequency. Figure 5.13 illustrates the effect of the particle mixing coefficient on solid-phase combustion.

As shown in Figure 5.13, an increase in the particle mixing coefficient leads to a decrease in both solid temperature and gas temperature, along with an increase in O_2 concentration and a decrease in CO_2 concentration. While a high mixing coefficient can facilitate short-term and intense combustion, our experimental findings suggest that neither the temperature nor CO_2 levels indicate violent combustion [28]. This implies that the particle mixing coefficient may already be at an optimal level under our grate design, and further increases could result in excessive mixing.

5.4.2.3.3 Emissivity

Emissivity indicates the ability of a material to radiate energy, and Figure 5.14 illustrates its influence on solid-phase combustion.

As shown in Figure 5.14, both solid and gas temperatures gradually increase with higher emissivity. Concurrently, the O_2 concentration decreases, while the CO_2 concentration increases. The flame and the wall primarily radiate energy to the solid material, exerting minimal influence on gas temperature. The intense combustion of solids leads to an increase in CO_2 emissions due to strong radiation effects. In summary, it is essential for the simulation to select an appropriate emissivity based on the furnace structure, material, and actual combustion conditions.

5.4.2.3.4 Moisture Content

Figure 5.15 illustrates the impact of moisture content on solid-phase combustion.

Figure 5.15 shows that the temperatures of solids and gases gradually decrease as moisture content increases. This aligns with the experimental curves presented in Refs. [29, 32]. High moisture content hinders MSW combustion, resulting in incomplete combustion that produces less CO_2 and an increase in O_2 content. Therefore, it

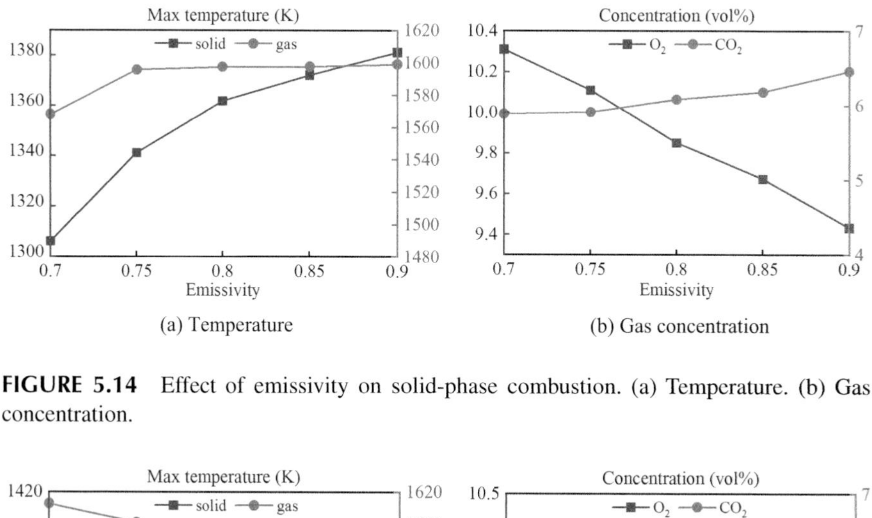

(a) Temperature (b) Gas concentration

FIGURE 5.14 Effect of emissivity on solid-phase combustion. (a) Temperature. (b) Gas concentration.

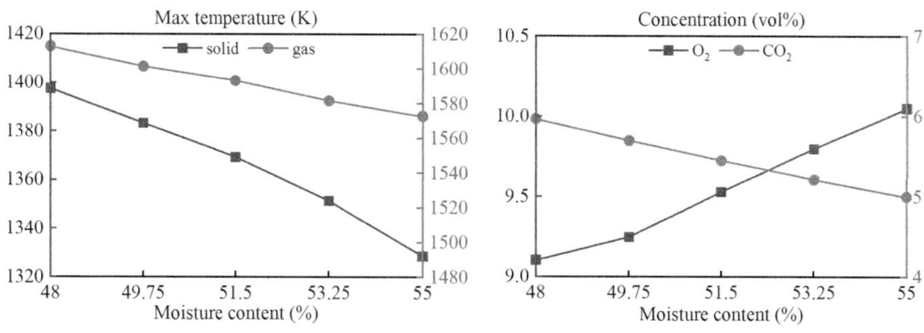

(a) Temperature (b) Gas concentration

FIGURE 5.15 Effect of moisture content on solid-phase combustion. (a) Temperature. (b) Gas concentration.

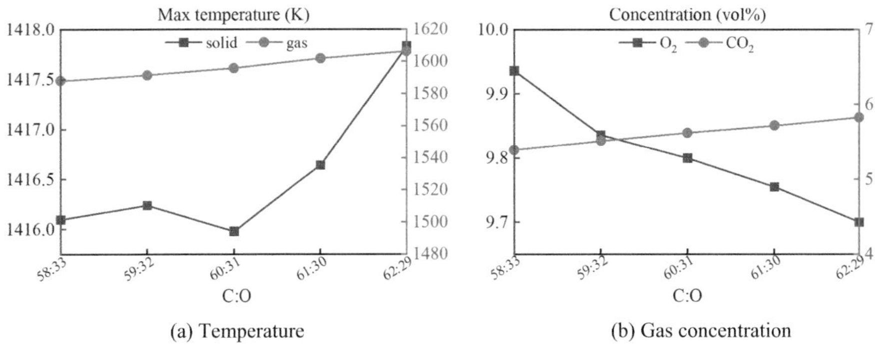

FIGURE 5.16 Effect of the C:O ratio on solid-phase combustion. (a) Temperature. (b) Gas concentration.

is essential to incorporate fermentation and dehydration processes in MSWI to ensure that the fuel attains a low calorific value.

5.4.2.3.5 C:O Ratio

Figure 5.16 illustrates the effect of the C:O ratio on solid-phase combustion.

Figure 5.16 shows that the C:O ratio has minimal impact on temperature (see). However, the concentration of CO_2 is directly proportional to carbon (C), while the concentration of O_2 is directly proportional to oxygen (O). In practice, the elemental composition of MSWs can vary significantly; however, the proportions of C and O remain relatively stable during the analysis. As a result, the influence on the outcomes is limited.

5.4.3 Comprehensive Analysis

The operating-condition benchmark illustrates the combustion state within the furnace of the MSWI power plant during stable operation, serving as a reference for analyzing various operating parameters individually. Multiple operating conditions reflect the influence of different manipulated variables, indicating potential changes in combustion behavior. Therefore, by utilizing a numerical simulation model that incorporates various operational modifications, it becomes feasible to adjust the manipulated variables based on theoretical knowledge and empirical expertise to achieve a stable state of normal combustion.

Non-manipulated variables, on the other hand, are uncontrollable through actions taken in the distributed control system. Among these non-manipulated variables, particle size and mixing coefficient indicate the accumulation of waste on the grate, while emissivity reflects the wall's ability to radiate heat. These factors affect the reaction outcomes from a microscopic perspective. Additionally, moisture content and the C:O ratio highlight the differences in the composition of MSWs during actual incineration, impacting the reaction results from the source of combustion. Thus, understanding non-manipulated variables is essential for subsequent optimal control, particularly in minimizing pollution emissions.

5.5 SUMMARY

A solid-phase numerical simulation model of the actual MSWI process was developed using FLIC software. A method for generating multi-operating condition mechanism data for the MSWI process was proposed based on a biorthogonal numerical simulation experiment. The solid-phase combustion conditions in the furnace were analyzed under the operating-condition benchmark and various multi-operating conditions, with a focus on the effects of changes in non-manipulated variables. The contributions are as follows:

(1) Under the operating-condition benchmark of the studied MSWI power plant, the maximum solid temperature reached 1,360 K, while the gas temperature peaked at 1,600 K in the main combustion zone. Water evaporation and volatile release occurred rapidly, and combustion was completed before reaching 10 m. O_2 and CO_2 were generated between 2.8 and 10 m at high solid temperatures. Conversely, low solid temperatures did not yield effective combustion results.

(2) The mechanism data for multi-operating conditions exhibited fluctuations. The maximum solid temperature varied between 842 and 1,467 K, while the maximum gas temperature ranged from 1,177 to 1,548 K. This mechanism data encompassed the incineration conditions under different operating scenarios, validating the design of multi-operating conditions based on the biorthogonal experimental method.

(3) For the non-manipulated variables studied, larger particles were found to be more conducive to combustion. The particle mixing coefficient and emissivity should be selected based on the specific circumstances. However, high moisture content proved detrimental to combustion, and the C:O ratio had minimal effect on combustion efficiency.

REFERENCES

1. Tang J, Xia H, Yu W Q J F. Research status and prospects of intelligent optimization control for municipal solid waste incineration process. *Acta Automatica Sinica*, 2023, 49(10): 2019–2059.
2. Qiao J, Guo Z, Tang J. Dioxin emission concentration measurement approaches for municipal solid wastes incineration process: a survey. *Acta Automatica Sinica*, 2020, 46(6): 1063–1089.
3. Pham T P T, Kaushik R, Parshetti G K, et al. Food waste-to-energy conversion technologies: Current status and future directions. *Waste Management*, 2015, 38: 399–408.
4. Xia H, Tang J, Aljerf L, et al. Investigation on dioxins emission characteristic during complete maintenance operating period of municipal solid waste incineration. *Environmental Pollution*, 2023, 318: 120949.
5. Alao M A, Popoola O M, Ayodele T R. Waste-to-energy nexus: an overview of technologies and implementation for sustainable development. *Cleaner Energy Systems*, 2022, 3: 100034.
6. Shin D, Choi S. The combustion of simulated waste particles in a fixed bed. *Combustion and Flame*, 2000, 121(1): 167–180.

7. Goh Y R, Yang Y B, Zakaria R, et al. Development of an incinerator bed model for municipal solid waste incineration. *Combustion Science and Technology*, 2001, 162(1): 37–58.
8. Yang Y B, Goh Y R, Zakaria R, et al. Mathematical modelling of MSW incineration on a travelling bed. *Waste Management*, 2002, 22(4): 369–380.
9. Yang Y B, Newman R, Sharifi V, et al. Mathematical modelling of straw combustion in a 38MWe power plant furnace and effect of operating conditions. *Fuel*, 2007, 86(1): 129–142.
10. Yu Z, Ma X, Liao Y. Mathematical modeling of combustion in a grate-fired boiler burning straw and effect of operating conditions under air- and oxygen-enriched atmospheres. *Renewable Energy*, 2010, 35(5): 895–903.
11. Ryu C, Yang Y B, Nasserzadeh V, et al. Thermal reaction modeling of a large municipal solid waste incinerator. *Combustion Science and Technology*, 2004, 176(11): 1891–1907.
12. Goddard C D, Yang Y B, Goodfellow J, et al. Optimisation study of a large waste-to-energy plant using computational modelling and experimental measurements. *Journal of the Energy Institute*, 2005, 78(3): 106–116.
13. Ismail T M, El-Salam M, El-Kady M A, et al. Three dimensional model of transport and chemical late phenomena on a MSW incinerator. *International Journal of Thermal Sciences*, 2014, 77: 139–157.
14. Yang Y B, Lim C N, Goodfellow J, et al. A diffusion model for particle mixing in a packed bed of burning solids. *Fuel*, 2005, 84(2): 213–225.
15. Yang Y B, Ryu C, Khor A, et al. Fuel size effect on pinewood combustion in a packed bed. *Fuel*, 2005, 84(16): 2026–2038.
16. Yang Y B, Yamauchi H, Nasserzadeh V, et al. Effects of fuel devolatilisation on the combustion of wood chips and incineration of simulated municipal solid wastes in a packed bed. *Fuel*, 2003, 82(18): 2205–2221.
17. Yang Y B, Sharifi V N, Swithenbank J. Effect of air flow rate and fuel moisture on the burning behaviours of biomass and simulated municipal solid wastes in packed beds. *Fuel*, 2004, 83(11): 1553–1562.
18. Yan M, Tian X, Antoni, et al. Influence of multi-temperature primary air on the characteristics of MSW combustion in a moving grate incinerator. *Journal of Environmental Chemical Engineering*, 2021, 9(6): 106690.
19. Yan M, Antoni, Wang J, et al. Numerical investigation of MSW combustion influenced by air preheating in a full-scale moving grate incinerator. *Fuel*, 2021, 285: 119193.
20. Aliman I, Pasek A D. The evaluation of experimental and numerical study of combustion process on mini traveling chain grate furnace (incinerator) by computational fluid dynamics method. In: *Proceedings of the International Conference on Thermal Science & Technology*, Helsinki, Finland, 19–20 July 2018. 2018.
21. Liang Z, Ma X. Mathematical modeling of MSW combustion and SNCR in a full-scale municipal incinerator and effects of grate speed and oxygen-enriched atmospheres on operating conditions. *Waste Management*, 2010, 30(12): 2520–2529.
22. Hu Z, Jiang E, Ma X. Numerical simulation on NO_x emissions in a municipal solid waste incinerator. *Journal of Cleaner Production*, 2019, 233: 650–664.
23. Lin T, Liao Y, Dai T, et al. Investigation on co-disposal technology of sludge and municipal solid waste based on numerical simulation. *Fuel*, 2023, 343: 127882.
24. Chen J, Tang J, Xia H, et al. Modelling the furnace temperature field of a municipal solid waste incinerator using the numerical simulation and the deep forest regression algorithm. *Fuel*, 2023, 347: 128511.

25. Tang J, Zhuang J, Aljerf L, et al. Numerical simulation modelling on whole municipal solid waste incineration process by coupling multiple software for the analysis of grate speed and air volume ratio. *Process Safety and Environmental Protection*, 2023, 176: 506–527.
26. Gu T, Yin C, Ma W, et al. Municipal solid waste incineration in a packed bed: a comprehensive modeling study with experimental validation. *Applied Energy*, 2019, 247: 127–139.
27. Sun R, Ismail T M, Ren X, et al. Influence of simulated MSW sizes on the combustion process in a fixed bed: CFD and experimental approaches. *Waste Management*, 2016, 49: 272–286.
28. Yang Y B, Swithenbank J. Mathematical modelling of particle mixing effect on the combustion of municipal solid wastes in a packed-bed furnace. *Waste Management*, 2008, 28(8): 1290–1300.
29. Sun R, Ismail T M, Ren X, et al. Numerical and experimental studies on effects of moisture content on combustion characteristics of simulated municipal solid wastes in a fixed bed. *Waste Management*, 2015, 39: 166–178.
30. Zhuang J, Tang J, Aljerf L. Comprehensive review on mechanism analysis and numerical simulation of municipal solid waste incineration process based on mechanical grate. *Fuel*, 2022, 320: 123826.
31. Yang Y B, Sharifi V N, Swithenbank J. Converting moving-grate incineration from combustion to gasification – numerical simulation of the burning characteristics. *Waste Management*, 2007, 27(5): 645–655.
32. Yang Y B, Sharifi V N, Swithenbank J. Numerical simulation of municipal solid waste incineration in a moving-grate furnace and the effect of waste moisture content. *Progress in Computational Fluid Dynamics: An International Journal*, 2007, 7(5): 261–273.
33. Arthur J R. Reactions between carbon and oxygen. *Transactions of the Faraday Society*, 1951, 47: 164–178.
34. Yang Y B, Goodfellow J, Sharifi V, et al. Investigation of biomass combustion systems using CFD techniques: a parametric study of packed-bed burning characteristics. *Progress in Computational Fluid Dynamics*, 2006, 6: 262–271.
35. Fernando N, Narayana M. A comprehensive two dimensional computational fluid dynamics model for an updraft biomass gasifier. *Renewable Energy*, 2016, 99: 698–710.
36. Smoot L D, Pratt D T. *Pulverized-Coal Combustion and Gasification*. Springer, 1979.

6 Unveiling Dioxin Dynamics

A Whole-Process Simulation Model

6.1 RELATED WORKS

6.1.1 Whole-Process Simulation Description

The MSWI technology has emerged as the preferred method for waste disposal due to its numerous advantages [1], including high reduction rates, significant waste volume reduction, and energy recovery [2]. DXN is an inevitable byproduct of nearly all high-temperature reactions and combustion processes [3], often occurring in conjunction with polychlorinated dibenzofurans (PCDFs) and polychlorinated dibenzo-*p*-dioxins (PCDDs). Currently, the proportion of the trace organic pollutant DXN emitted during the MSWI process has reached 37.6% [4, 5]. Due to its toxicity and bioaccumulation properties, DXN has become a major focus of scientific research [6].

Early research primarily focused on the pathways for the formation of DXN-like compounds. For example, Ballschmiter et al. experimentally identified four fundamental reaction pathways for the formation of PCDFs [7]: polychlorobiphenyl cyclization, polychlorodiphenyl ether cyclization, dibenzofuran chlorination, and octachlorodibenzofuran dechlorination. This work provided a theoretical foundation for subsequent studies on the numerical simulation modeling of DXN formation. Additionally, Weber and Hagenmaier summarized the existing theories of DXN formation and further elucidated the reaction processes involving chlorophenols (CPhs) and chlorobenzenes (CBzs) in the formation of PCDDs and PCDFs at various temperatures [8]. Peng et al. reviewed the role of chlorophenols (CPs) as a major precursor of DXNs in MSWI process [9], while Chen et al. investigated methods to suppress the formation pathways of DXNs in full-scale MSWI operations [10]. By this stage, a general consensus had emerged within the scientific community regarding the mechanisms underlying DXN formation, which involve two primary types of generation reactions: de novo synthesis and precursor synthesis [11].

Currently, research on DXNs primarily focuses on modeling, which can be categorized into two types: first-principles modeling and numerical simulation modeling [12]. The roots of first-principles modeling for DXNs can be traced back to 1983, when Shaub and Tsang developed a kinetic model known as the "ST model" to describe the formation of DXNs [13]. Experimental results indicated that the PCDD predictions made by the "ST model" are at least five orders of magnitude lower than

DOI: 10.1201/9781003634195-8

the actual PCDD detection values observed in MSWI processes. Subsequently, a growing number of researchers began scientific investigations based on the "ST model" [14–16]. Another kinetic model was developed for DXN formation in coal combustion processes [17], utilizing reaction rate information [18]. Taking into account the empirical relationships among the 17 congeners of DXNs, Stanmore constructed separate kinetic models [19], which is built for isomeric and non-isomeric reactions based on the Arrhenius equation [20]. Experimental results revealed a 20% discrepancy between this model and the actual measured values. Furthermore, Altarawneh et al. developed a PCDF generation model based on density functional theory [21], demonstrating its effectiveness in describing the oxidation processes by which chlorobenzenes (CBzs) and polychlorinated biphenyls (PCBs) form PCDFs.

Numerical simulation software modeling has become a focal point in recent years for research on the mechanisms of DXN formation [22]. Schwartz et al. developed a Windows desktop application for simulating DXN emissions using the.NET 2.0 framework [23]. This software provides a flexible platform for studying DXN emissions; however, its current unavailability for installation presents challenges in reproducing and applying it in practical engineering research. In [24], the Simulink tool within MATLAB was utilized to simulate the degradation reaction process of DXNs in flue gas containing activated carbon, oxygen, water, and hydroxyl radicals under electron beam irradiation. The research identified optimal conditions for degrading DXNs using electron beam irradiation. Sun et al. employed FLIC to simulate the solid-phase combustion of MSW [25], which is based on the kinetic model [19]. They calculated the distribution of DXN concentrations along the grate using customized HCl values. Experimental results demonstrated that the combination of FLIC and the kinetic model effectively captured the trend of DXN concentration distribution. However, verifying whether the experimental results align with the actual DXN distribution remains a challenge. In another study, Tungalag utilized Fortran subprograms within Aspen Plus software to simulate the volatile products from MSW combustion [26], incorporating DXN (specifically 2,3,7,8-TCDD) as one of the gaseous products. The experimental results indicated that Aspen Plus software effectively simulated the production of 2,3,7,8-TCDD, providing robust support for DXN simulation studies using this software.

As mentioned above, the simulation software currently available for DXN numerical modeling includes FLIC and Aspen Plus. However, existing numerical simulation research has largely overlooked the simulation of the 17 congeners of DXNs. Therefore, there is a pressing need to explore the design and construction of a comprehensive numerical simulation model (CNSM) that is specifically tailored to encompass the whole lifecycle of DXN concentration in the MSWI process.

6.1.2 Whole Process of DXN Concentration Description

The MSWI process, which includes feeding, combustion, heat exchange, gas cleaning, and emission stages, is illustrated in Figure 6.1. The system depicted employs a typical grate-type incinerator with a daily capacity of 800 tons of MSW located in Beijing, China. The flue gas cleaning stage is equipped with advanced air pollution

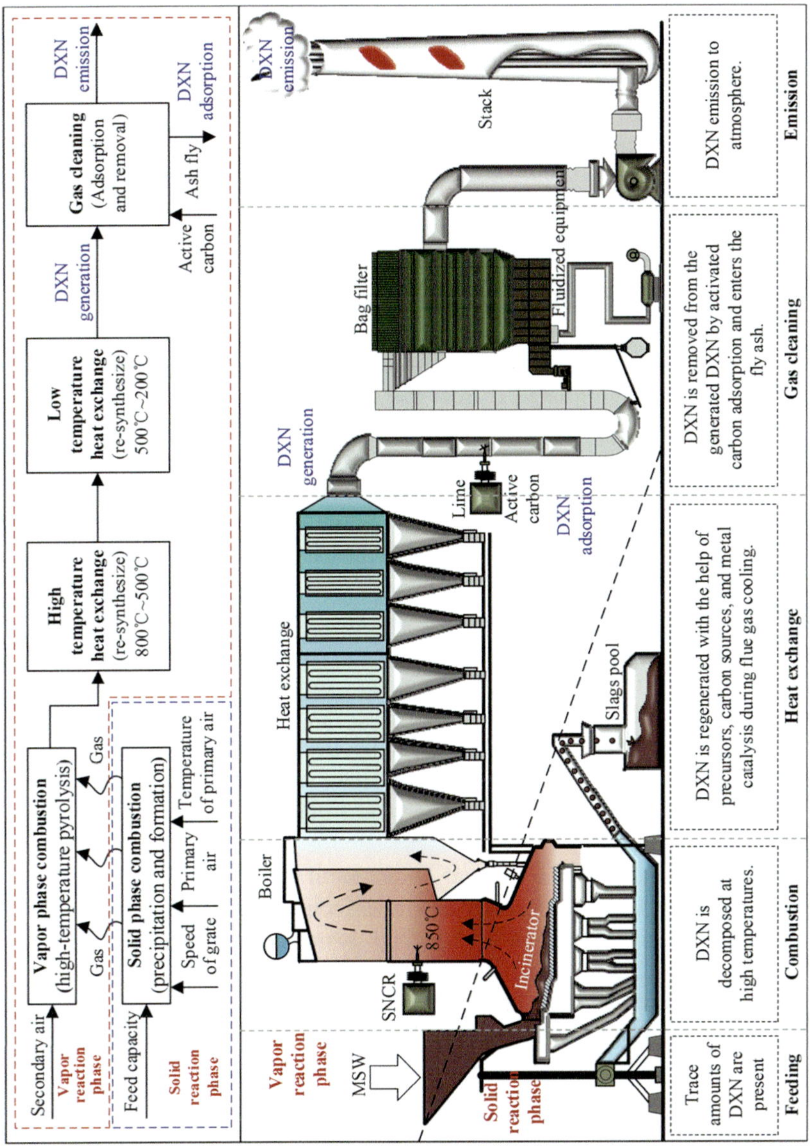

FIGURE 6.1 Schematic diagram of MSWI process and the whole lifecycle of DXN concentration.

control devices (APCDs), including selective non-catalytic reduction, activated carbon injection, semi-dry deacidification, and a bag filter system.

The model encompasses five distinct zones: MSW solid-phase combustion, MSW gas-phase combustion, high-temperature heat exchange, low-temperature heat exchange, and gas cleaning. The analysis of the manipulated variables in terms of DXN concentration within these five regions of the MSWI process is as follows.

(1) MSW Solid-Phase Combustion Zone: The core objective is to utilize "air distribution and material distribution" operations to regulate heat radiation and combustion within the furnace, thereby facilitating the drying, volatilization, and combustion conversion of MSW into high-temperature gas-phase substances. Consequently, the inputs for this zone include fermented material (i.e., MSW), bed control, and air volume operations. The manipulated variables can be represented by the MSW feed capacity, grate speed, primary air flow, and primary air temperature.
(2) MSW Gas-Phase Combustion Zone: Building on the high-temperature gas-phase substances generated in the solid-phase combustion zone, this process utilizes the remaining oxygen from the primary air, along with additional oxygen supplied by the secondary air, to further enhance the combustion of gas-phase substances. This approach ensures that the furnace temperature exceeds 850°C, which is essential for effective pyrolysis of DXNs. Consequently, the inputs for this zone include the outputs from the solid-phase combustion zone and the secondary air. The manipulated variables can be represented by the secondary air flow and the secondary air temperature.
(3) High-Temperature Heat Exchange Zone: This stage primarily involves the physical cooling of high-temperature flue gas based on the principle of heat exchange. The temperature typically ranges from 800°C to 500°C, which corresponds to the region where superheater devices are utilized. Assuming a constant heat exchange efficiency in the boiler system, both the input and output in this process flow consist of gas-phase substances. In the context of the CNSM, this portion does not involve any manipulated variables related to the MSWI process.
(4) Low-Temperature Heat Exchange Zone: Similar to the high-temperature heat exchange, this stage operates on the principle of heat exchange to further cool the flue gas output from the high-temperature zone. The temperature typically ranges from 500°C to 200°C, corresponding to the zone where economizer devices are utilized. Like the high-temperature heat exchange, this zone does not involve any manipulated variables related to the MSWI process.
(5) Gas Cleaning Zone: The core function of this zone is to utilize the combined technology of activated carbon and bag filters to effectively remove a significant amount of DXNs from the flue gas through physical adsorption. Consequently, the inputs for this zone include the flue gas output from the low-temperature heat exchange zone (below 200°C) and activated carbon. The operational variable in the MSWI process for this stage can be represented by the amount of activated carbon used.

6.1.3 Manipulated Variables from MSWI Process

The benchmark condition values and their actual ranges for various process parameters are presented in Table 6.1.

Based on operational experience from the actual MSWI process, the values of the selected manipulated variables and their corresponding levels are provided in Table 6.2.

TABLE 6.1
Benchmark Condition Values and Their Actual Ranges for Some Process Parameters

Parameters		Setting		Real Range	Unit
MSW feed capacity		800		600–800	tons/d
Speed of grate		7.5		5.5–7.5	m/h
Primary air	Drying	1,090	261.6 (24%)	1,017–1,175	m^3/min
	Burning 1		490.5 (45%)		
	Burning 2		218 (20%)		
	Burnout		119.9 (11%)		
Temperature of primary air (drying)		473		413–503	K
Temperature of primary air (burning)		453		393–473	K
Size of grate		11 (length); 12.9 (width)			m

TABLE 6.2
Values of the Selected Manipulated Variables and Their Levels

Factors		Units	Level						
MSW treatment capacity		tons/h	28,000	28,333	29,667	30,500	31,333	32,166	33,000
Speed of grate		m/h	5.5	5.9	6.2	6.5	6.8	7.1	7.5
Drying (24%) First burning (45%) Second burning (20%) Burnout (11%)	Primary air	Nm^3/min	1,017	1,043	1,070	1,096	1,122	1,149	1,175
Temperature of primary air (drying)		K	414	429	444	459	474	489	503
Temperature of primary air (burning)		K	394	407	421	434	447	461	473

The manipulated variables and their corresponding levels were carefully selected based on operational experience and literature reviews to represent a range of realistic scenarios for MSWI processes. These levels were chosen to reflect typical operating conditions and to facilitate the exploration of DXN dynamics under varying process conditions.

6.2 MODELING STRATEGY

In this chapter, we collaboratively utilize FLIC, Aspen Plus, and Aspen Adsorption software to establish a CNSM for the entire lifecycle of DXNs. The specific modeling process is illustrated in Figure 6.2.

As shown in Figure 6.2, the functions of the different modules of the CNSM model are described as follows:

(1) MSW Bed Combustion Model Module: This module is implemented using FLIC, where MSW is treated as spherical particles. It is assumed to undergo four stages on the grate: drying, volatile release, volatile combustion, and char combustion. These processes generate gas-phase substances such as C_mH_n, CO, CO_2, O_2, N_2, H_2O, H_2, and NO_x, along with a small amount of ash residue. The gas-phase substances and their molar fractions simulated by FLIC are then used as inputs for Aspen Plus.

(2) Solid–Gas-Phase Model Module: This module is implemented with Aspen software, with details as follows.

1. Solid–Gas-Phase Combustion Sub-module: The inputs for this module include MSW components, primary air, temperature of the primary air, and other relevant parameters. The output is based on the gas-phase substances and their molar fractions simulated by FLIC. Additionally, to account for the gas-phase substances produced by solid-phase MSW combustion, trace elements such as chlorine and copper are incorporated into the model.
2. Precipitation and Formation Sub-module: This part simulates the precipitation and rapid formation of DXN substances during the stage prior to the environmental temperature rising to 850°C in the combustion process.
3. High-Temperature Pyrolysis Sub-module: This part simulates the high-temperature pyrolysis process of DXNs during combustion. As the combustion temperature reaches 850°C and secondary air is injected, the gas-phase substances in the furnace continue to combust. The high-temperature flue gas remains within the 850°C range for at least 2 seconds, effectively simulating the pyrolysis process of DXNs.
4. High-Temperature Gas-Phase Synthesis Sub-module: This part section simulates the region where the superheater is located during the heat exchange stage, specifically the area where the flue gas temperature decreases from 850°C to 500°C. Based on the reactions for high-temperature gas-phase generation of DXNs, it implements the process in which precursor compounds such as CBzs and CPhs undergo condensation or chlorination to produce DXNs.
5. Low-Temperature Catalytic Synthesis Sub-module: This part simulates the region where the economizer is located during the heat exchange stage, specifically the area where the flue gas temperature decreases from 500°C to 200°C. The generation of DXNs involves both precursor catalytic generation and de novo synthesis. In the low-temperature

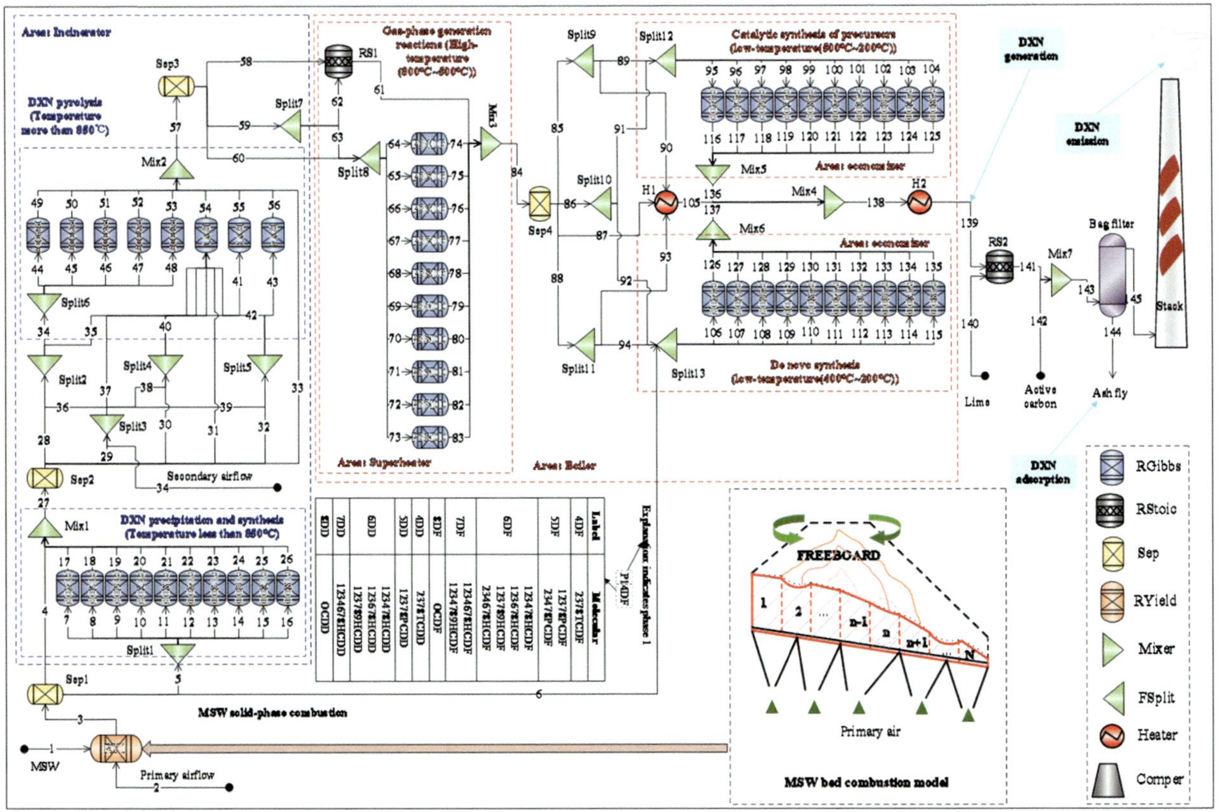

FIGURE 6.2 CNSM model for the whole lifecycle of DXN.

precursor catalytic generation reaction, precursor compounds such as CBzs and CPhs undergo catalytic action by $CuCl_2$ to produce DXN. The de novo synthesis reaction entails the formation of DXN from carbon through a series of elementary reactions, including chlorination and oxidation.

6. Activated Carbon Adsorption Modeling Sub-module: This part utilizes Aspen Adsorption to simulate the adsorption of 17 congeners by activated carbon. Using the DXN generation concentration from Aspen Plus as input, and maintaining a fixed configuration for the activated carbon adsorption bed, the efficiency of activated carbon in removing DXNs is determined.
7. Emissions to the Atmosphere Sub-module: This section simulates the emission of DXNs that have not been completely adsorbed and removed, releasing them from the stack into the atmosphere.

Note: During the simulation modeling process of the whole lifecycle of DXN generation, adsorption, and emission, the theoretical assumptions include the following aspects: (1) The model operates at thermodynamic equilibrium, meaning parameters such as temperature, pressure, and component concentrations within the system remain stable; (2) Gas-phase reaction components in the simulation are treated as ideal gases, implying that there are no intermolecular forces between gas molecules and volume effects are ignored; (3) The model operates under steady-state conditions, meaning internal parameters are assumed not to vary with time during the simulation; (4) The temperature field within the model system is assumed to be uniformly distributed, indicating that the internal temperature is considered equal throughout the same zone; (5) The pressure field within the model system is also assumed to be uniformly distributed, meaning that the internal pressure is regarded as equal across the same zone.

The input parameters for the CNSM model of the whole lifecycle of DXN generation, adsorption, and emission are presented in Table 6.3.

Table 6.3 displays the model input parameters. Considering the characteristics of the actual MSWI operating process and the theoretical assumptions, certain parameters (e.g., grate size, particle size distribution, particle mixing coefficient, and activated carbon dosage) only need to be set once. Additionally, the input for the Aspen Plus model is based on the MSW composition and the output from the FLIC model, with the 15 sets of MSW compositions determined in the previous section. Therefore, all parameters involved in the multi-condition experimental design ultimately originate from the FLIC model, namely the five key parameters. They are the MSW feed capacity, grate speed, primary air, drying temperature of primary air, and burning temperature of primary air.

Therefore, by leveraging existing knowledge on MSW combustion mechanisms and the processes of DXN generation, adsorption, and emission, FLIC and Aspen software are utilized to develop a multi-stage model that captures the entire DXN evolution process. The main contributions are as follows: (1) Introduce a comprehensive numerical simulation modeling strategy for DXNs and offers a novel perspective for DXN modeling research within the context of MSWI processes; (2) Provides a

TABLE 6.3
Input Parameters for CNSM Model

Actual Input Parameters Based on MSW and MSWI Processes		Numerical Simulation Input Parameters Based on Elemental or Thermal Energy Equivalence Assumptions		
MSW	**MSWI processes**	**FLIC**	**Aspen Plus**	**Aspen adsorption**
C	MSW treatment capacity	MSW	MSW	Aspen Plus
H	Speed of feed	MSWI processes	FLIC	Particle size of active carbon
O	Primary air	Particle size distribution	Secondary air	Parking density of active carbon
N	Temperature of primary air (drying)	Particle mixing coefficient	Temperature of secondary air	Void between activated carbon particles
S	Temperature of primary air (burning)	Thermal radiance	Pressure of secondary air	Void in activated carbon particles
Cl	Size of grate	Void ratio	Volume of calcium hydroxide injection	Activated carbon shape factor
Moisture		Density of each component	Volume of active carbon injection	Inside diameter of adsorption bed
Ash		Specific heat capacity of each component	Temperature of each reactor	Height of adsorption bed
Fixed carbon		Pre-exponential factor		Transfer coefficient
Volatile		Activation energy		Isotherm parameters 1
		Water evaporation temperature		Isotherm parameters 2
		Zone of bed		

comprehensive depiction of DXN evolution, significantly aiding in both the qualitative description and understanding of its progression; (3) Offers valuable insights into the mechanisms underlying DXN emission control, thereby laying a foundation for future research in DXN emission reduction and control.

6.3 MODELING IMPLEMENTATION

6.3.1 MSW Bed Combustion Model Module

The mechanistic model of MSW solid-phase combustion encompasses fundamental conservation equations, component transport dynamics, and radiation heat transfer equations. The fundamental conservation equations (e.g., comprising mass, momentum, and energy conservation) treat MSW as a porous medium undergoing combustion, which serves as the foundation for thermal reactions and mass exchange between the gas and solid phases. Therefore, the mass continuity equations for the gas and solid phases of MSW can be expressed as follows:

$$\frac{\partial\left(\rho_g \phi_g\right)}{\partial t}+\nabla \cdot\left(\rho_g \phi\left(V_g-V_B\right)\right)=S_{\mathrm{sg}} \tag{6.1}$$

$$\frac{\partial \rho_{\mathrm{sb}}}{\partial t}+\nabla \cdot\left(\rho_{\mathrm{sb}}\left(V_s-V_B\right)\right)=S_s \tag{6.2}$$

where ϕ_g represents the bed voidage; V_g and V_s are gas velocity and solid particle velocity, respectively; V_B indicates velocity of the moving boundary; S_{sg} denotes rate of conversion of solid to gas; S_s expresses the mass source term; and ρ_{sb} means the parking density of bed.

Taking into account the viscosity and friction effects at the interface between the gas and solid phases in MSW, which result in momentum losses within the flow, the momentum equations [27] for the gas and solid phases during the combustion process can be expressed as follows:

$$\frac{\partial\left(\rho_g \phi_g V_g\right)}{\partial t}+\nabla \cdot\left(\rho_g \phi\left(V_g-V_B\right) V_g\right)=-\nabla p_g+F(v) \tag{6.3}$$

$$\frac{\partial\left(\rho_{\mathrm{sb}} V_s\right)}{\partial t}+\nabla \cdot\left(\rho_{\mathrm{sb}}\left(V_s-V_B\right) V_s\right)=-\nabla \cdot \sigma-\nabla \cdot \tau+\rho_{\mathrm{sb}} g+A \tag{6.4}$$

where ρ_g represents the pressure of gas; $F(v)$ is the resistance of solid particles to fluid flow in porous media; σ and τ indicate normal stress tensor and shear stress tensor inside the bed respectively; and A denotes the random particle motion.

In the combustion process within the MSW bed, it is assumed that the temperature of the particles in the porous medium is uniformly distributed, with solid particles exchanging heat with adjacent particles through conduction and radiation. Consequently, the energy equations for both the gas phase and the solid phase can be expressed as follows:

$$\frac{\partial\left(\rho_g \phi_g H_g\right)}{\partial}+\nabla \cdot\left(\rho_g \phi_g\left(V_g-V_B\right) H_g\right)=\nabla \cdot\left(\lambda_g \nabla T_g\right)+S_a h'_s\left(T_s-T_g\right)+Q_h \tag{6.5}$$

$$\frac{\partial\left(\rho_{\mathrm{sb}} H_s\right)}{\partial t}+\nabla \cdot\left(\rho_{\mathrm{sb}}\left(V_s-V_B\right) H_s\right)=\nabla \cdot\left(\lambda_s \nabla T_s\right)-S_a h'_s\left(T_s-T_g\right)+\nabla \cdot q_r+Q_{\mathrm{sh}} \tag{6.6}$$

where H_g and H_s are the enthalpies of the gas phase and enthalpy of the solid phase, respectively; T_g and T_s are temperature of the solid-phase and temperature of the gas-phase, respectively; Q_h is heat loss/gain of the gas; Q_{sh} is the mass source term; q_r is radiative heat flux; λ_g is thermal dispersion coefficient composed of diffusion and turbulent contributions; and λ_s is the thermal conductivity of the bed layer.

The specific calculation process of λ_g and λ_s is as follows.

$$\lambda_g = \lambda_0 + 0.5 d_p V_g \rho_g C_{\mathrm{pg}} \tag{6.7}$$

$$\lambda_s = \lambda_{s0} + \lambda_{\mathrm{sm}} \tag{6.8}$$

where λ_0 is the thermal diffusivity, d_p is the particle diameter, C_{pg} is the heat capacity of the mixture gas, λ_{s0} is the thermal conductivity of solid materials, and λ_{sm} is the heat transfer caused by random particle motion.

Building on the previously mentioned conservation equations for mass, momentum, and energy, the component transport equations [28] for the gas and solid phases in the interaction among the combustion mixture components of MSW can be expressed as follows:

$$\frac{\partial\left(\rho_g Y_{i,g}\phi_g\right)}{\partial t} + \nabla\cdot\left(\rho_g\phi_g\left(V_g - V_B\right)Y_{i,g}\right) = \nabla\cdot\left(D_{ig}\nabla\left(\rho_g Y_{i,g}\right)\right) + S_{Y_{i,g}} \tag{6.9}$$

$$\frac{\partial\left(\rho_{\mathrm{sb}} Y_{i,s}\right)}{\partial t} + \nabla\cdot\left(\rho_{\mathrm{sb}}\left(V_s - V_B\right)Y_{i,s}\right) = \nabla\cdot\left(D_s\nabla\left(\rho_{\mathrm{sb}} Y_{i,s}\right)\right) + S_{Y_{i,s}} \tag{6.10}$$

where $Y_{i,g}$ is mass fraction of an individual gas-phase component, $S_{Y_{i,g}}$ is the mass source of volatile gases and char, $S_{Y_{i,s}}$ is the mass source of MSW particle composition, D_{ig} is fluid dispersion coefficient composed of diffusion and turbulent contributions, and D_s means mixing coefficient of particles in MSW bed.

Radiative heat transfer returns the heat generated by the combustion reaction to the MSW bed, supplying energy for moisture evaporation, MSW pyrolysis, and other endothermic solid-phase reactions. Typically, radiative heat transfer within the furnace is modeled using a four-directional radiation approach [29]. The specific calculation process is as follows:

$$\frac{dI_x^+}{dx} = -\left(k_a + k_s\right)I_x^+ + \frac{1}{4}K_a E_b + \frac{1}{4}k_s\left(I_x^+ + I_x^- + I_z^+ + I_z^-\right) \tag{6.11}$$

$$-\frac{dI_x^+}{dx} = -\left(k_a + k_s\right)I_x^+ + \frac{1}{4}K_a E_b + \frac{1}{4}k_s\left(I_x^+ + I_x^- + I_z^+ + I_z^-\right) \tag{6.12}$$

$$\frac{dI_z^+}{dz} = -\left(k_a + k_s\right)I_z^+ + \frac{1}{4}K_a E_b + \frac{1}{4}k_s\left(I_x^+ + I_x^- + I_z^+ + I_z^-\right) \tag{6.13}$$

$$-\frac{dI_z^-}{dz} = -\left(k_a + k_s\right)I_z^- + \frac{1}{4}K_a E_b + \frac{1}{4}k_s\left(I_x^+ + I_x^- + I_z^+ + I_z^-\right) \tag{6.14}$$

where I_x^+, I_x^-, I_z^+, and I_z^- are radiation intensity in four directions, respectively; k_a and k_s are absorption and scattering coefficients, respectively; and E_b is blackbody radiation.

When hypothesis k_s is determined to be zero, the calculation process for k_a can be written as follows,

$$k_a = -\frac{1}{d_p}\ln\left(\phi_g\right) \tag{6.15}$$

Based on the fundamental conservation equations, component transport dynamics, and radiation heat transfer equations, a mechanistic model of MSW solid-phase combustion is developed through four key processes such as moisture evaporation, volatile release, volatile combustion, and char combustion. Ultimately, this model generates data on the gaseous products and molar fractions resulting from MSW solid-phase combustion.

6.3.2 Solid–Gas-Phase Model Module

6.3.2.1 Solid–Gas-Phase Combustion Sub-Module

The solid-gas-phase combustion of MSW utilizes inputs such as MSW composition (including carbon, hydrogen, oxygen, nitrogen, sulfur, as well as moisture, volatile matter, fixed carbon, and ash), primary air, the temperature of the primary air, and the pressure of the primary air. The output is generated based on the MSW bed combustion model and includes trace elements of chlorine and metallic copper as byproducts. The specific details of the solid-gas-phase combustion model are presented in Table 6.4.

At this stage, Stream 1 is designated as the MSW composition, Stream 2 as the primary air, and Stream 3 as the gas-phase output products. The outputs from the RY1 reactor include the gas-phase products from the MSW bed combustion model, as well as chlorine elements and metallic copper. Consequently, the process represented by the RY1 reactor and the output of Stream 3 can be expressed as follows:

$$\begin{aligned} MSW + O_2 + N_2 \rightarrow\ & CH_4 + CO + CO_2 + O_2 + N_2 + H_2O + N_2O + H_2 + SO_2 + HCl \\ & + C + C_6H_6 + C_6Cl_5OH + C_6H_2Cl_3OH + C_6Cl_6 + C_6HCl_5 + CuCl_2 \end{aligned} \tag{6.16}$$

where C_6H_6 is benzene, C_6Cl_5OH is pentachlorophenol, $C_6H_2Cl_3OH$ is 2,4,6-Trichlorophenol, C_6Cl_6 is hexachlorobenzene, and C_6HCl_5 is pentachlorobenzene.

6.3.2.2 Precipitation and Formation Sub-Module

The products of the solid-gas-phase combustion model serve as inputs for constructing the DXN precipitation and formation reaction model. Specific details are presented in Table 6.5.

TABLE 6.4
Information of Solid-Gas-Phase Combustion Model

Model	Type	Number	Mark
Reaction model	RYield	1	RY1

TABLE 6.5
Information of DXN Precipitation and Formation Model

Model	Type	Number	Mark
Reaction model	RYield	10	P14DF, P15DF, P16DF, P17DF, P18DF, P14DD, P15DD, P16DD, P17DD, P18DD
	Sep	1	Sep1
	Mixer	1	Mix1
	FSplit	1	Split1

At this stage, the Sep1 separator plays a crucial role in dividing the gas-phase products from RY1 into three streams as follows. (1) Stream 4: This stream carries products such as CPhs and CBzs, which do not participate in the DXN precipitation and formation process; (2) Stream 5: This stream is specifically directed toward simulating the precipitation and formation of DXN at temperatures below 850°C; and (3) Stream 6: This stream contains particulate matter (in part), which includes carbon sources relevant to DXN formation.

In the DXN precipitation and formation reaction, reactors are established based on the number of chlorine (Cl) atom substitutions for PCDDs and PCDFs. The following reactors are configured accordingly: (1) P14DF and P14DD reactors for 4 Cl atom substitutions; (2) P15DF and P15DD reactors for 5 Cl atom substitutions; (3) P16DF and P16DD reactors for 6 Cl atom substitutions; (4) P17DF and P17DD reactors for 7 Cl atom substitutions; (5) P18DF and P18DD reactors for 8 Cl atom substitutions. All 10 RYield reactors are set to a temperature of 600°C and a pressure of 101 kPa. Additionally, other products generated include low-CPhs (which have fewer Cl atom substitutions than the produced DXN), low-CBzs (also with fewer Cl atom substitutions than the produced DXN), and conventional gases. Therefore, the DXN precipitation and formation reaction can be represented as follows:

$$\text{MSW} \xrightarrow{600^\circ\text{C}} \text{DXN} + \text{H}_2\text{O} + \cdots \tag{6.17}$$

$$\begin{cases} \text{①}(\text{CBzs} + \text{CPhz} + \text{O}_2 + \text{HCl} \rightarrow 2,3,7,8\text{TCDF} + \text{H}_2\text{O} \\ \text{②}(\text{CBzs} + \text{CPhz} + \text{O}_2 + \text{HCl} \rightarrow 1,2,3,7,8\text{PeCDF} + 2,3,4,7,8\text{PeCDF} + \text{H}_2\text{O} \\ \text{③}\begin{pmatrix} \text{CBzs} + \text{CPhz} + \text{O}_2 + \text{HCl} \rightarrow 1,2,3,4,7,8\text{HxCDF} + 1,2,3,6,7,8\text{HxCDF} \\ \quad + 1,2,3,7,8,9\text{HxCDF} + 2,3,4,6,7,8\text{HxCDF} + \text{H}_2\text{O} \end{pmatrix. \\ \text{④}(\text{CBzs} + \text{CPhz} + \text{O}_2 + \text{HCl} \rightarrow 1,2,3,4,6,7,8\text{HpCDF} + 1,2,3,4,7,8,9\text{HpCDF} + \text{H}_2\text{O} \\ \text{⑤}(\text{CBzs} + \text{CPhz} + \text{O}_2 + \text{HCl} \rightarrow \text{OCDF} + \text{H}_2\text{O} \\ \text{⑥}(\text{CBzs} + \text{CPhz} + \text{O}_2 + \text{HCl} \rightarrow 2,3,7,8\text{TCDD} + \text{H}_2\text{O} \\ \text{⑦}(\text{CBzs} + \text{CPhz} + \text{O}_2 + \text{HCl} \rightarrow 1,2,3,7,8\text{PeCDD} + \text{H}_2\text{O} \\ \text{⑧}\begin{pmatrix} \text{CBzs} + \text{CPhz} + \text{O}_2 + \text{HCl} \rightarrow 1,2,3,4,7,8\text{HxCDD} + 1,2,3,6,7,8\text{HxCDD} \\ \quad + 1,2,3,7,8,9\text{HxCDD} + \text{H}_2\text{O} \end{pmatrix. \\ \text{⑨}(\text{CBzs} + \text{CPhz} + \text{O}_2 + \text{HCl} \rightarrow 1,2,3,4,6,7,8\text{HpCDD} + \text{H}_2\text{O} \\ \text{⑩}(\text{CBzs} + \text{CPhz} + \text{O}_2 + \text{HCl} \rightarrow \text{OCDD} + \text{H}_2\text{O} \end{cases} \tag{6.18}$$

To ensure equilibrium in the DXN precipitation and formation reaction, the Split1 splitter allocates the appropriate proportions of reactants to the 10 RYield reactors based on the ratio of congeners. In the simulation process, aside from 2,3,7,8-Tetrachlorodibenzo-*p*-dioxin (1,2,3,7,8PeCDD), octachlorodibenzo-*p*-dioxin (OCDD), 1,2,3,4,7,8-hexachlorodibenzofuran (1,2,3,4,7,8HxCDF), and octachlorodibenzofuran (OCDF), the remaining 13 congeners of DXNs require molecular structure depictions to obtain complete substance information. Detailed information is provided in Table 6.6.

After the DXN precipitation and formation reaction, the outputs from the 10 RYield reactors are mixed with Stream 2 using Mixer 1 to obtain all gas-phase substances from the DXN precipitation and formation stage. Additionally, the output from Stream 6, which contains particles carrying carbon sources, directly serves as the input for the low-temperature de novo synthesis reaction stage of DXN.

6.3.2.3 High-Temperature Pyrolysis Sub-Module

In the DXN high-temperature pyrolysis stage, the products from the DXN precipitation and formation reaction are utilized as inputs to construct a DXN pyrolysis reaction model at 850°C. Specific information is provided in Table 6.7.

Following the mixing process in Mixer 1, the Sep2 separator is used to separate the resulting gas-phase products into six distinct streams, each with specific compositions and functions. The details are as follows: (1) Stream 28: This stream carries all DXNs formed during the previous stage, along with CBzs, a small amount of CO, and some residual O_2; (2) Stream 29: This stream primarily consists of particulate matter and the majority of CO, supplemented by secondary air input for further reactions; (3) Stream 30: This stream contains CBzs and a portion of the available oxygen; (4) Stream 31: This stream carries a small amount of CO, HCl, and some O_2; (5) Stream 32: Similar to Stream 30, this stream consists of CBzs and a portion of the available oxygen; and (6) Stream 33: This stream is composed of gas-phase components that do not participate in further reactions, including residual CO, CO_2, H_2O, and N_2.

In the DXN high-temperature pyrolysis reaction zone, 8 RYield reactors are established for various processes, including DXN pyrolysis, conventional gas combustion reactions, and the pyrolysis of CPhs and CBzs. Reactor RY2 is primarily designated for DXN pyrolysis and conventional gas reactions. Its inputs include Stream 35 (which contains most of the DXNs), Stream 31 (which contains CO, HCl, O_2, and other conventional gases), Stream 37 (which includes CO, O_2, N_2, and other conventional gases), Stream 40 (which consists of CO, O_2, N_2, and chlorobenzene), and Stream 42 (which contains CO, O_2, N_2, and chlorobenzene). The typical reaction products of DXN pyrolysis are CO_2 and HCl. The process can be represented as follows:

$$\text{DXN} + \text{CPhs} + \text{CBzs} + \text{PCBs} + \cdots + O_2 \xrightarrow{850^\circ\text{C}} CO_2 + \text{HCl} \quad (6.19)$$

In this zone, the remaining portion of DXN enters five reactors: P14DDF, P15DDF, P16DDF, P17DDF, and P18DDF. Each reactor is designed to target the thermal decomposition of DXN congeners based on their level of chlorine substitution. The

TABLE 6.6
Chemical Formulas and Names of 17 DXN Substances

Types	Congener	Molecular	Name	CAS Number
Polychlorinated dioxins	2,3,7,8TCDD	$C_{12}H_4Cl_4O_2$	2,3,7,8-tetrachlorodibenzo-*p*-dioxin	1746-01-6
	1,2,3,7,8PeCDD	$C_{12}H_3Cl_5O_2$	1,2,3,7,8-pentachlorodibenzo-*p*-dioxin	40321-76-4
	1,2,3,4,7,8HxCDD	$C_{12}H_2Cl_6O_2$	1,2,3,4,7,8-hexachlorodibenzo-*p*-dioxin	39227-28-6
	1,2,3,6,7,8HxCDD	$C_{12}H_2Cl_6O_2$	1,2,3,6,7,8-hexachlorodibenzo-*p*-dioxin	57653-85-7
	1,2,3,7,8,9HxCDD	$C_{12}H_2Cl_6O_2$	1,2,3,7,8,9-hexachlorodibenzo-*p*-dioxin	19408-74-3
	1,2,3,4,6,7,8HpCDD	$C_{12}HCl_7O_2$	1,2,3,4,6,7,8-heptachlorodibenzo-*p*-dioxin	35822-46-9
	OCDD	$C_{12}Cl_8O_2$	octachlorodibenzo-*p*-dioxin	3268-87-9
Polychlorinated dibenzofurans	2,3,7,8TCDF	$C_{12}H_4Cl_4O$	2,3,7,8-tetrachlorodibenzofuran	51207-31-9
	1,2,3,7,8PeCDF	$C_{12}H_3Cl_5O$	1,2,3,7,8-pentachlorodibenzofuran	57117-41-6
	2,3,4,7,8PeCDF	$C_{12}H_3Cl_5O$	2,3,4,7,8-pentachlorodibenzofuran	57117-31-4
	1,2,3,4,7,8HxCDF	$C_{12}H_2Cl_6O$	1,2,3,4,7,8-hexachlorodibenzofuran	70648-26-9
	1,2,3,6,7,8HxCDF	$C_{12}H_2Cl_6O$	1,2,3,6,7,8-hexachlorodibenzofuran	57117-44-9
	1,2,3,7,8,9HxCDF	$C_{12}H_2Cl_6O$	1,2,3,7,8,9-hexachlorodibenzofuran	72918-21-9
	2,3,4,6,7,8HxCDF	$C_{12}H_2Cl_6O$	2,3,4,6,7,8-hexachlorodibenzofuran	60851-34-5
	1,2,3,4,6,7,8HpCDF	$C_{12}HCl_7O$	1,2,3,4,6,7,8-heptachlorodibenzofuran	67562-39-4
	1,2,3,4,7,8,9HpCDF	$C_{12}HCl_7O$	1,2,3,4,7,8,9-heptachlorodibenzofuran	55673-89-7
	OCDF	$C_{12}Cl_8O$	Octachlorodibenzofuran	39001-02-0

TABLE 6.7
Information of High-Temperature Pyrolysis Model

Model	Type	Number	Mark
Reaction model	RYield	8	P24DDF, P25DDF, P26DDF, P27DDF, P28DDF RY2, RY3, RY4
	Sep	1	Sep2
	Mixer	1	Mix2
	FSplit	5	Split2, Split3, Split4, Split5, Split6

details are as follows: (1) P14DDF focuses on congeners with four chlorine atoms substituted, specifically 2,3,7,8-TCDD and 2,3,7,8-TCDF; (2) P15DDF targets congeners with five chlorine atoms substituted, including 1,2,3,7,8-PeCDD, 1,2,3,7,8-PeCDF, and 2,3,4,7,8-PeCDF; (3) P16DDF focuses on congeners with six chlorine atoms substituted, such as 1,2,3,4,7,8-HxCDD, 1,2,3,6,7,8-HxCDD, 1,2,3,7,8,9-HxCDD, 1,2,3,4,7,8-HxCDF, 1,2,3,6,7,8-HxCDF, 1,2,3,7,8,9-HxCDF, and 2,3,4,6,7,8-HxCDF; (4) P17DDF targets congeners with seven chlorine atoms substituted, including 1,2,3,4,6,7,8-HpCDD, 1,2,3,4,6,7,8-HpCDF, and 1,2,3,4,7,8,9-HpCDF; and (5) P18DDF focuses on congeners with eight chlorine atoms substituted, namely OCDD and OCDF.

The thermal decomposition reactions in these five reactors produce trace amounts of corresponding DXN congeners, which are the congeners that remain incompletely decomposed during the pyrolysis process. The specific process can be represented as follows:

$$\begin{cases} ① \begin{cases} 2,3,7,8\text{TCDF} + 2,3,7,8\text{TCDD} + O_2 + \text{CO} + \text{CBzs} \xrightarrow{850^\circ\text{C}} 2,3,7,8\text{TCDF} \\ + 2,3,7,8\text{TCDD} + \text{CBzs}' + \text{CPhz}' + CO_2 + H_2O \end{cases} \\ ② \begin{cases} 1,2,3,7,8\text{PeCDF} + 2,3,4,7,8\text{PeCDF} + 1,2,3,7,8\text{PeCDD} + O_2 + \text{CO} + \text{CBzs} \xrightarrow{850^\circ\text{C}} \\ 1,2,3,7,8\text{PeCDF} + 2,3,4,7,8\text{PeCDF} + 1,2,3,7,8\text{PeCDD} + \text{CBzs}' + \text{CPhz}' \\ + CO_2 + H_2O \end{cases} \\ ③ \begin{cases} 1,2,3,4,7,8\text{HxCDF} + 1,2,3,6,7,8\text{HxCDF} + 1,2,3,7,8,9\text{HxCDF} \\ + 2,3,4,6,7,8\text{HxCDF} + 1,2,3,4,7,8\text{HxCDD} + 1,2,3,6,7,8\text{HxCDD} \\ + 1,2,3,7,8,9\text{HxCDD} + O_2 + \text{CO} + \text{CBzs} \xrightarrow{850^\circ\text{C}} 1,2,3,4,7,8\text{HxCDF} \\ + 1,2,3,6,7,8\text{HxCDF} + 1,2,3,7,8,9\text{HxCDF} + 2,3,4,6,7,8\text{HxCDF} \\ + 1,2,3,4,7,8\text{HxCDD} + 1,2,3,6,7,8\text{HxCDD} + 1,2,3,7,8,9\text{HxCDD} + \text{CBzs}' \\ + \text{CPhz}' + CO_2 + H_2O \end{cases} \\ ④ \begin{cases} 1,2,3,4,6,7,8\text{HpCDF} + 1,2,3,4,7,8,9\text{HpCDF} + 1,2,3,4,6,7,8\text{HpCDD} + O_2 + \text{CO} \\ + \text{CBzs} \xrightarrow{850^\circ\text{C}} 1,2,3,4,6,7,8\text{HpCDF} \\ + 1,2,3,4,7,8,9\text{HpCDF} + 1,2,3,4,6,7,8\text{HpCDD} + \text{CBzs}' + \text{CPhz}' + CO_2 + H_2O \end{cases} \\ ⑤ \begin{cases} \text{OCDF} + \text{OCDD} + O_2 + \text{CO} + \text{CBzs} \xrightarrow{850^\circ\text{C}} \text{OCDF} + \text{OCDD} + \text{CBzs}' + \text{CPhz}' \\ \quad + CO_2 + H_2O \end{cases} \end{cases} \tag{6.20}$$

where CBzs′ is CBzs with low chlorine substitution, CPhz′ is CPhs with low chlorine substitution. In specific reactions, their lower chlorine content is reflected in fewer chlorine substitutions compared to DXN congeners.

Additionally, the RY3 and RY4 reactors correspond to the pyrolysis reactions of CPhs and CBzs, respectively. Their reaction mechanisms are similar to those of the P14DDF, P15DDF, P16DDF, P17DDF, and P18DDF reactors, meaning that they take inputs from CPhs and CBzs, with their pyrolysis products also containing trace amounts of these compounds.

After the high-temperature pyrolysis process, the outputs from the eight RYield reactors are mixed with Stream 33 using Mixer 2 to obtain all gas-phase substances from the DXN high-temperature pyrolysis stage. The final output is represented by Stream 57.

6.3.2.4 High-Temperature Gas-Phase Synthesis Sub-Module

Using the products from the high-temperature pyrolysis stage as input, a DXN gas-phase synthesis reaction model is constructed with a temperature range of 800°C to 500°C. Specific information is provided in Table 6.8.

This stage begins with the use of the Sep3 separator to divide the gaseous products from Mixer 2 into three streams as follows. (1) Stream 58: This stream contains trace amounts of DXN, CBzs, CO, the majority of HCl, and a significant portion of O_2; (2) Stream 59: This stream consists predominantly of HCl, CBzs, and CPhs; (3) Stream 60: This stream comprises CO_2, H_2O, minor amounts of O_2, and SO_2.

In the high-temperature gas-phase generation reaction zone, the main processes include DXN generation and the HCl oxidation reaction. The HCl oxidation reaction is conducted in the RS1 reactor, which also receives input from Stream 62 (containing HCl, CPhs, and CBzs). Therefore, the reaction process can be represented as follows:

$$4HCl + O_2 \rightarrow 2Cl_2 + 2HO_2 \tag{6.21}$$

TABLE 6.8
Information of High-Temperature Gas-Phase Formation Model for DXN

Model	Type	Number	Mark
Reaction model	RYield	10	P34DF, P35DF, P36DF, P37DF, P38DF, P34DD, P35DD, P36DD, P37DD, P38DD
	RStoic	1	RS1
	Sep	1	Sep3
	Mixer	1	Mix3
	FSplit	2	Split7, Split8

In the DXN re-synthesis reaction, 10 RYield reactors are employed, designated as P14DF, P15DF, P16DF, P17DF, P18DF, P14DD, P15DD, P16DD, P17DD, and P18DD. These reactors correspond to DXN congeners with 4, 5, 6, 7, and 8 chlorine atom substitutions, respectively. The temperature and pressure settings for all 10 RYield reactors are 750°C and 101 kPa.

Other products generated in this zone include lower CPhs (with fewer chlorine substitutions than the produced DXN), lower CBzs (also with fewer chlorine substitutions than the produced DXN), and conventional gases. Consequently, without considering product concentration, the DXN high-temperature gas-phase generation reaction process corresponds to Eq. (6.21). For a detailed explanation of the reaction mechanism, please refer to the high-temperature gas-phase reaction mechanism.

After the completion of the DXN high-temperature gas-phase generation reaction, the outputs from the 10 RYield reactors are combined with the output from the RS1 reactor using Mixer 3. This mixing process yields all gas-phase substances from the DXN high-temperature gas-phase generation stage, with the final output designated as Stream 84.

6.3.2.5 Low-Temperature Catalytic Synthesis Sub-Module

This sub-module mainly involves the low-temperature catalytic synthesis of DXN. The input primarily consists of precursors that have fully reacted during the DXN high-temperature gas-phase generation stage, metal catalysts that did not participate in the high-temperature gas-phase synthesis reaction, particulate matter, carbon sources, and other materials. The reaction temperature range is between 500°C and 200°C. Specific information is provided in Table 6.9.

In this zone, the gas-phase products from Mixer 3 are separated into four streams using the Sep4 separator. The separated streams are as follows: (1) Stream 85: This stream contains CBzs, CPhs, and a portion of $CuCl_2$; (2) Stream 86: This stream comprises common gases such as CO_2, H_2O, a portion of O_2, and SO_2; (3) Stream 87: This stream includes DXN generated in the high-temperature gas phase, along with CO, CO_2, HCl, the majority of Cl_2, and a portion of O_2; and (4) Stream 88: This stream contains a small portion of Cl_2 and a portion of $CuCl_2$.

TABLE 6.9
Information of Low-Temperature Precursor Catalytic Formation Model

Model	Type	Number	Mark
Reaction model	RYield	20	P44DF, P45DF, P46DF, P47DF, P48DF, P44DD, P45DD, P46DD, P47DD, P48DD, P54DF, P55DF, P56DF, P57DF, P58DF, P54DD, P55DD, P56DD, P57DD, P58DD
	Heater	2	H1, H2
	Sep	1	Sep4
	Mixer	3	Mix4, Mix5, Mix6
	FSplit	5	Split9, Split10, Split11, Split12, Split13

In the following sections, the modeling process of the DXN low-temperature catalytic generation reaction is described, focusing on the low-temperature precursor catalytic synthesis and de novo synthesis components.

6.3.2.5.1 Low-Temperature Precursor Catalytic Synthesis

This component utilizes 10 RYield reactors (P44DF, P45DF, P46DF, P47DF, P48DF, P44DD, P45DD, P46DD, P47DD, and P48DD) to simulate the catalytic synthesis of DXN from precursors. The input for these reactors is sourced from the Split12 splitter, which receives inputs from Stream 89 (containing CBzs, CPhs, and $CuCl_2$) and Stream 91 (containing common gases such as CO_2, H_2O, O_2, and SO_2). The temperature and pressure settings for all 10 RYield reactors are maintained at 350°C and 101 kPa. Other reaction products include lower CPhs (with fewer chlorine substitutions than the generated DXN), lower CBzs (also with fewer chlorine substitutions than the generated DXN), and conventional gases. Therefore, the reaction process can be represented as follows:

$$\begin{cases}
①(CBzs+CPhz+H_2O\xrightarrow{CuCl_2}2,3,7,8TCDF+CBzs'+CPhz'+HCl+H_2O\\
②\begin{cases}CBzs+CPhz+H_2O\xrightarrow{CuCl_2}1,2,3,7,8PeCDF+2,3,4,7,8PeCDF+CBzs'\\
\qquad +CPhz'+HCl+H_2O\end{cases}\\
③\begin{cases}CBzs+CPhz+H_2O\xrightarrow{CuCl_2}1,2,3,4,7,8HxCDF+1,2,3,6,7,8HxCDF\\
\qquad +1,2,3,7,8,9HxCDF+2,3,4,6,7,8HxCDF+CBzs'\\
\qquad +CPhz'+HCl+H_2O\end{cases}\\
④\begin{cases}CBzs+CPhz+H_2O\xrightarrow{CuCl_2}1,2,3,4,6,7,8HpCDF+1,2,3,4,7,8,9HpCDF\\
\qquad +CBzs'+CPhz'+HCl+H_2O\end{cases}\\
⑤(CBzs+CPhz+H_2O\xrightarrow{CuCl_2}OCDF+CBzs'+CPhz'+HCl+H_2O\\
⑥(CBzs+CPhz+H_2O\xrightarrow{CuCl_2}2,3,7,8TCDD+CBzs'+CPhz'+HCl+H_2O\\
⑦(CBzs+CPhz+H_2O\xrightarrow{CuCl_2}1,2,3,7,8PeCDD+CBzs'+CPhz'+HCl+H_2O\\
⑧\begin{cases}CBzs+CPhz+H_2O\xrightarrow{CuCl_2}1,2,3,4,7,8HxCDD+1,2,3,6,7,8HxCDD\\
\qquad +1,2,3,7,8,9HxCDD+CBzs'+CPhz'+HCl+H_2O\end{cases}\\
⑨(CBzs+CPhz+H_2O\xrightarrow{CuCl_2}1,2,3,4,6,7,8HpCDD+CBzs'+CPhz'+HCl+H_2O\\
⑩(CBzs+CPhz+H_2O\xrightarrow{CuCl_2}OCDD+CBzs'+CPhz'+HCl+H_2O
\end{cases} \tag{6.22}$$

After the completion of the DXN low-temperature precursor catalytic synthesis reaction, the outputs from the 10 RYield reactors contain 17 congeners of DXN.

Subsequently, these outputs are mixed using Mixer 5, yielding the synthesized DXN from precursors, with the final output represented as Stream 136.

6.3.2.5.2 De Novo Synthesis

This component also utilizes 10 RYield reactors (P54DF, P55DF, P56DF, P57DF, P58DF, P54DD, P55DD, P56DD, P57DD, and P58DD) to simulate the de novo synthesis of DXN. The input for these reactors comes from the Split13 splitter, which receives inputs from a stream containing particulate matter with carbon sources, Stream 92 (containing common gases such as CO_2, H_2O, O_2, and SO_2), and Stream 94 (containing Cl_2 and $CuCl_2$). The temperature and pressure settings for all 10 RYield reactors are maintained at 350°C and 101 kPa. Other reaction products generated include lower CPhs (with fewer chlorine substitutions than the produced DXN), lower CBzs (also with fewer chlorine substitutions than the produced DXN), and conventional gases. Consequently, the reaction process can be represented as follows:

$$
\begin{cases}
\text{①}(C + H_2 + H_2O + Cl_2 \xrightarrow{CuCl_2} 2,3,7,8TCDF + CBzs' + CPhz' + H_2 + H_2O \\
\text{②}\begin{cases} C + H_2 + H_2O + Cl_2 \xrightarrow{CuCl_2} 1,2,3,7,8PeCDF + 2,3,4,7,8PeCDF + CBzs' \\ \qquad + CPhz' + H_2 + H_2O \end{cases} \\
\text{③}\begin{cases} C + H_2 + H_2O + Cl_2 \xrightarrow{CuCl_2} 1,2,3,4,7,8HxCDF + 1,2,3,6,7,8HxCDF \\ \qquad + 1,2,3,7,8,9HxCDF + 2,3,4,6,7,8HxCDF + CBzs' \\ \qquad + CPhz' + H_2 + H_2O \end{cases} \\
\text{④}\begin{cases} C + H_2 + H_2O + Cl_2 \xrightarrow{CuCl_2} 1,2,3,4,6,7,8HpCDF + 1,2,3,4,7,8,9HpCDF \\ \qquad + CBzs' + CPhz' + H_2 + H_2O \end{cases} \\
\text{⑤}(C + H_2 + H_2O + Cl_2 \xrightarrow{CuCl_2} OCDF + CBzs' + CPhz' + H_2 + H_2O \\
\text{⑥}(C + H_2 + H_2O + Cl_2 \xrightarrow{CuCl_2} 2,3,7,8TCDD + CBzs' + CPhz' + H_2 + H_2O \\
\text{⑦}(C + H_2 + H_2O + Cl_2 \xrightarrow{CuCl_2} 1,2,3,7,8PeCDD + CBzs' + CPhz' + H_2 + H_2O \\
\text{⑧}\begin{cases} C + H_2 + H_2O + Cl_2 \xrightarrow{CuCl_2} 1,2,3,4,7,8HxCDD + 1,2,3,6,7,8HxCDD \\ \qquad + 1,2,3,7,8,9HxCDD + CBzs' + CPhz' + H_2 + H_2O \end{cases} \\
\text{⑨}(C + H_2 + H_2O + Cl_2 \xrightarrow{CuCl_2} 1,2,3,4,6,7,8HpCDD + CBzs' + CPhz' + H_2 + H_2O \\
\text{⑩}(C + H_2 + H_2O + Cl_2 \xrightarrow{CuCl_2} OCDD + CBzs' + CPhz' + H_2 + H_2O
\end{cases}
\tag{6.23}
$$

After the completion of the DXN low-temperature de novo synthesis reaction, the outputs from the 10 RYield reactors are mixed using Mixer 6. This mixing process results in the synthesized DXN from de novo reactions, with the final output designated as Stream 137

According to the flowchart of the MSWI process, Mixer 6 combines 3 streams to obtain the complete gas-phase output, designated as Stream 138. The 3 streams involved are Stream 136, Stream 137, and Stream 105. The details are as follows: (1) Stream 136 corresponds to the low-temperature precursor catalytic synthesis; (2) Stream 137 represents the low-temperature de novo synthesis; and (3) Stream 105 includes other gas-phase substances that did not participate in the DXN generation reaction.

Subsequently, heat exchanger H2 is employed to cool the flue gas to 200°C. This cooling process results in the gas-phase composition and concentration at the outlet of the waste heat exchange stage. The DXN concentration obtained at this point is equivalent to the DXN generation concentration at the actual sampling location.

6.3.2.6 Active Carbon Adsorption Modeling Module

In the actual MSWI process, gas cleaning encompasses denitrification, deacidification, DXN removal, heavy metal removal, and particulate matter removal. Since the research focus in this article utilizes selective non-catalytic reduction technology for denitrification, this aspect is not included in the CNSM modeling process. The gas cleaning stage constructed for the DXN CNSM model includes deacidification, DXN removal, heavy metal removal, and particulate matter removal. Using the products from the DXN generation stage as input, a gas cleaning reaction model is developed under temperature conditions below 200°C. Specific information regarding this model is provided in Table 6.10.

In this zone, the RS2 reactor is first utilized to perform desulfurization and deacidification treatment on the gas-phase products output from heat exchanger H_2. The input to the RS2 reactor also includes Stream 140, which primarily consists of $Ca(OH)_2$. The specific reaction process is as follows:

$$\begin{cases} ①(SO_2 + CA(OH)_2 \rightarrow CASO_2 + H_2O \\ ②(SO_3 + CA(OH)_2 \rightarrow CASO_3 + H_2O \\ ③(2HCl + CA(OH)_2 \rightarrow CACl_2 + 2H_2O \end{cases} \quad (6.24)$$

After the flue gas desulfurization and deacidification reaction is complete, the output is designated as Stream 141. Subsequently, Mixer 6 is used to combine Stream 141 with Stream 142 (which primarily consists of activated carbon) before inputting them into the bag filter.

TABLE 6.10
Information of Flue Gas Purification Stage Model

Model	Type	Number	Mark
Reaction model	RStoic	1	RS2
	Mixer	1	Mix7
	Aspen adsorption	1	Bag filter

TABLE 6.11
Theoretical Assumptions of Activated Carbon Adsorption Bed Model

Discretization Method	First-Order Upwind Scheme (UDS 1)
Number of nodes	20
Mass balance	Convection
Momentum balance	Kaman-Kozeny equation
Energy balance	Equal temperature
Kinetic modal	Lumped resistance
Transfer coefficient	Constant
Isothermal	Langmuir type 1

This chapter equates the activated carbon injection and bag filter processes in the actual MSWI process to an activated carbon adsorption bed for modeling the DXN adsorption mechanism. The activated carbon adsorption bed model is constructed using Aspen Adsorption. Table 6.11 presents the methods corresponding to the assumptions utilized in the model-building and calculation process.

The specific implementation methods and strategies for the activated carbon adsorption bed model are as follows: (1) Discretization method: A first-order upwind difference scheme is employed as the standard discretization method. This approach balances stability, rapid convergence, accuracy, and minimal oscillation; (2) Material balance: The diffusion term is omitted from the material balance, assuming that only convective transport occurs during the adsorption process; (3) Momentum balance: The Kaman-Kozeny equation is utilized for calculating momentum balance; (4) Temperature assumptions: It is assumed that gas and solid temperatures remain constant and equal throughout the adsorption process, allowing for the use of an isothermal mode in energy balance calculations; (5) Kinetic model: A linear lumped resistance model is applied as the kinetic model; (6) Mass transfer coefficient: The mass transfer coefficient is considered constant during the adsorption process; and (7) Adsorption isotherm: The Langmuir type 1 isotherm is used to describe the adsorption isotherm. This isotherm assumes monolayer adsorption on a surface with a uniform distribution of adsorption sites, constant adsorption energy, and negligible interactions between adsorbate molecules. Consequently, the adsorption isotherm represents the relationship between the amount of DXN adsorbed per gram of activated carbon and the concentration of DXN at adsorption equilibrium. The equation is as follows:

$$q = \frac{q_m b c_e}{\left(1 + b c_e\right)} \tag{6.25}$$

where q is adsorption capacity of the adsorbent (mg/g), c_e is concentration of the adsorbate (mg/L), q_m is maximum adsorption capacity when the adsorbate fully covers the surface (mg/g), and b is adsorption equilibrium constant (L/mg).

After the activated carbon adsorption bed model is constructed in Aspen Adsorption, the adsorption process [30], based on the initial DXN concentration provided by Stream 141 and the activated carbon from Stream 142, can be represented as follows:

$$\frac{dc_{\mathrm{DXN}}^{\mathrm{Generation}}}{dt} = -k_i A_a c_{\mathrm{DXN}}^{\mathrm{Generation}} \tag{6.26}$$

where $c_{\mathrm{DXN}}^{\mathrm{Generation}}$ is concentration of DXN in flue gas (ng/Nm^3), t is the time, k_i indicates adsorption coefficient, and A_a is effective adsorption sites.

Therefore, the removal efficiency of DXN can be expressed as:

$$\eta_i = \frac{c_{\mathrm{DXN}}^{\mathrm{Generation}} - c_{\mathrm{DXN}}^{\mathrm{Emission}}}{c_{\mathrm{DXN}}^{\mathrm{Generation}}} = 1 - \exp\left(-k_i A_a t\right) \tag{6.27}$$

where $c_{\mathrm{DXN}}^{\mathrm{Emission}}$ represents emission concentration of DXN in flue gas.

After determining the removal efficiency of the 17 congeners based on the activated carbon adsorption bed model, the bag filter reactor separates the gas-phase DXN and solid particles from the flue gas according to these removal efficiency results. Consequently, the outputs are designated as Stream 144 (DXN removed by adsorption) and Stream 145 (DXN not removed by adsorption).

6.3.2.7 Emission to the Atmosphere Sub-Module

In the DXN emission stage, the gas-phase products output from the bag filter reactor are used as input to construct a DXN emission model. Specific information regarding this model is provided in Table 6.12.

This stage is a single-input, single-output model that does not involve any chemical reaction processes. Its purpose is to simulate the emission of trace gas-phase DXN into the atmosphere alongside flue gas after DXN removal. In this context, assuming the emitted DXN mass concentration is represented as $c_{\mathrm{DXN}}^{\mathrm{Emission}}$, the final TEQ concentration of DXN emissions can be expressed as:

$$c_{\mathrm{DXNTEQ}}^{\mathrm{Emission}}\left(\mathrm{ng\,TEQ}/\mathrm{m}^3\right) = \frac{21 - \varphi_{\mathrm{normal}}\left(O_2\right)}{21 - \varphi_{\mathrm{working}}\left(O_2\right)} \times c_{\mathrm{DXN}}^{\mathrm{Emission}}\left(\mathrm{ng}/\mathrm{m}^3\right) \tag{6.28}$$

6.3.3 Implementation Process Description

The implementation process of the CNSM model using FLIC and Aspen software is illustrated in Figure 6.3.

TABLE 6.12
Information of DXN Emission Model

Model	Type	Number	Mark
Reaction model	Compr	1	Stack

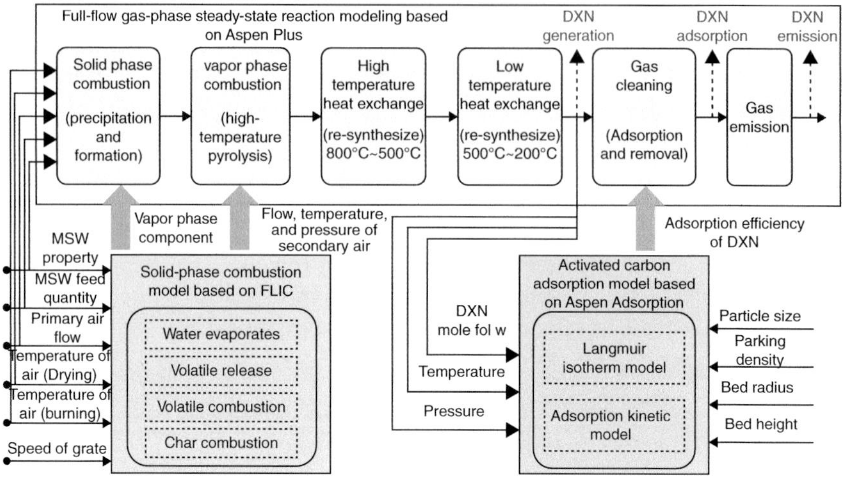

FIGURE 6.3 Implementation process of the CNSM model.

In general, the implementation steps for the CNSM model are as follows:

(1) Obtain elemental and industrial analysis data of MSW by collecting actual MSW samples entering the furnace. Based on the feasible parameter domain of the actual MSWI process, employ orthogonal experimental designs to derive multiple input conditions for the model.
(2) Build a comprehensive gas-phase reaction mechanism model for DXN using Aspen Plus, taking into account the division of five regions and the DXN evolution mechanism.
(3) With fixed parameters for the MSWI process and assumed parameters for MSW, input the previously obtained information from the multiple conditions to generate gas-phase products from MSW combustion. This will yield the comprehensive generation concentrations of DXN under various conditions.
(4) Construct an activated carbon adsorption model within Aspen Adsorption based on the principles of activated carbon adsorption beds. Utilize the comprehensive generation concentrations of DXN to simulate the adsorption efficiency of 17 congeners, subsequently obtaining the adsorption and emission concentrations of DXN under multiple conditions.
(5) These steps collectively establish a comprehensive simulation mechanism model for DXN, allowing for the evaluation of DXN generation, adsorption, and emission across multiple conditions.

Remark: This study assumes that the facility operates at thermodynamic equilibrium, with stable temperature, pressure, and component concentrations throughout the process. Reconciling our model with the reality that MSWI facilities do not maintain thermodynamic equilibrium presents a significant challenge that requires further investigation. This is an important direction for future research, and there are two potential approaches to address this limitation. The first approach is to adjust the model parameters using a larger dataset of real DXN detection data. By doing so, the model can be refined to better capture the dynamic nature of MSWI facilities. The second approach involves exploring the application of data-driven models by leveraging AI technology to guide and modify the parameters of this model. This would enable the model to learn patterns and relationships that may not be easily captured by traditional thermodynamic equilibrium assumptions. As a result, the ability to simulate real-world MSWI scenarios would be enhanced, leading to more accurate predictions of DXN emissions.

6.4 EXPERIMENTAL STUDY

6.4.1 Data Description

During the simulation process, both the solid-phase combustion and gas-phase reaction processes adhere to the principles of thermodynamic equilibrium, mass conservation, energy conservation, and momentum conservation. Thus, the industrial analysis of MSW forms the foundation for combustion process calculations, while the elemental analysis of MSW provides the basis for understanding combustion products and gas-phase reaction processes. To this end, this study conducts sampling and laboratory analyses of the MSW entering the furnace during the actual MSWI process.

In this study, the quartering method is employed for sampling the waste fed into the furnace. This involves dividing the mixed waste sample into four equal parts and selecting one part as the representative sample, which helps to minimize variability. The MSW industrial analysis includes four components: moisture, volatile matter, fixed carbon, and ash, in accordance with the standards GB/T 211-2017 [31] and GB/T 212-2008 [32] in China. The MSW elemental analysis comprises six elements: carbon (C), hydrogen (H), oxygen (O), nitrogen (N), sulfur (S), and chlorine (Cl), following the standards GB/T 476-2008 [33], GB/T 31391-2015 [34], and CJ/T 96-213 [35] in China. The results of these analyses are presented in Table 6.13.

The benchmark experiment for the CNSM model utilizes the municipal solid waste (MSW) analysis results obtained on March 11, 2021, as input. The data from the industrial and elemental analyses are presented in Table 6.14.

6.4.2 Benchmark Condition Case

6.4.2.1 Parameter Setting

Table 6.15 displays the benchmark settings and the actual ranges for each parameter in the FLIC-based bed combustion model.

Table 6.16 presents the benchmark settings and the actual ranges for each parameter in the Aspen Plus-based solid-gas-phase combustion model.

Table 6.17 displays the temperature and pressure parameter settings for the reactors at each stage of the Aspen Plus-based model.

Table 6.18 presents the benchmark settings for each parameter in the Aspen Adsorption-based simulation model.

TABLE 6.13
Results of MSW Industrial Analysis and Elemental Analysis

		MSW Analysis									
		Industrial Analysis (Average)						Elemental Analysis (Average)			
Date	Counts	C (%)	H (%)	O (%)	N (%)	S (%)	Cl (%)	Moisture (%)	Volatile (%)	Ash (%)	Fixed Carbon (%)
Mar. 2021	1	65.62	8.09	24.93	1.12	0.11	4.27	49.7	32.2	10.3	7.8
May. 2021	1	64.73	8.38	26.26	0.37	0.25	0.64	40.6	40.3	14.9	4.1
Jul. 2021	3	61.91	8.06	28.68	1.12	0.23	7.00	43.8	38.5	11.1	6.5
Sep. 2021	2	63.41	8.24	27.22	0.88	0.25	0.51	42.2	39.9	11.3	6.6
Oct. 2021	4	64.66	8.58	26.14	0.39	0.22	1.31	45.9	33.0	14.2	6.9
Dec. 2021	1	64.57	8.31	26.64	0.31	0.18	0.55	35.8	40.4	14.1	9.7
Jan. 2022	3	64.24	9.11	25.81	0.48	0.37	0.49	39.8	39.8	12.5	7.9

TABLE 6.14
MSW Industrial Analysis and Elemental Analysis Data for Benchmark Experiment

	Items	Account	Unit
Industrial analysis	Moisture	49.7	wt%
	Volatile	32.22	wt%
	Fixed carbon	7.82	wt%
	Ash	10.26	wt%
Elemental analysis	C	65.62	wt%
	H	8.09	wt%
	O	24.93	wt%
	N	1.12	wt%
	S	0.11	wt%

TABLE 6.15
Parameter Settings Values of FLIC-Based Simulation Model

Parameters		Setting	Real Range	Unit
Particle size distribution		100	20–150	Mm
Particle mixing coefficient		2E–06		
Thermal radiance		0.8	0.7–0.9	
Void ratio		0.58	0.5–0.65	%
Density of each component	Moisture	1,000		kg/m^3
	Volatile	1,700		kg/m^3
	Fixed carbon	2,260		kg/m^3
	Ash	2,300		kg/m^3
Specific heat capacity of each component	Moisture	4,180		J/kg °C
	Volatile	3,432		J/kg °C
	Fixed carbon	1,434		J/kg °C
	Ash	1,230		J/kg °C
Pre-exponential factor		5,160,000		
Activation energy of volatile		10,700		
Water evaporation temperature		373		K
Zone of bed		11		m

TABLE 6.16
Parameter Settings Values of Aspen Plus-Based Simulation Model

Parameters	Setting	Real Range	Unit
Secondary air	1,078	0–30,000	m^3/h
Temperature of secondary air	398		°C
Pressure of secondary air	1,090	0–3	kPa
Volume of calcium hydroxide injection	270		kg/h
Volume of active carbon injection	17		kg/h
Temperature of each reactor	See Table 6.17		°C

TABLE 6.17
Temperature and Pressure Parameter Settings Values of Aspen Plus-Based Simulation Model

Phase	Type	Name	Temperature (°C)	Pressure (kPa)
MSW solid–gas-phase combustion	RYield	RY1	850	101
Precipitation and formation	RGibbs	P14DF, P15DF, P16DF, P17DF, P18DF, P14DD, P15DD, P16DD, P17DD, P18DD	400	101
High-temperature pyrolysis	RGibbs	P24DDF, P25DDF, P26DDF, P27DDF, P28DDF RY2, RY3, RY4	900	101
High-temperature gas-phase synthesis	RGibbs	P34DF, P35DF, P36DF, P37DF, P38DF, P34DD, P35DD, P36DD, P37DD, P38DD	750	101
	RStoic	RS1	800	101
Low-temperature precursor catalytic synthesis	RGibbs	P44DF, P45DF, P46DF, P47DF, P48DF, P44DD, P45DD, P46DD, P47DD, P48DD, P54DF, P55DF, P56DF, P57DF, P58DF, P54DD, P55DD, P56DD, P57DD, P58DD,	350	101
Gas cleaning	RStoic	RS2	170	101
	Aspen adsorption	Bag filter	170	101.2
Emission	Compr	Stack	170	101.4

TABLE 6.18
Parameter Settings Values of Aspen Adsorption-Based Simulation Model

Parameters		Setting	Unit
Particle size of active carbon		1.05	mm
Parking density of active carbon		400	kg/m^3
Void between activated carbon particles		0.4	kPa
Void in activated carbon particles		0	m^3 void/m^3
Activated carbon shape factor		1	N/A
Inside diameter of adsorption bed		1E–04	m
Height of adsorption bed		1E–03	m
Transfer coefficient		0.45	
Isotherm parameters 1	PCDD congener	0.0936	N/A
	PCDF congener	0.00936	N/A
Isotherm parameters 2		3.5	N/A

6.4.2.2 Results and Discussions

6.4.2.2.1 Bed Combustion Simulation Based on FLIC

Based on the parameter settings provided in Tables 6.14, 6.1, and 6.15, the results of MSW solid-phase combustion in the FLIC model are illustrated in Figure 6.4.

In Figure 6.4, the horizontal axis represents the length of the grate, while the vertical axis indicates the stacking height of the MSW solids. The drying grate extends from 0 to 2.84 m, the burning1 grate spans from 2.84 to 5.84 m, the burning2 grate ranges from 5.84 to 8.68 m, and the burnout grate covers from 8.68 to 11 m. Based on the experimental results, the specific analysis is as follows:

(1) Figure 6.4(a) and (b) illustrate the temperature distribution results for MSW solid-phase and gas-phase combustion over a total duration of 90 minutes. The movement direction of the MSW on the grate is from left to right, while primary air is injected from below the grate, passing through the MSW bed and entering the furnace to provide sufficient oxygen for combustion. The primary air temperature in the drying section is 473 K, while in the combustion section, it is 453 K. On the drying grate, the combination of high-temperature primary air and heat radiation from the furnace causes the moisture in the MSW to gradually evaporate, resulting in a decrease in mass. As the MSW moves onto the burning1 grate, combustion begins, leading to a rapid decline in stacking height and indicating a significant reduction in mass. At the end of the burning1 grate, the MSW solid temperature reaches a maximum of 1,220 K, signifying the peak of combustion intensity. The highest temperature in the MSW gas phase occurs at the front end of the burning2 grate, reaching 1,460 K. As the combustion process continues, the MSW solid completely burns out, converting to ash, and its temperature eventually drops to approximately 850.15 K.

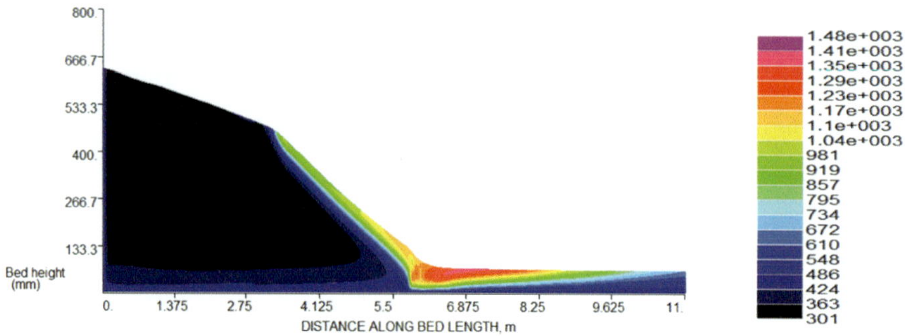

a) Schematic diagram of MSW solid-phase temperature distribution

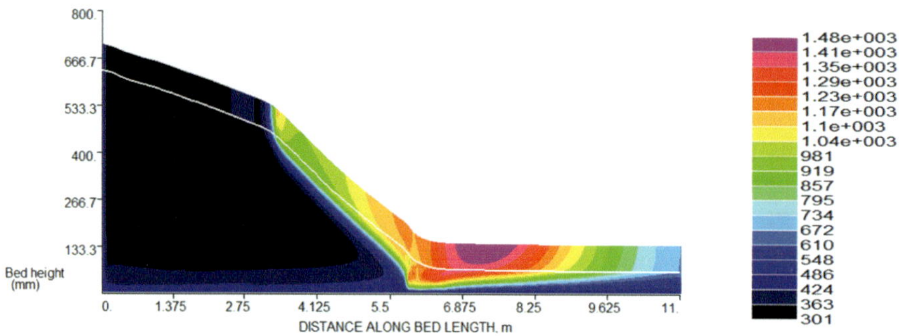

b) Schematic diagram of MSW gas-phase temperature distribution

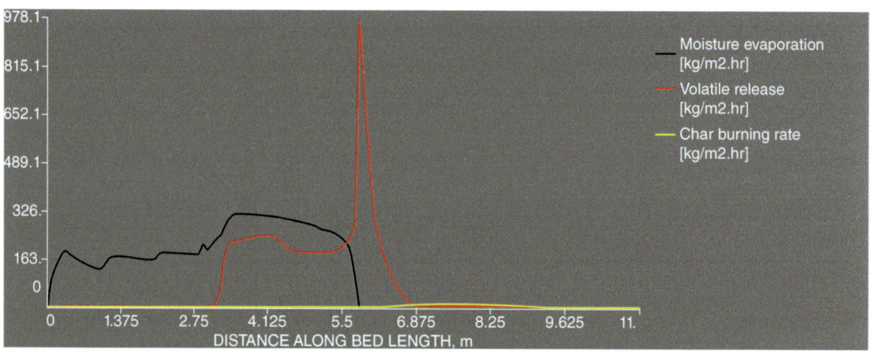

c) Schematic diagram of the distribution of reaction rates in different stages

FIGURE 6.4 Results of MSW solid-phase combustion simulation. (a) Schematic diagram of MSW solid-phase temperature distribution. (b) Schematic diagram of MSW gas-phase temperature distribution. (c) Schematic diagram of the distribution of reaction rates in different stages.

(Continued)

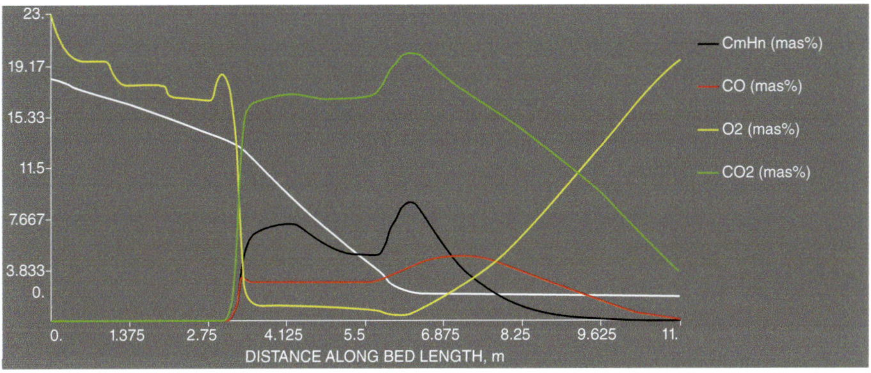

d) Schematic diagram of the distribution of gas concentrations in solid-phase combustion

FIGURE 6.4 (Continued) (d) Schematic diagram of the distribution of gas concentrations in solid-phase combustion.

(2) Figure 6.4(c) presents the reaction rate distribution results for various substances, including three key indicators: the moisture evaporation rate (the mass of MSW moisture evaporated due to heat over a specific period), the volatile release rate (the mass of C_mH_n, CO, CO_2, and H_2O released from MSW due to heat within a certain timeframe), and the char combustion rate (the mass of CO and CO_2 produced from char combustion in MSW during a defined period). From Figure 6.4(c), the following observations can be made: Moisture evaporation occurs between 0 and 5.8 m, encompassing both the drying grate and the first combustion grate; Volatile release takes place between 3.17 and 7.08 m, covering the burning1 and burning2 grates; Char combustion occurs between 6.16 and 10 m, spanning the burning2 combustion grate and the burnout grate. These reaction processes are consistent with the MSW combustion mechanism. Furthermore, under the configured settings for an 800-ton processing capacity, it has been demonstrated that this incinerator can completely combust MSW within 90 minutes.

(3) Figure 6.4(d) presents the distribution results of gas products, specifically illustrating the volume fraction distribution of C_mH_n, O_2, CO, and CO_2. From the results in Figure 6.4(d), several observations can be made: In the drying grate, the primary process is moisture evaporation, characterized by minimal oxygen consumption and nearly zero gas-phase products; Starting from the burning1 grate, the oxygen concentration displays a linear downward trend, indicating that MSW is undergoing combustion and releasing significant amounts of C_mH_n, CO, and CO_2. Between the burning1 and burning2 grates, the trend shows that combustion gradually intensifies, reaches a peak, and then slowly diminishes. As the combustion of MSW weakens, the concentrations of C_mH_n, CO, and CO_2 gradually decrease, while the O_2 concentration steadily increases, ultimately approaching 23%.

Therefore, based on the FLIC solid-phase combustion model, the final molar mass fraction results for the output gas-phase components are presented in Table 6.19.

6.4.2.2.2 Solid–Gas-Phase Simulation Based on Aspen Plus

Based on the gas-phase composition from the FLIC solid-phase combustion model output, as shown in Table 6.19, and utilizing the parameter settings from Tables 6.16 and 6.17, the DXN gas-phase reaction results are presented in Tables 6.20–6.22.

TABLE 6.19
Gas-Phase Results of MSW Solid-Phase Combustion

Gas-Phase Component	Volume Fraction (Vol%)	Molar Mass Fraction (mol%)
C_mH_n	2.293	0.022907093
CO	2.179	0.021768232
CO_2	7.49	0.074825175
O_2	6.676	0.066693307
N_2	38.3	0.426935664
H_2O	35.12	0.350849151
NO_2	5.758	0.013204196
H_2	2.184	0.021818182

TABLE 6.20
Aspen Plus Gas-Phase Simulation Results 1

DXN		Unit	DXN Precipitation and Formation	DXN High-Temperature Pyrolysis
Gas flow rate		mg/h	1,061.66	1,103.69
Temperature		K	673.15	1,173.15
Pressure		kPa	101	101
O_2		%	5.8	6.3
H_2O		%	27	26
PCDD	2,3,7,8TCDD	mg/Nm³	3.2694E−13	4.2665E−14
	1,2,3,7,8PCDD	mg/Nm³	7.2715E−20	1.4609E−10
	1,2,3,4,7,8HCDD	mg/Nm³	2.1142E−25	2.3197E−13
	1,2,3,6,7,8HCDD	mg/Nm³	2.1142E−25	2.3197E−13
	1,2,3,7,8,9HCDD	mg/Nm³	1.4234E−24	1.9395E−12
	1,2,3,4,6,7,8HCDD	mg/Nm³	1.3746E−17	3.5576E−11
	OCDD	mg/Nm³	5.9786E−30	3.3673E−11
PCDF	2,3,7,8TCDF	mg/Nm³	4.4857E−06	3.2357E−19
	1,2,3,7,8PCDF	mg/Nm³	5.1370E−18	1.7674E−16
	2,3,4,7,8PCDF	mg/Nm³	4.1869E−18	2.2496E−16
	1,2,3,4,7,8HCDF	mg/Nm³	6.3244E−18	6.3978E−19
	1,2,3,6,7,8HCDF	mg/Nm³	1.0650E−17	2.3559E−18
	2,3,4,6,7,8HCDF	mg/Nm³	9.8426E−18	1.8509E−18
	1,2,3,7,8,9HCDF	mg/Nm³	1.1781E−17	2.9988E−18

(Continued)

TABLE 6.20 (Continued)
Aspen Plus Gas-Phase Simulation Results 1

DXN		Unit	DXN Precipitation and Formation	DXN High-Temperature Pyrolysis
	1,2,3,4,6,7,8HCDF	mg/Nm³	4.0735E−27	6.5079E−16
	1,2,3,4,7,8,9HCDF	mg/Nm³	3.7233E−27	5.1128E−16
	OCDF	mg/Nm³	1.6182E−22	2.0672E−16
DXNs		mg/Nm³	4.4859E−06	2.1779E−10

TABLE 6.21
Aspen Plus Gas-Phase Simulation Results 2

DXN		Unit	DXN High-Temperature Gas-Phase Synthesis	Low-Temperature Catalytic Synthesis: Low-Temperature Precursor Catalytic Synthesis	Low-Temperature Catalytic Synthesis: Low-Temperature De Novo Synthesis
Gas flow rate		mg/h	1,101.42	1,095.59	1,095.59
Temperature		K	1,023.15	623.15	623.15
Pressure		kPa	101	101	101
O_2		%	5.7	4.6	4.6
H_2O		%	26	26	26
PCDD	2,3,7,8TCDD	mg/Nm³	2.9221E−09	1.2922E−07	4.4972E−08
	1,2,3,7,8PCDD	mg/Nm³	3.2745E−09	1.1444E−07	4.1458E−08
	1,2,3,4,7,8HCDD	mg/Nm³	8.7521E−10	3.7824E−08	1.3703E−08
	1,2,3,6,7,8HCDD	mg/Nm³	8.7521E−10	3.7824E−08	1.3703E−08
	1,2,3,7,8,9HCDD	mg/Nm³	7.2270E−09	2.3810E−07	8.6257E−08
	1,2,3,4,6,7,8HCDD	mg/Nm³	2.7915E−09	9.7557E−08	3.5342E−08
	OCDD	mg/Nm³	2.6157E−09	9.2261E−08	3.3424E−08
PCDF	2,3,7,8TCDF	mg/Nm³	2.0249E−13	1.2280E−07	1.7055E−16
	1,2,3,7,8PCDF	mg/Nm³	2.7902E−09	1.0538E−07	3.8175E−08
	2,3,4,7,8PCDF	mg/Nm³	3.4649E−09	1.1323E−07	4.1019E−08
	1,2,3,4,7,8HCDF	mg/Nm³	1.0540E−09	7.0951E−08	2.5013E−08
	1,2,3,6,7,8HCDF	mg/Nm³	3.4216E−09	1.0990E−07	3.8743E−08
	2,3,4,6,7,8HCDF	mg/Nm³	2.7553E−09	1.0223E−07	3.6058E−08
	1,2,3,7,8,9HCDF	mg/Nm³	4.2491E−09	1.1808E−07	4.1628E−08
	1,2,3,4,6,7,8HCDF	mg/Nm³	2.9758E−09	9.7243E−08	3.5228E−08
	1,2,3,4,7,8,9HCDF	mg/Nm³	2.3963E−09	9.0502E−08	3.2786E−08
	OCDF	mg/Nm³	2.3496E−12	8.9051E−08	3.2261E−08
DXNs		mg/Nm³	4.3691E−08	1.7666E−06	5.8977E−07

TABLE 6.22
Aspen Plus Gas-Phase Simulation Results 3

DXN		Unit	DXN Generation Concentration	DXN Adsorption Concentration	DXN Emission Concentration
Gas flow rate		kmol/h	1,095.59	1,157.32	1,157.32
Temperature		K	453.15	432.15	432.15
Pressure		kPa	101	101	101.4
O_2		%	4.6	7.65	7.65
H_2O		%	26	19	19
PCDD	2,3,7,8TCDD	mg/Nm^3	7.5278E–08	1.1232E–06	5.1567E–09
	1,2,3,7,8PCDD	mg/Nm^3	6.7370E–08	1.0052E–06	4.6150E–09
	1,2,3,4,7,8HCDD	mg/Nm^3	2.2267E–08	0	0
	1,2,3,6,7,8HCDD	mg/Nm^3	2.2267E–08	0	0
	1,2,3,7,8,9HCDD	mg/Nm^3	1.4017E–07	2.0914E–06	9.6019E–09
	1,2,3,4,6,7,8HCDD	mg/Nm^3	5.7432E–08	8.5690E–07	3.9342E–09
	OCDD	mg/Nm^3	5.4314E–08	8.1039E–07	3.7206E–09
PCDF	2,3,7,8TCDF	mg/Nm^3	5.3069E–08	7.9181E–07	3.6353E–09
	1,2,3,7,8PCDF	mg/Nm^3	6.2036E–08	9.2560E–07	4.2496E–09
	2,3,4,7,8PCDF	mg/Nm^3	6.6656E–08	9.9454E–07	4.5661E–09
	1,2,3,4,7,8HCDF	mg/Nm^3	4.1471E–08	6.1876E–07	2.8408E–09
	1,2,3,6,7,8HCDF	mg/Nm^3	6.4234E–08	9.5839E–07	4.4001E–09
	2,3,4,6,7,8HCDF	mg/Nm^3	5.9781E–08	8.9196E–07	4.0951E–09
	1,2,3,7,8,9HCDF	mg/Nm^3	6.9018E–08	1.0298E–06	4.7276E–09
	1,2,3,4,6,7,8HCDF	mg/Nm^3	5.7247E–08	8.5414E–07	3.9215E–09
	1,2,3,4,7,8,9HCDF	mg/Nm^3	5.3279E–08	7.9494E–07	3.6497E–09
	OCDF	mg/Nm^3	5.2424E–08	7.8219E–07	3.5912E–09
DXNs		mg/Nm^3	1.0183E-06	1.4529E–05	6.6706E–08

In the benchmark experiment of the DXN gas-phase reaction, Table 6.19 presents the concentration results for DXN precipitation and formation. During the MSW pyrolysis and gasification process (when the temperature is below 1,123.15 K), the mass concentration of DXN from precipitation and formation is measured at 4.4859E–06 mg/Nm^3. However, when the combustion temperature rises to 1,173.15 K, a significant amount of DXN in the flue gas undergoes pyrolysis, resulting in its concentration reducing to 2.1779E–10 mg/Nm^3, achieving a pyrolysis efficiency of approximately 99.995%.

The next stage is the DXN generation process, with its concentration distribution results presented in Table 6.20. The reaction temperature for the high-temperature gas-phase generation stage of DXN is set to 750°C, resulting in a generated DXN mass concentration of 4.3691E–08 mg/Nm^3. The low-temperature catalytic synthesis process for DXN consists of two sub-processes: the low-temperature precursor catalytic synthesis, yielding a DXN mass concentration of 1.7666E–06 mg/Nm^3, and the low-temperature de novo synthesis, which produces a DXN mass concentration of 5.8977E–07 mg/Nm³. Consequently, during the DXN generation stage, the low-temperature precursor catalytic synthesis contributes the most to the overall generated concentration, while the high-temperature gas-phase generation contributes the least. The final generated DXN mass concentration is 1.0183E–06 mg/Nm³.

6.4.2.2.3 Active Carbon Adsorption Simulation Based on Aspen Adsorption

The generated DXN concentration obtained above is then input into the activated carbon adsorption bed model developed in Aspen Adsorption to simulate DXN removal. The parameters for the activated carbon adsorption bed are detailed in Table 6.18. Based on these parameters, the amount of activated carbon used in the bed is 9.46E–03 kg/kmol, and the adsorption curves for the 17 DXN congeners are illustrated in Figure 6.5.

Considering the actual process design and flue gas flow rate of the MSWI process, and based on the activated carbon adsorption curves shown in Figure 6.4, the adsorption time for the 17 DXN congeners is determined to be 3 seconds. Using Eq. (6.41) for calculation, the adsorption efficiency for all 17 congeners is found to be 95%. Consequently, the final DXN removal efficiency is established at 95%.

Based on this removal efficiency, the concentration distribution results for DXN generation, adsorption, and emission are presented in Table 6.21. The mass concentration of adsorbed DXN is measured at 1.4529E–05 mg/Nm³, while the mass concentration of emitted DXN is 6.6706E–08 mg/Nm³. Subsequently, by employing Eq. (6.42), the DXN TEQ concentration under standard conditions is calculated, yielding

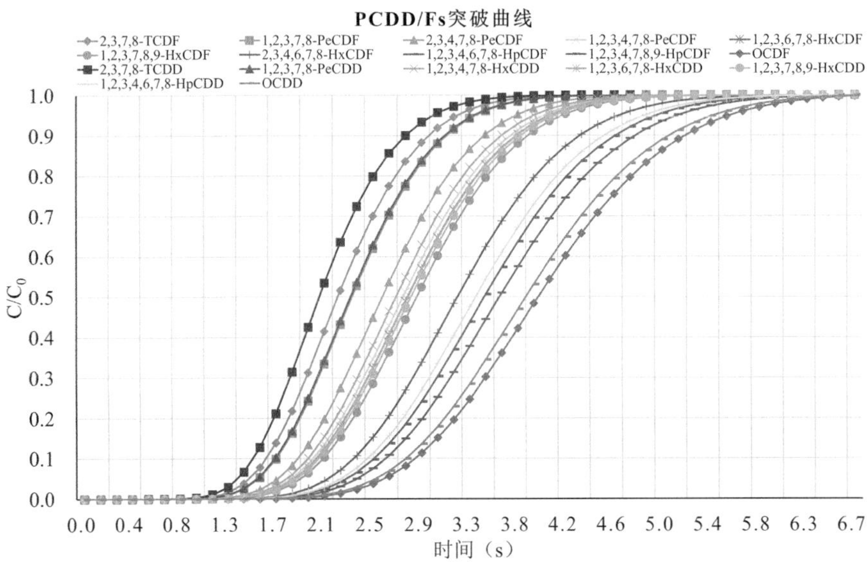

FIGURE 6.5 Adsorption curves of 17 DXN homologs on activated carbon. (a) PCDD. (b) PCDF.

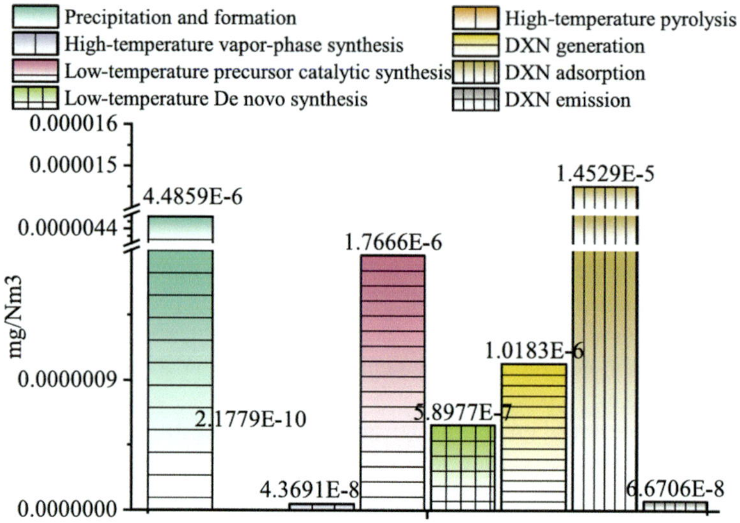

FIGURE 6.6 Whole-process DXN concentration distribution.

a value of 0.0502 ng TEQ/Nm^3. This DXN emission concentration complies with the emission limit specified in the GB 18485-2014 standard.

Figure 6.6 illustrates the DXN concentration distribution results throughout the entire process. The evolution can be summarized as follows: (1) Before reaching 1,123.15 K, the concentration of DXN generated from MSW combustion is relatively high; (2) At 1,123.15 K, the DXN concentration is nearly completely decomposed; (3) As the flue gas temperature gradually decreases, the DXN concentration begins to rise again; (4) A significant amount of the generated DXN is adsorbed and removed by activated carbon, resulting in only a trace amount of DXN that meets emission standards being released into the atmosphere. These experimental results demonstrate the effectiveness of the proposed CNSM model.

6.4.3 Multiple Conditions Case

6.4.3.1 MSW Components

The analysis results of the 15 sets of MSW components are presented in Figure 6.7.

Based on the MSW analysis results presented above, several key findings can be noted:

(1) Seasonal Variation: The mean statistical results indicate that winter MSW exhibits the lowest moisture content (below 40%), which implies a higher heating value. In contrast, spring MSW has the highest moisture content, reaching up to 49.7%, indicating a relatively lower heating value. Elemental analysis shows that the ratios and proportions of C, H, and O remain relatively stable, while chlorine content exhibits significant variation, reaching as high as 7%.

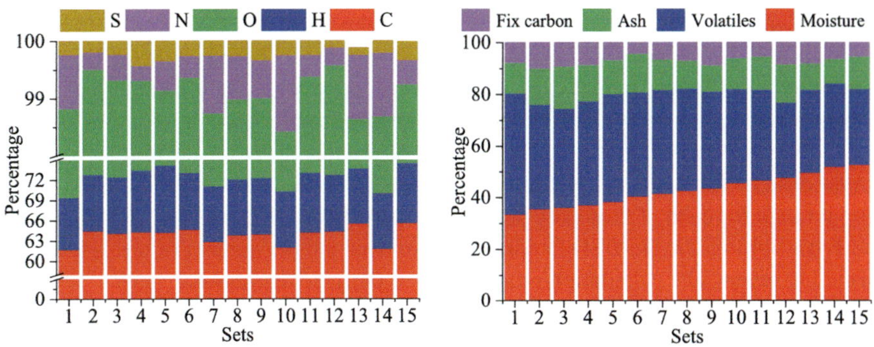

(a) Elemental analysis (b) Industrial analysis

FIGURE 6.7 MSW industrial analysis and elemental analysis results. (a) Elemental analysis. (b) Industrial analysis.

(2) Elemental composition: The analysis data from the 15 sets of MSW (as shown in Figure 6.6) demonstrate that the most stable elements in MSW are C, H, and O, with their combined proportion exceeding 98%. The elements N, S, and Cl collectively account for less than 2%. Additionally, the average moisture content across the 15 sets is approximately 40%, with volatile matter accounting for about 37.3%, ash content around 12.6%, and fixed carbon constituting approximately 7.1%. Under these proportions, the lower heating value of MSW is estimated to be approximately 9,131 KJ/kg.

These findings suggest that waste sorting and disposal effectively reduce the proportions of N, S, and Cl elements in MSW, while simultaneously increasing the heating value of the MSW entering the furnace. This is beneficial for the stable operation of the MSWI process and helps to reduce pollutant emissions.

In summary, the elemental composition of MSW varies across different sampling times, and the distribution of moisture, volatile matter, fixed carbon, and ash also shows differences. Overall, these variations contribute to an increase in the heating value of MSW and a decrease in the proportions of nitrogen (N), sulfur (S), and chlorine (Cl) elements. However, MSW with different compositions can lead to varying combustion conditions, which significantly impact pollutant emissions. Therefore, this article utilizes the 15 sets of MSW component data as representative of 15 different types of MSW for numerical simulation experiments. These experiments investigate the effects of different types of MSW on DXN generation, adsorption, and emission.

6.4.3.2 Experimental Design

The CNSM model for the entire lifecycle of DXN concentration, based on a multi-software coupling approach, can be represented as follows:

$$\boldsymbol{y} = F_{\text{SMM}}\left(\boldsymbol{x}_{\text{Actual}}, \boldsymbol{x}_{\text{Hypothesis}}\right) \tag{6.29}$$

where $F_{\mathrm{SMM}}(\cdot)$ represents the CNSM model; $\boldsymbol{y}$ is the simulated DXN concentration in terms of whole lifecycle; $\boldsymbol{x}_{\mathrm{Actual}}$ indicates input parameters based on the actual MSW and MSWI processes; and $\boldsymbol{x}_{\mathrm{Hypothesis}}$ denotes input parameters for the numerical simulation model based on material or thermal energy equivalent assumptions.

Orthogonal experimental design (OED) is employed to create multi-condition input designs for the five key parameters mentioned above. Table 6.23 displays the input range for each of these key parameters during the OED process.

From Table 6.23, the OED factors are set to five, and the number of levels is established at seven. The calculation process can be outlined as follows:

$$\boldsymbol{X}_{\mathrm{ODE}} = f_{\mathrm{OED}}^{1}\left(\boldsymbol{x}_{\mathrm{FV}}, \boldsymbol{x}_{\mathrm{GS}}, \boldsymbol{x}_{\mathrm{PA}}, \boldsymbol{x}_{\mathrm{PAD}}, \boldsymbol{x}_{\mathrm{PAB}}\right) \in R^{72\times 5} \tag{6.30}$$

where,

$$\begin{cases} \boldsymbol{x}_{\mathrm{FV}} = \left[28{,}000 \quad 28{,}333 \quad 29{,}667 \quad 30{,}500 \quad 31{,}333 \quad 32{,}166 \quad 33{,}000\right] \in R^{1\times 5} \\ \boldsymbol{x}_{\mathrm{PA}} = \left[5.5 \quad 5.9 \quad 6.2 \quad 6.5 \quad 6.8 \quad 7.1 \quad 7.5\right] \in R^{1\times 5} \\ \boldsymbol{x}_{\mathrm{GS}} = \left[1{,}017 \quad 1{,}043 \quad 1{,}070 \quad 1{,}096 \quad 1{,}122 \quad 1{,}149 \quad 1{,}175\right] \in R^{1\times 5} \\ \boldsymbol{x}_{\mathrm{PAD}} = \left[414 \quad 429 \quad 444 \quad 459 \quad 474 \quad 489 \quad 503\right] \in R^{1\times 5} \\ \boldsymbol{x}_{\mathrm{PAB}} = \left[394 \quad 407 \quad 421 \quad 434 \quad 447 \quad 461 \quad 473\right] \in R^{1\times 5} \end{cases} \tag{6.31}$$

where $\boldsymbol{X}_{\mathrm{ODE}} \in R^{72\times 5}$ represents the OED matrix from a 5-factor 7-level design; $f_{\mathrm{OED}}^{1}(\cdot)$ is an orthogonal experimental calculation model; and $\boldsymbol{x}_{\mathrm{FV}}$, $\boldsymbol{x}_{\mathrm{GS}}$, $\boldsymbol{x}_{\mathrm{PA}}$, $\boldsymbol{x}_{\mathrm{PAD}}$, and $\boldsymbol{x}_{\mathrm{PAB}}$ denote MSW feed capacity, grate speed, primary air, primary air (drying) temperature, and primary air (burning) temperature, respectively.

In general, under the 5-factor, 7-level condition, a 72 × 5 orthogonal experimental table is generated. By combining this orthogonal table with the 15 sets of MSW components, the multi-condition design for the simulation mechanism model is established as follows: each set of MSW components is treated as a specific type of MSW. With the input from the orthogonal experimental table, the concentration distribution of DXN can be examined under various operating conditions. Consequently, the total number of inputs for the CNSM model amounts to 1,080 groups.

6.4.3.3 Results and Discussions

Using the MSW composition data and parameter setting ranges from Tables 6.13 and 6.23 as the foundation, a sample space comprising 1,080 input groups with various parameter settings is ultimately obtained. Taking the first set of MSW components as the basis, Figure 6.7 illustrates the distribution space of 72 groups of data under OED.

Figure 6.8 presents a two-dimensional distribution space of 72 groups of orthogonal experimental data generated using the OED model, with the first set of MSW components as the basis. A visual examination of the figure reveals that the data points are dispersed throughout the parameter space, indicating that the OED model effectively captures the variability in inputs corresponding to different operating conditions. This capability is crucial for the simulation study, as it allows us to

TABLE 6.23
Orthogonal Experimental Design for Inputs of CNSM Model

Regions	Factors		Unit	Level						
MSW component	15 groups MSW component									
	MSW feed capacity		tons/h	28,000	28,333	29,667	30,500	31,333	32,166	33,000
	Speed of grate		m/h	5.5	5.9	6.2	6.5	6.8	7.1	7.5
Operation parameters	Drying (24%)	Primary air	Nm³/min	1,017	1,043	1,070	1,096	1,122	1,149	1,175
	First burning (45%)									
	Second burning (20%)									
	Burnout (11%)									
	Temperature of primary air (drying)		K	414	429	444	459	474	489	503
	Temperature of primary air (burning)		K	394	407	421	434	447	461	473

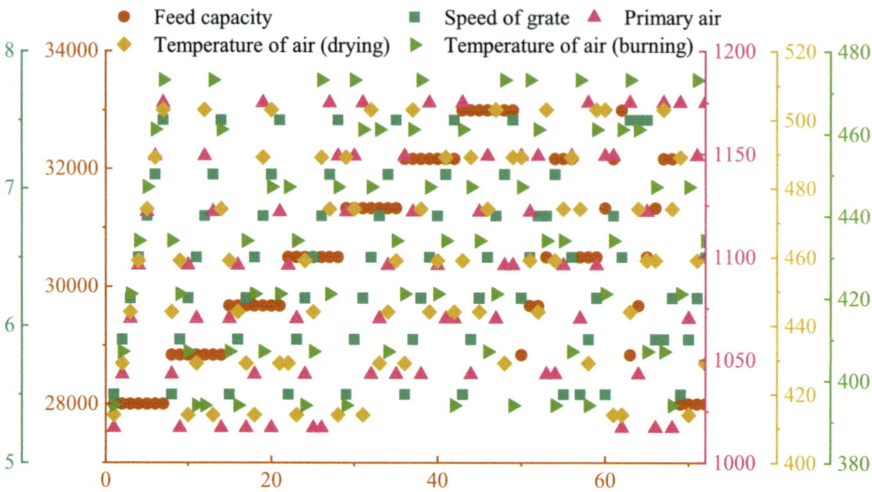

FIGURE 6.8 Distribution space of 72 sets of orthogonal experimental data.

investigate the responses of DXN generation, adsorption, and emission to changes in the operating conditions of the MSWI process.

Furthermore, the distribution of the data points in Figure 6.8 suggests that the OED model can adequately represent multiple operating conditions within the parameter range defined in Table 6.23. This finding is significant, as it demonstrates the potential of the OED model to simulate the behavior of the incineration process across a wide array of operating conditions. By leveraging this capability, we can conduct numerous simulations to analyze the sensitivity of DXN generation, adsorption, and emission to variations in operating conditions.

In light of these findings, we have utilized the 1,080 sets of multi-condition results generated by the OED model as input to investigate the concentration response curves of DXN generation, adsorption, and emission under various operating conditions. The results of this analysis are presented in the subsequent sections of the manuscript, providing new insights into the dynamics of DXN generation and emission during the whole MSWI process.

6.4.3.3.1 Bed Combustion Simulation Based on FLIC

Based on the 1,080 multi-condition experimental designs, the simulation results of the FLIC solid-phase combustion model are presented in Figure 6.9.

As depicted in Figure 6.9, the temperature distribution in the solid-phase combustion and the trends in gas-phase composition offer valuable insights into the dynamics of DXN formation in the whole MSWI process. The solid temperature of the MSW varies within a range of 842–1,467 K, playing a crucial role in the thermal decomposition of organic matter and the subsequent formation of DXNs. The gas-phase temperature on the MSW surface ranges from 1,177 to 1,548 K, significantly influencing the rate of chemical reactions and the distribution of gas-phase species.

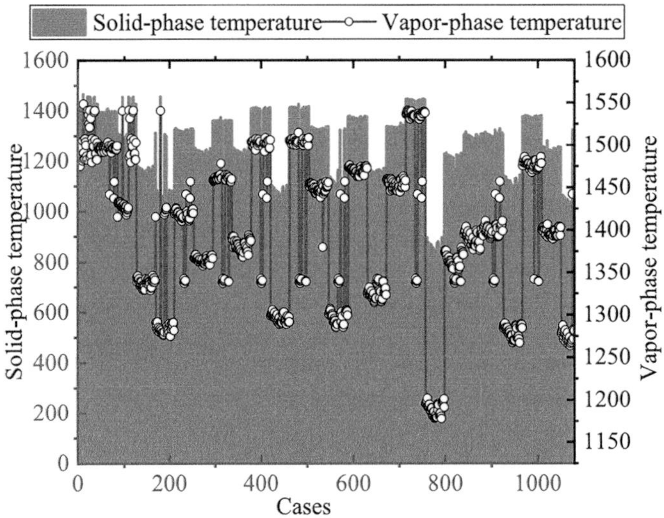

(a) Different temperature distribution viewpoint

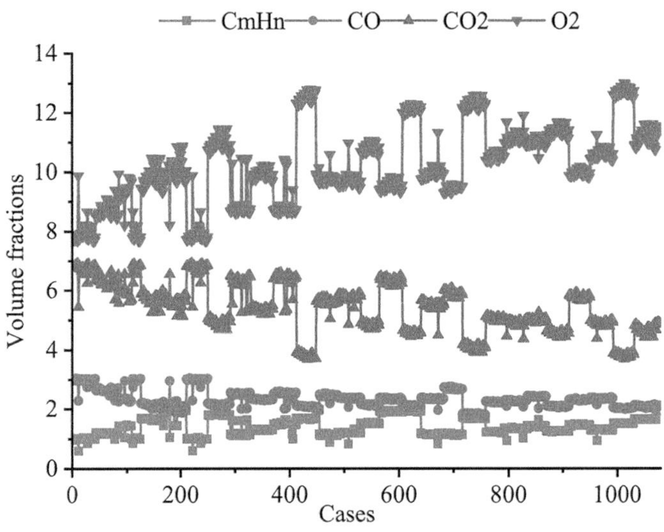

(b) Different gas component viewpoint

FIGURE 6.9 Temperature distribution in solid-phase combustion and trends in gas-phase composition. (a) Different temperature distribution viewpoint. (b) Different gas component viewpoint.

The volume fractions of volatile compounds (C_mH_n), CO, CO_2, and O_2 range from 0.603% to 1.989%, 1.61% to 3.054%, 3.559% to 6.911%, and 7.671% to 13.13%, respectively. These values provide a detailed picture of the gas-phase composition during MSW combustion. These results are particularly significant, as they illustrate the complex interplay between the solid-phase combustion process and the gas-phase chemistry. For instance, the variations in CO and CO_2 concentrations suggest that the combustion process may be incomplete, leading to the formation of partial oxidation products that can influence DXN formation. Furthermore, fluctuations in O_2 concentration imply that the combustion process is oxygen-limited, which can impact both the rate of DXN formation and its destruction.

Simultaneously considering these factors is essential for understanding the dynamics of DXN formation in the whole MSWI process. The validation of the multi-condition OED model is further supported by the reasonable ranges of temperature and gas-phase product compositions obtained from the FLIC solid-phase combustion model. This outcome underscores the effectiveness of the simulation approach in capturing the complexities of MSW combustion and DXN formation.

The 1,080 sets of gas-phase component data derived from MSW solid-phase combustion provide a rich dataset for further analysis and modeling. These data can be utilized to develop more accurate models of DXN formation and destruction, informing strategies to minimize DXN emissions during MSW incineration. In conclusion, the results presented in Figure 6.9 offer a comprehensive understanding of the dynamics of DXN formation during MSW incineration. By expanding on the interpretation of these results, we have provided a detailed picture of the complex interactions between solid-phase combustion and gas-phase chemistry, which is essential for developing effective strategies to reduce DXN emissions.

6.4.3.3.2 Solid–Gas-Phase Simulation Based on Aspen Plus and Adsorption

Using the 1,080 sets of gas-phase component outputs from the FLIC solid combustion model as input, the multi-condition experimental results of the DXN gas-phase reaction process based on Aspen PLUS and adsorption are presented in Figure 6.10.

According to the results presented in Figure 6.10, the following observations can be made:

(1) DXN Precipitation and Formation Stage: The center of the concentration distribution is 0.0073 mg/Nm^3, with a variance of 5.9476E–06. The minimum value is 9.8551E–04 mg/Nm^3, while the maximum value reaches 0.0161 mg/Nm^3.
(2) DXN High-Temperature Pyrolysis Stage: The center of the concentration distribution is 9.11734E–07 mg/Nm^3, with a variance of 9.3052E–14. The minimum value is 4.1170E–08 mg/Nm^3, and the maximum value is 2.0068E–06 mg/Nm^3.
(3) DXN High-Temperature Gas-Phase Synthesis Stage: The center of the concentration distribution is 5.3113E–07 mg/Nm^3, with a variance of 4.5729E–14. The minimum value is 0 mg/Nm^3, while the maximum value is 1.2179E–06 mg/Nm^3.
(4) DXN Low-Temperature Catalytic Synthesis Stage: The center of the DXN generation concentration distribution is 3.1987E–07 mg/Nm^3, with a variance of 1.2974E–14. The minimum value is 2.390E–08 mg/Nm^3, and the maximum value is 7.0236E–07 mg/Nm^3.

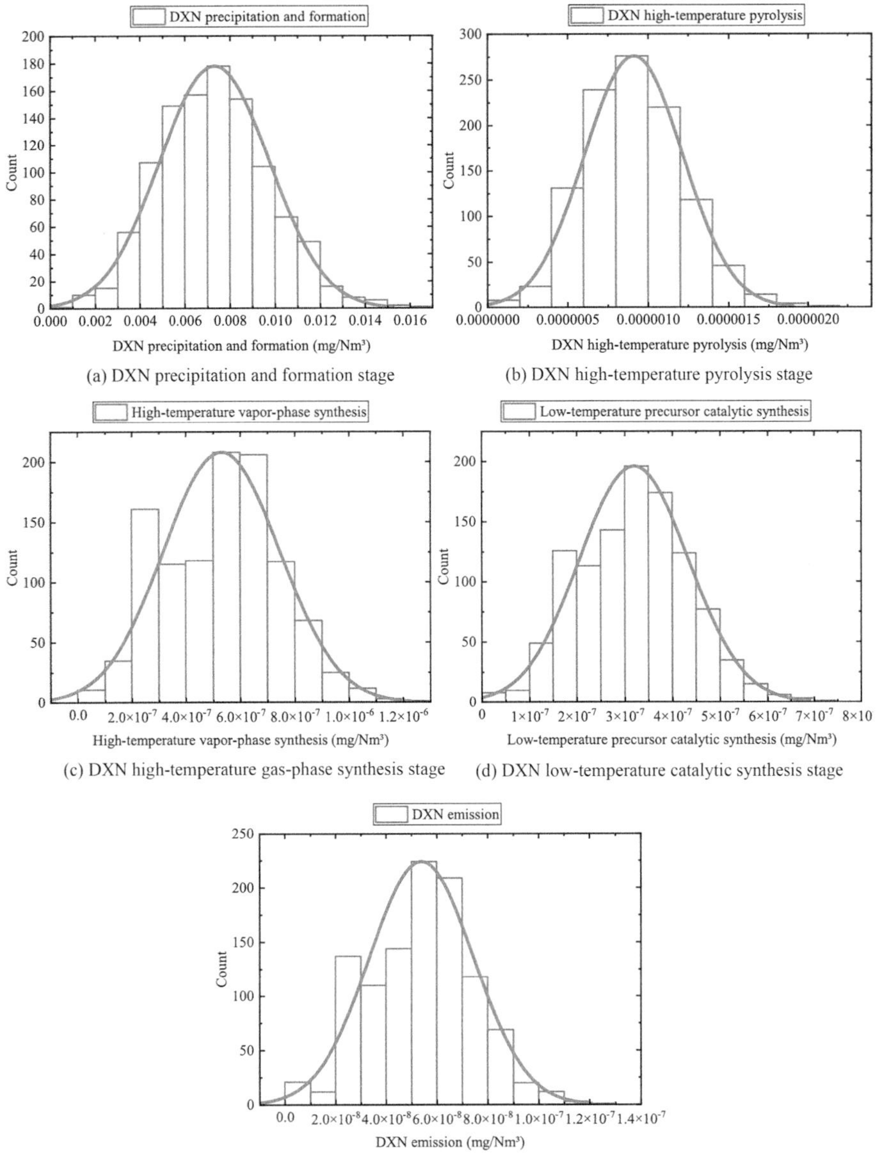

(a) DXN precipitation and formation stage

(b) DXN high-temperature pyrolysis stage

(c) DXN high-temperature gas-phase synthesis stage

(d) DXN low-temperature catalytic synthesis stage

(e) DXN emission stage

FIGURE 6.10 Probability distribution of the whole-process DXN concentration under multiple operating conditions. (a) DXN precipitation and formation stage. (b) DXN high-temperature pyrolysis stage. (c) DXN high-temperature gas-phase synthesis stage. (d) DXN low-temperature catalytic synthesis stage. (e) DXN emission stage.

(5) DXN Emission Stage: The center of the concentration distribution is 5.3916E–08 mg/Nm^3, with a variance of 4.2114E–16. The minimum value is 2.2129E–13 mg/Nm^3, while the maximum value is 1.2010E–07 mg/Nm^3.

Among the 1,080 sets of data representing the whole lifecycle DXN concentration distribution process, only 2.6% of emissions exceed 0.1 ng TEQ/Nm^3. Therefore, it can be concluded that the DXN CNSM model demonstrates high simulation performance for the actual MSWI process, and the operating conditions covered by the orthogonal experiment are diverse.

6.4.3.3.3 *Influence of Manipulated Variables on DXN Concentration*

Based on the comprehensive DXN concentration data obtained under multiple operating conditions, this section further analyzes the influence of operating parameters on the concentration distribution to describe and characterize the potential relationships between the operating process and the whole lifecycle DXN concentration. The selected manipulated variables include feed capacity, grate speed, and primary air flow. The analysis is conducted by examining pairs of manipulated variables corresponding to a single DXN concentration. The experimental results are presented in Figure 6.11.

As shown in Figure 6.11, for different stages (e.g., precipitation and formation, high-temperature pyrolysis, high-temperature gas-phase synthesis, low-temperature catalytic synthesis, and emission) the trajectory of the influence of operating parameter changes on DXN concentration is largely consistent, with differences primarily in concentration values. For variations in a single operating parameter, the fluctuations in DXN concentration at each stage do not exhibit a significant linear correlation. These experimental results indicate that changes in DXN concentration are difficult to describe and characterize using a single operating parameter, further highlighting the strong uncertainty associated with DXN concentrations.

Based on the above experimental findings, there are the following three risks as follows:

(1) Risk 1: When the grate speed is fast and the primary air flow is high, there is an increased risk of rising DXN generation and emission concentrations during the MSWI process;
(2) Risk 2: When the feed capacity remains constant and the primary air flow is low, the concentration of DXN generation and emission also faces an increased risk of rising;
(3) Risk 3: When the feed capacity is low and the grate speed is fast, there is similarly an increased risk of elevated DXN generation and emission concentrations during the MSWI process.

Based on the identified risks, the following operational recommendations are proposed:

(1) Addressing Risk 1: The current high DXN concentration is attributed to fast grate speed and increased primary air volume, as indicated by the CNSM model. Under these operating conditions, there is a risk of incomplete

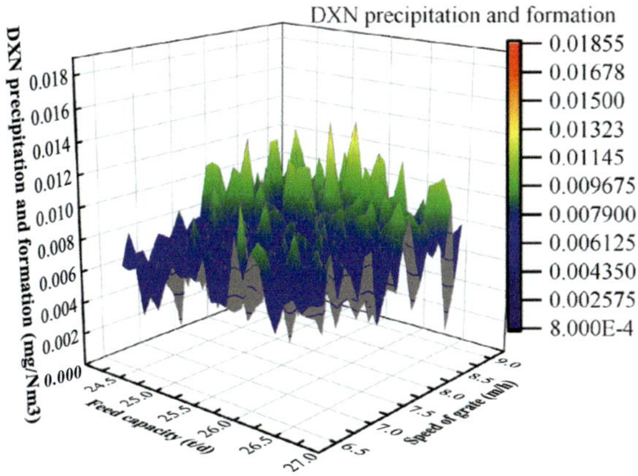

(a1)

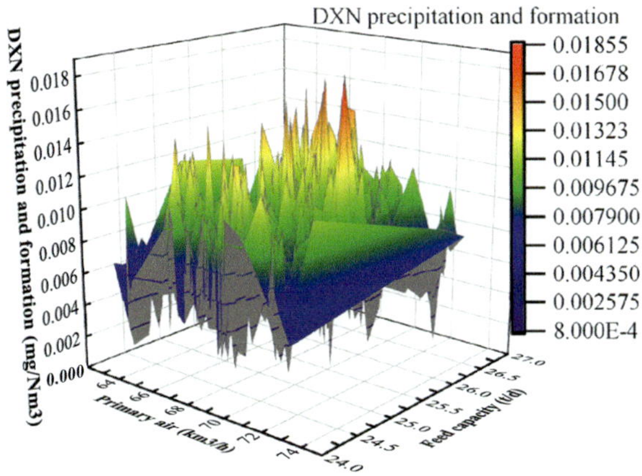

(a2)

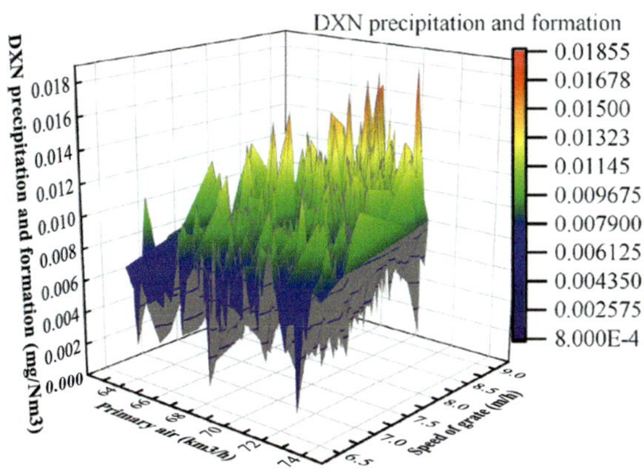

(a3)

(a) Impact on precipitation and formation

FIGURE 6.11 Influence of operating parameters on the distribution of the whole-process DXN concentration. (a) Impact on precipitation and formation.

(Continued)

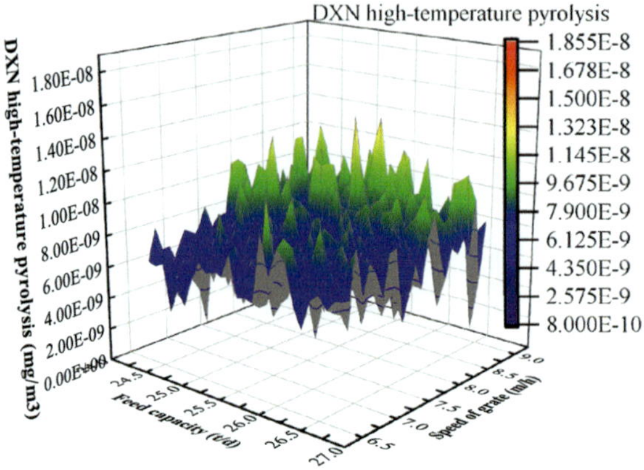

(b1)

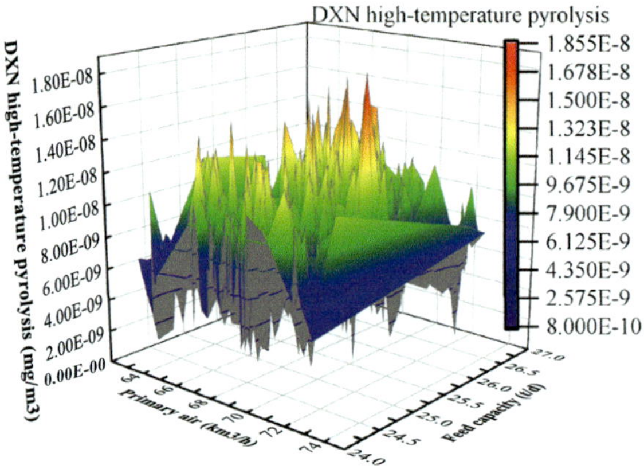

(b2)

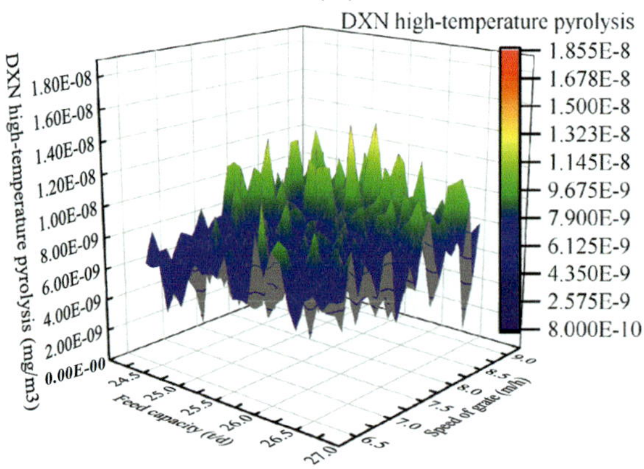

(b3)

(b) Impact on high-temperature pyrolysis

FIGURE 6.11 (Continued) (b) Impact on high-temperature pyrolysis.

(Continued)

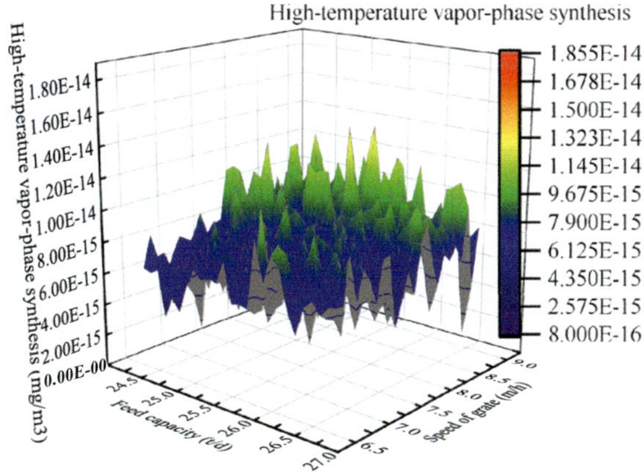

(c1)

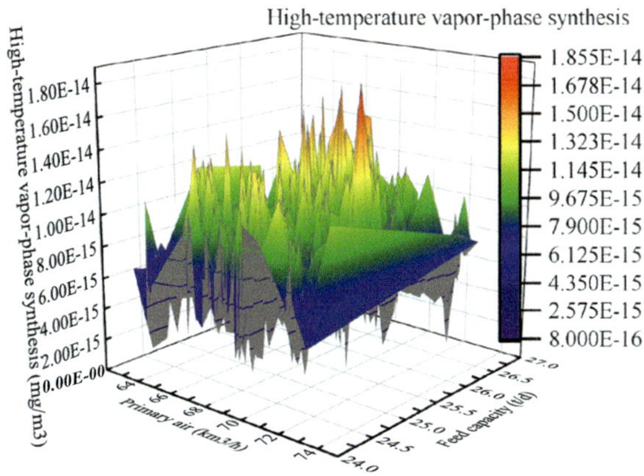

(c2)

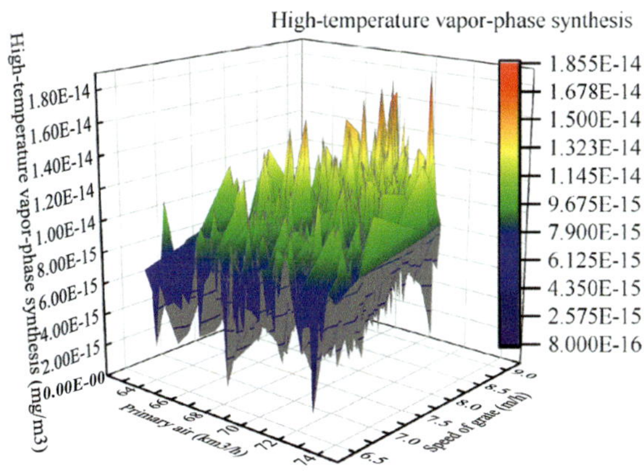

(c3)

(c) Impact on high-temperature gas-phase synthesis

FIGURE 6.11 (Continued) (c) Impact on high-temperature gas-phase synthesis.

(Continued)

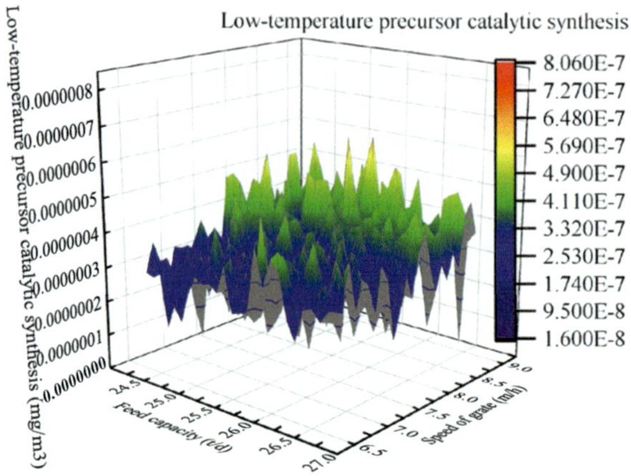

(d1)

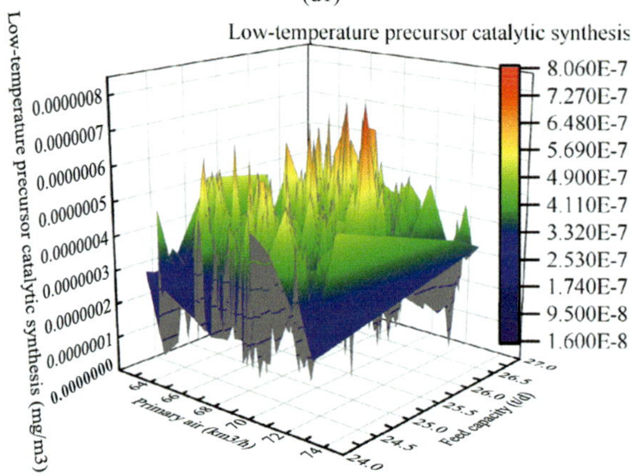

(d2)

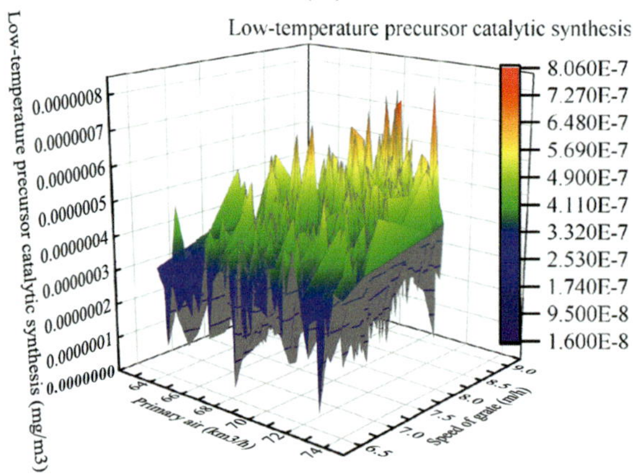

(d3)

(d) Impact on low-temperature catalytic synthesis

FIGURE 6.11 (Continued) (d) Impact on low-temperature catalytic synthesis.

(Continued)

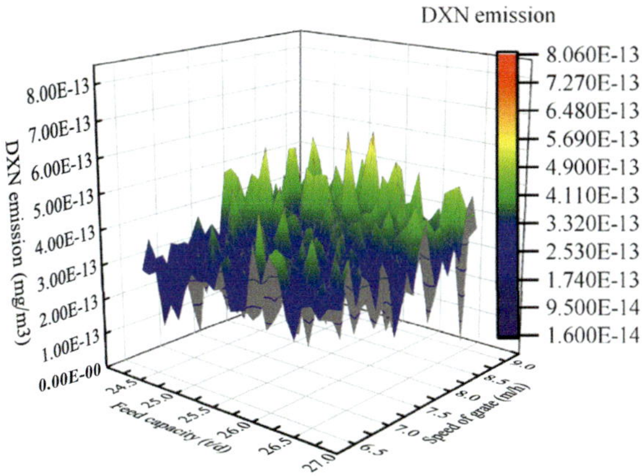

(e1)

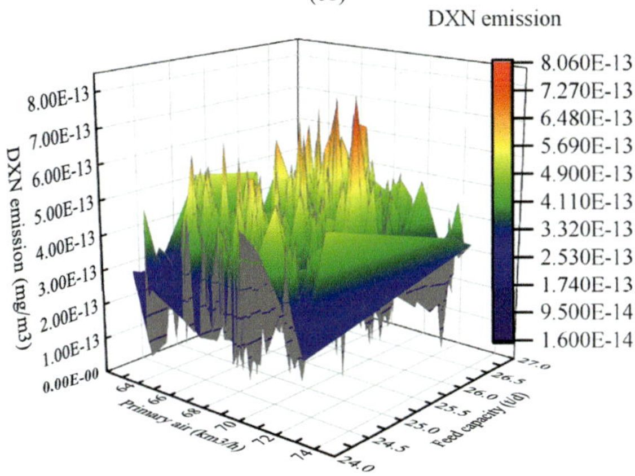

(e2)

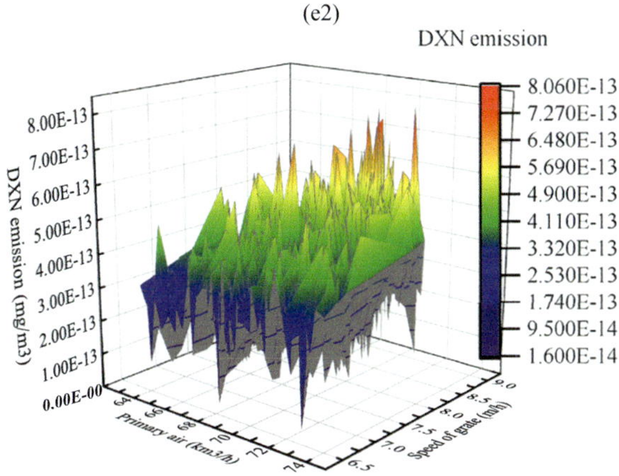

(e3)

(e) Impact on DXN emission

FIGURE 6.11 (Continued) (e) Impact on DXN emission.

combustion of MSW, leading to a significant increase in the production of incomplete combustion products, which may elevate DXN concentrations. This observation aligns with the findings in [36]. Therefore, it is recommended to reduce the grate speed and optimize the primary air volume to minimize DXN formation and emissions.

(2) Addressing Risk 2: The high DXN concentration in this scenario is due to constant feed capacity and low primary air volume, as represented by the CNSM model. With continuous feeding of MSW, a smaller primary air volume limits the available oxygen for combustion, resulting in anaerobic conditions. This ultimately leads to incomplete combustion and increased DXN generation and emissions. Hence, it is advisable to reduce the feed rate, increase the primary air volume, or decrease the grate speed to enhance the combustion efficiency of MSW, ultimately aiming to lower DXN concentrations.

(3) Addressing Risk 3: The elevated DXN concentration is linked to small feed capacity and fast grate speed, as indicated by the CNSM model. Although the MSW undergoes complete combustion within the furnace under current conditions, the small feed capacity may result in insufficient heat load, risking a temperature drop. According to the research in [37], a decrease in furnace temperature can lead to the reformation of DXN. Therefore, it is recommended to increase the feed capacity, reduce the grate speed, and decrease the primary air volume to mitigate DXN formation and emissions.

6.4.4 Potential Applications and Limitations

This work presents a comprehensive CNSM model for MSWI processes. Developed through the integration of FLIC and Aspen software, the CNSM model effectively captures the evolution of DXN, encompassing several crucial stages: (1) DXN Precipitation and Formation Stage: This stage simulates the initial formation and precipitation of DXN within the combustion environment; (2) DXN High-Temperature Pyrolysis Stage: This stage models the thermal decomposition of DXN at temperatures exceeding 1,123.15 K; (3) DXN High-Temperature Gas-Phase Synthesis Stage: This stage characterizes the formation of DXN through gas-phase reactions occurring at temperatures between 1,073.15 and 773.15 K; (4) DXN Low-Temperature Catalytic Synthesis Stage: This stage evaluates the contribution of catalytic reactions to DXN formation at temperatures ranging from 773.15 to 473.15 K; (5) DXN Activated Carbon Adsorption: This stage simulates the adsorption of DXN onto an activated carbon bed; (6) DXN Emission to the Atmosphere Stage: This stage determines the final emission levels of DXN resulting from the MSWI process. Experimental validation has demonstrated that the CNSM model accurately represents DXN concentrations throughout the whole MSWI process. This validated model serves as a powerful tool for understanding the complex formation and emission mechanisms of DXN.

Building upon the strengths of the developed CNSM model, several advantages and potential applications are evident: (1) Scientific Research View: The CNSM model offers a new framework and modeling approach for understanding the mechanisms of DXN formation, effectively advancing the development of simulation models for professional researchers in this field. For newcomers, the CNSM model serves as a comprehensive demonstration of the entire process of steady-state reactions, enabling them to quickly grasp relevant concepts. Furthermore, the introduction of the CNSM model supports the advancement of DXN digital twin technology, facilitates intelligent optimization and control for emission reduction, and enhances artificial intelligence-based online detection technologies. The current version of the CNSM model can also function as a qualitative evaluation tool for assessing changes in DXN concentrations throughout the MSWI process; (2) Engineering Application View: The CNSM model is grounded in the DXN evolution mechanism, ensuring that its representation of the entire DXN concentration process addresses the needs of most MSWI plants (excluding those employing selective catalytic reduction technology). Engineers can derive theoretical whole-process DXN concentrations by inputting the composition of the MSW and specific operational details. By adjusting operational parameters, they can simulate DXN concentration trends under various conditions and evaluate the likelihood of emission exceedances. Therefore, it can be concluded that the CNSM model holds significant engineering application value and can facilitate advancements in environmental management and DXN control.

The current version of the CNSM model has several limitations that need to be addressed in future research: (1) Improving Uncertainty and Nonlinearity: The analysis results indicate that changes in operational parameters exhibit a significant linear trend concerning the whole lifecycle concentrations of DXN. This linearity can distort the outcomes generated by the CNSM model. To enhance the model's rationality and authenticity, it is essential to introduce concepts of uncertainty and nonlinearity, thereby incorporating dynamic characteristics into the existing framework; (2) Optimizing and Adjusting the Model: While the CNSM model effectively characterizes the whole lifecycle concentrations of DXN, its outputs often align more closely with theoretical values than actual measurements. A critical challenge is to utilize empirical detection results to optimize and adjust the CNSM model. This adjustment would improve the accuracy of the model's concentration representations, ensuring they reflect real-world conditions more accurately; (3) Co-Modeling Multi-Pollutant Emissions: Although the CNSM model currently estimates DXN concentrations effectively, its applicability could be expanded by incorporating simulations of other relevant pollutants. Specifically, integrating the concentrations of CO, CO_2, NO_x, SO_2, and HCl into the DXN emission stage would enhance the model's comprehensiveness and credibility. This expansion would require a refined analysis and potential modifications to the intermediate processes within the model to ensure alignment with the complex nature of MSWI emissions. By addressing these limitations in future research, the CNSM model can be significantly enhanced, thereby increasing its utility and effectiveness in predicting and managing emissions from MSWI processes.

6.5 SUMMARY

To address the knowledge gaps regarding the generation, adsorption, and emission mechanisms of DXN in actual MSWI processes, this article presents a CNSM. This model integrates multiple software platforms to simulate the whole lifecycle of DXN concentration, offering mechanistic insights into the distribution of DXN concentrations throughout the MSWI process. The CNSM model encompasses several stages, including precipitation formation, high-temperature pyrolysis, re-synthesis, activated carbon adsorption, and emission. It effectively describes the evolution of DXN from combustion to emission, providing a theoretical foundation for online monitoring and optimal control strategies related to DXN.

Based on the proposed model and supporting experimental results, the following conclusions can be drawn: (1) Effective Simulation of Trends: Under steady-state conditions, the model successfully simulates the trends of DXN formation, adsorption, and emission in response to temperature changes, closely aligning with the actual dynamics of the MSWI process; (2) Optimal Furnace Temperature Range: Simulation results under stable conditions reveal that furnace temperatures between 850°C and 900°C are sufficient to completely decompose DXN. Therefore, maintaining furnace temperatures above 1,000°C in practical MSWI operations does not yield additional benefits for DXN decomposition; (3) Influence of Operational Parameters: The MSW composition, along with feed capacity, grate speed, and primary air volume, significantly impacts DXN emissions. Different control strategies can produce varying emission outcomes, highlighting the potential risk of exceeding DXN emission limits even under stable combustion conditions.

While the development of the CNSM model represents a significant advancement, several key challenges remain that warrant further investigation to enhance its robustness and applicability. These include improving the treatment of uncertainty and nonlinearity, optimizing and adjusting the model based on real-world data, and co-modeling the emissions of multiple pollutants. Addressing these challenges will unlock the full potential of the CNSM model as a powerful tool for analyzing, measuring, and reducing emissions throughout the whole MSWI process.

REFERENCES

1. Gómez-Sanabria A, Kiesewetter G, Klimont Z, et al. Potential for future reductions of global GHG and air pollutants from circular waste management systems. *Nature Communications*, 2022, 13(106): 1–12.
2. Kumar A, Samadder S R. A review on technological options of waste to energy for effective management of municipal solid waste. *Waste Management*, 2020, 69: 407–422.
3. Simpson A K, Drysdale M, Gamberg M, et al. Human biomonitoring of dioxins, furans, and non-ortho dioxin-like polychlorinated biphenyls (PCBs) in blood plasma from Old Crow, Yukon, Canada (2019). *Science of the Total Environment*, 2024, 923: 171222.
4. Wang B, Wang P, Xie L H, et al. A stable zirconium based metal-organic framework for specific recognition of representative polychlorinated dibenzo-p-dioxin molecules. *Nature Communications*, 2019, 10(1): 1–8.
5. Kanan S, Samar F. Dioxins and furans: a review from chemical and environmental perspectives. *Trends in Enviromental analytical Chemistry*, 2018, 17: 1–13.

6. McKay G. Dioxin characterisation, formation and minimisation during municipal solid waste (MSW) incineration: review. *Chemical Engineering Journal*, 2002, 86(3): 343–368.
7. Ballschmiter K, Zoller W, Buchert H, Class T. Correlation between substitution pattern and reaction pathway in the formation of polychlorodibenzofurans. *Fresenius' Zeitschrift für analytische Chemie*, 1985, 322(6): 587–594.
8. Weber R, Hagenmaier H. Mechanism of the formation of polychlorinated dibenzo-p-dioxins and dibenzofurans from chlorophenols in gas phase reactions. *Chemosphere*, 1999, 38(3): 529–549.
9. Peng Y Q, Chen J H, Lu S Y, et al. Chlorophenols in municipal solid waste incineration: a review. *Chemical Engineering Journal*, 2016, 292: 398–414.
10. Chen Z L, Lin X Q, Lu S Y, et al. Suppressing formation pathway of PCDD/Fs by S-N-containing compound in full-scale municipal solid waste incinerators. *Chemical Engineering Journal*, 2019, 359: 1391–1399.
11. Peng Y, Lu S, Li X, et al. Formation, measurement, and control of dioxins from the incineration of municipal solid wastes: recent advances and perspectives. *Energy & Fuels*, 2020, 34(11): 13247–13267.
12. Qiao J F, Guo Z H, Tang J. Dioxin emission concentration measurement approaches for municipal solid wastes incineration process: a survey. *Acta Automatica Sinica*, 2020, 46(6): 1063–1089.
13. Shaub W M, Tsang W. Dioxin formation in incinerators. *Environmental Science & Technology*, 1983, 17(12): 721–730.
14. Sidhu S S, Maqsud L, Dellinger B, et al. The homogeneous, gas-phase formation of chlorinated and brominated dibenzo-p-dioxin from 2, 4, 6-trichloro-and 2, 4, 6-Tribromophenols. *Combustion and Flame*, 1995, 100(1–2): 11–20.
15. Huang H, Buskens A. Comparison of dioxin formation levels in laboratory gas-phase flow reactors with those calculated using the SHAUB-Tsang mechanism. *Chemosphere*, 1999, 38(7): 1595–1602.
16. Babushok V I, Tsang W. Gas-phase mechanism for dioxin formation. *Chemosphere*, 2003, 51(10): 1023–1029.
17. Gerasimov G Y. Formation of dioxins by incineration of chlorine–containing fuels. *Combustion Explosion, and Shock Eaves*, 2001, 37(2): 148–152.
18. Altwicker E R. Relative rates of formation of polychlorinated dioxins and furans from precursor and de novo reactions. *Chemosphere*, 1996, 33(10): 1897–1904.
19. Stanmore B R. Modeling the formation of PCDD/F in solid waste incinerators. *Chemosphere*, 2002, 47(6): 565–573.
20. Laidler K J. The development of the Arrhenius equation. *Journal of Chemical Education*, 1984, 61(6): 494.
21. Altarawneh M, Dlugogorski B Z, Kennedy E M, Mackie J C. Mechanisms for PCDF and PCB formation from fires: pathways from oxidation of chlorobenzenes. In: *Proceedings 7th Asia-Oceania Symposium on Fire Science and Technology*, 2007.
22. Alobaid F, Almohammed N, Farid M M, et al. Progress in CFD simulations of fluidized beds for chemical and energy process engineering. *Progress in Energy and Combustion Science*, 2022, 91: 1–213.
23. Schwartz J R, Mark S, Wolfson A. A first-order simulator to control dioxin emissions: NMCRC-ATMOS. *Waste Management & Research: The Journal for a Sustainable Circular Economy*, 2009, 28(5): 461–471.
24. Zhong W R. Matlab-simulink simulation study on electron beam irradiation on dioxins in waste incineration flue gas. *Environmental Science & Technology*, 2013, 36(6L): 236–240.

25. Sun R, Ismail T M, Ren X, Abd El-Salam M. Numerical simulation of gas concentration and dioxin formation for MSW combustion in a fixed bed. *Journal of Environmental Management*, 2015, 157: 111–117.
26. Tungalag A, Lee B, Yadav M, Akande O. Yield prediction of MSW gasification including minor species through ASPEN plus simulation. *Energy*, 2020, 198: 117296.
27. Yang Y B, Goodfellow J, Sharifi V N, Swithenbank J. Investigation of biomass combustion systems using CFD techniques: a parametric study of packed-bed burning characteristics. *Progress in Computational Fluid Dynamics*, 2006, 6(4/5): 262.
28. Wakao N, Kaguei S. Heat and mass transfer in packed beds. *AICHE Journal*, 1983, 1(2): 193–199.
29. Smoot L D, Pratt D T. *Pulverized Coal Combustion and Gasification*. Springer, 1979.
30. Shin D, Choi S, Oh J E, Chang O S. Evaluation of polychlorinated dibenzo-p-dioxin/dibenzofuran (PCDD/F) emission in municipal solid waste incinerators. *Environmental Science & Technology*, 1999, 33(15): 2657–2666.
31. National Standard of the People's Republic of China. Determination of total moisture in coal. GB/T 211-2017. China, 2017.
32. National Standard of the People's Republic of China. Proximate analysis of coal. GB/T 212-2008. China, 2008.
33. National Standard of the People's Republic of China. Determination of carbon and hydrogen in coal. GB/T 476-2008. China, 2008.
34. National Standard of the People's Republic of China. Ultimate analysis of coal. GB/T 31391-2015. China, 2015.
35. Ministry of Housing and Urban-Rural Development of the People's Republic of China. General detecting methods for the chemical characteristic of domestic refuse. CJ/T 96-2013. China, 2013.
36. Hunsinger H, Jay K, Vehlow J. Formation and destruction of PCDD/F inside a grate furnace. *Chemosphere*, 2002, 46(9–10): 1263–1272.
37. Tejima H, Nishigaki M, Fujita Y, et al. Characteristics of dioxin emissions at startup and shutdown of MSW incinerators. *Chemosphere*, 2007, 66(6): 1123–1130.

Part III

Dioxins with AI Driven Detection Modeling and Risk Warning

7 Multi-Scale Modeling Using Generative Adversarial Network and Deep Forest Regression for Dioxin Emission Concentration

7.1 RELATED WORKS

7.1.1 Data Missing Filling Description

Complete modeling data serves as the foundation for building effective soft sensing models [1]. Due to sensor malfunctions, unforeseen disturbances, or human errors, these data often exhibit varying degrees of missing information. For example, an analysis of recorded data from an incineration plant in Beijing between 2009 and 2020 reveals a missing data rate of 2%–3%. This level of missing data is clearly detrimental to uncovering the underlying patterns in operational data and establishing an effective soft sensing model for DXN emissions.

Currently, the primary methods for handling missing data include simple deletion, weighted methods, and filling methods [2–4]. The simple deletion method removes rows containing missing data and is suitable for situations where the proportion of missing data in the overall sample is small. However, when a significant amount of data is missing, this approach can result in the loss of key information, compromising the quality of the dataset.

The weighted method assigns weights to missing attributes based on complete attributes through logistic or probit regression, increasing the weight of complete data to minimize the bias introduced by missing values. However, when multiple attributes are missing, this method can increase computational complexity and may reduce prediction accuracy. The filling method employs statistical techniques and machine learning algorithms to predict missing values. Statistical methods, such as expectation maximization, multiple imputation, and regression analysis, rely on assumptions derived from the dataset. Machine learning methods classify missing data and construct models to predict missing values by learning patterns from the sample distribution. Techniques include K-nearest neighbor imputation, K-means imputation, and Bayesian networks. This chapter primarily focuses on the filling method.

DOI: 10.1201/9781003634195-10

Regarding data-filling methods, Kabir et al. used the average value of available data to fill in missing features in water network databases [5], which is only suitable for small datasets with a low proportion of missing data. Deng et al. proposed an expectation maximization filling algorithm based on random processes to address abnormal data in the remote monitoring of concrete pump trucks [6]. They employed a random process approximation method to enhance the filling effect, significantly improving model performance. However, this approach does not account for local data similarities, and its convergence speed slows down as the ratio of missing data increases. Zhao et al. introduced a missing data-filling method based on an improved K-means algorithm [7], which aims to identify the most similar data points in the complete dataset for filling. This method offers advantages such as low computational complexity and fast processing speed, but it is limited by the challenge of defining similarity criteria applicable to the dataset. In the context of municipal sewage treatment processes, Lu et al. introduced a data-filling method based on a radial basis function (RBF) neural network [8], while Han et al. developed a filling model based on an improved support vector machine (SVM) [9]. The core of these approaches involves performing nonlinear fitting on the missing data. However, all of these filling algorithms rely on a single model, which introduces limitations such as uncertainty and difficulties in application, especially when dealing with datasets containing numerous missing attributes.

In response to this issue, Matteo et al. proposed a multi-level filling strategy that incorporates model selection [10]. Gondara et al. introduced a multi-fill model based on an ultra-deep denoising autoencoder [11], capable of handling missing data with varying types, patterns, proportions, and distributions. The essence of these methods lies in inferring multiple predicted values for the missing data initially. Subsequently, they comprehensively analyze these multiple complete datasets based on the predicted values, ultimately determining the final filling result according to specific criteria. This approach goes beyond selecting estimates from a single filling model.

The generation methods for filling missing data include variational autoencoders (VAEs) and generative adversarial networks (GANs). The significant success of deep learning has made data filling based on generative methods a prominent area of research. For instance, Sánchez-Morales et al. proposed a denoising superimposed autoencoder with an auxiliary classifier to enhance the quality of filled data through classification results [12]. Seo et al. utilized an iterative denoising autoencoder to address missing values in dissolved gas analysis data [13], which is crucial for diagnosing transformer health status. Additionally, Li et al. introduced a multi-source importance-weighted adversarial VAE filling method specifically designed for industrial missing data [14], employing domain adaptive training to minimize model mismatch. However, a challenge remains in effectively removing noise from the filled data, as these methods often struggle to filter out high-quality filled information.

In the realm of filling methods based on GANs, Yoon et al. proposed a GAN-based filling framework where the generator estimates missing values by leveraging real data vectors to produce complete vectors [15]. The discriminator, in turn, is tasked with distinguishing between real data and the filled data. Li et al. introduced a GAN-based interpolator aimed at filling high-quality data for missing entries characterized by complex and high-dimensional features [16]. Zhang et al. developed a GAN model for photovoltaic forecasting to address missing data, with results validated using solar data [17]. Sarda et al. employed GANs to fill in missing data within the steel industry [18]; however, this approach did not account for the modeling aspects involved. Yao et al. concentrated on downstream tasks following data filling and integrated semi-supervised learning techniques for industrial processes [19].

These studies collectively demonstrate the efficiency and applicability of GANs as a data-filling method. Nevertheless, further exploration is needed to determine how to effectively utilize GANs for filling missing data while considering the specific characteristics of processes and the requirements of downstream tasks.

7.1.2 Multi-scale Modeling Problem Description

Currently, it is essential to measure and analyze key operational indices that are crucial for managing complex industrial processes and ensuring environmental protection [20]. However, certain indices (e.g., production parameters and environmental indicators) are challenging to measure directly due to the complexities of the technology involved and the associated high costs [21]. To address this issue, data-driven soft sensor models based on process variables have been developed to quickly and accurately predict these operational indices [22]. These models play a vital role in monitoring operational states, achieving stable control, enhancing product quality, and reducing pollutant emissions [23, 24]. In recent years, deep learning (DL) has demonstrated significant potential and superior predictive performance in the application of soft sensors for industrial parameters, outperforming traditional statistical machine learning approaches [25]. Yao et al. introduced a locally weighted DL method for soft sensor modeling [26], validating its effectiveness through real industrial applications. Liu et al. developed a stacked multiple manifold autoencoder to predict the distillation temperature index in a practical hydrocracking process [27]. Subsequently, Wang et al. created a stacked autoencoder that incorporated residual information, achieving remarkable accuracy [28]. While DL methods excel in efficiently extracting deep features, they often struggle with missing data and small sample sizes.

Dioxin (DXN) exhaust emissions are a primary contributor to the "not in my backyard" phenomenon associated with the construction of municipal solid waste incineration (MSWI) power plants. Consequently, developing a DXN emission soft sensor model represents an effective approach for online measurement and real-time control of DXN emissions, thereby facilitating their reduction [29]. However, the modeling data for DXN emissions face challenges related to small sample sizes and missing values, and the input features collected may not directly correspond to DXN emission concentration data due to differing measurement methodologies. Therefore, addressing these issues is essential for effectively implementing a DXN soft sensor model using DL.

Existing research defines the small sample problem in engineering applications as having fewer than 50 samples, while in academic research, this threshold is set at fewer than 30 samples [30]. Small sample issues may arise from various factors [31], including: (1) difficulties in obtaining useful process data due to noise and missing values, and (2) high measurement costs associated with collecting difficult-to-measure key operational indices. Traditional single-model algorithms, such as decision trees (DTs) and neural networks, often experience overfitting and exhibit poor generalization performance under these conditions [32]. Consequently, ensemble learning has emerged as a prominent area of study for small sample modeling, with random forest (RF) serving as a significant branch of this approach [33].

For instance, Cao et al. developed a soft sensor model for wide and thick plate crowns using RF [34]. Zhang et al. applied RF to predict the moisture content of biomass in fluidized bed dryers [35], fine-tuning the RF parameters through a Bayesian optimization algorithm. Following this, Zhou et al. introduced a deep forest (DF) algorithm that leverages the concept of layer-by-layer learning from deep neural networks to address classification problems [36], thereby implementing a DL model without requiring a neural network structure. While DF exhibits capabilities similar to deep feature extraction, it is primarily designed for classification tasks. Subsequently, Tang et al. proposed a DFR algorithm to construct a soft sensor model for complex industrial processes [37]. However, DFR employs only two types of forest algorithms, which restricts the feature characterization ability of its cascade layers.

Complex industrial process data often exhibits missing and abnormal values due to mutual interference between devices and improper data storage practices [38, 39]. Constructing an accurate and robust soft sensor model using traditional DL neural networks is challenging, particularly when operational indices are based on small sample sizes [40]. Consequently, having sufficient complete data and employing DL algorithms that are well-suited for small samples are essential for developing soft sensor models with high generalization performance [41]. Researchers typically use the 3σ criteria to identify abnormal and missing data, subsequently removing the corresponding samples or features [42]. While this approach effectively mitigates the impact of problematic data on modeling, it also results in information loss and a reduction in sample size [43, 44].

The modeling data for certain difficult-to-measure operational indices exhibit multi-scale characteristics. For instance, the DXN emission concentration is a key operational index in the MSWI process. Typically, the sampling process for DXN emission concentration requires a duration of at least two hours (hour-time scale); however, the process data are generated in seconds (second-time scale) during this interval [45]. As a result, the input and output data, collected using different methods, operate on varying time scales, leading to the presence of multiple time-scale characteristics in the modeling data. It is important to note that the multi-scale process data discussed in this study differ from the multi-time-scale issues commonly encountered in time series prediction [46].

7.1.3 Generative Adversarial Network Description (GAN)

A GAN consists of two main components: a generator and a discriminator. The generator is responsible for producing virtual samples, while the discriminator's role is to differentiate between real samples and those generated by the generator.

The generator is structured as a multi-layer perceptron (MLP), comprising an input layer, N_{Gen} hidden layers, and an output layer. The hidden layers utilize the Relu activation function, while the output layer employs a linear activation function. The training process for the generator is outlined as follows:

$$\tilde{X}_{\text{vir}} = f_{\text{Generator}}\left(Z_{\text{noise}}, W_{\text{Gen}}, b_{\text{Gen}}\right) \tag{7.1}$$

where $f_{\text{Generator}}(\cdot)$ is the generator; W_{Gen} and b_{Gen} denote the weight and bias of the generator; Z_{noise} is the random noise; and $\tilde{X}_{\text{vir}}$ are the virtual samples.

The discriminator is also structured as an MLP, comprising an input layer, N_{Disc} hidden layers, and an output layer. The hidden layers utilize the Relu activation function, while the output layer employs a sigmoid activation function. The output of the discriminator provides the distinction between real and generated samples, as follows:

$$\left[\xi_{\text{vir}}, \xi_{\text{real}}\right] = f_{\text{Discriminator}}\left(\left[\tilde{X}_{\text{vir}}, X_{\text{real}}\right], W_{\text{Disc}}, b_{\text{Disc}}\right) \tag{7.2}$$

where $f_{\text{Discriminator}}(\cdot)$ is the discriminator, X_{real} is the real sample, W_{Disc} and b_{Disc} denote the weight and bias of the discriminator, ξ_{vir} and ξ_{real} denote the distinguished results of X_{vir} and X_{real}.

The GAN is trained through a competitive game between the generator and the discriminator. Its objective function can be formulated as follows:

$$\begin{cases} L_{\text{Disc}} = -\dfrac{1}{N_{\text{real}}} \displaystyle\sum_{n_{\text{real}}=1}^{N_{\text{real}}} \ln\left(\xi_{n_c}^{\text{real}}\right) - \dfrac{1}{N_{\text{vir}}} \displaystyle\sum_{n_{\text{vir}}=1}^{N_{\text{vir}}} \left(1 - \ln\left(\xi_{n_c^{\text{vir}}}^{\text{vir}}\right)\right) \\ L_{\text{Gen}} = -\dfrac{1}{N_{\text{vir}}} \displaystyle\sum_{n_{\text{vir}}=1}^{N_{\text{vir}}} \ln\left(\xi_{n_c^{\text{vir}}}^{\text{vir}}\right) \end{cases} \tag{7.3}$$

where L_{Disc} and L_{Gen} denote the loss values of the discriminator and generator, N_{real} and N_{vir} are the number of X_{real} and X_{vir}.

7.1.4 Deep Forest Regression Description (DFR)

DFR is a non-neural network DL model developed based on the principles of ensemble learning, with its core logic centered around "layered feature abstraction + ensemble decision-making." In the initial layer, the input data is processed through a "forest module" consisting of multiple base forest models (such as random forests or completely random forests). Each base model generates prediction results or high-dimensional feature representations, which are then concatenated to form the input for the subsequent layer. Through multiple rounds of hierarchical iterations, the model evolves from raw features to higher-order abstract features, ultimately producing regression predictions via the top-level ensemble module.

Compared to traditional single-tree models, DFR enhances the ability to fit complex nonlinear relationships through "hierarchical stacking." Unlike neural networks, it provides benefits such as strong model interpretability, high robustness to small sample sizes, and resistance to overfitting. This makes DFR particularly suitable for regression tasks characterized by limited domain knowledge, small data size, and complex feature dimensions.

Introducing DFR into the DXN emission concentration modeling process offers several advantages. It can directly handle high-dimensional input features without the need for complex feature selection and extraction preprocessing. Additionally, by employing a multi-layered ensemble structure, DFR automatically learns the relationships among features in terms of "raw material characteristics—combustion state—flue gas cleaning," thereby enhancing prediction accuracy. Currently, DFR has become an essential modeling tool for parameters that are difficult to measure in complex industrial processes, effectively addressing the performance limitations of traditional soft sensing models.

7.2 MODELING STRATEGY

The proposed improved GAN-DFR DXN soft sensor for handling multi-scale missing data comprises four key modules: missing data division, adversarial generative data filling, input and output data matching, and improved DFR (IDFR) modules. The methodology is illustrated in Figure 7.1.

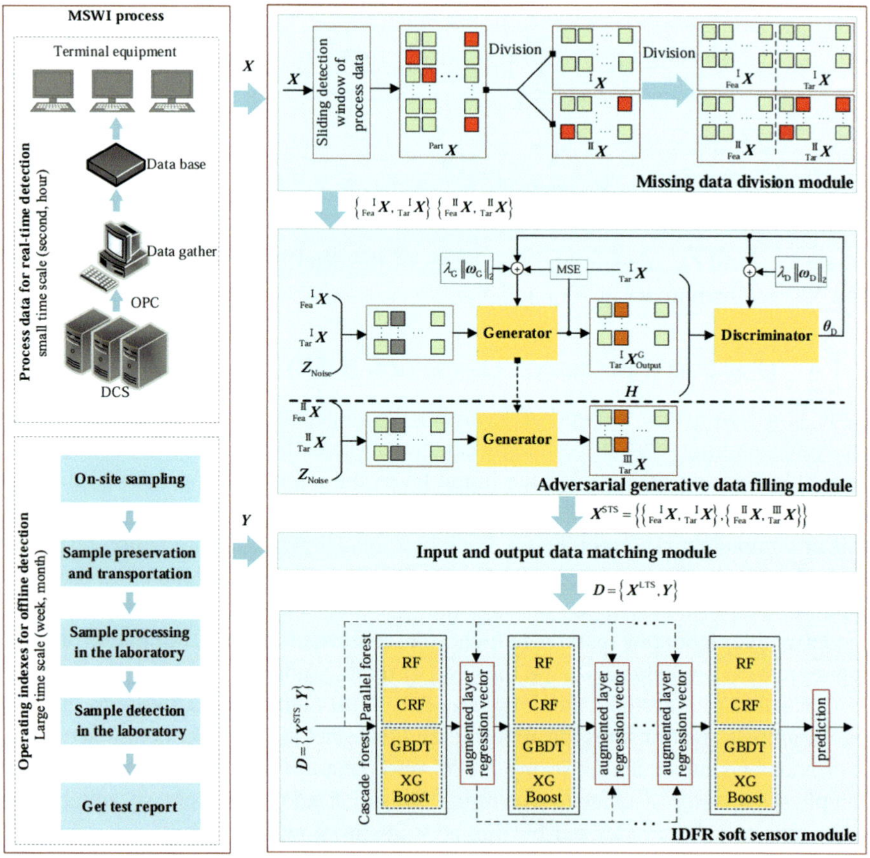

FIGURE 7.1 Strategy of improved GAN-DFR DXN soft sensor for multi-scale missing data.

In Figure 7.1, $\boldsymbol{X}$ and $\boldsymbol{Y}$ denote the input and output of the original dataset, respectively; ${}^{\text{Part}}\boldsymbol{X}$ is the input data to be filled; ${}^{\mathbf{I}}X$ and ${}^{\mathbf{II}}X$ denote complete input data (recorded as class I data) and missing input data (recorded as class II data), respectively; ${}^{\mathbf{I}}_{\text{Fea}}\boldsymbol{X}$ and ${}^{\mathbf{I}}_{\text{Tar}}\boldsymbol{X}$ denote the input and target feature of class I data, respectively; ${}^{\mathbf{II}}_{\text{Fea}}\boldsymbol{X}$ and ${}^{\mathbf{II}}_{\text{Tar}}\boldsymbol{X}$ denote the input and target feature of class II data, respectively; $\boldsymbol{Z}_{\text{Noise}}$ is random noise; $\lambda_G \parallel \omega_G \parallel_2$ and $\lambda_D \parallel \omega_D \parallel_2$ denote the model complexity penalty term of the generator and the discriminator, respectively; ω_G and ω_D are the respective weights of the generator and discriminator; λ_G and λ_D are the respective weights of the complexity penalty for the generator and discriminator; ${}^{\mathbf{I}}_{\text{Tar}}\boldsymbol{X}^{G}_{\text{Output}}$ is the result of the generator in training; $\boldsymbol{H}$ is the hint matrix; θ_D is the output of the discriminator; ${}^{\text{III}}_{\text{Tar}}\boldsymbol{X}$ is the filling results of missing data (recorded as class III data); $\boldsymbol{X}^{\text{STS}}$ is the small-time-scale input data after filling; $\boldsymbol{X}^{\text{LTS}}$ is the large time scale input data; and $\boldsymbol{D}$ is the modeling data after matching the input and output.

The functions of the different modules are as follows:

(1) Missing Data Division Module: This module partitions short-time-scale data based on the presence of missing values in the features.
(2) Adversarial Generative Data-Filling Module: This module utilizes the complete data to train the IGAN, allowing the trained generator to fill in the missing data. To ensure the quality and stability of the filled data, it incorporates mean squared error (MSE) constraints and model complexity penalty terms.
(3) Input and Output Data Matching Module: This module compresses the short-time-scale input data to align with the long-time-scale output data.
(4) IDFR Soft Sensor Module: This module develops a DXN emission soft sensor model using a multi-layer DF algorithm. Each layer employs fully connected parallel and cascade forest algorithms, along with a weighting strategy to enhance performance.

The functions of the different modules are as follows.

(1) Missing Data Division Module: It partitions short time-scale data according to whether there are missing values in features.
(2) Adversarial Generative Data-Filling Module: It uses the complete data to train the IGAN, and the missing data are filled by the trained generator. The MSE constraint and model complexity penalty terms are used to ensure the quality and stability of filling.
(3) Input and Output Data Matching Module: It compresses short-time-scale input data to match the long-time-scale output data.
(4) IDFR Soft Sensor Module: It builds a DXN emission soft sensor model with a multi-layer DF algorithm. Each layer uses fully connected parallel and cascade forest algorithms, and the weighting strategy is employed.

7.3 ALGORITHM IMPLEMENTATION

7.3.1 Missing Data Division Module

In this study, the proposed missing data-filling method specifically addresses cases of missing completely at random (MCAR). A sliding detection window is employed to identify the missing values and their surrounding data, denoted as dataset ${}^{\text{Part}}\boldsymbol{X}$. The dataset is divided into complete input data ${}^{\mathbf{I}}\boldsymbol{X}$ (designated as class I data) and missing input data ${}^{\mathbf{II}}\boldsymbol{X}$ (designated as class II data) based on the presence of missing values in the samples, as follows:

$$ {}^{\text{Part}}\boldsymbol{X} = \left\{ \left[{}^{\mathbf{I}}\boldsymbol{X} \right] \in \mathbb{R}^{N_{\text{I}} \times D}, \left[{}^{\mathbf{II}}\boldsymbol{X} \right] \in \mathbb{R}^{N_{\text{II}} \times D} \right\} \in \mathbb{R}^{N \times D} \tag{7.4} $$

$$ {}^{\mathbf{I}}\boldsymbol{X} = \left[{}^{\mathbf{I}}\boldsymbol{x}_1; {}^{\mathbf{I}}\boldsymbol{x}_2; \ldots; {}^{\mathbf{I}}\boldsymbol{x}_{N_{\text{I}}} \right] \in \mathbb{R}^{N_{\text{I}} \times D} \tag{7.5} $$

$$ {}^{\mathbf{II}}\boldsymbol{X} = \left[{}^{\mathbf{II}}\boldsymbol{x}_1; {}^{\mathbf{II}}\boldsymbol{x}_2; \ldots; {}^{\mathbf{II}}\boldsymbol{x}_{N_{\text{II}}} \right] \in \mathbb{R}^{N_{\text{II}} \times D} \tag{7.6} $$

where ${}^{\text{I}}N$ and ${}^{\text{II}}N$ denote the sample sizes of ${}^{\mathbf{I}}\boldsymbol{X}$ and ${}^{\mathbf{II}}\boldsymbol{X}$, respectively; and N and D denote the sample size and dimension of ${}^{\text{Part}}\boldsymbol{X}$, respectively.

Further, ${}^{\mathbf{II}}\boldsymbol{X}$ is divided into the input feature ${}^{\mathbf{II}}_{\text{Fea}}\boldsymbol{X}$ and target feature ${}^{\mathbf{II}}_{\text{Tar}}\boldsymbol{X}$ according to the missing values in the features, and ${}^{\mathbf{I}}\boldsymbol{X}$ is also divided into the input feature ${}^{\mathbf{I}}_{\text{Fea}}\boldsymbol{X}$ and target feature ${}^{\mathbf{I}}_{\text{Tar}}\boldsymbol{X}$. They are denoted as follows:

$$ {}^{\mathbf{I}}_{\text{Fea}}\boldsymbol{X} = \left[{}^{\mathbf{I}}_{\text{Fea}}\boldsymbol{x}_1; {}^{\mathbf{I}}_{\text{Fea}}\boldsymbol{x}_2; \ldots; {}^{\mathbf{I}}_{\text{Fea}}\boldsymbol{x}_{N_{\text{I}}} \right] \in \mathbb{R}^{N_{\text{I}} \times D_{\text{Fea}}} \tag{7.7} $$

$$ {}^{\mathbf{I}}_{\text{Tar}}\boldsymbol{X} = \left[{}^{\mathbf{I}}_{\text{Tar}}\boldsymbol{x}_1; {}^{\mathbf{I}}_{\text{Tar}}\boldsymbol{x}_2; \ldots; {}^{\mathbf{I}}_{\text{Tar}}\boldsymbol{x}_{N_{\text{I}}} \right] \in \mathbb{R}^{N_{\text{I}} \times D_{\text{Tar}}} \tag{7.8} $$

$$ {}^{\mathbf{II}}_{\text{Fea}}\boldsymbol{X} = \left[{}^{\mathbf{II}}_{\text{Fea}}\boldsymbol{x}_1; {}^{\mathbf{II}}_{\text{Fea}}\boldsymbol{x}_2; \ldots; {}^{\mathbf{II}}_{\text{Fea}}\boldsymbol{x}_{N_{\text{II}}} \right] \in \mathbb{R}^{N_{\text{II}} \times D_{\text{Fea}}} \tag{7.9} $$

$$ {}^{\mathbf{II}}_{\text{Tar}}\boldsymbol{X} = \left[{}^{\mathbf{II}}_{\text{Tar}}\boldsymbol{x}_1; {}^{\mathbf{II}}_{\text{Tar}}\boldsymbol{x}_2; \ldots; {}^{\mathbf{II}}_{\text{Tar}}\boldsymbol{x}_{N_{\text{II}}} \right] \in \mathbb{R}^{N_{\text{II}} \times D_{\text{Tar}}} \tag{7.10} $$

where D_{Fea} and D_{Tar} denote the number of input and target features, respectively.

7.3.2 Adversarial Generative Data-Filling Module

In this module, ${}^{\mathbf{I}}_{\text{Fea}}\boldsymbol{X}$ and ${}^{\mathbf{I}}_{\text{Tar}}\boldsymbol{X}$ are used to train the adversarial generative data-filling model, and ${}^{\mathbf{II}}_{\text{Tar}}\boldsymbol{X}$ is filled when the generator training is completed. The proposed framework of the adversarial generative data-filling model is shown in Figure 7.2.

The adversarial generative data-filling process is detailed as follows.

First, random noise $\boldsymbol{Z}_{\text{Noise}}$ is utilized to replace a random feature in ${}^{\mathbf{I}}_{\text{Tar}}\boldsymbol{X}$, which is then fed along with ${}^{\mathbf{I}}_{\text{Fea}}\boldsymbol{X}$ to the generator. The process is as follows:

$$ {}^{\mathbf{I}}\boldsymbol{X}^{\text{G}}_{\text{Input}} = \left\{ {}^{\mathbf{I}}_{\text{Fea}}\boldsymbol{X}, {}^{\mathbf{I}}_{\text{Tar}}\boldsymbol{X} \odot \boldsymbol{H} + \boldsymbol{Z}_{\text{Noise}} \odot \left(\boldsymbol{1} - \boldsymbol{H} \right) \right\} \tag{7.11} $$

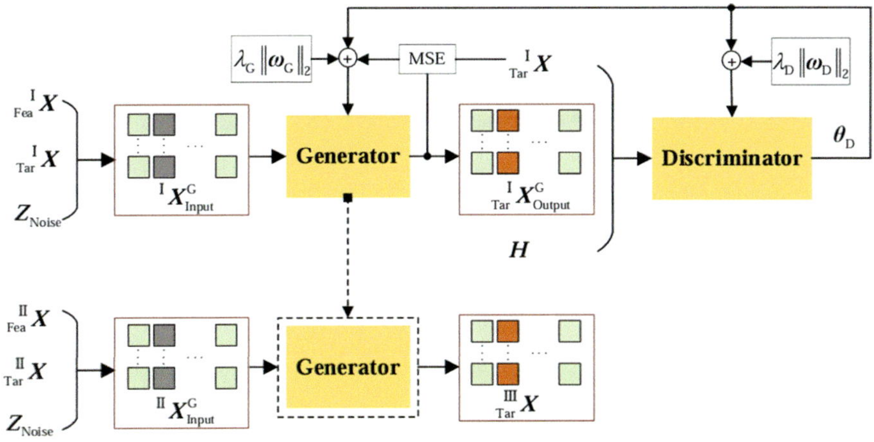

FIGURE 7.2 Adversarial generative data-filling framework.

where $\boldsymbol{H}$ is the hint matrix, and the d th feature of $\boldsymbol{H}$ is as follows:

$$h_{i,j} = \begin{cases} 0, & j = d \\ 1, & j \neq d \end{cases} \tag{7.12}$$

where $h_{i,j}$ is the row i and column j element of $\boldsymbol{H}$.

The generator consists of N_G-layer fully connected neural networks, comprising one input layer, one output layer, and $N_G - 2$ hidden layers. The output of the first hidden layer of the generator is calculated as follows:

$$\boldsymbol{H}_{G,1}^{\text{hidden}} = f_{G,1}\left({}^{\text{I}}\boldsymbol{X}_{\text{Input}}^{G} \cdot \omega_{G,1} + \boldsymbol{b}_{G,1}\right) \tag{7.13}$$

where $\omega_{G,1} \in \mathbb{R}^{D_{\text{Tar}} \times D_{G,1}}$ and $\boldsymbol{b}_{G,1} \in \mathbb{R}^{N_{\text{I}} \times D_{G,1}}$ denote the weights and biases between the input layer and first hidden layer of the generator, respectively; $D_{G,1}$ is the dimension of the first hidden layer; $f_{G,1}(\cdot)$ is the Relu activation function; and $\boldsymbol{H}_{G,1}^{\text{hidden}}$ is the output of the first hidden layer.

Thus, the output of the n_Gth hidden layer of the generator is calculated by

$$\boldsymbol{H}_{G,n_G}^{\text{hidden}} = f_{G,n_G}\left(\boldsymbol{H}_{G,n_G-1}^{\text{hidden}} \cdot \omega_{G,n_G} + \boldsymbol{b}_{G,n_G}\right), n_G = 2, \ldots, N_G - 2 \tag{7.14}$$

The output of the generator in training is calculated as

$${}_{\text{Tar}}^{\text{I}}\boldsymbol{X}_{\text{Output}}^{G} = f_{\text{Output}}^{G}\left(\boldsymbol{H}_{G,N_G-2}^{\text{hidden}} \cdot \omega_{G,N_G-1} + \boldsymbol{b}_{G,N_G-1}\right) \tag{7.15}$$

where ω_{G,N_G-1} and $\boldsymbol{b}_{G,N_G-1}$ denote the weights and biases between the $(N_G - 2)$ th hidden layer and the output layer, respectively; and $f_{\text{Output}}^{G}(\cdot)$ is the linear activation function.

Next, the filled data and real data are discriminated by the discriminator. The input to the discriminator is expressed as

$$ {}^{\mathbf{I}}\boldsymbol{X}_{\text{Input}}^{D} = \left\{ {}_{\text{Tar}}^{\mathbf{I}}\boldsymbol{X}, {}_{\text{Tar}}^{\mathbf{I}}\boldsymbol{X}_{\text{Output}}^{G} \odot \boldsymbol{H} + {}_{\text{Tar}}^{\mathbf{I}}\boldsymbol{X} \odot \left(1-\boldsymbol{H}\right), \boldsymbol{H} \right\} \tag{7.16} $$

The discriminator consists of N_D-layer fully connected neural networks, which includes one input layer, one output layer, and $N_D - 2$ hidden layers. The output of the first hidden layer of the discriminator is calculated as follows:

$$ \boldsymbol{H}_{D,1}^{\text{hidden}} = f_{D,1}\left({}^{\mathbf{I}}\boldsymbol{X}_{\text{Input}}^{D} \cdot \omega_{D,1} + \boldsymbol{b}_{D,1} \right) \tag{7.17} $$

where $\omega_{D,1}$ and $\boldsymbol{b}_{D,1}$ denote the weights and biases between the input layer and the first hidden layer of the discriminator, respectively; and $\boldsymbol{H}_{D,1}^{\text{hidden}}$ is the output of the first hidden layer.

Thus, the output of the n_Dth hidden layer of the discriminator is calculated by

$$ \boldsymbol{H}_{D,nG}^{\text{hidden}} = f_{D,nG}\left(\boldsymbol{H}_{D,nG-1}^{\text{hidden}} \cdot \omega_{D,nG} + \boldsymbol{b}_{D,nG} \right),\ n_D = 2,\ldots,N_D - 2 \tag{7.18} $$

The output of the discriminator is calculated as

$$ \theta_D = f_{\text{Output}}^{D}\left(\boldsymbol{H}_{D,N_D-2}^{\text{hidden}} \cdot \omega_{D,N_D-1} + \boldsymbol{b}_{D,N_D-1} \right) \tag{7.19} $$

where ω_{D,N_D-1} and $\boldsymbol{b}_{D,N_D-1}$ denote the weights and biases between the $\left(N_D - 2\right)$ th hidden layer and the output layer, respectively; and $f_{\text{Output}}^{D}\left(\cdot\right)$ is the sigmoid activation function.

The MSE constraint term and model complexity penalty terms are incorporated into the original GAN to enhance the quality of the filled data and mitigate overfitting. The proposed loss function is calculated as follows:

$$ \begin{aligned} \min_G \max_D V\left(D,G\right) &= E_{P\left({}^{\mathbf{I}}\boldsymbol{X}\right)}\left(\log {}^{\mathbf{I}}\theta_D\right) + E_{P(\boldsymbol{Z}_{\text{noise}})}\left[\log\left(1 - {}^{\mathbf{III}}\theta_D\right)\right] \\ &+ \lambda_{\text{MSE}} f_{\text{MSE}}\left({}_{\text{Tar}}^{\mathbf{I}}\boldsymbol{X}, {}_{\text{Tar}}^{\mathbf{I}}\boldsymbol{X}_{\text{Output}}^{G} \right) + \lambda_G \| \omega_G \|_2 + \lambda_D \| \omega_D \|_2 \end{aligned} \tag{7.20} $$

where $P\left({}^{\mathbf{I}}\boldsymbol{X}\right)$ is the distribution of ${}^{\mathbf{I}}\boldsymbol{X}$; $P\left(\boldsymbol{Z}_{\text{noise}}\right)$ is the distribution of $\boldsymbol{Z}_{\text{noise}}$; ${}^{\mathbf{I}}\theta_D$ is the output of the discriminator for real data; ${}^{\mathbf{III}}\theta_D$ is the output of the discriminator for filling data (recorded as class III data); λ_{MSE} is the weight of the MSE constraint term; $f_{\text{MSE}}\left(\cdot\right)$ is the MSE function; λ_G and λ_D are the weights of the model complexity penalty term of the generator and discriminator, respectively.

Bivariate cross entropy is used as the loss function of the original GAN and is calculated as follows:

$$ \text{Loss}_{\text{BCE}} = -\frac{1}{N'} \sum_{a=1}^{N'} \left[y_a \log\left(f\left(\boldsymbol{x}_a\right)\right) + \left(1 - y_a\right) \log\left(1 - f\left(\boldsymbol{x}_a\right)\right) \right] \tag{7.21} $$

where $\boldsymbol{x}_a$ and $f\left(\boldsymbol{x}_a\right)$ denote the input and output of the model, respectively; y_a is the binary label; and N' is the number of data.

The generator parameters are fixed while the discriminator is being trained, with real data labeled as 1 (i.e., $y_a = 1$). The loss for the discriminator is calculated by:

$$\mathrm{Loss}_D^1 = -\frac{1}{{}^{\mathrm{I}}N}\sum_{a=1}^{{}^{\mathrm{I}}N}\log\left({}^{\mathrm{I}}\theta_D\right) \tag{7.22}$$

Then, the filled data are marked as 0 (i.e., $y_a = 0$). The loss for the discriminator in this case is calculated by:

$$\mathrm{Loss}_D^0 = -\frac{1}{{}^{\mathrm{I}}N}\sum_{a=1}^{{}^{\mathrm{I}}N}\log\left(1 - {}^{\mathrm{III}}\theta_D\right) \tag{7.23}$$

Further, Eqs. (7.19) and (7.20), along with the complexity penalty, are combined to formulate the loss function of the discriminator as follows:

$$\mathrm{Loss}_D = -\left[\frac{1}{{}^{\mathrm{I}}N}\sum_{a=1}^{{}^{\mathrm{I}}N}\left[\log\left({}^{\mathrm{I}}\theta_D\right) + \log\left(1 - {}^{\mathrm{III}}\theta_D\right)\right] + \lambda_D \parallel \omega_D \parallel_2\right] \tag{7.24}$$

The discriminator parameters are fixed while the generator is being trained, with filling data labeled as 1 (i.e., $y_a = 1$). The loss for the generator is calculated by:

$$\mathrm{Loss}_G^1 = -\frac{1}{{}^{\mathrm{I}}N}\sum_{a=1}^{{}^{\mathrm{I}}N}\log\left({}^{\mathrm{III}}\theta_D\right) \tag{7.25}$$

For Eq. (7.22), the MSE constraint and complexity penalty are combined to form the loss function of the generator as follows:

$$\mathrm{Loss}_G = -\left[\frac{1}{{}^{\mathrm{I}}N}\sum_{a=1}^{{}^{\mathrm{I}}N}\log\left({}^{\mathrm{III}}\theta_D\right) + \lambda_{\mathrm{MSE}} f_{\mathrm{MSE}}\left({}^{\mathrm{I}}_{\mathrm{Tar}}\boldsymbol{X}, {}^{\mathrm{I}}_{\mathrm{Tar}}\boldsymbol{X}^G_{\mathrm{Output}}\right) + \lambda_G \parallel \omega_G \parallel_2\right] \tag{7.26}$$

The generator and discriminator are trained for N_{epoch} epochs, which brings the filling data increasingly close to the real data. This iterative process allows both models to improve their performance over time, enhancing the quality of the generated data.

Finally, $\boldsymbol{Z}_{\mathrm{Noise}}$ is used to replace the missing data in ${}^{\mathrm{II}}_{\mathrm{Tar}}\boldsymbol{X}$ with ${}^{\mathrm{II}}_{\mathrm{Fea}}\boldsymbol{X}$ as input to the generator. Thus, the filling data ${}^{\mathrm{III}}_{\mathrm{Tar}}\boldsymbol{X}$ are generated.

7.3.3 Input and Output Data Matching Module

As the process data for real-time detection operate on a small time scale, while the key operational indexes for offline detection by laboratories are measured on a larger time scale, a direct match between the two is not possible. Therefore, multiple short-time-scale process data points are averaged to create an input that can be aligned with the operational indexes from the long time scale, resulting in the final modeling data.

The input data for the short time scale, after filling, and the output data for the long time scale are represented as follows:

$$\boldsymbol{X}^{\text{STS}} = \left\{ \left\{ {}_{\text{Fea}}^{\ \ \text{I}}\boldsymbol{X}, {}_{\text{Tar}}^{\ \ \text{I}}\boldsymbol{X} \right\}, \left\{ {}_{\text{Fea}}^{\ \ \text{II}}\boldsymbol{X}, {}_{\text{Tar}}^{\ \ \text{III}}\boldsymbol{X} \right\} \right\} = \left[\boldsymbol{x}_1^{\text{STS}}; \boldsymbol{x}_2^{\text{STS}}; \ldots; \boldsymbol{x}_N^{\text{STS}} \right] \in \mathbb{R}^{N \times D} \tag{7.27}$$

$$\boldsymbol{Y} = \left[y_1, y_2, \ldots, y_{N_{\text{large}}} \right] \in \mathbb{R}^{N_{\text{large}} \times 1} \tag{7.28}$$

Each N_{scale} sample in $\boldsymbol{X}^{\text{STS}}$ is averaged to obtain input data for the long time-scale, as follows:

$$\boldsymbol{x}_i^{\text{LTS}} = \frac{1}{N_{\text{scale}}} \sum_{j=(i-1)\cdot N_{\text{scale}}+1}^{i \cdot N_{\text{scale}}} \boldsymbol{x}_j^{\text{STS}}, i = 1, 2, \ldots, N_{\text{large}} \tag{7.29}$$

$$\boldsymbol{X}^{\text{LTS}} = \left[\boldsymbol{x}_1^{\text{LTS}}; \boldsymbol{x}_2^{\text{LTS}}; \ldots; \boldsymbol{x}_{N_{\text{large}}}^{\text{LTS}} \right] \in \mathbb{R}^{N_{\text{large}} \times D} \tag{7.30}$$

where $N_{\text{large}} = N / N_{\text{scale}}$ is the number of data for a long time scale.

The input and output data for the long time-scale are matched to obtain the final modeling data, as follows:

$$\boldsymbol{D} = \left\{ \boldsymbol{X}^{\text{LTS}}, \boldsymbol{Y} \right\} \in \mathbb{R}^{N_{\text{large}} \times (D+1)} \tag{7.31}$$

7.3.4 Improved DFR(IDFR) Soft Sensor Modeling Module

The proposed IDFR module is shown in Figure 7.3.

Compared to the existing DFR, the proposed IDFR introduces the cascade forest algorithm at each layer. As a result, IDFR incorporates two types of ensemble learning methods: cascade forest (XGBoost and GBDT) and parallel forest (RF and completely RF), thereby enhancing its diversity. The augmented layer regression vector from each forest layer, along with the input data, is fully connected as the input for the subsequent layer, which enhances its feature characterization ability at each level. To improve predictive performance, the output of the IDFR is obtained by performing a weighted summation of the outputs from the last layer.

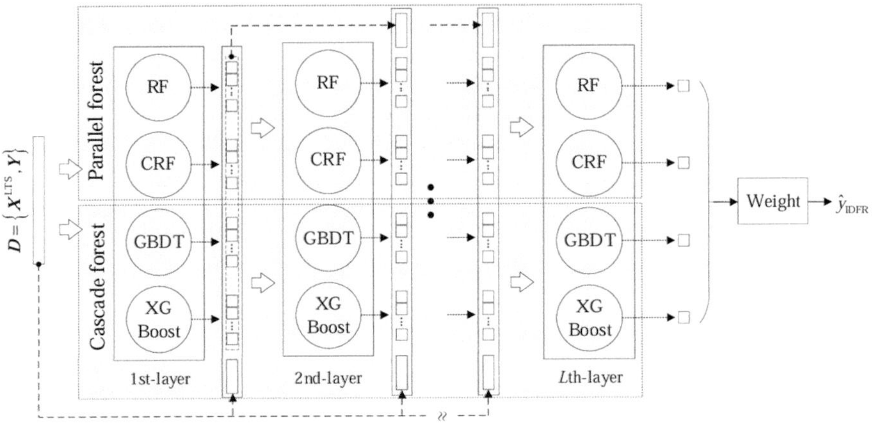

FIGURE 7.3 Structure of the IDFR module.

7.3.4.1 Parallel Forest

Parallel forest consists of RF and completely RF (CRF).

RF consists of N_{DT} parallel DTs as follows:

$$f_{\text{RF}}\left(\boldsymbol{x}\right)=\frac{1}{N_{\text{DT}}}\sum_{n_{\text{DT}}=1}^{N_{\text{DT}}}h_{n_{\text{DT}}}\left(\boldsymbol{x},\theta_{\text{sample}}\right) \tag{7.32}$$

where $h_{n_{\text{DT}}}\left(\boldsymbol{x},\theta_{\text{sample}}\right)$ is the jth DT.

The modeling process of DT is referred to in [32]. CRF and RF belong to the same category of learners. The only difference between them is that the segmentation criteria for CRF are completely random, rather than based on a loss function. Therefore, CRF can be expressed as $f_{\text{CRF}}\left(\boldsymbol{x}\right)$.

7.3.4.2 Cascade Forest

Cascade forest consists of GBDT and XGBoost, which utilize the iterative decision forest algorithm to minimize residual errors. The regularization coefficient λ is incorporated into the loss function $\text{L}_{\text{XGBoost}}\left(\boldsymbol{y},\hat{\boldsymbol{y}}\right)=\left(\boldsymbol{y}-\hat{\boldsymbol{y}}\right)^2/2$ of XGBoost as follows:

$$\text{L}_{\text{XGBoost}}\left(\boldsymbol{y},\hat{\boldsymbol{y}}\right)=\sum_{i=1}^{N_{\text{Train}}}\text{L}_{\text{XGBoost}}\left(y_i,\hat{y}_i\right)+\frac{1}{2}\lambda\left(c^{\text{XGBoost}}\right)^2 \tag{7.33}$$

where $\hat{y}_i$ is the prediction value of the ith DT; and c^{XGBoost} is the prediction value of the next DT model, as follows:

$$c^{\text{XGBoost}}=\frac{-\sum_{i=1}^{N_{\text{Train}}}\frac{d\text{L}_{\text{XGBoost}}\left(\boldsymbol{y},\hat{y}_i\right)}{d\hat{y}_i}}{\sum_{i=1}^{N_{\text{Train}}}\frac{d^2\text{L}_{\text{XGBoost}}\left(\boldsymbol{y},\hat{y}_i\right)}{d^2\hat{y}_i}+\lambda} \tag{7.34}$$

Therefore, GBDT and XGBoost are recorded as $f_{\text{GBDT}}(\boldsymbol{x})$ and $f_{\text{XGBoost}}(\boldsymbol{x})$, respectively. Further, the first layer of IDFR is as follows:

$$F_1(\cdot)=\left\{f_{\text{RF}}(\cdot),f_{\text{CRF}}(\cdot),f_{\text{GBDT}}(\cdot),f_{\text{XGBoost}}(\cdot)\right\} \tag{7.35}$$

7.3.4.3 Weighted Output

The predictions of the four forest algorithms are calculated, and the first-layer augmented layer regression vector is denoted as follows:

$$\hat{\boldsymbol{y}}_1^{\text{regvec}}=\left\{\hat{y}_1^{\text{RF}},\hat{y}_1^{\text{CRF}},\hat{y}_1^{\text{GBDT}},\hat{y}_1^{\text{XGBoost}}\right\} \tag{7.36}$$

Hence, the input of lth-layer of IDFR $(l=2,\ldots,L)$ is calculated using a fully connected approach as follows:

$$\boldsymbol{D}_l^{\text{IDFR}}=\left\{\hat{\boldsymbol{y}}_1^{\text{regvec}},\ldots,\hat{\boldsymbol{y}}_{l-1}^{\text{regvec}},\boldsymbol{D}\right\} \tag{7.37}$$

The training process of IDFR stops when the prediction accuracy of the validation set declines as the number of layers increases. This process is illustrated as follows:

$$F_{\text{IDFR}}(\cdot)=\begin{cases}\text{continue}, & f_{\text{RMSE}}\left(\boldsymbol{y}^{\text{Val}},\hat{\boldsymbol{y}}_L^{\text{Val}}\right)<f_{\text{RMSE}}\left(\boldsymbol{y}^{\text{Val}},\hat{\boldsymbol{y}}_{L-1}^{\text{Val}}\right)\\ \left\{F_l(\cdot)\right\}_{l=1}^{L}, & f_{\text{RMSE}}\left(\boldsymbol{y}^{\text{Val}},\hat{\boldsymbol{y}}_L^{\text{Val}}\right)\geq f_{\text{RMSE}}\left(\boldsymbol{y}^{\text{Val}},\hat{\boldsymbol{y}}_{L-1}^{\text{Val}}\right)\end{cases} \tag{7.38}$$

where $F_{\text{IDFR}}(\cdot)$ is the IDFR model; $\hat{\boldsymbol{y}}_{L-1}^{\text{Val}}$ and $\hat{\boldsymbol{y}}_L^{\text{Val}}$ denote the validation prediction accuracy of the $(L-1)$th-layer and Lth-layer of IDFR, respectively; $\boldsymbol{y}^{\text{Val}}$ is the validation set truth value; and $f_{\text{RMSE}}(\cdot)$ is the function of RMSE.

The final prediction value of IDFR $\hat{y}_{\text{IDFR}}$ is calculated as

$$\hat{y}_{\text{IDFR}}=\omega_1\hat{y}_L^{\text{RF}}+\omega_2\hat{y}_L^{\text{CRF}}+\omega_3\hat{y}_L^{\text{GBDT}}+\omega_4\hat{y}_L^{\text{XGBoos}} \tag{7.39}$$

where $\left\{\omega_j\right\}_{j=1}^{4}$ are the output weights of the corresponding forests, defined as follows:

$$\omega_j=\frac{1}{3}\left(1-\frac{r_j}{\sum_{j=1}^{4}r_j}\right) \tag{7.40}$$

where $\{r_j\}_{j=1}^{4}$ are the validation RMSE values of corresponding forests, defined as follows:

$$\begin{cases} r_1 = f_{\text{RMSE}}\left(\boldsymbol{y}^{\text{Val}}, \hat{\boldsymbol{y}}_L^{\text{ValRF}}\right) \\ r_2 = f_{\text{RMSE}}\left(\boldsymbol{y}^{\text{Val}}, \hat{\boldsymbol{y}}_L^{\text{ValCRF}}\right) \\ r_3 = f_{\text{RMSE}}\left(\boldsymbol{y}^{\text{Val}}, \hat{\boldsymbol{y}}_L^{\text{ValGBDT}}\right) \\ r_4 = f_{\text{RMSE}}\left(\boldsymbol{y}^{\text{Val}}, \hat{\boldsymbol{y}}_L^{\text{ValXGBoos}}\right) \end{cases} \tag{7.41}$$

where $\hat{\boldsymbol{y}}_L^{\text{ValRF}}$, $\hat{\boldsymbol{y}}_L^{\text{ValCRF}}$, $\hat{\boldsymbol{y}}_L^{\text{ValGBDT}}$, and $\hat{\boldsymbol{y}}_L^{\text{ValXGBoos}}$ denote the validation prediction accuracies of the Lth-layer RF, CRF, GBDT, and XGBoost of IDFR, respectively.

7.3.5 Pseudocode

The pseudocode of multi-scale missing data filling using GAN is demonstrated in Algorithm 1.

The pseudocode of soft sensor modeling using IDFR is demonstrated in Algorithm 2.

Algorithm 1: Multi-scale missing data filling using GAN

Input: Original sample $\{\boldsymbol{X}, \boldsymbol{Y}\}$, epoch of training N_{epoch}, random noise $\boldsymbol{Z}_{\text{Noise}}$, model complexity penalty terms of GAN λ_{G} and λ_{D}, weight of the MSE constraint term λ_{MSE}, layer number of the generator and discriminator N_{G} and N_{D}.

Output: Input data for small time scale after filling $\boldsymbol{X}^{\text{STS}}$.

${}^{\text{Part}}\boldsymbol{X}$ is obtained from $\boldsymbol{X}$ through sliding detection window;

${}^{\text{Part}}\boldsymbol{X}$ is divided into ${}^{\text{I}}\boldsymbol{X}$ and ${}^{\text{II}}\boldsymbol{X}$;

${}^{\text{II}}\boldsymbol{X}$ is divided into ${}^{\text{II}}_{\text{Fea}}\boldsymbol{X}$ and ${}^{\text{II}}_{\text{Tar}}\boldsymbol{X}$;

${}^{\text{I}}\boldsymbol{X}$ is divided into ${}^{\text{I}}_{\text{Fea}}\boldsymbol{X}$ and ${}^{\text{I}}_{\text{Tar}}\boldsymbol{X}$;

for $e \leftarrow 1$ **to** N_{epoch} **do**

  The generator and discriminator are trained based on Eqs.(8)-(23);

end

The generator is used to fill in missing values of ${}^{\text{II}}_{\text{Tar}}\boldsymbol{X}$ to obtain ${}^{\text{III}}_{\text{Tar}}\boldsymbol{X}$;

${}^{\text{I}}_{\text{Fea}}\boldsymbol{X}$, ${}^{\text{III}}_{\text{Tar}}\boldsymbol{X}$, ${}^{\text{II}}_{\text{Fea}}\boldsymbol{X}$, and ${}^{\text{II}}_{\text{Tar}}\boldsymbol{X}$ are combined and become $\boldsymbol{X}^{\text{STS}}$.

Algorithm 2: Soft sensor modeling using IDFR

Input: Input data for small time scale after filling $\boldsymbol{X}^{\mathrm{STS}}$, matching scale N_{scale}, DT number of IDFR N_{DT}, minimum samples number of IDFR θ_{sample}, feature selection number IDFR N_{feature}, maximum layer number of IDFR $L_{\max}$, forest algorithm number of IDFR N_{forest}, regularization coef cient of IDFR λ, learning rate of IDFR α_{IDFR}.

Output: Final prediction of IDFR $\hat{y}_{\mathrm{IDFR}}$

Each N_{scale} sample in $\boldsymbol{X}^{\mathrm{STS}}$ is averaged to obtain input data for large time scale $\boldsymbol{X}^{\mathrm{LTS}}$;
$\boldsymbol{X}^{\mathrm{LTS}}$ and $\boldsymbol{Y}$ are matched to obtain the final modeling data $\boldsymbol{D} = \{\boldsymbol{X}^{\mathrm{LTS}}, \boldsymbol{Y}\}$;
for $l \leftarrow 1$ **to** $L_{\max}$ **do**
    if $l == 1$ **then**
        The input of IDFR is $\boldsymbol{D}$;
    else
        The input of IDFR is $\boldsymbol{D}_l^{\mathrm{IDFR}}$;
    end
    The RF model $f_{\mathrm{RF}}(\cdot)$ and CRF model are built based on Eq.(29);
    The GBDT model $f_{\mathrm{GBDT}}(\cdot)$ and XGBoost model $f_{\mathrm{XGBoost}}(\cdot)$ are built based on Eq.(30)-(31);
    The l th-layer of IDFR is built as $F_l(\cdot) = \{f_{\mathrm{RF}}(\cdot), f_{\mathrm{CRF}}(\cdot), f_{\mathrm{GBDT}}(\cdot), f_{\mathrm{XGBoost}}(\cdot)\}$;
    The l th-layer augmented layer regression vector $\hat{\boldsymbol{y}}_l^{\mathrm{regvec}} = \{\hat{y}_l^{\mathrm{RF}}, \hat{y}_l^{\mathrm{CRF}}, \hat{y}_l^{\mathrm{GBDT}}, \hat{y}_l^{\mathrm{XGBoost}}\}$;
    if $f_{\mathrm{RMSE}}(\boldsymbol{y}^{\mathrm{Val}}, \hat{\boldsymbol{y}}_L^{\mathrm{Val}}) \geq f_{\mathrm{RMSE}}(\boldsymbol{y}^{\mathrm{Val}}, \hat{\boldsymbol{y}}_{L-1}^{\mathrm{Val}})$ **then**
        $F_{\mathrm{IDFR}}(\cdot) = \{F_l(\cdot)\}_{l=1}^{L}$;
        break;
    end
end
The output weights of IDFR $\left\{\omega_j\right\}_{j=1}^{4}$ are calculated based on Eqs.(37)-(38);
$\hat{y}_{\mathrm{IDFR}} = \omega_1 \hat{y}_L^{\mathrm{RF}} + \omega_2 \hat{y}_L^{\mathrm{CRF}} + \omega_3 \hat{y}_L^{\mathrm{GBDT}} + \omega_4 \hat{y}_L^{\mathrm{XGBoos}}$.

7.4 APPLICATION RESULTS

7.4.1 Experimental Environment and Evaluation Indicators

In this chapter, the proposed method is applied to the MSWI process to verify its effectiveness. The configurations of the experimental platform are as follows: CPU—Intel Core i9-11900K (3.50 GHz), GPU—NVIDIA GeForce RTX 3060 Ti,

OS—Windows 11 (64-bit), and RAM—32 GB. The developed code is based on MATLAB R2022a.

The DXN data from an MSWI power plant in Beijing includes 109 days of process data without true values and 58 days of modeling data with true values, spanning from 2009 to 2020. The input variable consists of 116 dimensions, while the output variable represents the DXN emission concentration [19]. The sampling frequency of the process variables for each group of data is hourly (i.e., there are 24 samples of process data per day). The sliding detection window selects process data from the same year to fill in the missing values through the adversarial generative data-filling module, addressing the randomness and uncertainty of the DXN data. The input variables of the 58-day dataset with true values, after filling, are used to obtain the final modeling data of IDFR, resulting in a total of 141 samples. The DXN dataset is then randomly divided into training, validation, and testing sets in a ratio of 2:1:1, respectively.

In this study, root mean square error (RMSE) and coefficient of determination (R^2) are chosen as evaluation metrics to compare the performance of different methods. They are expressed as follows:

$$\text{RMSE} = \sqrt{\frac{\sum_{i=1}^{N} \left(y_i - \hat{y}_i\right)^2}{N-1}} \tag{7.42}$$

$$R^2 = 1 - \frac{\sum_{i=1}^{N} \left(y_i - \hat{y}_i\right)^2}{\sum_{i=1}^{N} \left(y_i - \overline{y}_i\right)^2} \tag{7.43}$$

where y_i is the true value, $\hat{y}_i$ is the predicted value, $\overline{y}_i$ is the mean value, and N is the number of samples.

7.4.2 Experimental Results

7.4.2.1 Missing Data Division Results

The abnormal and missing values in the datasets, both with and without true values, are identified based on the 3σ criteria. The number of abnormal and missing values for each feature is illustrated in Figure 7.4.

Figure 7.4 indicates that it is essential to address abnormal and missing values in the dataset to ensure the accuracy and reliability of the modeling process.

7.4.2.2 Data-Filling Results

The structure of the adversarial generative data-filling model and the experimental settings are presented, respectively, in Tables 7.1 and 7.2.

The input features 9–12 and 17–20 are selected to demonstrate the filling effect, with their specific meanings detailed in Table 7.3. Additionally, the relationship between MSE and training epochs for these eight features during the training process is illustrated in Table 7.4.

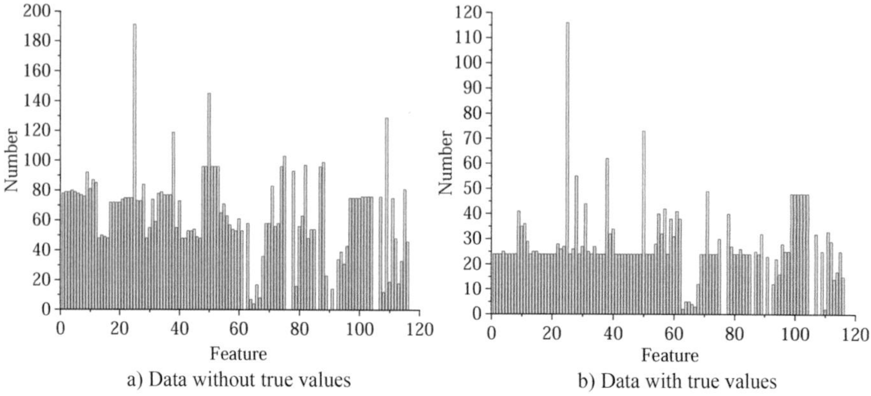

FIGURE 7.4 Number of missing and abnormal values. (a) Data without true values. (b) Data with true values.

TABLE 7.1
Structure of Adversarial Generative Data-Filling Model

Model	Layer	Number of Neurons	Activation Function
Generator	Input layer	116	\
	First hidden layer	580	Relu
	Second hidden layer	1,160	Relu
	Output layer	D_{Tar}	Linear
Discriminator	Input layer	D_{Tar}	\
	First hidden layer	$D_{\text{Tar}}{}^{*}5$	Relu
	Second hidden layer	$D_{\text{Tar}}{}^{*}10$	Relu
	Output layer	1	Sigmoid

TABLE 7.2
Training Settings of Adversarial Generative Data-Filling Model

Epoch	Learning Rate	Batch Size	λ_{MSE}	λ_G	λ_D
800	1E–04	$N_{\text{large}}/10$	1E–01	1E–02	1E–02

The 9–12 and 17–20 input features are selected to show the filling effect, and their specific meanings are shown in Table 7.3. The relationship between MSE and the training epoch of eight features in the training process is shown in Table 7.3. Figure 7.5 shows that the MSE of the filled data and real data decreases with the increase in training epochs.

Figure 7.5 demonstrates that the filling effect stabilizes when the training epochs reach around 700–800. To evaluate the effectiveness of the proposed approach, comparisons are made between adversarial generative data filling, linear interpolation

TABLE 7.3
Input Feature Information for the DXN Dataset

NO.	Feature	Unit
9	Left inlet temperature of drying grate	°C
10	Left outlet temperature of drying grate	°C
11	Right inlet temperature of drying grate	°C
12	Right outlet temperature of drying grate	°C
17	Left inner temperature of combustion grate 1–1	°C
18	Left outer temperature of combustion grate 1–1	°C
19	Right inner temperature of combustion grate 1–1	°C
20	Right outer temperature of combustion grate 1–1	°C

TABLE 7.4
Statistical Results of the Deviations of the Center of Distribution

Feature	Target	IGAN	Linear	VAE
9	150.75094	150.89168	149.99085	151.40899
		0.14074	0.76009	0.65805
10	161.19991	160.87257	160.98927	160.90406
		0.32734	0.21064	0.29585
11	151.30802	151.78028	151.35608	151.80259
		0.47226	0.04806	0.49457
12	160.92614	160.42409	161.93792	163.65651
		0.50205	1.01178	2.73037
17	261.94267	266.65615	259.21737	260.91481
		4.71348	2.7253	1.02786
18	263.95105	265.71244	261.75005	262.08545
		1.76139	2.201	1.8656
19	271.58861	272.88601	268.44923	271.59175
		1.2974	3.13938	0.00314
20	260.46775	263.91291	255.04496	258.82863
		3.44516	5.42279	1.63912

filling, and variational autoencoder (VAE) filling methods. The parameters for the VAE are configured as follows: the number of neurons in the encoder is set to 116-580-1160-20, the number of neurons in the decoder is 20-1160-580-116, with a total of 800 epochs, a learning rate of 1×10^{-4}–1×10^{-4}, and a batch size of $N_{\text{large}} / 10$. The distributions of the missing data and the filled data from the three methods are illustrated in Figure 7.6, while the deviations of the distribution centers are summarized in Table 7.4.

Figure 7.6 and Table 7.4 indicate that the filling distribution of the proposed method is reasonable. The IGAN demonstrates excellent filling results for features 9, 12, and 18, while the linear interpolation filling method performs better for features

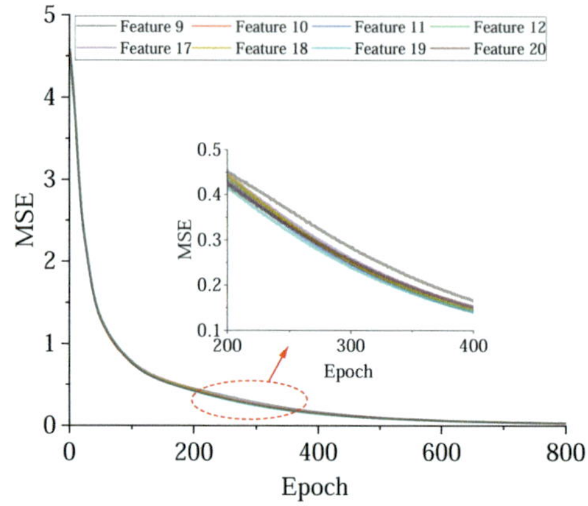

FIGURE 7.5 MSE of filling data for the DXN dataset.

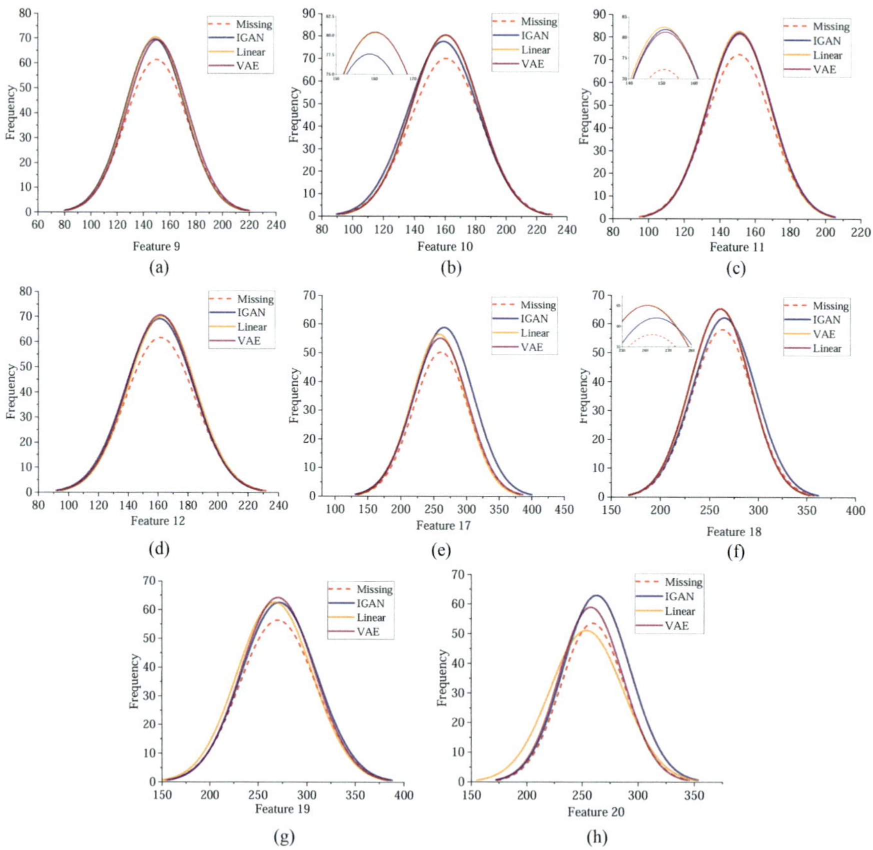

FIGURE 7.6 (a–h) Distribution contrast of filling data for the DXN dataset.

10 and 11. The VAE exhibits superior results for features 17, 19, and 20. However, IGAN shows significant deviations in the filling results for features 17, 18, and 20.

From the perspective of the distribution center, the filling results from the linear interpolation and VAE methods are similar, whereas the filling center of IGAN is relatively distant from the center. It is important to note that the distribution of the missing data remains unknown since the missing data are classified as MCAR type. Therefore, the filling results for features 17, 18, and 20 can still be considered acceptable despite the deviations.

Moreover, the distribution ranges of IGAN are broader compared to those of the linear interpolation and VAE methods, suggesting a greater likelihood that the filling results encompass the actual unknown distribution. This analysis highlights that different filling methods exhibit their advantages across various features, underscoring the importance of selecting an appropriate filling method based on the specific requirements of the industrial process.

7.4.2.3 Input and Output Data Matching Results

The missing and abnormal data of the 58-day truth values have been filled, and the input variables are aligned according to the sampling time of the DXN true values. The statistical results of the 141 modeling samples are presented in Table 7.5.

7.4.2.4 Soft Sensor Modeling Results

The DFR soft sensor model is built based on the training set and validation set after filling. The parameters of IDFR are set as follows: θ_{sample} is 7, N_{DT} is 50, N_{feature} is $\sqrt{D}$, $L_{\max}$ is 50, N_{forest} is 4, λ is 1, and α_{IDFR} is 0.3. The output weights of the IDFR are presented in Table 7.6. To account for randomness, the IDFR soft sensor modeling experiments are repeated 30 times, and the statistical results of these experiments are summarized in Table 7.7.

Table 7.7 indicates that the average and variance of RMSE for the DXN dataset are as follows: for the training dataset, it is 0.0101 ± 2.4148E–06; for the validation dataset, it is 0.0200 ± 6.9659E–07; and for the testing dataset, it is 0.0157 ±

TABLE 7.5
Statistical Results of the DXN Dataset

Dataset	Dimension of Input	Dimension of Output	Number of Samples
DXN	116	1	141

TABLE 7.6
Output Weights of IDFR for the DXN Dataset

ω_1	ω_2	ω_3	ω_4
0.2640	0.2529	0.2330	0.2501

TABLE 7.7
Statistical Results of IDFR Modeling Experiments for the DXN Dataset

	RMSE		
Dataset	**Training**	**Validation**	**Testing**
DXN	0.0101 ± 2.4148E−06	0.0200 ± 6.9659E−07	0.0157 ± 9.1818E−07

TABLE 7.8
Interval Setting of Hyper-Parameters

Hyper-Parameter	DXN
θ_{sample}	[1, 30]
$N_{feature}$	[1, 30]
N_{DT}	[10, 300]
λ	[0.1, 3]
α_{IDFR}	[0.05, 1]

9.1818E−07. These results demonstrate that the proposed method exhibits good accuracy and stability across all datasets.

7.4.3 Analysis of Hyper-Parameter Sensitivity

In the proposed IDFR, five hyper-parameters, namely, minimum sample number θ_{sample}, feature selection number $N_{feature}$, DT number of IDFR N_{DT}, regularization coefficient λ, and learning rate α_{IDFR}, are analyzed. The intervals of the hyper-parameters are detailed in Table 7.8, providing a comprehensive overview of the ranges used for analysis. The results of the hyper-parameter analysis on the testing set are illustrated in Figure 7.7, showcasing how variations in the hyper-parameters affect model performance and aiding in identifying optimal configurations for improved predictive accuracy.

Figure 7.7 shows the following:

(1) Figure 7.7(a) shows that the RMSE of the DXN dataset decreases first and then increases, and the minimum value is obtained at $\theta_{sample} = 7$. Thus, setting θ_{sample} to 1/20–1/10 of the number of training samples is suitable.
(2) Figure 7.7(b) shows that the RMSE of the DXN dataset decreases with the increase in $N_{feature}$ and tends to stabilize at [3, 43]. Thus, $N_{feature} = D/4$ is suitable for the DXN dataset.
(3) Figure 7.7(c) shows that the lower RMSE is obtained at [40, 60] for DXN datasets. Thus, N_{DT} at [40, 60] is reasonable for the DXN dataset.
(4) Figure 7.7(d) shows that the RMSE of the DXN dataset fluctuates with the increase in λ. Thus, the DXN dataset is sensitive to the change of λ, and $\lambda = 1$ is suitable for the DXN modeling data.
(5) Figure 7.7(e) shows that the RMSE of the DXN dataset decreases first and then increases, and the minimum values are obtained at 0.5. Thus, α_{IDFR} at [0.4, 0.6] can ensure model performance.

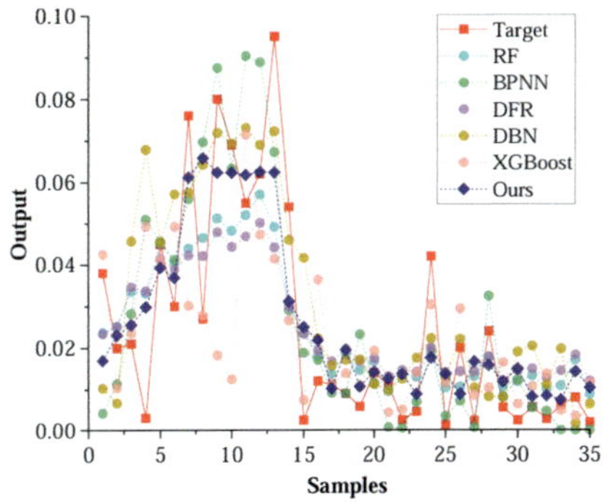

a) No filling and our method

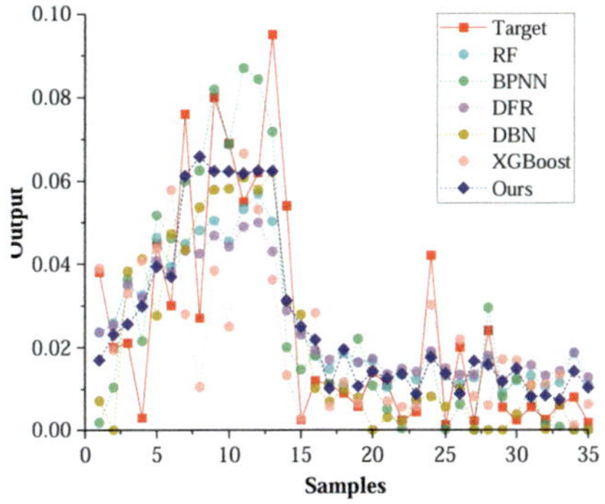

b) Linear filling and our method

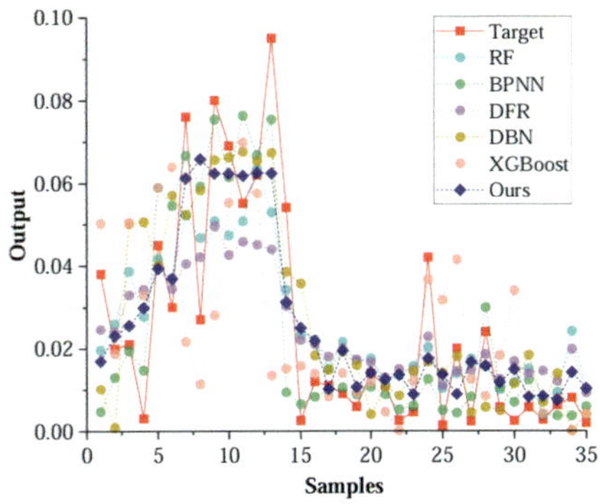

c) VAE filling and our method

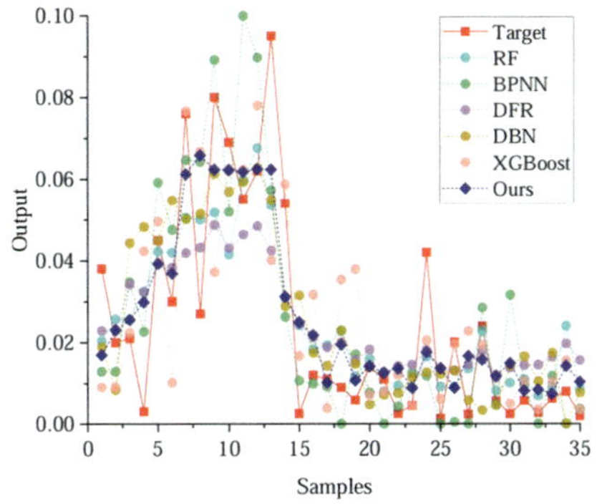

d) IGAN filling and our method

FIGURE 7.7 (a–d) Parameter sensitivity curves for the DXN dataset.

7.4.4 Comparative Results

The proposed method is compared with several existing methods, including RF, BPNN, DFR, deep belief network (DBN), and XGBoost, under three conditions: unprocessed data, linear interpolation, and VAE interpolation. This comparison aims to verify the effectiveness of the proposed method. The parameter settings for the comparable methods are either the same or similar.

In particular:

For RF, the minimum sample number of DTs is set to 7, the number of features selected is specified as $\sqrt{D}$, and the number of DTs is set to 50.

For BPNN, the architecture includes neurons arranged in layers as 116-580-1160-116-1, with the activation functions of the hidden layer being sigmoid and the output layer being linear. The training consists of 1,000 epochs with a learning rate of 1×10^{-3}–1×10^{-3}.

For DFR, the cascade forests consist of 50 layers, with 4 forests comprising 2 CRFs and 2 RFs. The minimum sample number of DTs is 7, the number of features selected is $\sqrt{D}$, and the number of DTs is 50.

For DBN, the configuration mirrors that of BPNN, with 116-580-1160-116-1 neurons, sigmoid and linear activation functions, 1000 training epochs, and a learning rate of 1×10^{-3}–1×10^{-3}.

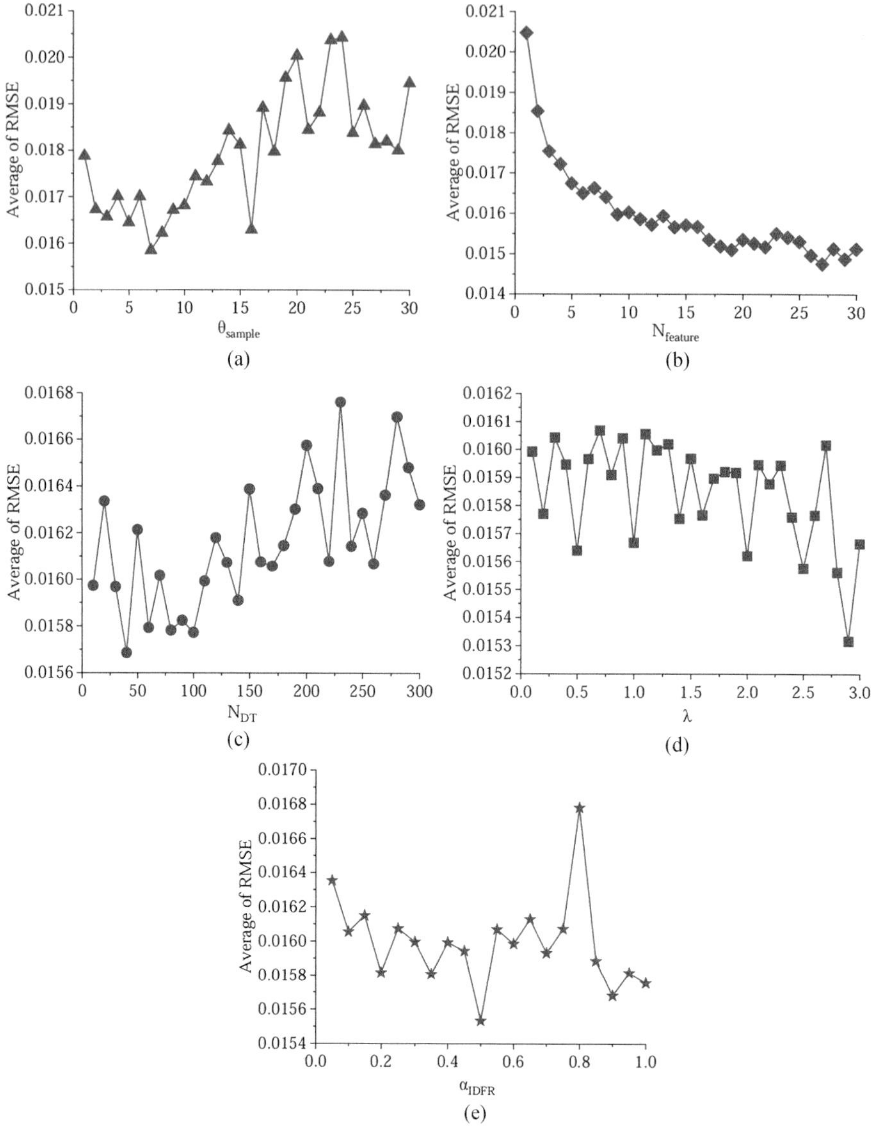

FIGURE 7.8 Comparative experimental fitting curves for the DXN dataset. (a) No filling and our method. (b) Linear filling and our method. (c) VAE filling and our method. (d) IGAN filling and (e) our method.

For XGBoost, the minimum sample number of DTs is 7, the feature selection number is $\sqrt{D}$, the number of DTs is 50, the maximum depth is 1, and the learning rate is set to 0.3.

The specific settings of the comparison experiments are detailed in Table 7.9, and the respective testing prediction curves are illustrated in Figure 7.8. The statistical results of the comparisons are presented in Tables 7.9–7.11.

TABLE 7.9
Comparative Experimental Settings for DXN Datasets

No.	Filling Method	Modeling Method				
1		RF	BPNN	DFR	DBN	XGBoost
2	Linear interpolation filling	RF	BPNN	DFR	DBN	XGBoost
3	VAE filling	RF	BPNN	DFR	DBN	XGBoost
4	Adversarial generative filling	RF	BPNN	DFR	DBN	XGBoost
5	Adversarial generative filling	IDFR				

TABLE 7.10
Comparative Experimental Statistical Results for RMSE

Filling Method	Modeling Method	RMSE		
		Training	Validation	Testing
	RF	0.0116	0.0195	0.0163
	BPNN	0.0056	0.0235	0.0206
	DFR	0.0134	0.0203	0.0173
	DBN	0.0176	0.0211	0.0191
	XGBoost	0.0113	0.0223	0.0233
	RF	0.0117	0.0195	0.0164
	BPNN	0.0070	0.0222	0.0175
Linear filling	DFR	0.0133	0.0203	0.0174
	DBN	0.0178	0.0214	0.0181
	XGBoost	0.0118	0.0222	0.0232
	RF	0.0115	0.0194	0.0161
	BPNN	**0.0051**	0.0219	0.0174
VAE filling	DFR	0.0134	0.0204	0.0173
	DBN	0.0176	0.0214	0.0179
	XGBoost	0.0117	0.0219	0.0230
	RF	0.0116	**0.0186**	0.0160
	BPNN	0.0094	0.0220	0.0172
IGAN filling	DFR	0.0135	0.0201	0.0171
	DBN	0.0188	0.0229	0.0178
	XGBoost	0.0094	0.0248	0.0212
IGAN filling	IDFR	0.0101	0.0200	**0.0157**

The bold ones are the best values in terms of different datasets.

Tables 7.10 and 7.11 present the following findings:

(1) The performance of the models improves with the filling of missing data; however, not all filling methods are suitable for every predictive model. For example, the linear filling method is not appropriate for RF.

TABLE 7.11
Comparative Experimental Statistical Results for R^2

Filling Method	Modeling Method	R^2 Training	R^2 Validation	R^2 Testing
	RF	0.8387 ± 1.4147E−04	0.5126 ± 1.3320E−03	0.6097 ± 1.7427E−03
	BPNN	0.9494 ± 2.9514E−03	0.2920 ± 1.9092E−02	0.3518 ± 7.8034E−02
	DFR	0.7867 ± 9.7932E−05	0.4741 ± 2.5408E−04	0.5604 ± 4.0587E−04
	DBN	0.6319 ± 9.3700E−04	0.4306 ± 1.3513E−03	0.4626 ± 1.8440E−03
	XGBoost	0.8035 ± 3.2437E−03	0.3443 ± 1.4902E−02	0.4053 ± 1.9457E−02
Linear filling	RF	0.8354 ± 1.3755E−04	0.5160 ± 6.4896E−04	0.6036 ± 9.1621E−04
	BPNN	0.9302 ± 3.3567E−03	0.3688 ± 1.6876E−02	0.5406 ± 1.7222E−02
	DFR	0.7879 ± 1.0656E−04	0.4732 ± 2.6169E−04	0.5538 ± 2.5400E−04
	DBN	0.6243 ± 1.2384E−03	0.4174 ± 3.3196E−03	0.5142 ± 1.8887E−03
	XGBoost	0.7819 ± 5.1635E−03	0.3522 ± 1.1391E−02	0.4140 ± 1.1776E−02
VAE filling	RF	0.8426 ± 1.0200E−04	0.5188 ± 1.3586E−03	0.6184 ± 7.9443E−04
	BPNN	**0.9596 ± 1.6929E−03**	0.3882 ± 5.7031E−03	0.5494 ± 5.6910E−03
	DFR	0.7847 ± 1.4719E−04	0.4699 ± 3.3881E−04	0.5586 ± 1.9766E−04
	DBN	0.6302 ± 6.4243E−04	0.4140 ± 1.8622E−03	0.5255 ± 9.3953E−04
	XGBoost	0.7884 ± 3.2437E−03	0.3660 ± 1.4902E−02	0.4165 ± 1.9457E−02
IGAN filling	RF	0.8395 ± 9.4346E−05	**0.5591 ± 7.4708E−04**	0.6224 ± 7.7071E−04
	BPNN	0.8479 ± 4.0888E−02	0.3802 ± 1.1844E−02	0.5547 ± 2.0767E−02
	DFR	0.7837 ± 6.9694E−05	0.4840 ± 1.2084E−04	0.5686 ± 1.3878E−04
	DBN	0.5802 ± 1.4969E−04	0.3303 ± 5.6390E−04	0.5336 ± 1.3637E−04
	XGBoost	0.8900 ± 1.9473E−03	0.2091 ± 1.3717E−02	0.3206 ± 3.8632E−02
IGAN filling	IDFR	0.8757 ± 1.2475E−03	0.4883 ± 1.8435E−03	**0.6374 ± 1.9317E−03**

The bold ones are the best values in terms of different datasets.

(2) The proposed IGAN filling method outperforms other filling methods on the testing set, although the improvement observed on the validation set is minimal.
(3) Neural network-based methods, such as BPNN and DBN, are not effective for small sample modeling. Specifically, BPNN exhibits significant overfitting issues and lacks stability compared to other methods.
(4) The performance of IDFR surpasses that of other single ensemble learning methods on both training and testing sets, indicating that the proposed method enhances prediction accuracy compared to traditional ensemble learning methods like RF, DFR, and XGBoost. However, its stability is only superior to that of BPNN and XGBoost. Future research should focus on improving both the accuracy and stability of the model.
(5) Figure 7.8 illustrates that the proposed method does not perfectly match the "Target" output, and in some cases, it performs worse than other methods. Nonetheless, statistical results indicate that the proposed method achieves

TABLE 7.12
Statistical Results of the Ablation Experiment

Experiment	RMSE	R^2
Our method	0.0157 ± 9.1818E–07	0.6374 ± 1.9317E–03
Without A1	0.0167 ± 3.1072E–07	0.5894 ± 7.5149E–04
Without A2	0.0164 ± 1.0443E–06	0.6001 ± 2.5670E–03
Without A3	0.0173 ± 2.5104E–07	0.5576 ± 6.6046E–04
Without A4	0.0175 ± 2.6291E–06	0.5460 ± 7.0370E–03

the highest R^2 value among all compared methods. With the best R^2 for the testing data being only 0.6374, there is a clear need for further improvement in the fitting performance of the proposed method. This represents an ongoing challenge for modeling long-time-scale industrial process data, highlighting the necessity for additional research to enhance fitting performance.

7.4.5 Ablation Experiment

In the ablation study, experiments are conducted to evaluate the impact of various components on model performance, specifically focusing on the following elements: MSE-based constraint term (A1), model complexity penalties (A2), cascade forests (A3), and parallel forests (A4). The results of these ablation experiments on the testing set are presented in Table 7.12. This table provides insights into how each component contributes to the overall performance of the model, helping to identify which features are most critical for achieving optimal results.

Table 7.12 reveals several important insights: (1) Each of the four components has a distinct influence on the performance of the proposed method. (2) The MSE-based constraint term enhances the filling results, while the model complexity penalties are designed to improve stability. (3) Both types of forests exert a positive impact, with cascade forests demonstrating greater improvement than parallel forests.

The different modules exhibit varying levels of enhancement in the soft sensor's performance based on the ablation experiments conducted. In future research, incorporating anomaly detection as an additional module could further enhance the method's performance. There are two potential approaches: one is to implement anomaly detection prior to the soft sensor model to identify drifted operating conditions and unreliable measurements; the other is to conduct such detection after the soft sensor model using expert knowledge or lagged true values. Consequently, integrating anomaly detection with the soft sensor for effective updating will require further investigation.

7.5 SUMMARY

This chapter presents a multi-scale missing data modeling approach using an improved GAN-DFR for soft sensor modeling of DXN emission concentration. The main contributions are as follows: (1) A modeling framework is proposed that combines an improved GAN data-filling method for short-time-scale process data with an IDFR soft sensor method for long-time-scale operational indices. (2) MSE constraint and model complexity penalty terms are incorporated to enhance and stabilize the training process of the GAN. (3) Fully connected parallel and cascade forest algorithms for IDFR are employed to improve diversity, while a weight adaptive strategy is implemented to enhance generalization performance. The effectiveness of the proposed method is validated using actual DXN datasets. However, challenges remain regarding the high-dimensional filling issues associated with GAN and the lightweight nature of the IDFR depth structure, which require further optimization. Additionally, there are still numerous unlabeled samples that could be utilized more efficiently.

Future work will focus on several key issues. First, the model structure of the GAN will be improved to enhance its performance in filling high-dimensional missing data. Second, the IDFR structure will be optimized to reduce redundancy and increase processing speed. Finally, semi-supervised learning techniques will be introduced to better leverage unlabeled samples.

REFERENCES

1. Tang J, Cui C L, Xia H, Qiao J F. A survey of virtual sample generation for complex industrial processes. *Acta Automatica Sinica*, 2024, 50(4): 688–718.
2. Liu S. Incomplete big data filling algorithm based on machine learning. *International Core Journal of Engineering*, 2021, 7(12): 461–466.
3. Wu S, Feng X D, Shan Z G. Missing data imputation approach based on incomplete data clustering. *Chinese Journal of Computers*, 2012, 35(8): 1726–1738.
4. Tlamelo E, Thabiso M, Dimane M. A survey on missing data in machine learning. *Journal of Big Data*, 2021, 8(1): 140.
5. Kabir G, Tesfamariam S, Hemsing J. Handling incomplete and missing data in water network database using imputation methods. *Sustainable and Resilient Infrastructure*, 2020, 5(6): 365–377.
6. Deng Z W, Tang Z H, Zhu H Q. An improved expectation maximization algorithm for missing data management of concrete pump truck. *Journal of Central South University (Science and Technology)*, 2021, 52(2): 443–449.
7. Zhao X, Zhang Y, Yin B C. Imputation of incomplete bus arrival time based on the improved k-means algorithm. *Journal of Beijing University of Technology*, 2018, 44(01): 135–143.
8. Lu S W, Wu X L, Zheng J. Data-cleaning method based on dynamic fusion LOF for municipal wastewater treatment process. *Control and Decision*, 2022, 37(5): 1231–1240.
9. Han H G, Lu SW, Wu X L. Abnormal data cleaning method for municipal wastewater treatment based on improved support vector machine. *Journal of Beijing University of Technology*, 2021, 47(9): 1011–1020.
10. Matteo Q, James R. Multiple imputation for discrete data: evaluation of the joint latent normal model. *Biometrical Journal*, 2019, 61(4): 1003–1019.
11. Gondara L, Wang K. MIDA: multiple imputation using denoising autoencoders. In: *Pacific-Asia Conference on Knowledge Discovery and Data Mining*, Melbourne, VIC, Australia, June 3–6, 2018. Springer, 2018, 260–272.
12. Sánchez-Morales A, Sancho-Gómez J L, Figueiras-Vidal A R. Complete autoencoders for classification with missing values. *Neural Computing and Applications*, 2020, 33(6): 1951–1957.

13. Seo B, Shin J, Kim T, Youn B D. Missing data imputation using an iterative denoising autoencoder (IDAE) for dissolved gas analysis. *Electric Power Systems Research*, 2022, 212: 1–9.
14. Li B X, Yao Z J, Zhao C H. Variational imputation model of multi-source industrial data based on domain adaptation. *Control Engineering of China*, 2022, 29(4): 627–636.
15. Yoon J, Jordon J, Schaar M. GAIN: missing data imputation using generative adversarial nets. In: *International Conference on Machine Learning (ICML)*, Stockholm, Sweden, 10–15 July 2018. 2018, 5689–5698.
16. Li S C X, Jiang B, Marlin B. Misgan: learning from incomplete data with generative adversarial networks. 2019, arXiv: 1902.09599.
17. Zhang W, Luo Y, Zhang Y, Srinivasan D. SolarGAN: multivariate solar data imputation using generative adversarial network. *IEEE Transactions on Sustainable Energy*, 2021, 12(1): 743–746.
18. Sarda K, Yerudkar A, Vecchio C D. Missing data imputation for real time-series data in a steel industry using generative adversarial networks. In: *IECON 2021–47th Annual Conference of the IEEE Industrial Electronics Society*, Toronto, ON, Canada, 13–16 October 2021. 2021, 1–6.
19. Yao Z, Zhao C. FIGAN: a missing industrial data imputation method customized for soft sensor application. *IEEE Transactions on Automation Science and Engineering*, 2022, 19(4): 3712–3722.
20. Liu J, He J Z, Tang Z H, et al. Frame-dilated convolutional fusion network and GRU-based self-attention dual-channel network for soft-sensor modeling of industrial process quality indexes. *IEEE Transactions on Systems, Man, and Cybernetics: Systems*, 2021, 52(9): 5989–6002.
21. Jiang X, Ge Z. Augmented multidimensional convolutional neural network for industrial soft sensing. *IEEE Transactions on Instrumentation and Measurement*, 2021, 70: 1–10.
22. Kadlec P, Gabrys B, Strandt S. Data-driven soft sensors in the process industry. *Computers & Chemical Engineering*, 2009, 33(4): 795–814.
23. Zhong K, Han M, Han B. Data-driven based fault prognosis for industrial systems: a concise overview. *IEEE/CAA Journal of Automatica Sinica*, 2019, 7(2): 330–345.
24. Maschler B, Weyrich M. Deep transfer learning for industrial automation: a review and discussion of new techniques for data-driven machine learning. *IEEE Industrial Electronics Magazine*, 2021, 15(2): 65–75.
25. Sun Q, Ge Z. A survey on deep learning for data-driven soft sensors. *IEEE Transactions on Industrial Informatics*, 2021, 17(9): 5853–5866.
26. Yao L, Ge Z. Dynamic features incorporated locally weighted deep learning model for soft sensor development. *IEEE Transactions on Instrumentation and Measurement*, 2021, 70: 1–11.
27. Liu C, Wang K, Wang Y, Yuan X. Learning deep multimanifold structure feature representation for quality prediction with an industrial application. *IEEE Transactions on Industrial Informatics*, 2021, 18(9): 5849–5858.
28. Wang Y, Luo J, Liu C, et al. Layer-wise residual-guided feature learning with deep learning networks for industrial quality prediction. *IEEE Transactions on Instrumentation and Measurement*, 2022, 71: 1–11.
29. Qiao J F, Guo Z H, Tang J. Dioxin emission concentration measurement approaches for municipal solid wastes incineration process: a survey. *Acta Automatica Sinica*, 2020, 46(6): 1063–1089.
30. Zhu Q X, Hou K R, Chen Z S, et al. Novel virtual sample generation using conditional GAN for developing soft sensor with small data. *Engineering Applications of Artificial Intelligence*, 2021, 106: 1–9.

31. Tian Y, Xu Y, Zhu Q X, He Y L. Novel virtual sample generation using target-relevant autoencoder for small data-based soft sensor. *IEEE Transactions on Instrumentation and Measurement*, 2021, 70: 1–10.
32. Sun J, Yu H, Zhong G, et al. Random shapley forests: cooperative game-based random forests with consistency. *IEEE Transactions on Cybernetics*, 2022, 52(1): 205–214.
33. Xia H, Tang J, Qiao J F, et al. DF classification algorithm for constructing a small sample size of data-oriented DF regression model. *Neural Computing and Applications*, 2022, 34(4): 2785–2810.
34. Cao J, Jiang J, Zhao Q, et al. Wide and heavy plate crown control based on data mining. *Journal of Central South University (Science and Technology)*, 2019, 50(11): 2743–2752.
35. Zhang W, Cheng X, Hu Y, Yan Y. Online prediction of biomass moisture content in a fluidized bed dryer using electrostatic sensor arrays and the Random Forest method. *Fuel*, 2019, 239: 437–445.
36. Zhou Z H, Feng J. Deep forest: towards an alternative to deep neural networks. In: *International Joint Conference on Artificial Intelligence (IJCAI)*, Melbourne, VIC, Australia, 19–25 August, 2017. 2017, 3553–3559.
37. Tang J, Xia H, Zhang J, et al. Deep forest regression based on cross-layer full connection. *Neural Computing and Applications*, 2021, 33(15): 9307–9328.
38. Lin J Q, Wu H C, Chan S C. A new regularized recursive dynamic factor analysis with variable forgetting factor and subspace dimension for wireless sensor networks with missing data. *IEEE Transactions on Instrumentation and Measurement*, 2021, 70: 1–13.
39. Hu W, Wang T, Chu F. Fault feature recovery with Wasserstein generative adversarial imputation network with gradient penalty for rotating machine health monitoring under signal loss condition. *IEEE Transactions on Instrumentation and Measurement*, 2022, 71: 1–12.
40. Guo F, Bai W T, Huang B. Output-relevant variational autoencoder for just-in-time soft sensor modeling with missing data. *Journal of Process Control*, 2020, 92: 90–97.
41. Jiang Y, Yin S, Dong J, Kaynak O. A review on soft sensors for monitoring, control, and optimization of industrial processes. *IEEE Sensors Journal*, 2021, 21(11): 12868–12881.
42. Wang K, Liu X J, Liu E H, Qing L S. Core temperature prediction model based on AdaBoost algorithm. *Journal of Iron and Steel Research*, 2020, 32(5): 363–369.
43. Zhu J, Ge Z, Song Z, Gao F. Review and big data perspectives on robust data mining approaches for industrial process modeling with outliers and missing data. *Annual Reviews in Control*, 2018, 46: 107–133.
44. Chen L, Li G, Huang G, Shi P. A missing type-aware adaptive interpolation framework for sensor data. *IEEE Transactions on Instrumentation and Measurement*, 2021, 70: 1–15.
45. Xia H, Tang J, Aljerf L, et al. Assessment of PCDD/Fs formation and emission characteristics at a municipal solid waste incinerator for one year. *Science of the Total Environment*, 2023, 883: 163705.
46. Zou Q, Xiong Q, Li Q, et al. A water quality prediction method based on the multi-time scale bidirectional long short-term memory network. *Environmental Science and Pollution Research*, 2020, 27: 16853–16864.

8 Hybrid Simulator-Based Mechanism and Data-Driven Modeling for Multi-Demand Dioxin Emissions Concentration

8.1 RELATED WORKS

8.1.1 Multi-demand Dioxin Emissions Concentration Description

Accurately measuring the emission concentration of dioxin (DXN) is challenging due to the limitations of current detection technologies [1, 2]. As a result, the use of artificial intelligence modeling to predict the MSWI process has emerged as a promising solution. Previous research in this area can be categorized into three primary approaches: those based on statistical methods [3], classical machine learning techniques [4], and those utilizing deep learning and broader learning methodologies [5–7]. The aforementioned research demonstrates effectiveness but also highlights several unresolved demands that require further attention:

(1) The importance of adopting a hybrid modeling strategy tailored to the specific nuances of the modeling process. A primary challenge is to develop an interpretable framework for both mechanism-driven (MD) and data-driven (DD)-driven modeling. Furthermore, DXN analysis typically spans several hours, while optimal control operates on a timescale of seconds. Addressing the multifaceted requirements for accurate DXN emission prediction is crucial.

(2) The mechanisms governing DXN generation, emission, and adsorption remain unclear, which hinders the development of an accurate mathematical model. The concentrations of CO and CO_2, which serve as indicators of the combustion state, are directly linked to DXN concentration. While several commercial software options are available for simulating the MSWI process, there is a secondary challenge in acquiring multi-condition virtual mechanism data to construct an MD surrogate model that effectively represents the combustion state.

(3) There is a limited amount of actual DXN modeling data that utilizes easily measurable variables as inputs. Consequently, the challenge arises in constructing a DD model based on real historical data, along with the necessity to establish a relationship between the representation variables of the

DOI: 10.1201/9781003634195-11

combustion state and DXN concentration in relation to virtual mechanical data for developing the MD model. Additionally, exploring the integration of DD and MD models should be considered.

Three model prediction demands concerning various time scales in the actual MSWI power plant are identified as follows: second-scale, hourly, and daily predictions.

8.1.2 DXN Emission and Combustion State Description

DXN is produced in nearly all thermal and combustion processes. However, its emission mechanism is highly complex, making the development of accurate mathematical models exceedingly challenging. Numerous studies have shown a strong correlation between the concentration of DXN emissions and the combustion state observed in incinerators [1].

The mechanisms of DXN formation primarily include homogeneous synthesis, precursor synthesis, and de novo synthesis [8]. Hunsinger et al. emphasized the strong relationship between DXN emissions and products of incomplete combustion (PICs) [9]. Additionally, Peng et al. suggested that CO content could serve as an indicator of the combustion state [10], rather than relying solely on PICs. Previous mapping modeling efforts underscored the effectiveness of using CO content as an indicator. Research has discussed the impact of excess air on DXN emission levels, while Chang et al. observed a linear correlation between DXN concentration, combustion temperature, and CO concentration [3].

Research indicates that the concentrations of CO and O_2 significantly influence the combustion state, directly affecting DXN emissions. Additionally, CO_2 plays a crucial role in characterizing combustion states, as outlined in the combustion efficiency standards (e.g., GB18484-2020 in China). Therefore, a comprehensive assessment of the combustion state requires considering CO, CO_2, and O_2. As a result, developing a mechanistic model of the combustion state to map DXN emissions emerges as a viable solution.

8.1.3 Semi-Supervised Learning Modeling

The core principle of semi-supervised learning is to enhance the limited set of original labeled samples by generating pseudo-labels for unlabeled samples, thereby improving the learner's generalization performance. The critical aspect of this approach is assessing the confidence of the pseudo-labeled samples [11]. Consequently, when a substantial number of unlabeled samples are available, it is essential to filter them to determine their positive impact on the constructed soft measurement model.

A semi-supervised support vector regression (SS-SVR) model based on a self-training strategy was established [12], which estimates the pseudo-label distribution of unlabeled samples using a probabilistic local reconstruction model. This approach expands the training dataset by resampling both the unlabeled samples and their associated pseudo-labels. Cheng et al. proposed a semi-supervised regression algorithm featuring a dual preference mechanism [13], which initially selects unlabeled samples through similarity calculations with the centers of dense regions of labeled samples. They then optimize the labeled samples based on their pairwise similarities and train a learner using Gaussian process regression to generate pseudo-labels for the dual-preferred unlabeled samples. Shi et al. developed a Gaussian process regression soft measurement model based on an improved self-training algorithm [14]. It first estimates the missing dominant variable values of unlabeled samples using similarity measures. Then, it filters the estimated samples and subsequently incorporates the high-generalization pseudo-labeled samples into the original labeled sample set to construct the soft measurement model.

However, a drawback of the aforementioned methods is that selecting unreliable unlabeled samples for labeling may hinder the model's ability to correct accumulated bias during the training process. Furthermore, these methods utilize traditional single learners to construct soft measurement models during the labeling process, and their generalization capabilities and interpretability require further enhancement.

8.1.4 Data Mechanism Hybrid-Driven Modeling

The MD model, often referred to as white-box modeling, is based on understanding the internal mechanisms of industrial processes and employs fundamental principles such as the conservation of momentum, heat, mass, and reaction kinetics [15, 16]. However, constructing an effective MD model presents challenges due to the complexity and variability inherent in the reaction mechanisms of modern industrial processes [17]. With the emergence of industrial big data [18], machine learning and deep learning modeling methods have advanced rapidly [19]. DD models establish a mapping relationship between independent and dependent variables. However, the performance of DD models can be adversely affected by insufficient data samples or incomplete coverage ranges [20]. To address these issues, a hybrid strategy that integrates both MD and DD approaches is proposed. This strategy leverages the strengths of both methods, resulting in a higher-performing model through mutual compensation between MD and DD [21].

From a structural perspective, the hybrid approach combining MD and DD can be categorized into three types: cascade, parallel, and cascade-parallel fusion [22]. Cascade models can be further divided into two main categories: MD model front and MD model post. In the former, the output of the MD model serves as input for the DD model, as seen in applications such as estimating the capacity of depleted $LiFePO_4$ batteries [23]. The latter constructs the DD model using historical data, exemplified by predicting key parameters in cane sugar crystallization [24]. Typically, constructing an MD model is preferred when the underlying mechanism is well understood. However, practical MD models often experience reduced accuracy due to frequent changes in operating conditions. To address this, a parallel structure is proposed where the DD model compensates for the errors of the MD model. Existing studies include multi-variable predictions for underflow concentration in thickeners [25] and monitoring the contents of salvianolic acid [26]. Nonetheless, these parallel models encounter limitations related to the effective range of historical data and struggle to provide reliable outputs for unknown data, resulting in poor generalization performance.

To overcome these challenges, significant efforts have been made to enhance parallel structures. For example, Ren et al. [27] applied a cascade-parallel ensemble modeling strategy to the Czochralski process. Other researchers have focused on fusion methods, such as the Kalman filter linear fusion [28] and linear fusion based on correlation matrixes [29]. Furthermore, Zhang et al. [30] demonstrated that the ensemble strategy combining MD and DD models outperforms both cascade and parallel structures. However, fusion strategies that integrate MD and DD models have not received sufficient attention, leading to a lack of related studies. Additionally, implementing MD and DD fusion models should be tailored to the specific characteristics of different applications.

8.2 MODELING STRATEGY

The proposed hybrid modeling strategy comprises five modules: a combustion state model based on mechanisms and numerical simulations, a white-box multiple-input multiple-output (MIMO) surrogate model, a semi-supervised transfer learning mapping model, a Bagging LRDT model, and a CIRWNN fusion model. The workflow of these five modules is illustrated in Figure 8.1. Within the five modules, modules 1, 2, and 3 are associated with the MD model, module 4 represents the DD model, and module 5 pertains to the fusion model.

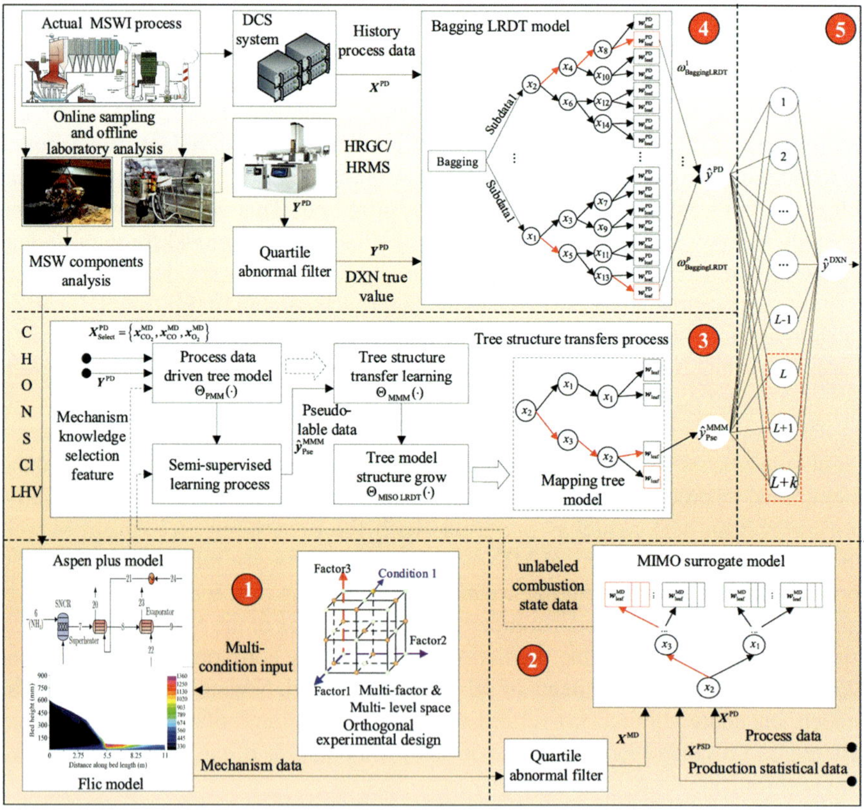

FIGURE 8.1 Mechanism- and DD modeling strategy.

The functions of the five modules in Figure 8.1 are as follows:

(1) Combustion State Modeling Based on Mechanisms and Numerical Simulation Module: This module integrates orthogonal experimental design with multiple soft-coupling numerical simulations. The orthogonal design generates multi-condition input data, including the low heat value (LHV) of municipal solid waste (MSW), primary airflow (PA), and feeding capacity (FC). The coupling involves FLIC and Aspen Plus, where FLIC simulates the solid-phase combustion of MSW, while Aspen Plus manages the vapor-phase processes.
(2) White-Box MIMO LRDT Surrogate Modeling Module: This module incorporates a quartile abnormal filter (QAF) and a MIMO LRDT model. The QAF is used to eliminate outlier data points within a fixed time window. The MIMO LRDT model is trained using LHV, PA, and FC as inputs to predict CO_2, CO, and O_2 emissions.
(3) Bagging LRDT Modeling Module: Process data is collected from the distribution control system (DCS), while DXN emission concentrations are obtained through online sampling and offline laboratory analysis (HRGC/HRMS). In this module, LRDT serves as the base learner, with Bagging applied to enhance stability, and iterated Tikhonov regularization utilized to improve performance.
(4) Semi-supervised Transfer Learning Mapping Modeling Module: This module includes LRDT training using process data, tree structure transfer, and the development of a DXN mapping model based on semi-supervised mechanism data. Initially, an LRDT is constructed using process data focused on three key elements: CO_2, CO, and O_2. The LRDT structure is then transferred to incorporate knowledge from mechanism data. Finally, a semi-supervised approach utilizing unlabeled mechanism data establishes a DXN mapping model that links the combustion state of the MSWI process with DXN concentration.
(5) CIRWNN Modeling Module: This module consists of a random weighted neural network (RWNN) and employs a constrained increment learning strategy. The RWNN takes input from the outputs of both the Bagging LRDT and semi-supervised LRDT. Additionally, the constrained increment learning strategy [31] is employed to regulate the model's learning performance.

The motivation and relationships among the various modules within the proposed hybrid modeling architecture are outlined as follows:

(1) DXN Emission Mechanism: The DXN emission mechanism involves phases of generation, adsorption, and emission, making it challenging to construct a direct DXN mechanism model. Thus, this chapter aims to adopt an indirect strategy that utilizes the combustion state as a proxy for representing DXN concentration. This approach addresses two key issues: establishing a combustion state model and bridging the relationship between the combustion state and DXN concentration.

(2) Challenges in Historical Data-based DD Models: The small sample sizes and high-dimensional features present in DXN modeling data pose significant challenges that need to be addressed. To tackle this, this chapter proposes a DD model based on a tree structure characterized by high-generalization capability and strong interpretability.
(3) Hybrid Data-Driven Modeling: In the context of hybrid data-driven modeling, the challenge lies in effectively balancing the contributions of predictive values from both MD and DD models. Recognizing the increasing ability of domain experts to integrate mechanistic knowledge with historical experience, this chapter employs a fast increment neural network to fuse the outputs of the MD and DD models.
(4) Establishment of a Robust Prediction System: By utilizing the aforementioned five distinct modules along with various time-scaled inputs, this chapter effectively establishes a robust system for predicting DXN emissions that meets multiple demands.

The significant novelties and contributions of this chapter include:

(1) A hybrid modeling strategy that fuses mechanistic (MD) and DD approaches is proposed, which integrates simulated mechanism data with real historical data for multi-demand prediction modeling. This approach primarily utilizes interpretable tree-structured methods, effectively addressing common challenges such as small sample sizes and the complexities associated with high-dimensional data learning.
(2) The virtual mechanical data are used to develop a novel MIMO linear regression decision tree (LRDT) surrogate model. Our research introduces a strategy that combines numerical simulation and agent-based modeling to assist in modeling key parameters that are difficult to measure, particularly when constructing the mechanism model proves complex.
(3) A semi-supervised transfer learning mapping model is used to establish correlations between easy-to-measure process variables and difficult-to-measure key parameters, thereby facilitating the construction of a virtual mechanism DD prediction model. Additionally, A novel Bagging LRDT method is designed to model difficult-to-measure key parameters using real historical data, addressing challenges posed by small sample sizes and high dimensionality. Following this, a neural network method is employed to balance the contributions of prediction values derived from both MD and DD models.

8.3 ALGORITHM IMPLEMENTATION

In the module concerning the combustion state modeling based on mechanism and numerical simulation, a coupling of numerical simulation technologies and an orthogonal experimental design was applied, which is the same as that of Chapter 6.

8.3.1 White-Box MIMO Surrogate Model Module

The original multi-condition mechanistic data are denoted as $\mathbf{D}_{\text{Orig}}^{\text{MD}} \in \mathbb{R}^{2058\times(11\times13)}$. To characterize the combustion state, LHV, PA, and FC are selected as input variables, while CO_2, CO, and O_2 are defined as output variables. Therefore, the multi-condition mechanistic data can be expressed as $\mathbf{D}^{\text{MD}} \in \mathbb{R}^{2058\times(3\times3)}$.

First, the moving time window QAF is introduced to eliminate outlier data points. The length of the data set $\boldsymbol{D}_1^{\text{MD}}$ is denoted as $T_{D_1^{\text{MD}}} = 2058$, and the fixed sampling time window is set to t_{Wind}. For the data within the time window t_{Wind}, the quartile method is employed to identify and remove outliers. The removal rule is as follows:

$$\begin{cases} x^{\text{AB}} > x^{\text{UB}} = 2.5Q_{3/4}(\boldsymbol{x}) - 1.5Q_{1/4}(\boldsymbol{x}) \\ x^{\text{AB}} < x^{\text{DB}} = 2.5Q_{1/4}(\boldsymbol{x}) - 1.5Q_{3/4}(\boldsymbol{x}) \end{cases} \tag{8.1}$$

where x^{AB} denotes the abnormal data point; x^{UB} and x^{DB} indicate the up and down boundary; $Q_{3/4}(\cdot)$ and $Q_{1/4}(\cdot)$ is the 1st and 3rd quartile of $\boldsymbol{x}$.

Thus, a data point within the time window is considered an abnormal point and is deleted if it is greater than x^{UB} or less than x^{DB}. After processing the data through the QAF method, the resulting dataset $\boldsymbol{D}_2^{\text{MD}} \in \mathbb{R}^{N_{D_2^{\text{MD}}}\times(3\times3)}$ is obtained, which can be used for surrogate modeling.

Considering the multivariate input and output characteristics inherent in mechanistic data $\mathbf{D}_2^{\text{MD}} \in \mathbb{R}^{N_{D_2^{\text{MD}}}\times(3\times3)}$, a MIMO LRDT algorithm is introduced for surrogate modeling. Generally, the MISO LRDT algorithm represents an enhanced version of the classical MISO CART method.

The motivation for proposing the LRDT method for DD modeling arises from the limitation of the CART model, where the leaf nodes utilize only the target information. Specifically, the prediction value at the leaf node is the statistical expectation of the target variable. Furthermore, the CART model neglects feature space information when making predictions at the leaf nodes. It reduces its generalization capacity in response to significant changes in data distribution, as it primarily focuses on the average of the training data. To address these limitations and effectively leverage information from both feature and target spaces, a linear regression method is employed to compute multiple target-predicted outputs at the leaf nodes. This improvement strategy enhances both learning and generalization performance compared to the traditional CART model.

The MIMO LRDT represents an advancement over the MISO LRDT, allowing both its non-leaf nodes and leaf nodes to simultaneously process multiple target outputs. Let x_{nonleaf}^1, x_{nonleaf}^2, and x_{nonleaf}^3 denote the 1st, 2nd, and 3rd non-leaf nodes, while t_{leaf}^1 and $t_{\text{leaf}}^{T/2}$ represent the 1st and $(T/2)$th multiple-output leaf nodes, respectively. Additionally, $\{x_{\text{nonleaf}}^1, x_{\text{nonleaf}}^3, \ldots, x_{\text{nonleaf}}^t\}$ signifies the feature set associated with the nodes along the $(T/2)$th path t. The MIMO mechanistic data is re-denoted as $\mathbf{D}_2^{\text{MD}} = \{\mathbf{x}_i^{\text{MD}}, \mathbf{y}_i^{\text{MD}}\}_{i=1}^{N_{D_2^{\text{MD}}}} \in \mathbb{R}^{N_{D_2^{\text{MD}}}\times(3\times3)}$, where $\mathbf{x}_i^{\text{MD}} = [x_1^{\text{MD}}; x_2^{\text{MD}}; x_3^{\text{MD}}]^{\text{T}} \in \mathbb{R}^{1\times3}$ and

$\mathbf{y}_i^{\text{MD}} = \left[y_1^{\text{MD}}; y_2^{\text{MD}}; y_3^{\text{MD}} \right]^{\text{T}} \in \mathbb{R}^{1\times 3}$ represent the input features and output targets, respectively.

The learning process of the MIMO LRDT method is as follows. Firstly, the modeling data are divided as

$$\begin{cases} \mathbf{D}_{2,\text{left}}^{\text{MD}} : \left\{ \mathbf{D}_{2,\text{left}}^{\text{MD}} \in \mathbb{R}^{N_{\text{Left}}^{\text{Model2}} \times (3\times 3)} \,|\, \mathbf{x}_i^{\text{MD}} \geq x_{1,\text{nonleaf}}^{\text{MD}} \right\} \\ \mathbf{D}_{2,\text{right}}^{\text{MD}} : \left\{ \mathbf{D}_{2,\text{right}}^{\text{MD}} \in \mathbb{R}^{N_{\text{Right}}^{\text{Model2}} \times (3\times 3)} \,|\, \mathbf{x}_i^{\text{MD}} < x_{1,\text{nonleaf}}^{\text{MD}} \right\} \end{cases} \tag{8.2}$$

where $N_{\text{Left}}^{\text{Model2}}$ and $N_{\text{Right}}^{\text{Model2}}$ denote the number of samples in $\mathbf{D}_{2,\text{left}}^{\text{MD}}$ and $\mathbf{D}_{2,\text{right}}^{\text{MD}}$, respectively; $\mathbf{x}_i^{\text{MD}}$ is the ith feature vector; and $x_{1,\text{nonleaf}}^{\text{MD}}$ represents the value of vector $\mathbf{x}_i^{\text{MD}}$.

To find the optimal parameters $x_{1,\text{nonleaf}}^{\text{MD}}$, all points in the dataset D_2^{MD} are traversed to calculate the mean squared error (MSE) of the target values. The multiple outputs of MSE can be expressed as follows:

$$\begin{aligned} L_k^{\text{MSE}}(n,i) &= f_{\text{MSE}}\left(\mathbf{D}_{2,\text{left}}^{\text{MD}}\right) + f_{\text{MSE}}\left(\mathbf{D}_{2,\text{right}}^{\text{MD}}\right) \\ &= \frac{1}{3}\sum_{\text{Targ}=1}^{3} \left\{ \begin{aligned} &\left(\left(\mathbf{y}_{\text{Targ,Left}}^{\text{MD}} - \overline{y}_{\text{Targ,Left}}^{\text{MD}}\right) I\left(x_i^{\text{MD}} \in \mathbf{D}_{2,\text{left}}^{\text{MD}}\right)\right)^2 \\ &+ \left(\left(\mathbf{y}_{\text{Targ,Right}}^{\text{MD}} - \overline{y}_{\text{Targ,Right}}^{\text{MD}}\right) I\left(x_i^{\text{MD}} \in \mathbf{D}_{2,\text{right}}^{\text{MD}}\right)\right)^2 \end{aligned} \right\} \end{aligned} \tag{8.3}$$

where $L_k^{\text{MSE}}(n,i)$ represents the loss value of MSE, and (n,i) denotes the ith feature value of the nth sample from kth iteration.

The 1st non-leaf node $x_{1,\text{nonleaf}}^{\text{MD}}$ is determined by minimizing the MSE across all points, as follows:

$$x_{1,\text{nonleaf}}^{\text{MD}} = L_{\text{Best}}^{\text{MSE}}(n,i) = \min\left\{ L_k^{\text{MSE}}(n,i) \right\}_{k=1}^{K} \tag{8.4}$$

where $L_{\text{Best}}^{\text{MSE}}(n,i)$ expresses the optimal coordinate of the dataset, and K is the maximum of iteration.

Based on Eqs. (8.1) and (8.2), $(T/2-1)$ middle nodes $\left\{ x_{\text{nonleaf}}^{1}, x_{\text{nonleaf}}^{3}, \ldots, x_{\text{nonleaf}}^{T/2-1} \right\}$ can be obtained. There are T routes from the root node to all the leaf nodes, and each path contains node information that exhibits significant differences. Therefore, the input features of the leaf nodes are selected based on the nodes along the route, as follows:

$$\begin{aligned} \mathbf{D}_{t,\text{leaf}}^{\text{MD}} &= \left\{ \text{Route}_{t,\text{leaf}}^{\text{MD}}, \mathbf{Y}_{t,\text{leaf}}^{\text{MD}} \right\} \in \mathbb{R}^{N_{t,\text{leaf}}^{\text{Model2}} \times \left(M_{t,\text{leaf}}^{\text{MD}} + 3\right)} \\ &= \left\{ x_{1,\text{nonleaf}}^{\text{MD}}, \ldots, x_{t,\text{nonleaf}}^{\text{MD}}, \mathbf{Y}_{t,\text{leaf}}^{\text{MD}} \right\} \in \mathbb{R}^{N_{t,\text{leaf}}^{\text{Model2}} \times \left(M_{t,\text{leaf}}^{\text{MD}} + 3\right)} \end{aligned} \tag{8.5}$$

where $\mathbf{D}_{t,\text{leaf}}^{\text{MD}}\left(\mathbf{D}_{t,\text{leaf}}^{\text{MD}}=\left\{\mathbf{X}_{t,\text{leaf}}^{\text{MD}},\mathbf{Y}_{t,\text{leaf}}^{\text{MD}}\right\}\in\mathbb{R}^{N_{t,\text{leaf}}^{\text{Model2}}\times\left(M_{t,\text{leaf}}^{t}+3\right)}\right)$ is the training dataset of the tth leaf node, $\text{Route}_{t,\text{leaf}}^{\text{MD}}$ is the tth path of the tth leaf node, $N_{t,\text{leaf}}^{\text{Model2}}$ is the number of sample of $\mathbf{D}_{t,\text{leaf}}^{\text{MD}}$, and $M_{t,\text{leaf}}^{\text{MD}}$ is the dimension of $\mathbf{D}_{t,\text{leaf}}^{\text{MD}}$.

Further, the linear regression function is provided as follows,

$$\hat{\mathbf{Y}}_{t,\text{leaf}}^{\text{MD}}=\mathbf{X}_{t,\text{leaf}}^{\text{MD}}\mathbf{W}_{t,\text{leaf}}^{\text{MD}}=\mathbf{X}_{t,\text{leaf}}^{\text{MD}}\left[\mathbf{w}_{1,\text{leaf}}^{\text{MD}},\mathbf{w}_{2,\text{leaf}}^{\text{MD}},\mathbf{w}_{3,\text{leaf}}^{\text{MD}}\right] \tag{8.6}$$

where $\hat{\mathbf{Y}}_{t,\text{leaf}}^{\text{MD}}$ is multiple output of the tth leaf node, $\mathbf{X}_{t,\text{leaf}}^{\text{MD}}$ is the feature space data, and $\mathbf{W}_{t,\text{leaf}}^{\text{MD}}$ is the weight matrix and needs to be solved.

Then, the regularized least squares cost function $L\left(\mathbf{W}_{t,\text{leaf}}^{\text{MD}}\right)$ is used to calculate the weight matrix $\mathbf{W}_{t,\text{leaf}}^{\text{MD}}$, as follows,

$$L\left(\mathbf{W}_{t,\text{leaf}}^{\text{MD}}\right)=\begin{bmatrix}\frac{1}{2}\left(\|\mathbf{X}_{t,\text{leaf}}^{\text{MD}}\mathbf{w}_{t,\text{leaf}}^{1}-\mathbf{y}_{t,\text{leaf}}^{1}\|_2^2+\lambda\|\mathbf{w}_{t,\text{leaf}}^{1}\|_2^2\right)\\ \frac{1}{2}\left(\|\mathbf{X}_{t,\text{leaf}}^{\text{MD}}\mathbf{w}_{t,\text{leaf}}^{2}-\mathbf{y}_{t,\text{leaf}}^{2}\|_2^2+\lambda\|\mathbf{w}_{t,\text{leaf}}^{2}\|_2^2\right)\\ \frac{1}{2}\left(\|\mathbf{X}_{t,\text{leaf}}^{\text{MD}}\mathbf{w}_{t,\text{leaf}}^{3}-\mathbf{y}_{t,\text{leaf}}^{3}\|_2^2+\lambda\|\mathbf{w}_{t,\text{leaf}}^{3}\|_2^2\right)\end{bmatrix}^{\text{T}} \tag{8.7}$$

where λ is the regularization parameters and $\lambda \geq 0$.

The gradient descent method is applied to solve the weight matrix $\mathbf{W}_{t,\text{left}}^{\text{MD}}$, and the calculation is as follows,

$$\frac{\partial L\left(\mathbf{W}_{t,\text{leaf}}^{\text{MD}}\right)}{\partial\left(\mathbf{W}_{t,\text{leaf}}^{\text{MD}}\right)}=\left[\frac{\partial L\left(\mathbf{W}_{t,\text{leaf}}^{\text{MD}}\right)}{\partial\left(\mathbf{w}_{t,\text{leaf}}^{1}\right)^{\text{T}}},\frac{\partial L\left(\mathbf{W}_{t,\text{leaf}}^{\text{MD}}\right)}{\partial\left(\mathbf{w}_{t,\text{leaf}}^{2}\right)^{\text{T}}},\frac{\partial L\left(\mathbf{W}_{t,\text{leaf}}^{\text{MD}}\right)}{\partial\left(\mathbf{w}_{t,\text{leaf}}^{3}\right)^{\text{T}}}\right]^{\text{T}} \tag{8.8}$$

Using Targth as an example, the solve process can be written as follows,

$$\begin{aligned}\frac{\partial L\left(\mathbf{W}_{t,\text{leaf}}^{\text{MD}}\right)}{\partial\left(\mathbf{w}_{t,\text{leaf}}^{\text{Targ}}\right)^{\text{T}}}&=\frac{\partial}{\partial\left(\mathbf{w}_{t,\text{leaf}}^{\text{Targ}}\right)^{\text{T}}}\left(\begin{aligned}&\left(\mathbf{X}_{t,\text{leaf}}^{\text{MD}}\mathbf{w}_{t,\text{leaf}}^{\text{Targ}}-\mathbf{y}_{t,\text{leaf}}^{\text{Targ}}\right)^{\text{T}}\left(\mathbf{X}_{t,\text{leaf}}^{\text{MD}}\mathbf{w}_{t,\text{leaf}}^{\text{Targ}}-\mathbf{y}_{t,\text{leaf}}^{\text{Targ}}\right)\\&+\lambda\left(\mathbf{w}_{t,\text{leaf}}^{\text{Targ}}\right)^{\text{T}}\left(\mathbf{w}_{t,\text{leaf}}^{\text{Targ}}\right)\end{aligned}\right)\\&=\left(\mathbf{X}_{t,\text{leaf}}^{\text{MD}}\right)^{\text{T}}\mathbf{X}_{t,\text{leaf}}^{\text{MD}}\mathbf{w}_{t,\text{leaf}}^{\text{Targ}}-\left(\mathbf{X}_{t,\text{leaf}}^{\text{MD}}\right)^{\text{T}}\mathbf{y}_{t,\text{leaf}}^{\text{Targ}}+\lambda\left(\mathbf{w}_{t,\text{leaf}}^{\text{Targ}}\right)\end{aligned} \tag{8.9}$$

Then, let $\frac{\partial L\left(\mathbf{W}_{t,\text{leaf}}^{\text{MD}}\right)}{\partial\left(\mathbf{w}_{t,\text{leaf}}^{\text{Targ}}\right)^{\text{T}}}=0$, the Targth, $\mathbf{w}_{t,\text{leaf}}^{\text{Targ}}$ can be obtained as,

$$\mathbf{w}_{t,\text{leaf}}^{\text{Targ}}=\left(\left(\mathbf{X}_{t,\text{leaf}}^{\text{MD}}\right)^{\text{T}}\mathbf{X}_{t,\text{leaf}}^{\text{MD}}+\lambda\mathbf{I}\right)^{-1}\left(\mathbf{X}_{t,\text{leaf}}^{\text{MD}}\right)^{\text{T}}\mathbf{y}_{t,\text{leaf}}^{\text{Targ}} \tag{8.10}$$

Therefore, the weight matrix $\mathbf{W}_{t,\text{leaf}}^{\text{MD}}$ is obtained, and the prediction value of the tth leaf node can be calculated using the linear regression method.

Finally, the MIMO LRDT model can be expressed as,

$$\hat{\mathbf{y}}^{\text{MD}} = f_{\text{MIMOLRDT}}\left(\mathbf{x}^{\text{MD}}\right) \in \mathbb{R}^{1\times 3} \tag{8.11}$$

8.3.2 Semi-Supervised Transfer Mapping Model Module

The MIMO LRDT model produces outputs of CO_2, CO, and O_2, which allows for the characterization of the combustion state within the MSWI process. This methodology consists of four sequential steps. First, the model develops the combustion state and the DXN emission concentration model using process data. Next, the unlabeled virtual mechanism data is labeled. Following this, the structure of the process mapping model (PMM) is transferred, and pseudo-labeled data are employed to construct the mechanism mapping model (MMM). Finally, the structure of the MMM is updated. The details of these steps are as follows.

Step (1): The combustion state and the DXN emission concentration model are constructed using the process data $\mathbf{X}^{\text{PD}}$. CO_2, CO, and O_2 are selected as input features, while the DXN emission concentration $\mathbf{y}^{\text{MD}}$ serves as the target variable to obtain the learning data $\mathbf{D}^{\text{PDM}} = \left\{\mathbf{x}_{\text{CO}_2}^{\text{MD}}, \mathbf{x}_{\text{CO}}^{\text{MD}}, \mathbf{x}_{\text{O}_2}^{\text{MD}}, \mathbf{y}^{\text{MD}}\right\} \in \mathbb{R}^{N^{\text{Model3}}\times(3+1)}$ for the PMM. The modeling procedure follows the same approach as that of the MISO LRDT, and the resulting PMM is denoted as $\Theta_{\text{PMM}}(\cdot)$.

Step (2): The unlabeled virtual mechanism data, denoted as $\hat{\mathbf{Y}}^{\text{MD}} = \left\{\hat{\mathbf{y}}_{\text{CO}_2}^{\text{MD}}, \hat{\mathbf{y}}_{\text{CO}}^{\text{MD}}, \hat{\mathbf{y}}_{\text{O}_2}^{\text{MD}}\right\} \in \mathbb{R}^{N^{\text{MD}}\times 3}$, is the output of the MIMO surrogate model. The labeling principle in the PMM method consists of two primary steps. First, feeding the unlabeled mechanism data (CO_2, CO, and O_2) into the PMM model. Second, computing prediction values based on these inputs. Finally, these predicted values serve as pseudo-labels for the unlabeled virtual mechanism data (CO_2, CO, and O_2). That is to say, the trained PMM model predicts values for the DXN concentration corresponding to the unlabeled virtual mechanism data. When $\hat{\mathbf{Y}}^{\text{MD}}$ is used as an input for $\Theta_{\text{PMM}}(\cdot)$, the labeled process is illustrated as follows.

$$\hat{\mathbf{y}}_{\text{Pse}}^{\text{MMM}} = \Theta_{\text{PMM}}\left(\hat{\mathbf{Y}}^{\text{MD}}\right) \in \mathbb{R}^{N^{\text{MD}}\times 1} \tag{8.12}$$

Therefore, the pseudo-labeled dataset is generated by combining $\hat{\mathbf{Y}}^{\text{MD}}$ and $\hat{\mathbf{y}}_{\text{Pse}}^{\text{MMM}}$, as follows:

$$\mathbf{D}^{\text{MMM}} = \left\{\hat{\mathbf{y}}_{\text{CO}_2}^{\text{MD}}, \hat{\mathbf{y}}_{\text{CO}}^{\text{MD}}, \hat{\mathbf{y}}_{\text{O}_2}^{\text{MD}}, \mathbf{y}_{\text{Pse}}^{\text{MMM}}\right\} \in \mathbb{R}^{N^{\text{MD}}\times(3+1)} \tag{8.13}$$

Step (3): Subsequently, the trained PMM model is transferred to become the MMM model. The rationale for employing a transfer learning mechanism is twofold, i.e., to minimize model training costs and to retain knowledge from the source domains. In terms of training expenses, the volume of unlabeled virtual mechanistic data significantly exceeds that of the labeled process data, leading to substantially higher costs associated with retraining the MMM model using pseudo-labeled virtual mechanism data, compared to the cost of transferring the PMM model to the MMM model. Additionally, the pseudo-labeling of mechanistic data relies on the PMM model, and the structure of the source domain model enhances the reliability of the pseudo-labels. During the transfer learning process, the information within each non-leaf node of the PMM model remains unchanged and is directly incorporated as node information within the MMM model. This process is illustrated in Figure 8.2.

As shown in Figure 8.2, the pseudo-labeled data are fed into the PMM structure, where each non-leaf node acts as prior knowledge to divide the pseudo-labeled dataset $\mathbf{D}^{\text{MMM}}$ into two sub-nodes, ultimately resulting in all the sub-nodes $\left\{\text{Route}_{t,\text{leaf}}^{\text{PMM}}\right\}_{t=1}^{T}$ in T routes. The leaf nodes within the MMM model undergo retraining using the provided pseudo-labeled data to improve their understanding of the unlabeled targets. As a result, the MMM model $\Theta_{\text{MMM}}(\cdot)$ is implemented.

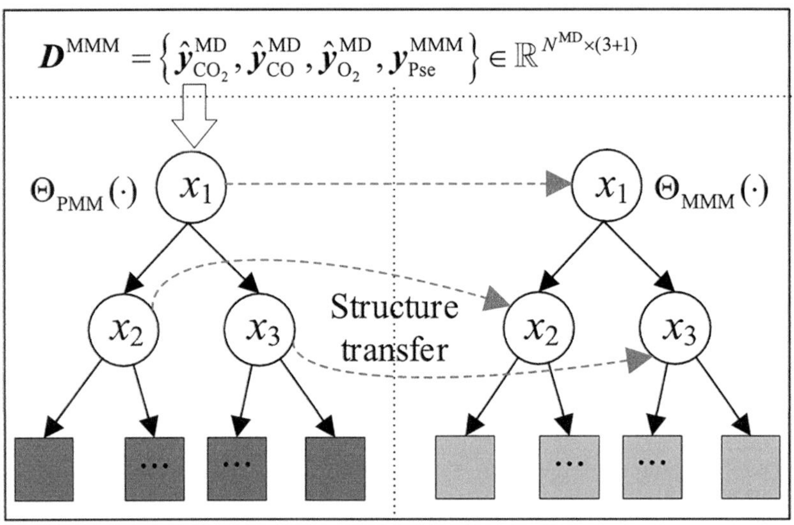

FIGURE 8.2 Graphic of tree structure transfer.

Step (4): Since the virtual mechanism data N^{MD} significantly outnumber the real process data N^{Model3}, the sample size of the leaf nodes in the transfer model greatly exceeds the threshold for standard training. Therefore, a tree structure update algorithm is employed for the MMM model.

First, the virtual mechanism data $\mathbf{D}^{\mathrm{MMM}}$ are fed into the MMM model, and the prediction performance of each leaf node is evaluated. This performance can be described as follows:

$$\mathbf{e}_{\mathrm{MMM}} = \left(\Theta_{\mathrm{MMM}}\left(\mathbf{D}^{\mathrm{MMM}}\right) - \hat{\mathbf{y}}_{\mathrm{Pse}}^{\mathrm{MMM}}\right)^2 \in \mathbb{R}^{1\times T} \tag{8.14}$$

where $\mathbf{e}_{\mathrm{MMM}}$ is the MSE of all T leaf nodes in MMM model.

Second, the mean of the MSE vector is calculated, and the leaf nodes that exceed this mean are identified as needing further updates.

$$\left\{e_{\mathrm{MMM}}^{t},\ldots\right\}_{t}^{t\ll T} \Leftarrow \left(e_{\mathrm{MMM}}^{t} > \bar{\mathbf{e}}_{\mathrm{MMM}}\right) \tag{8.15}$$

Third, the selected leaf nodes are updated by continuing to grow until the number of samples in each leaf node is less than or equal to the initially set threshold. The updating process follows the same procedure as the training process of the MISO LRDT.

8.3.3 Bagging LRDT Model Module

The real high-dimensional MSWI process data is collected from the actual power plant and is denoted as follows,

$$\mathbf{X}_{\mathrm{Ori}}^{\mathrm{PD}} = \left\{\mathbf{x}_1^{\mathrm{PD}}, \mathbf{x}_2^{\mathrm{PD}}, \ldots, \mathbf{x}_{M^{\mathrm{PD}}}^{\mathrm{PD}}\right\} \in \mathbb{R}^{N^{\mathrm{PD}}\times M^{\mathrm{PD}}} \tag{8.16}$$

The true values of the DXN emission concentrations are designated as $\mathbf{y}^{\mathrm{PD}} \in \mathbb{R}^{N^{\mathrm{PD}}\times 1}$. Subsequently, abnormal data points are removed using the QAF method, resulting in the acquisition of high-dimensional small-sample training data, denoted as follows,

$$\mathbf{D}^{\mathrm{PD}} = \left\{\mathbf{X}^{\mathrm{PD}}, \mathbf{y}^{\mathrm{PD}}\right\} \in \mathbb{R}^{N^{\mathrm{Model3}}\times\left(M^{\mathrm{PD}}+1\right)} \tag{8.17}$$

A novel bagging ensemble method based on MISO LRDT for process data modeling tasks is employed. First, bootstrap sampling is utilized to generate multiple subsets from the training data $\mathbf{D}^{\mathrm{PD}}$. This process can be expressed as follows:

$$\mathbf{D}^{\text{PD}} \overset{\text{Bagging}}{\Rightarrow} \begin{cases} \mathbf{D}_1^{\text{PD}} = \left\{\mathbf{X}_1^{\text{PD}}, \mathbf{y}_1^{\text{PD}}\right\} \in \mathbb{R}^{N^{\text{Model3}} \times \left(M^{\text{PD}}+1\right)} \\ \mathbf{D}_2^{\text{PD}} = \left\{\mathbf{X}_2^{\text{PD}}, \mathbf{y}_2^{\text{PD}}\right\} \in \mathbb{R}^{N^{\text{Model3}} \times \left(M^{\text{PD}}+1\right)} \\ \cdots \\ \mathbf{D}_P^{\text{PD}} = \left\{\mathbf{X}_P^{\text{PD}}, \mathbf{y}_P^{\text{PD}}\right\} \in \mathbb{R}^{N^{\text{Model3}} \times \left(M^{\text{PD}}+1\right)} \end{cases} \tag{8.18}$$

where $\mathbf{D}_p^{\text{PD}}$ denotes the pth training subset.

Second, the MISO LRDT models with number P are constructed using the aforementioned subsets $\left\{\mathbf{D}_p^{\text{PD}}\right\}_{p=1}^{P}$. The training procedure for MISO LRDT is similar to that of MIMO LRDT, with the primary differences lying in the determination of intermediate nodes and the computation of leaf nodes.

To search for an optimal non-leaf node $x_{1,\text{nonleaf}}^{\text{PD},p}$, the single output of MSE can be denoted as,

$$\begin{aligned} x_{1,\text{nonleaf}}^{\text{PD},p} &= \min L_k^{\text{MSE}}(n,i) = \varphi_{\text{MSE}}\left(\mathbf{D}_{2,\text{left}}^{\text{PD},p}\right) + \varphi_{\text{MSE}}\left(\mathbf{D}_{2,\text{right}}^{\text{PD},p}\right) \\ &= \left(\left(\mathbf{y}_{\text{Left}}^{\text{PD},p} - \overline{y}_{\text{Left}}^{\text{PD},p}\right) I\left(x_i \in \mathbf{D}_{2,\text{left}}^{\text{PD},p}\right)\right)^2 \\ &+ \left(\left(\mathbf{y}_{\text{Right}}^{\text{PD},p} - \overline{y}_{\text{Right}}^{\text{PD},p}\right) I\left(x_i \in \mathbf{D}_{2,\text{right}}^{\text{PD},p}\right)\right)^2 \end{aligned} \tag{8.19}$$

Based on the MIMO LRDT, the tth path of the tth leaf node of pth MISO LRDT is denoted as

$$\text{Route}_{t,\text{leaf}}^{\text{PD},p} = \left\{x_{1,\text{nonleaf}}^{\text{PD},p}, \ldots, x_{t,\text{nonleaf}}^{\text{PD},p}\right\} \in \mathbb{R}^{N_{t,\text{leaf}}^{\text{Model3}} \times M_{t,\text{leaf}}^{\text{PD}}} \tag{8.20}$$

Further, the input data of the tth leaf node of pth MISO LRDT is denoted as

$$\mathbf{D}_{t,\text{left}}^{\text{PD},p} = \left\{\text{Route}_{t,\text{left}}^{\text{PD},p}, \mathbf{y}_{t,\text{left}}^{\text{PD},p}\right\} \in \mathbb{R}^{N_{t,\text{left}}^{\text{Model3},p} \times \left(M_{t,\text{leaf}}^{\text{MD}}+1\right)} \tag{8.21}$$

Finally, the linear regression of the tth leaf node is represented as follows.

$$\hat{\mathbf{y}}_{t,\text{leaft}}^{\text{PD},p} = \mathbf{X}_{t,\text{leaf}}^{\text{PD},p} \mathbf{w}_{t,\text{leaft}}^{\text{PD},p} \tag{8.22}$$

where $\mathbf{w}_{t,\text{leaft}}^{\text{PD},p}$ is calculated as follows:

$$\mathbf{w}_{t,\text{leaft}}^{\text{PD},p} = \left(\left(\mathbf{X}_{t,\text{leaft}}^{\text{PD},p}\right)^{\text{T}} \mathbf{X}_{t,\text{leaft}}^{\text{PD},p} + \lambda \mathbf{I}\right)^{-1} \left(\mathbf{X}_{t,\text{leaft}}^{\text{PD},p}\right)^{\text{T}} \mathbf{y}_{t,\text{leaft}}^{\text{PD},p} \tag{8.23}$$

The weight $\mathbf{w}_{\text{BaggingLRDT}}^{p}$ is assigned to each MISO LRDT model to improve the performance of the Bagging LRDT model. For all the MISO LRDT model with number P, it can be shown as follows.

$$\begin{aligned}\hat{\mathbf{y}}^{\text{PD}} &= \sum_{p=1}^{P} \Theta_{\text{BaggingLRDT}}^{p}(\cdot)\, w_{\text{BaggingLRDT}}^{p} \\ &= \left[\Theta_{\text{BaggingLRDT}}^{1}\left(\mathbf{X}^{\text{MD}}\right),\ldots,\Theta_{\text{BaggingLRDT}}^{P}\left(\mathbf{X}^{\text{MD}}\right)\right]\mathbf{w}_{\text{BaggingLRDT}} \\ &= \mathbf{A}\mathbf{w}_{\text{BaggingLRDT}}\end{aligned} \tag{8.24}$$

where $\Theta_{\text{BaggingLRDT}}^{p}(\cdot)$ is the pth MISO LRDT model, and $\mathbf{w}_{\text{BaggingLRDT}}^{p}$ denotes the pth weight of $\Theta_{\text{BaggingLRDT}}^{p}(\cdot)$.

The iterative Tikhonov regularization is applied to solve for each weight of the MISO LRDT model, which can be expressed as follows:

$$\begin{cases}\mathbf{w}_{\text{BaggingLRDT}}^{p} = \mathbf{w}_{\text{BaggingLRDT}}^{p-1} + \left(\mathbf{A}^{\text{T}}\mathbf{A} + \lambda\mathbf{I}\right)\mathbf{A}^{\text{T}}\mathbf{B}^{p-1} \\ \mathbf{B}^{p-1} = \mathbf{C} - \mathbf{A}\mathbf{w}_{\text{BaggingLRDT}}^{p}\end{cases}, p = 1,2,\ldots \tag{8.25}$$

where $\mathbf{w}_{\text{BaggingLRDT}}^{p}$ is the weight vector, and $\mathbf{B}$ represents the residual vector. In the iterative solving process, the initial values are established as follows:

$$\begin{cases}\mathbf{w}_{\text{BaggingLRDT}}^{p} = \mathbf{0} \in \mathbb{R}^{p\times 1} \\ \mathbf{C} = \mathbf{B}^{0}\end{cases} \tag{8.26}$$

Finally, the MISO Bagging LRDT model can be expressed as,

$$\hat{y}^{\text{PD}} = f_{\text{BaggingLRDT}}\left(\mathbf{x}^{\text{PD}}\right) \in \mathbb{R}^{1\times 1} \tag{8.27}$$

8.3.4 CIRWNN Fusion Model Module

This CIRWNN fusion modeling strategy utilizes the outputs of the MD model (denoted as $f_{\text{MIMOLRDT}}(\cdot)$ and $\Theta_{\text{MMM}}(\cdot)$) and DD model $\left(f_{\text{BaggingLRDT}}(\cdot)\right)$ models as inputs to obtain the final soft sensing output. Given the training dataset, which $\hat{\mathbf{y}}_{\text{Pse}}^{\text{MMM}}$ is the output for the MMM model, $\hat{\mathbf{y}}^{\text{PD}}$ is the output of the MISO Bagging LRDT model, and $\mathbf{y}^{\text{DXN}}$ is the true value of the DXN emission concentration.

In nature, CIRWNN is a single-hidden-layer feed-forward neural network algorithm. The input weights ω_{In} and biases σ are randomly generated based on a uniform distribution (usually set to $[-\mu,\mu]$ with $\mu > 0$). The system input is denoted as $\mathbf{D}^{\text{FM}} \in \mathbb{R}^{N^{\text{PD}}\times(2+1)}$, and suppose the number of hidden layer neurons is $L-1$. The predicted output can be denoted as follows:

$$\begin{aligned}\hat{\mathbf{y}}^{\mathrm{DXN}} &= f_{\mathrm{CIRWNN}}^{L-1}\left(\left[\hat{\mathbf{y}}_{\mathrm{Pse}}^{\mathrm{MMM}}, \hat{\mathbf{y}}^{\mathrm{PD}}\right]\right) \overset{\mathbf{H}=\left[\hat{\mathbf{y}}_{\mathrm{Pse}}^{\mathrm{MMM}}, \hat{\mathbf{y}}^{\mathrm{PD}}\right]}{=} f_{\mathrm{CIRWNN}}^{L-1}(\mathbf{H}) \\ &= \sum_{i=1}^{L-1} \psi^{i}\left(\mathbf{H}\left(\omega_{\mathrm{In}}^{i}\right)^{\mathrm{T}} + \sigma^{i}\right)\omega_{\mathrm{Out}}^{i}\end{aligned} \tag{8.28}$$

where $f_{\mathrm{CIRWNN}}^{L-1}(\cdot)$ is the basic RWNN model with $L-1$ hidden layer neurons, $\psi(\cdot)$ is the activation function, and ω_{Out} is the weight for $L-1$ hidden layer neurons.

The least squares algorithm is used to obtain the output weight ω_{Out}, and it can be expressed as follows:

$$\begin{aligned}\omega_{\mathrm{Out}}^{*} &= \arg\min_{\omega_{\mathrm{Out}}} \frac{1}{2}\left\| \hat{\mathbf{y}}^{\mathrm{DXN}} - \Phi_{L-1}(\mathbf{H})\omega_{\mathrm{Out}} \right\|_2^2 \\ &= \left(\left[\Phi_{L-1}(\mathbf{H})\right]^{\mathrm{T}} \Phi_{L-1}(\mathbf{H})\right)^{-1} \left[\Phi_{L-1}(\mathbf{H})\right]^{\mathrm{T}} \hat{\mathbf{y}}^{\mathrm{DXN}}\end{aligned} \tag{8.29}$$

where $\Phi_{L-1}(\cdot)$ is a matrix of the output of the hidden layer.

During the increment of hidden layer neurons, the constrained inequality can be written as follows:

$$\zeta = \frac{\left\langle \left(\mathbf{e}_{L-1}^{\mathrm{CIRWNN}}\right)^{\mathrm{T}}, \Phi_L \right\rangle^2}{\langle \Phi_L^{\mathrm{T}}, \Phi_L \rangle} - (1 - r - \gamma_L) \times \left\langle \left(\mathbf{e}_{L-1}^{\mathrm{CIRWNN}}\right)^{\mathrm{T}}, \Phi_L \right\rangle > 0 \tag{8.30}$$

where $\langle \cdot,\cdot \rangle$ represents the scalar product; r and γ_L are the positive constants which $0 < r < 1$ and $0 < \gamma_L < 1 - r$ $\left(\lim_{L\to\infty} \gamma_L = 0\right)$; $\mathbf{e}_{L-1}^{\mathrm{CIRWNN}}$ and Φ_L are calculated as follows:

$$\mathbf{e}_{L-1}^{\mathrm{CIRWNN}} = \mathbf{y}^{\mathrm{DXN}} - f_{\mathrm{CIRWNN}}^{L-1}\left(\left[\hat{\mathbf{y}}_{\mathrm{Pse}}^{\mathrm{MMM}}, \hat{\mathbf{y}}^{\mathrm{PD}}\right]\right) \tag{8.31}$$

$$\Phi_L = \left[\phi^{L}\left(\mathbf{h}^{1}\left(\omega_{\mathrm{In}}^{L}\right)^{\mathrm{T}} + \sigma^{L}\right), \ldots, \phi^{L}\left(\mathbf{h}^{N^{\mathrm{MD}}}\left(\omega_{\mathrm{In}}^{L}\right)^{\mathrm{T}} + \sigma^{L}\right)\right]^{\mathrm{T}} \tag{8.32}$$

The new output of the hidden layer is denoted as $\left[\Phi_{L-1}(\cdot), \Phi_L(\cdot)\right]$, and the output weight ω_{Out} is calculated using Eq. (8.29). Based on the aforementioned constraint, CIRWNN is incrementally learned until the learning accuracy reaches a satisfactory level.

During this process, adjustments to the number of hidden layer neurons and their corresponding weights are made iteratively, ensuring that the constraints are satisfied while optimizing the model's performance. The goal is to achieve a balance between model complexity and predictive accuracy. Finally, the CIRWNN fusion model is obtained as follows.

$$\hat{y}^{\mathrm{DXN}} = f_{\mathrm{CIRWNN}}\left(\left[\hat{\mathbf{y}}_{\mathrm{Pse}}^{\mathrm{MMM}}, \hat{\mathbf{y}}^{\mathrm{PD}}\right]\right) \in \mathbb{R}^{1\times 1} \tag{8.33}$$

8.4 MULTI-DEMAND PREDICTION ANALYSIS

Three model prediction demands concerning various time scales in the actual MSWI power plant are outlined as follows:

(1) Second Scale (Process Data Input Only): Real-time DXN emission concentrations can be obtained by employing the Bagging LRDT model using real-time process data.
(2) Hourly Scale (Process Data Input Only): This strategy involves utilizing the MMM model, Bagging LRDT model, and CIRWNN fusion model. Initially, real-time process data is input into the Bagging LRDT model to acquire the initial DXN emission concentration. Subsequently, the MMM processes CO_2, CO, and O_2 from the process data to derive the secondary DXN emission concentration. Finally, both the initial and secondary DXN emission concentrations are input into the CIRWNN fusion model to obtain the ultimate DXN emission concentration.
(3) Daily Scale (Process Data and Production Data Input): All constructed models are employed at this scale. First, production data is fed into the MIMO surrogate model to derive outputs for CO_2, CO, and O_2. These values are then processed by the MMM model to determine a secondary DXN emission concentration. Next, process data is input into the Bagging LRDT model to establish the first DXN emission concentration. Lastly, both the first and second DXN emission concentrations are fed into the CIRWNN fusion model to obtain the final DXN concentration.

Therefore, the hybrid MD and DD model constructed in this chapter is applicable for the multi-demand task of predicting DXN emissions in the actual MSWI process.

Therefore, the MD and DD hybrid-driven model constructed in this chapter applies to the multi-demand task of predicting DXN emissions in the MSWI process.

8.5 APPLICATION RESULTS

8.5.1 Operating Conditions and Model Setting

The detailed parameters related to the numerical simulation model and the orthogonal experimental design are the same as those presented in Chapter 6. After manual processing, a virtual simulation mechanism dataset comprising 1,960 multi-condition virtual mechanism data points was generated. The DXN data obtained from the MSWI plant contains 116 process variables and one DXN truth value, with a total sample size of 82, covering the time from 2016 to 2020.

The hyperparameter settings for the various methods in the modeling comparison experiment are as follows:

(1) Decision Tree (DT): Number of samples = 3;
(2) Random Decision Tree (RDT): Number of samples = 3;
(3) Ridge Regression (RR): Regularization term = 0.9;

(4) SVR: Kernel function = Gaussian; iteration count = 1,000; cache size = 1,000;
(5) Back-Propagation Neural Network (BPNN): Number of neurons in a single hidden layer = $2M + 1$; iteration count = 1,500; convergence error = 0.001; learning rate = 0.01;
(6) Random Forest (RF): Number of samples = 3; feature selection = root number of M; number of trees = 500;
(7) Extreme Gradient Boosting (XGBoost): Number of samples = 3; number of trees = 30; learning rate = 0.3; regularization term = 0.1;
(8) Gradient Boosting Decision Tree (GBDT): Number of samples = 3; number of trees = 30; learning rate = 0.1;
(9) PMM Based on LRDT: Number of samples = 3; regularization term = 0.1;
(10) MMM Based on LRDT: Number of samples = 3; regularization term = 0.1;
(11) MIMOLRDT: Number of samples = 15; regularization term = 0.1;
(12) BaggingLRDT: Number of samples = 15; first regularization term = 0.1; number of trees = 500; second regularization term = 0.9;
(13) CIRWNN: Maximum number of hidden nodes = 100; maximum number of candidate nodes = 100; batch size = 1; tolerance error = 0.001.

Two evaluation indicators, namely the root mean square error (RMSE) and the coefficient of determination (R^2), are employed to assess the performance of the proposed method. Their calculations are as follows:

$$\text{RMSE} = \sqrt{\sum_{i=1}^{N}\left(y_i - \hat{y}_i\right)^2 \Big/ N-1} \tag{8.34}$$

$$R^2 = 1 - \sum_{i=1}^{N}\left(y_i - \hat{y}_i\right)^2 \Big/ \sum_{i=1}^{N}\left(y_i - \bar{y}_i\right)^2 \tag{8.35}$$

8.5.2 Modeling Results

8.5.2.1 White-Box MIMO Surrogate Model Results

The results of outlier removal from the 1,960 multi-condition virtual mechanism data are presented in Figure 8.3.

Figure 8.3 shows that the significant elimination of these outliers demonstrates the effectiveness of the QAF method. After applying the QAF method to remove the outliers, a total of 1,846 samples were obtained.

From this dataset, the final selection of input parameters included the LHV of MSW, PA, and FC. The output parameters consisted of CO_2, CO, and O_2, based on data obtained from the operational MSWI power plant. The modeling performance indicators for CO_2, CO, and O_2 are detailed in Table 8.1.

Table 8.1 highlights the impressive modeling performance of DT across all three targets: CO_2, CO, and O_2. Notably, DT outperforms several comparison methods, including RDT, RR, RF, GBDT, XGBoost, BPNN, BLSNN, IT2FNN, and DPNN,

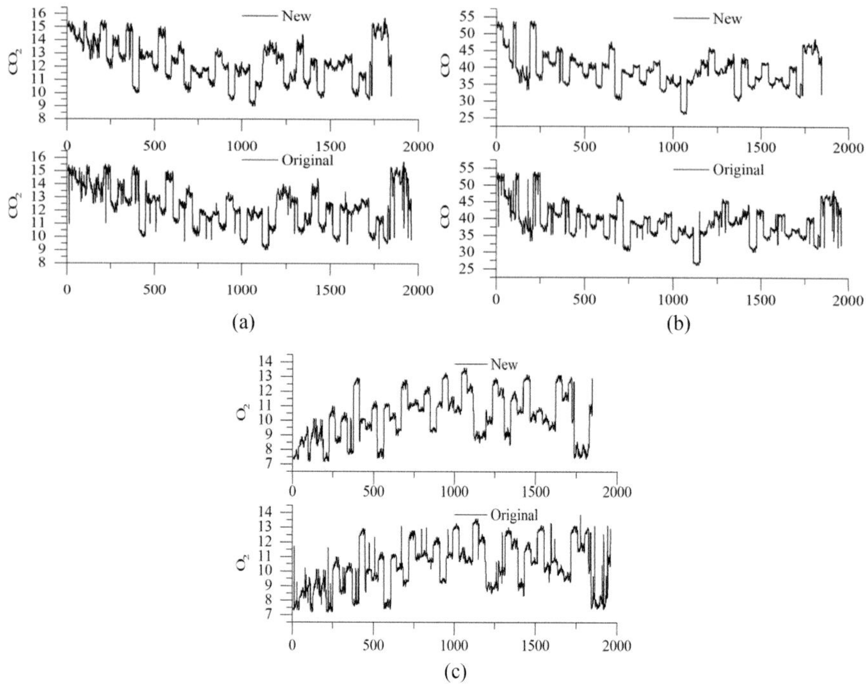

FIGURE 8.3 (a–c) Results of outlier removal in mechanism data.

TABLE 8.1
Different Method Performance Comparison Results of Virtual Mechanism Data

Methods	Target	Training		Testing	
		RMSE	R^2	RMSE	R^2
DT	CO_2	0.2688	0.9702	0.6153	0.8457
	CO	0.9519	0.9659	1.8184	0.8751
	O_2	0.2811	0.9710	0.6536	0.8446
RDT	CO_2	0.5400	0.8798	0.6237	0.8414
	CO	2.2356	0.8117	2.5335	0.7575
	O_2	0.6752	0.8325	0.7862	0.7751
RR	CO_2	1.40025	0.1918	1.3945	0.2072
	CO	4.9986	0.0586	4.9738	0.0652
	O_2	1.4306	0.2481	1.4233	0.2630
RF	CO_2	0.8127	0.7278	0.9900	0.6004
	CO	2.7334	0.7185	3.2301	0.6057
	O_2	0.8638	0.7259	1.0538	0.5960
GBDT	CO_2	0.1744	0.9875	0.5708	0.8672
	CO	0.6498	0.9841	1.7530	0.8839
	O_2	0.2153	0.9830	0.6073	0.8658

(Continued)

TABLE 8.1 (Continued)
Different Method Performance Comparison Results of Virtual Mechanism Data

		Training		Testing	
Methods	**Target**	**RMSE**	R^2	**RMSE**	R^2
XGBoost	CO_2	0.2567	0.9728	0.7699	0.7583
	CO	0.7982	0.9760	2.4466	0.7738
	O_2	0.2808	0.9710	0.8560	0.7334
BPNN	CO_2	1.1161	0.4866	1.1094	0.4982
	CO	3.2271	0.6076	3.1263	0.6307
	O_2	1.0850	0.5676	1.0687	0.5845
BLSNN	CO_2	1.0193	0.5718	1.0254	0.5713
	CO	3.6266	0.5044	3.5463	0.5248
	O_2	0.9717	0.6531	0.9571	0.6667
IT2FNN	CO_2	0.9775	0.6062	0.9572	0.6265
	CO	3.1347	0.6298	3.0363	0.6516
	O_2	1.0301	0.6102	1.0016	0.6350
DPNN	CO_2	0.6679	0.8161	0.6821	0.8103
	CO	2.5370	0.7575	2.5127	0.7614
	O_2	0.7264	0.8062	0.7051	0.8191
MISO LRDT	CO_2	0.4138	0.9294	0.5894	0.8584
	CO	1.3046	0.9359	1.7057	0.8901
	O_2	0.4282	0.9326	0.5487	0.8905
MIMO LRDT	CO_2	0.1645	0.9357	0.3089	0.8869
	CO	0.2220	0.9357	0.2991	0.8867
	O_2	0.4056	0.9812	0.5558	0.9747

indicating that a single learner is sufficient for virtual mechanism data. An analysis of the testing set indicators for the three targets shows that the proposed MISO LRDT method surpasses the DT model, underscoring the importance of using linear regression to derive predicted values from feature information at the leaf nodes. Furthermore, the statistical index R^2 results (0.8869 for CO_2, 0.8867 for CO, and 0.9747 for O_2) and the corresponding fitting curves indicate that the MIMO LRDT method exhibits superior fitting performance. Thus, these experiments validate the effectiveness of the proposed MIMO method.

8.5.2.2 Semi-Supervised Transfer Mapping Model Results

To construct a mapping model between CO_2, CO, O_2, and DXN concentrations, it is essential to first label the 1,846 unlabeled data points. Therefore, the prior MISO LRDT (PMM) model will be developed using the process data with labeled samples. The inputs for the model are CO_2, CO, and O_2, while the output is the DXN content. Based on the specified hyperparameter settings, the PMM model was trained, achieving an RMSE of 0.0013 and an R^2 of 0.9188. The fitting curve is presented in Figure 8.4.

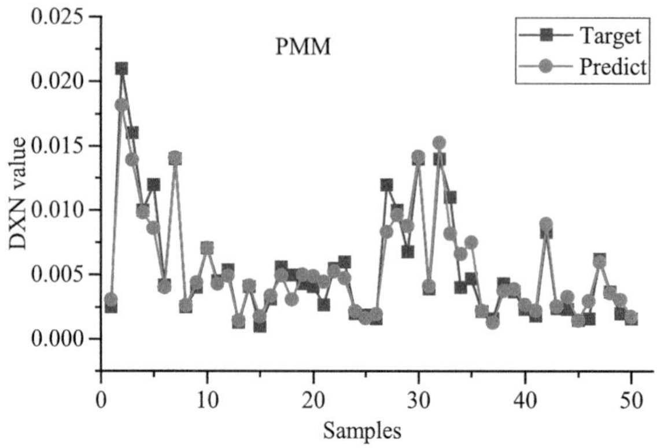

FIGURE 8.4 Result of the PMM model in DXN data.

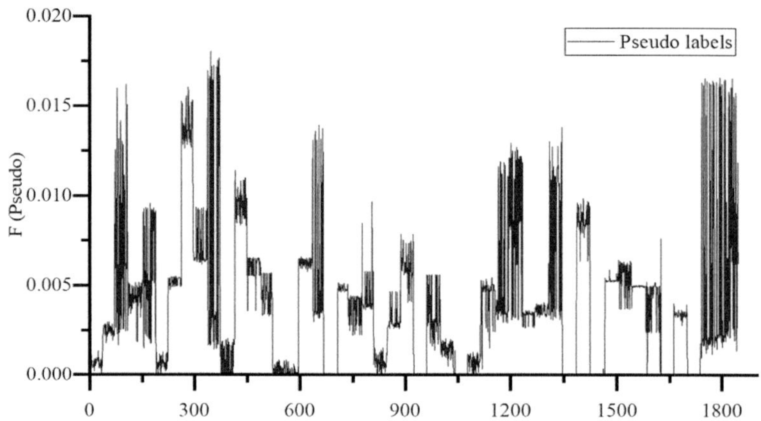

FIGURE 8.5 Curve of pseudo-labeled data.

The PMM model is then utilized to label 1,846 previously unmarked mechanism data points. As a result, 1,846 pseudo-labeled data points are generated, and their distribution is illustrated in Figure 8.5.

The 1,846 pseudo-labeled data points are divided into three equal parts, with the first two portions used for the training set (1,231 data points) and the last portion for the test set (615 data points). In the next step, these 1,231 training data points are input into the PMM model to facilitate tree structure migration and generate the MMM1 model. Subsequently, the MMM1 model is updated based on the error at the leaf nodes to produce the MMM2 model. The performances of these models are presented in Table 8.2 and illustrated in Figure 8.5.

As shown in Table 8.2 and Figure 8.6, the performance of the PMM model in both training and testing is commendable. The results indicate that the R^2 value for the training set is 0.8021, while the R^2 value for the testing set is 0.8072. After implementing the tree structure transfer, the performance of MMM1 improved by 11.07% for the training set and 10.48% for the testing set compared to PMM in terms of the

TABLE 8.2
Statistical Results of Pseudo-Labeled Mechanism Data

	Training		Testing	
Methods	**RMSE**	R^2	**RMSE**	R^2
PMM	0.0020	0.8021	0.0020	0.8072
MMM1	0.0015	0.8909	0.0015	0.8918
MMM2	0.0014	0.8960	0.0015	0.8965

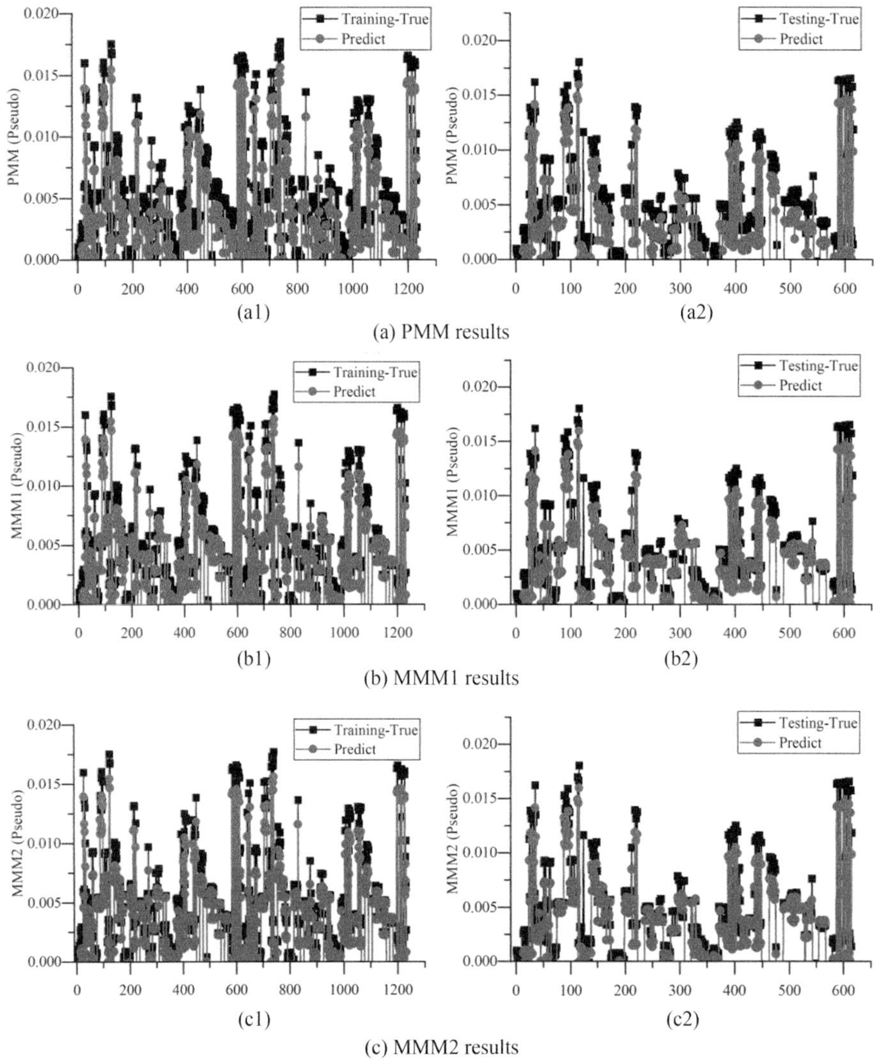

FIGURE 8.6 Training and testing curves of different methods in pseudo-labeled data. (a) PMM results. (b) MMM1 results. (c) MMM2 results.

R^2 indicator. Furthermore, MMM2 demonstrated an improvement of 0.57% for the training set and 0.53% for the testing set when compared to MMM1, calculated based on the R^2 indicators. These results confirm the effectiveness of the tree structure transfer and leaf node update strategies employed in the experiments.

8.5.2.3 Bagging LRDT Model Results

The QAF method, employing a sampling window of 15, was applied to manage outliers in the DXN content. This process resulted in a total of 74 DXN samples, successfully removing 8 outliers. The processing results are illustrated in Figure 8.7.

In Figure 8.7, the data distribution of DXN emission concentrations is clearer after the removal of outliers. Consequently, a total of 74 DXN samples were obtained, with 8 outliers successfully eliminated. These 74 samples are divided into three equal parts: the first two parts comprise the training set (50 samples), while the last part serves as the testing set (24 samples). Each method is repeated 30 times, and the best result from these iterations is selected as the statistical outcome. The modeling performance is presented in Table 8.3, and the corresponding fitting curve is illustrated in Figure 8.8.

Figure 8.8 and Table 8.3 illustrate the following findings:

(1) Among the single learners, DT (RMSE: 0.0008, R^2: 0.9689 in training), RDT (RMSE: 0.0033, R^2: 0.4765 in training), and MISO LRDT (RMSE: 0.0007, R^2: 0.9758 in training) demonstrate relatively lower performance compared to RR (RMSE: 0.0002, R^2: 0.9976 in training) and GBDT (RMSE: 0.0003, R^2: 0.9947 in training) concerning DXN data;

(2) The MISO LRDT (RMSE: 0.0029, R^2: 0.1450 in training) exhibits weaker generalization performance compared to the DT model (RMSE: 0.0026, R^2: 0.3321 in training), though it shows stronger learning capabilities than the DT model;

(3) In comparison with the results of existing literature methods, the Bagging LRDT method proposed in this chapter achieves the best performance in both training and testing (with an RMSE of 0 in the training set and 0.0017 in the testing set);

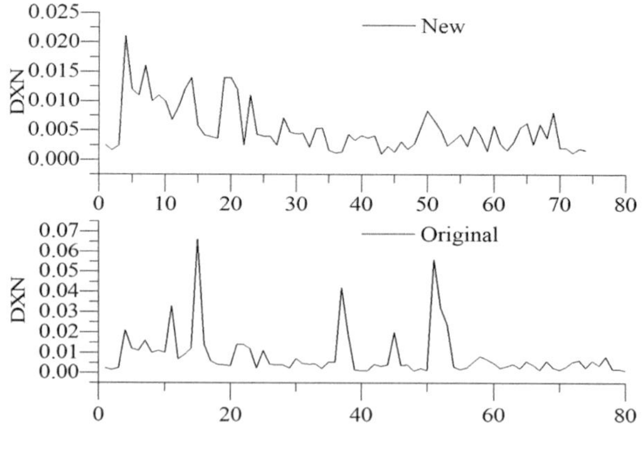

FIGURE 8.7 Results of outlier removal in DXN data.

TABLE 8.3
Different Method Performance Comparison Results of DXN Data

	Training		Testing	
Methods	**RMSE**	R^2	**RMSE**	R^2
DT	0.0008	0.9689	0.0026	0.3321
RDT	0.0033	0.4765	0.0019	0.6383
RR	0.0002	0.9976	0.0031	0.0663
SVR	0.0028	0.6035	0.0031	0.0232
BPNN	0.0001	0.9990	0.0021	0.5402
RF	0.0018	0.8320	0.0022	0.5173
GBDT	0.0003	0.9947	0.0022	0.5259
XGBoost	0.0020	0.7997	0.0030	0.0786
MISO LRDT	0.0007	0.9758	0.0029	0.1450
ImDFR	0.0077	0.9242	0.0179	0.6636
EnTSFRT	0.0149	0.7390	0.0151	0.6740
BaggingLRDT	0.0000	1.0000	0.0017	0.7202

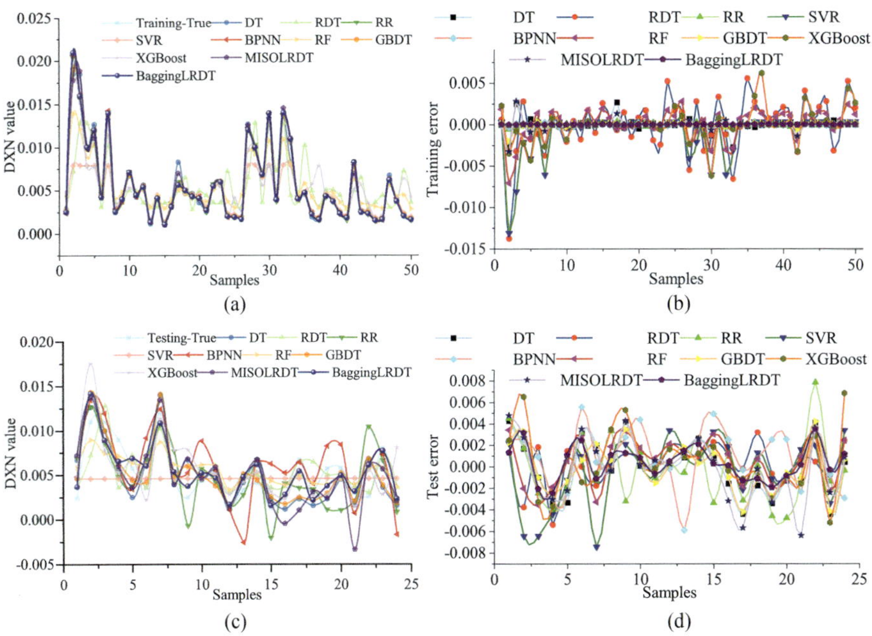

FIGURE 8.8 Curve of different methods in real DXN data. (a) Training sample prediction curves. (b) Training sample error curves. (c) Testing sample prediction curves. (d) Testing sample error curves.

(4) The Bagging LRDT method also excels in terms of fitting performance, achieving an R^2 of 1 in the training set and 0.7207 in the testing set. When compared with the highest accuracy model (EnTSFRT), Bagging LRDT shows an improvement of approximately 7% in R^2. Therefore, the Bagging LRDT stands out as an effective model for predicting DXN emission concentrations.

8.5.2.4 CIRWNN Fusion Model Results

The outputs from the Bagging LRDT and MMM2 LRDT models serve as inputs for the CIRWNN in the fusion model. The proposed method is categorized as a cascade-parallel fusion approach based on its workflow. Conducting comparative experiments with hybrid models can be challenging due to their specific design alignment with target characteristics. The training and testing flowchart for the fusion models is illustrated in Figure 8.9.

Results from the training and testing of the BaggingLRDT, MIMO LRDT, and MMM2 LRDT models have been presented in previous sections. This section focuses on the performance results of training and testing the CIRWNN-based fusion model.

Various state-of-the-art (SOTA) fusion methods are employed for performance comparisons, including long short-term memory neural networks (LSTM), SVR, RWNN, and partial least squares regression (PLSR). The hyperparameters for each method are specified as follows: (1) LSTM fusion model: The learning rate is set to 0.01, the dropout rate is 0.3, the number of iterations is 500, and the parameter update method used is Adam; (2) SVR fusion model: The kernel function is Gaussian, with the number of iterations set to 1000, a cache size of 1,000, and hyperparameter optimization performed using Bayesian methods; (3) RWNN fusion model: The number of hidden layer neurons is set to 10, and the activation function utilized is sigmoid; (4) PLSR fusion model: The number of latent variables is equal to the rank of the training data; and (5) CIRWNN fusion model: The maximum number of hidden neurons is set to 10, the maximum number of candidate nodes is 100, lambda is fixed at 0.6, and all other hyperparameters are set to their default values.

The experimental results are shown in Table 8.4, Figures 8.10 and 8.11.

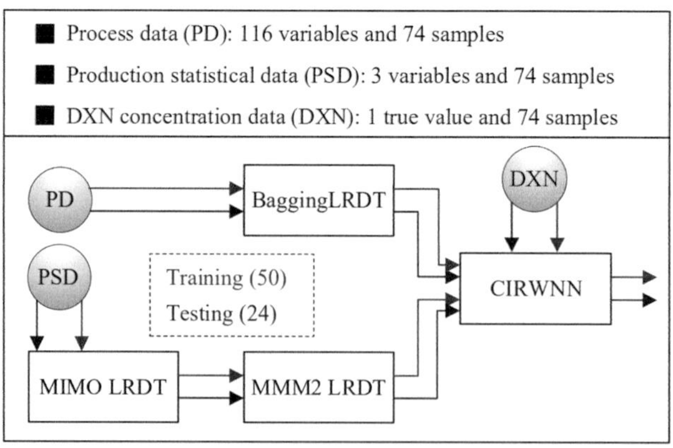

FIGURE 8.9 Flow chart of training and testing of fusion models.

TABLE 8.4
Comparison Results of Different Fusion Models

	Training		Testing	
Methods	**RMSE**	**R^2**	**RMSE**	**R^2**
LSTM	0.0000	0.9999	0.0017	0.7094
SVR	0.0001	0.9997	0.0016	0.7417
RWNN	0.0000	0.9999	0.0016	0.7411
PLSR	0.0004	0.9908	0.0016	0.7292
CIRWNN	0.0000	1.0000	0.0016	0.7437

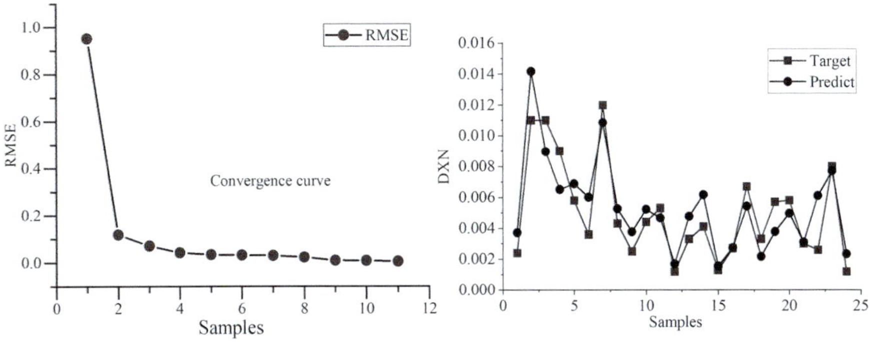

(a) Convergence curve (b) Testing set

FIGURE 8.10 Results of the CIRWNN fusion model. (a) Convergence curve. (b) Testing set.

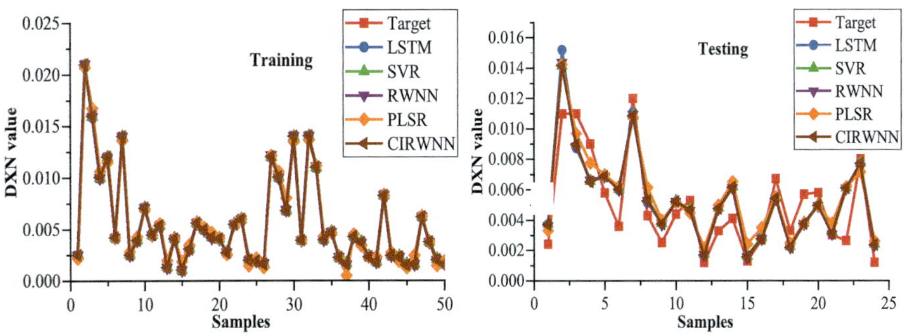

(a) (b)

FIGURE 8.11 Fitting curve of different fusion models. (a) Training sample prediction curves. (b) Testing sample prediction curves.

Table 8.4 presents the RMSE and R^2 values of all fusion models, which approach 0 and 1 within the training set, indicating robust learning capabilities across these models. However, during testing, the LSTM fusion model (RMSE: 0.0017, R^2: 0.7094) exhibits relatively weaker performance compared to the other models. Notably, SVR, RWNN, PLSR, and CIRWNN demonstrate identical RMSE values in the testing set.

Among these, the CIRWNN fusion model distinguishes itself by achieving superior prediction accuracy with an R^2 of 0.7437, making it the best performer among existing SOTA fusion methods based on this indicator. When compared to the highest accuracy model (SVR), CIRWNN shows an improvement of approximately 0.3% in terms of R^2. Figure 8.11 illustrates the close alignment between the predicted and actual distributions, further validating the efficacy of the CIRWNN fusion model.

8.5.3 Interpretability of the Proposed Hybrid Model

The visualization includes the MIMO LRDT, MMM, Bagging LRDT, and CIRWNN models. The MIMO LRDT model's visualization encompasses the tree structure, information on non-leaf nodes, and the weights of the leaf nodes. Similarly, the MMM model's visualization displays the tree structure along with details regarding non-leaf nodes and the weights of the leaf nodes. For the Bagging LRDT model, the visualization highlights the subtree structure, characteristics of non-leaf nodes, weights of the leaf nodes, and the weight distribution of each subtree. Finally, the CIRWNN model's visualization reveals the neural network structure, along with weight information for both the input and output layers. The resulting visual representations are presented in Figure 8.12.

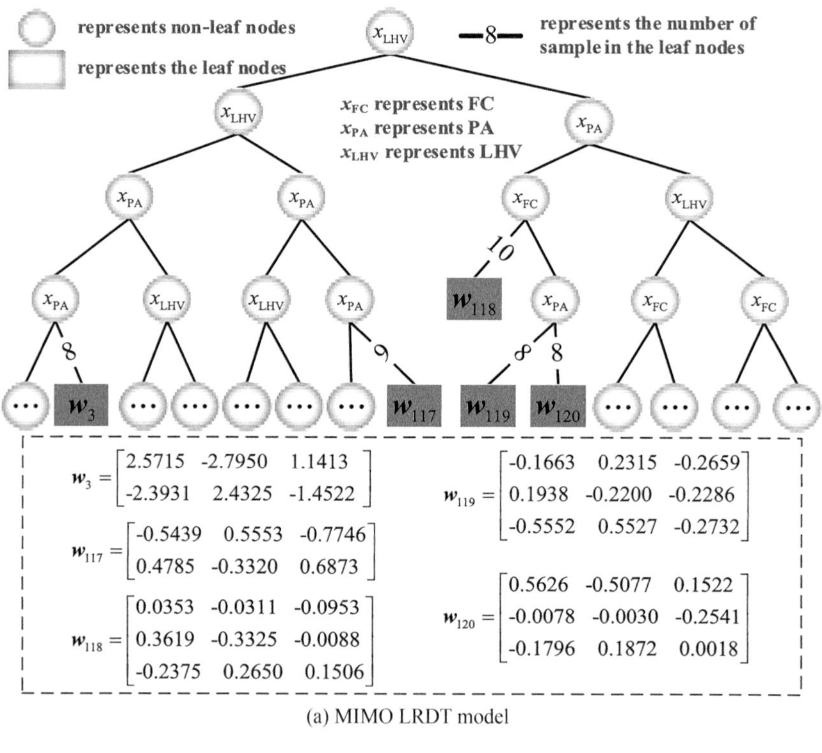

(a) MIMO LRDT model

FIGURE 8.12 Visualization of the proposed hybrid model. (a) MIMO LRDT model.

(*Continued*)

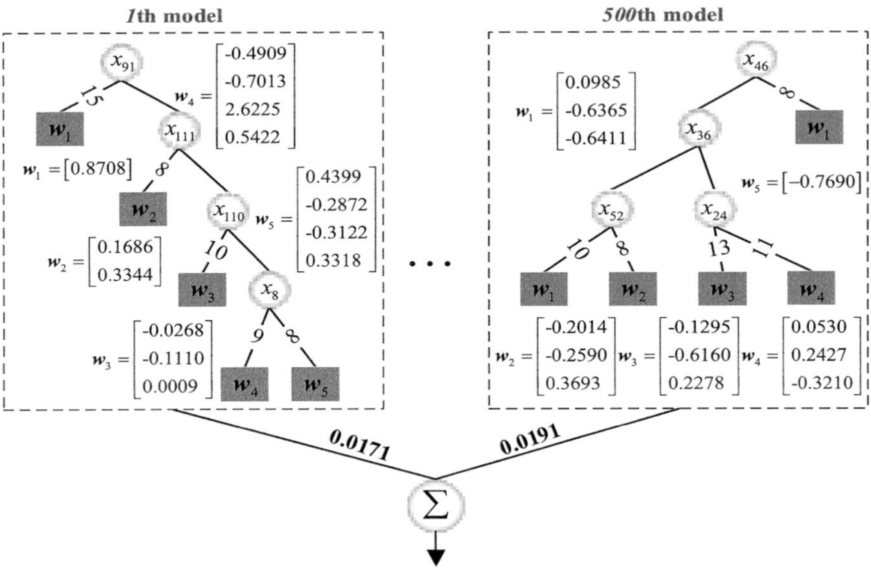

(b) MISO MMM model

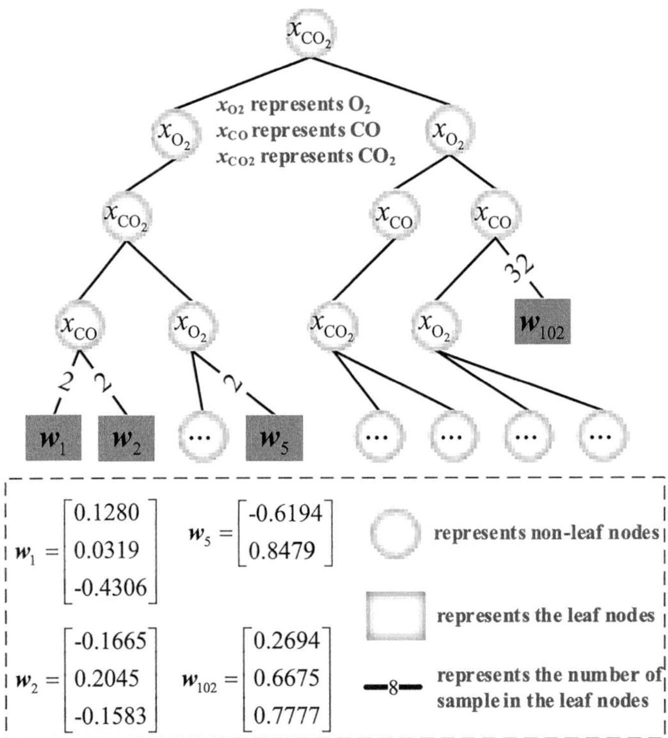

(c) Bagging LRDT model

FIGURE 8.12 (Continued) (b) MISO MMM model. (c) Bagging LRDT model.

(Continued)

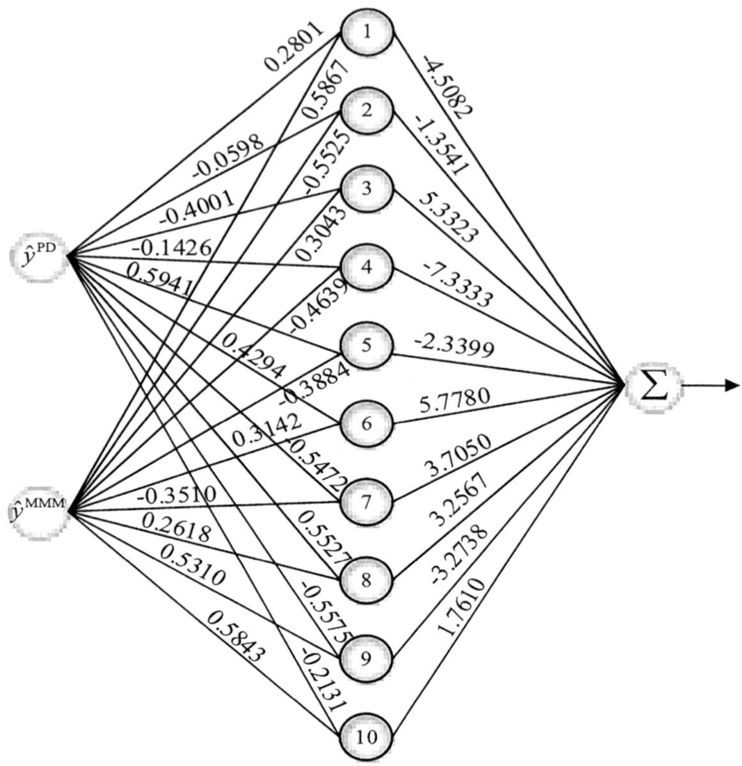

(d) CIRWNN model

FIGURE 8.12 (Continued) (d) CIRWNN model.

Figure 8.12 demonstrates the high interpretability of the proposed hybrid MD and DD modeling strategy. In the multi-demand prediction process, each model utilizes distinct input information, features a transparent calculation methodology, and produces clearly defined outputs. This interpretable hybrid model not only enhances the understanding of the prediction process but also assists in identifying potential issues, contributing to more effective decision-making and problem-solving in the context of demand forecasting.

8.5.4 Multi-Demand Modeling Results with Hardware-in-loop Platform

This section presents a DXN online prediction software system developed on the hardware-in-loop platform (HLP) for MSWI processes in the laboratory [32]. Initially, the fusion model is compiled into DLL files, which are then utilized to develop the DXN online prediction software system using the C# programming language. This system encompasses an OPC client, a human-computer interface, and fusion model calculations. The experimental setup of the DXN online prediction system on the HLP is illustrated in Figure 8.13.

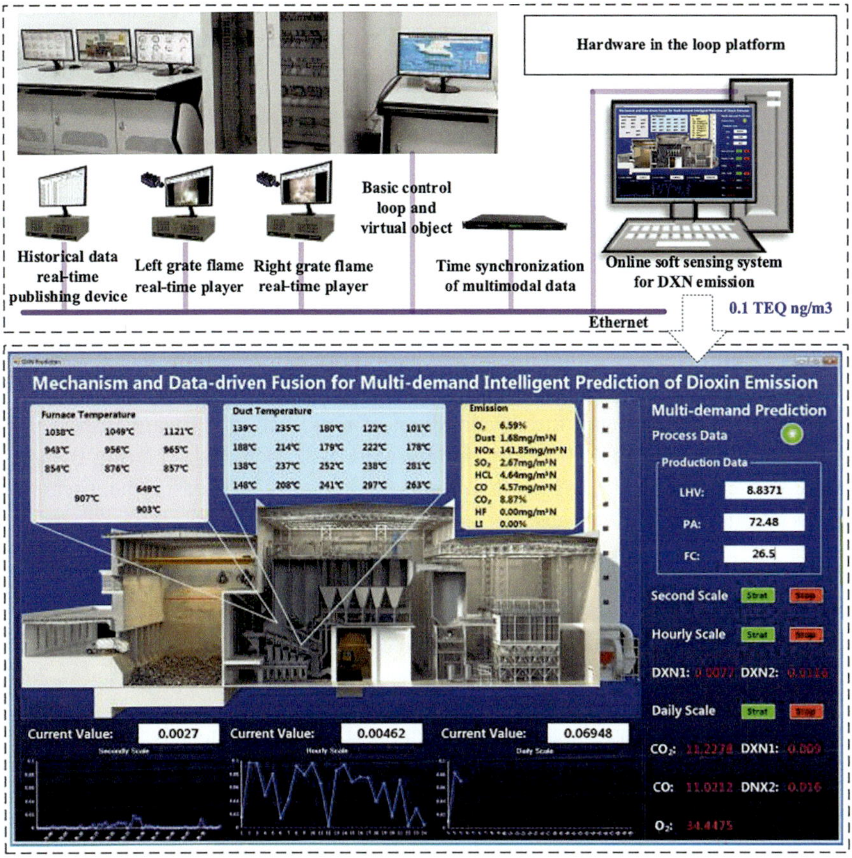

FIGURE 8.13 DXN online prediction software system.

The software system testing process consists of three parts:

(1) Seconds Scale Test: The real-time process data generated by the actual MSWI process can be simulated within the HLP. By clicking the "Start" button in the seconds scale section, the software retrieves the real-time process data from the platform using the OPC client. At this stage, only the BaggingLRDT model is utilized in the background.

(2) Hourly Scale Test: By clicking the "Start" button in the hourly scale section, the software employs the BaggingLRDT, MMM2 LRDT, and CIRWNN models to predict the DXN concentration. The predicted values from the BaggingLRDT and MMM2 LRDT models are displayed in the hourly scale area.

(3) Daily Scale Test: By clicking the "Start" button in the daily scale section, the software uses the BaggingLRDT, MIMO LRDT, MMM2 LRDT, and CIRWNN models to predict the DXN concentration. The predicted values from the BaggingLRDT, MIMO LRDT, and MMM2 LRDT models are also shown in the daily scale area.

The predictive curves representing these three strategies are displayed in the bottom-left section of the software interface in Figure 8.13. These results strongly affirm the efficacy of the DXN prediction software system.

8.6 SUMMARY

This chapter introduces a novel modeling strategy that combines MD approaches using numerical simulations with DD techniques utilizing an improved regression decision tree. This hybrid modeling strategy integrates virtual mechanism data with real historical data, facilitating intelligent multi-demand modeling for DXN emission concentration. It allows for rapid single-process DD predictions within seconds, process DD predictions within hours, and hybrid-driven predictions leveraging both process and production data within days. The effectiveness of this modeling strategy has been validated through the development of a prediction software system integrated into a HLP in the laboratory. Future work will focus on exploring dynamic updating methods for the hybrid model and applying online measurement technology in engineering applications.

REFERENCES

1. Qiao J F, Guo Z H, Tang J. Dioxin emission concentration measurement approaches for municipal solid wastes incineration process: a survey. *Acta Automatica Sinica*, 2020, 46(6): 1063–1089.
2. Xia H, Tang J, Aljerf L, et al. Assessment of DXNs formation and emission characteristics at a municipal solid waste incinerator for one year. *The Science of the Total Environment*, 2023, 883: 163705.
3. Chang N, Huang S. Statistical modelling for the prediction and control of PCDDs and PCDFs emissions from municipal solid waste incinerators. *Waste Management & Research*, 1995, 13: 379–400.
4. Bunsan S, Chen W, Chen H, et al. Modeling the dioxin emission of a municipal solid waste incinerator using neural networks. *Chemosphere*, 2013, 92(3): 258–264.
5. Tang J, Xia H, Zhang J, et al. Deep forest regression based on cross-layer full connection. *Neural Computing and Applications*, 2021, 33(15): 9307–9328.
6. Xia H, Tang J, Yu W, and Qiao J. Online measurement of dioxin emission in solid waste incineration using fuzzy broad learning. *IEEE Transactions on Industrial Informatics*, 2024, 20(1): 358–368.
7. Xia H, Tang J, Cui C, Qiao J. Soft sensing method of dioxin emission in municipal solid waste incineration process based on broad hybrid forest regression. *Acta Automatica Sinica*, 2023, 49(2): 343–365.
8. Altarawneh M, Dlugogorski B, Kennedy E, Mackie J. Mechanisms for formation, chlorination, dechlorination and destruction of polychlorinated dibenzo-*p*-dioxins and dibenzofurans (DXNs). *Progress in Energy and Combustion Science*, 2009, 35(3): 245–274.
9. Hunsinger H, Jay K, Vehlow J. Formation and destruction of DXN inside a grate furnace. *Chemosphere*, 2002, 46(9): 1263–1272.

10. Peng Y, Lu S, Li X, et al. Formation, measurement, and control of dioxins from the incineration of municipal solid wastes: recent advances and perspectives. *Energy & Fuels*, 2020, 34: 13247–13267.
11. Hu Z, Yang Z, Hu X, et al. SimPLE: similar pseudo label exploitation for semi-supervised classification. In: *2021 IEEE/CVF Conference on Computer Vision and Pattern Recognition (CVPR)*. 2021, 15094–15103.
12. Kang P, Kim D, Cho S. Semi-supervised support vector regression based on self-training with label uncertainty: An application to virtual metrology in semiconductor manufacturing. *Expert Systems with Applications*, 2016, 51: 85–106.
13. Cheng K M, Xiong W L. A dual-optimal semi-supervised regression algorithm. *CAAI Transactions on Intelligent Systems*, 2019, 14(4): 689–696.
14. Shi X D, Xiong W L. Semi-supervised Gaussian process regression modeling based on improved self-training algorithm. *Control Engineering of China*, 2020, 27(3): 451–455.
15. Wyper P, Antiochos S, DeVore C. A universal model for solar eruptions. *Nature*, 2017, 544: 452–455.
16. Liu Z, Han H G, Yang H, Qiao J F. Knowledge-aided and data-driven fuzzy decision making for sludge bulking. *IEEE Transactions on Fuzzy Systems*, 2023, 31(4): 1189–1201.
17. Sun Q, Ge Z. A survey on deep learning for data-driven soft sensors. *IEEE Transactions on Industrial Informatics*, 2021, 17(9): 5853–5866.
18. Yang Z, Ge Z Q. On paradigm of industrial big data analytics: from evolution to revolution. *IEEE Transactions on Industrial Informatics*, 2022, 18(12): 8373–8388.
19. Lu S, Chai T Y. Mesoscale particle size predictive model for operational optimal control of bauxite ore grinding process. *IEEE Transactions on Industrial Informatics*, 2020, 16(12): 7714–7721.
20. Zhou P, Zhang R, Xie J, et al. Data-driven monitoring and diagnosing of abnormal furnace conditions in blast furnace ironmaking: an integrated PCA-ICA method. *IEEE Transactions on Industrial Electronics*, 2021, 68(1): 622–631.
21. Ren J, Liu D, Wan Y. Modeling and application of Czochralski silicon single crystal growth process using hybrid model of data-driven and mechanism-based methodologies. *Journal of Process Control*, 2021, 104: 74–85.
22. Guo D, Wu Z W, Chai T Y, et al. A hybrid prediction model of energy consumption per ton for fused magnesia. *Control Theory and Technology*, 2019, 17(1): 24–36.
23. Ni Y, Xu J, Zhu C, Pei L. Accurate residual capacity estimation of retired $LiFePO_4$ batteries based on mechanism and data-driven model. *Applied Energy*, 2022, 305: 1–14.
24. Meng Y, Yu S, Zhang J, et al. Hybrid modeling based on mechanistic and data-driven approaches for cane sugar crystallization. *Journal of Food Engineering*, 2019, 257: 44–55.
25. Dong X, Xie H, Jiang L, et al. Research on a method for predicting the underflow concentration of a thickener based on the hybrid model. *Engineering Applications of Computational Fluid Mechanics*, 2020, 14: 13–26.
26. Dong X, Yan X, Qu H. Advanced process control for salvianolic acid a conversion reaction based on data-driven and mechanism-driven model. *Process Biochemistry*, 2022, 118: 1–10.
27. Ren J, Liu D, Wan Y. Data-driven and mechanism-based hybrid model for semiconductor silicon monocrystalline quality prediction in the czochralski process. *IEEE Transactions on Semiconductor Manufacturing*, 2022, 35(4): 658–669.

28. Zhu P, Xia L, Pan H. Multi-model fusion modeling method based on improved kalman filtering algorithm. *CIESC Journal*, 2015, 66(4): 1388–1394.
29. Zhuang Y, Liu Y, Ahmed A, et al. A hybrid data-driven and mechanistic model soft sensor for estimating CO_2 concentrations for a carbon capture pilot plant. *Computers in Industry*, 2022, 143: 1–14.
30. Zhang M, Liu H, Wang M, et al. Intelligence hybrid modeling method and applications in chemical process. *Chemical Industry and Engineering Progress*, 2021, 40(5): 1765–1776.
31. Wang D, Ming L. Stochastic configuration networks: fundamentals and algorithms. *IEEE Transactions on Cybernetics*, 2017, 47: 3466–3479.
32. Tang J, Wang T Z, Tian H, Yu W. AI-and security-empowered end-edge-cloud modular platform in complex process industries: a case study on municipal solid waste incineration. *Sensors*, 2025, 25(22): 6973.

9 Multimodal Data-Driven Interpretable Deep Modeling Approach of Dioxins Generation Concentration

9.1 RELATED WORKS

9.1.1 Dioxins Generation Concentration Modeling Description

DXN, categorized as highly toxic pollutants, comprise a group of complex chemicals including polychlorinated dibenzo-*p*-dioxins and polychlorinated dibenzofurans, which are challenging to measure and monitor in real-time [1]. Currently, the standard method for measuring DXN emission concentration involves offline direct detection, which requires online gas sampling followed by laboratory analysis. This approach presents limitations, such as significant lag times and high costs [2]. In light of these challenges, online measurements of DXN emission concentrations during the MSWI- process using artificial intelligence approaches have emerged as a significant area of research. There are two primary methodologies: online indirect detection and soft sensing direct detection, both of which utilize available input feature information to estimate DXN emission concentrations.

Within online indirect detection methods, data-driven models are constructed using associated organic compounds as input features. These associated compounds include 1,2,4-trichlorobenzene, 1,2,4-trichlorobenzeneas [3], and 1,3-dichlorobenzene, 1,4-dichlorobenzene, 1,2-dichlorobenzene, 1,3,5-trichlorobenzene, and 1,2,3-trichlorobenzene [4]. These compounds serve as indicators of DXN emission concentrations, providing valuable information for estimating pollution levels in the MSWI process.

In soft sensing direct detection approaches, process numerical data collected from the distribution control system (DCS) serve as inputs to map DXN emission concentrations [5]. Existing studies have focused on applying [6, 7], machine learning [8, 9], and deep learning [10–12] to estimate DXN emission concentrations using a data-driven approach. Additionally, DXN modeling utilizing transfer learning strategies has also been reported [13]. This study conducts data-based transfer learning research using DXN data from furnaces of the same type as the source domain. Transfer learning strategies reduce the excessive dependence of the modeling process on target domain data and can effectively lower the costs associated with model retraining.

DOI: 10.1201/9781003634195-12

Currently, transfer learning is widely applied in key parameter modeling, such as predicting equipment remaining useful life [14, 15], cancer prediction [16], and quality prediction [17]. However, a primary limitation of these studies is their focus on detecting DXN emission concentrations at the stack location, neglecting the crucial aspect of DXN generation concentrations at the boiler outlet. Understanding these concentrations is essential for effective emission reduction control related to their production sources.

According to the MSWI process flow, DXN undergoes five key transformations: (1) introduction with incoming municipal solid waste (MSW), which contains trace amounts of DXN [18]; (2) decomposition at high temperatures; (3) resynthesis under suitable temperature conditions; (4) adsorption and removal within the plant; and (5) emissions to the atmosphere [19]. Among these stages, numerous studies have indicated that a significant source of DXN within the MSWI process is its resynthesis during the cooling of flue gas [20]. There are three pathways for DXN generation: homogeneous synthesis (500°C–800°C) [21, 22], precursor synthesis (200°C–400°C) [23, 24], and de novo synthesis (200°C–400°C). Based on this mechanism, the DXN generation point is identified as the outlet of the boiler, prior to the gas cleaning stage. The most effective control methods for the adsorption and removal of generated DXN include activated carbon injection and bag filtration. When the DXN generation concentration is unknown, it is necessary to use additional activated carbon to reduce DXN levels to comply with emission standards. Consequently, online measurement of DXN generation concentration becomes a crucial step for effective DXN emission reduction and control.

Data-driven artificial intelligence (AI) approaches offer a promising solution for managing sustainable energy and for online measurement of DXN emission concentrations in MSWI power plants. However, MSWI processes generate a diverse dataset comprising multimodal data, including combustion flame videos, recorded statistical data, and commonly used process numerical data. Existing DXN modeling research has not yet constructed models that leverage the multimodal characteristics of the MSWI process. Thus, designing a DXN generation model utilizing such multimodal data presents a significant challenge. Additionally, the small sample size of available training data for DXN generation concentration is another critical issue to consider during the modeling process. Therefore, the specific challenges in measuring DXN generation concentration include the strong correlation with multimodal data, the difficulties associated with modeling small sample sizes, and the lack of research precedents.

The rapid development of computers and information technology has resulted in an increasing volume of data available in the industrial sector. Leveraging the advantages of big data, deep learning technology has emerged as a major research focus, particularly for modeling and predictive tasks in complex scenarios [25]. Among the various deep learning approaches, research based on deep neural networks (DNN) has gained significant prominence. For instance, He et al. utilized an enhanced long short-term memory (LSTM) approach to develop a soft sensing model for measuring hydrogen sulfide [26], demonstrating superior performance compared to classical methods. Similarly, Yao et al. introduced a dynamic mixture variational autoencoder regression technique specifically designed for soft sensing applications involving multimodal data [27]. Experimental results have shown that this innovative model is highly effective in accurately predicting product quality. Further, Guo et al. developed a deep multiple attention model for predicting thermal deformation in air preheater rotors [28]. While this model demonstrates computational efficiency and accuracy, its performance is heavily reliant on a large training dataset. Although the DNN-based modeling method shows strong capabilities in terms of performance, the training process requires a substantial amount of labeled data. Furthermore, the interpretability of the prediction results is relatively low, and tuning the model's parameters presents a significant challenge.

To enhance the interpretability of black-box models, an end-to-end deep learning method known as the deep neural decision forest has been proposed [29]. This approach utilizes a DNN to extract features, followed by the application of a random forest (RF) to build the model. End-to-end deep learning methods deeply extract information and then apply ensemble learning techniques for decision-making. This decision-making process aligns closely with human cognition and offers improved interpretability. Such methods have been widely applied and developed in the field of vision modeling tasks. For instance, Shen et al. proposed deep regression forests for age estimation [30], which leverage convolutional neural networks (CNN) to extract facial feature representations and employ the RF approach to construct the estimation model. Experimental results indicate that the predictive accuracy of this model significantly surpasses that of traditional deep networks and classical machine learning methods. Addressing the challenge of facial age estimation, Kuang et al. designed a new feature extraction network called the EfficientRF network [31]. Experimental findings demonstrate that enhancements to this network contribute to improved model accuracy. To tackle the positioning problem of mobile devices, Zhang et al. reported a deep fuzzy forest approach [32]. This method integrates fuzzy decision trees with a four-layer neural network within an end-to-end trainable architecture. Additionally, Cui et al. developed a novel deep neural forest that comprises multiple base decision tree models [33]. The deep neural forest method demonstrates an average relative improvement of 12.7% compared to the best-performing baseline.

While the end-to-end deep forest method enhances interpretability through its use of decision trees, offering greater transparency than traditional DNNs, its performance heavily depends on the availability of extensive training data—particularly when incorporating DNN units into its architecture. In many real-world scenarios, acquiring large volumes of labeled data can be both costly and challenging. Therefore, designing interpretable models that can perform well with limited labeled data presents a pressing practical challenge. To address this challenge, Zhou and Feng proposed a deep forest [34], a novel ensemble of decision trees with a cascade structure. Experiments across various domains, including image categorization, face recognition, music classification, and sentiment analysis, demonstrate that deep forest can achieve results comparable to, or even surpassing, those of DNNs. Capitalizing on the strong interpretability of deep forest, numerous researchers have begun applying it in industrial process domains, such as [35], forecast the corrosion rate of low-alloy steels [36], COVID-19 classification [37], and hyperspectral image classification [38]. Previous studies have highlighted the effectiveness of deep forest in a variety of industrial applications.

However, the current deep forest approach does not directly address regression modeling problems, particularly where class distribution probabilities are utilized as a representation strategy. To address this limitation, Tang et al. proposed a deep forest regression method called DFR-clfc [10]. Their approach modifies the feature representation strategy between layers in the deep forest by introducing cross-layer full connections. Experimental results indicate that DFR-clfc demonstrates strong modeling performance across high-dimensional, medium-dimensional, and low-dimensional datasets, even with a limited number of training samples (approximately 100). Recognizing the strengths of the DFR-clfc method in handling small-sample data, Xia et al. developed an improved DFR model to predict DXN emission concentrations in MSWI processes [11]. By enhancing the diversity of the layer forest model, they effectively modeled DXN emissions. Furthermore, Xia et al. designed a simplified DFR model for DXN emissions [12], showing that this streamlined approach maintains a high level of modeling accuracy. These studies provide a theoretical and modeling foundation for DXN generation concentration modeling in the MSWI process, demonstrating that DFR is well-suited for the DXN generation modeling task within this context.

9.1.2 Image Depth Feature Extraction

Deep feature extraction technology is increasingly replacing traditional surface features, establishing itself as a core approach for overcoming interference in complex industrial environments and enhancing the accuracy of combustion state recognition. Unlike surface features (e.g., color, texture, and geometry), which are manually engineered and carry specific physical meanings, deep feature extraction leverages deep learning models to automatically identify more representative abstract features from raw flame images or segmented flame region images through multi-layer nonlinear transformations. Theoretically, these deep features not only encompass visual details of flames but also capture implicit correlations that surface features often fail to represent. This process simulates the perceptual mechanisms of the cerebral cortex, enabling a more accurate reflection of the complexity and dynamics inherent in combustion processes within industrial settings. By utilizing deep feature extraction, the recognition of combustion states can be significantly improved, leading to better monitoring and control of combustion systems.

Research on deep feature extraction techniques for process industry images began relatively recently but has progressed rapidly. Deep learning (DL) technology utilizes multi-layer networks to automatically learn features from flame images [39], effectively overcoming the limitations associated with superficial feature representations and eliminating the necessity for complex manual feature extraction processes.

Among the various DL approaches, CNNs have emerged as particularly effective for extracting deep features [40]. This effectiveness stems from their unique convolutional operations, which allow CNNs to identify and learn spatial hierarchies of features from the input images. By employing multiple layers of convolutional filters, CNNs can capture intricate patterns and relationships within the data, making them well-suited for applications in combustion state recognition and other tasks within the process industry. The ability of CNNs to automatically learn relevant features significantly enhances the accuracy and efficiency of image analysis in complex industrial environments. For example, Reference [41] selected LeNet-5 as the recognition model for identifying flame combustion states. The study designed and tested the influence of multiple CNN architectures on flame image features, successfully achieving automatic extraction of deep features from flame images by optimizing the convolutional layers and activation functions. Reference [42] addresses the challenge of combustion state recognition under varying operating conditions in a swirl chamber. This study uses CNNs to extract spatial features from flame images and LSTM networks to capture temporal correlations, allowing for rapid detection of combustion instability and presenting a novel method for fusing spatiotemporal features. Reference [43] collected flame images from a 4.2 MW heavy oil combustion boiler operating under various combustion conditions. The researchers utilized Deep CNN (DCNN) to extract deep features from these images and implemented an SVM for predicting NO_x concentration based on the extracted features. This approach highlights the effectiveness of DL techniques in improving the accuracy of combustion state recognition and pollutant prediction in industrial applications.

The autoencoder, an unsupervised learning neural network model, can perform tasks such as data compression, feature extraction, and dimensionality reduction. For instance, reference [44] introduces an unsupervised feature extraction and classification framework that integrates a convolutional autoencoder (CAE), principal component analysis (PCA), and a hidden Markov model (HMM) to recognize the combustion flame states of coal powder. This framework emphasizes the automatic extraction of deep features from flame images using advanced DL techniques. Reference [45] employs a convolutional sparse autoencoder (CSAE) to extract deep features from coal-fired boiler flame images and utilizes a Minimum Support Vector Machine for state prediction. This approach demonstrates improved performance compared to existing neural network models. Reference [46] utilizes an adversarial denoising autoencoder (ADAE) for extracting deep features from heavy oil boiler flame images. The incorporation of adversarial mechanisms, denoising encoding, and dropout techniques enhances the model's capability to extract relevant deep features. Reference [47] introduces a cascaded stack autoencoder (SAE) model designed to integrate prior knowledge and hidden information for imbalanced data in rotary kiln sintering images. This model includes FE-SAE for extracting deep features and DF-SAE for fusing these features with prior knowledge, such as dynamic and static characteristics. Reference [48] addresses flame combustion state recognition in heavy oil boiler industrial environments. It utilizes a denoising autoencoder (DAE) to extract representative features from flame images. The DAE network is trained using the adversarial learning mechanism of generative adversarial networks (GANs) to mitigate issues of gradient vanishing or exploding during training, which can arise due to multiple hidden layers and a large number of parameters. Additionally, Reference

[49] focuses on applying flame combustion image recognition for the unsupervised quantitative assessment of combustion status in a 660 MW opposed coal powder combustion boiler furnace. This study constructs an unsupervised DL model that integrates a convolutional block attention module (CBAM) and a stacked convolutional autoencoder (SCAE) to effectively extract combustion features from flame images.

In summary, deep feature extraction technology for flame combustion in process industries has developed into a diverse system centered on CNN and their variants, while also integrating temporal models and Transformer architecture. This approach has yielded significant improvements in feature representation capabilities and adaptability to complex industrial scenarios. By incorporating domain adaptive learning and data augmentation techniques to enhance the physical interpretability of deep features, it is anticipated that this technology will facilitate the large-scale and stable application of deep feature extraction for combustion state recognition and soft sensing modeling in the process industry.

9.1.3 Multimodal Data Modeling

In actual industrial sites, domain experts identify key process parameters using multimodal data, including process data and flame videos. Several studies have constructed predictive models leveraging this multimodal data. For instance, a main steam temperature prediction model that integrates flame images and process data using neural networks. The results indicate that fusing image features can significantly enhance model accuracy. However, the static fusion method merely performs superficial splicing of multimodal information and fails to facilitate in-depth interaction and feature reconstruction between modalities, which depend on specific sample content [50]. Furthermore, when one modality provides more valuable information in a particular context, this rigid fusion mechanism struggles to adaptively amplify the contribution of key modalities, ultimately impacting model performance.

In contrast, the fields of image processing and natural language processing have widely embraced multimodal data for pattern recognition and semantic transformation in complex tasks [51, 52]. Currently, there are studies exploring the integration of image data and process data in industrial combustion processes, with the goal of constructing more accurate and robust models. For example, Lin et al. [53] incorporated flame images into traditional soft sensing frameworks to enhance model accuracy; Tóth et al. extracted 20 features from flame images and built a temperature prediction model using an additional 7 process features [54]; Zhou et al. utilized deep CNNs to extract image features and established an SVM based on multi-source information fusion [55].

Therefore, integrating multimodal data from process data and flame images to develop soft sensing models for parameters that are difficult to detect in complex industrial processes is a current and significant research focus. This integration involves leveraging both quantitative data (such as temperature, pressure, and flow rates) and qualitative data (like visual representations of flames) to create a multimodal data-driven model.

9.2 MSWI PROCESS DESCRIPTION IN TERMS OF DIOXINS GENERATION CONCENTRATION

The MSWI process is a comprehensive system that typically consists of several key stages, including: (1) Feeding: This initial stage involves the loading of MSW into the incineration system. The waste is often preprocessed to remove non-combustible materials and ensure uniformity; (2) Incineration: In this stage, the waste is subjected to high temperatures in the combustion chamber, where it is burned. This process reduces the volume of waste and generates heat; (3) Heat exchange: The heat generated during incineration is utilized to produce steam or hot water, which can then be used for energy generation or district heating applications; (4) Gas cleaning: After incineration, the flue gases produced must be cleaned to remove harmful pollutants, including dioxins, particulate matter, and other toxic substances. This stage typically involves various filtration and scrubbing technologies; and (5) Emission: The treated flue gases are released into the atmosphere through a stack, ensuring that they meet environmental regulations and standards for air quality.

A typical grate-type incinerator, designed to handle a daily capacity of 800 tons of MSW, is depicted in Figure 9.1, highlighting its concentration of dioxin generation.

In Figure 9.1, The flue gas cleaning stage is equipped with advanced air pollution control devices (APCDs), including selective non-catalytic reduction, activated carbon injection, semi-dry deacidification, and a bag filter system. Based on the MSWI process, it is evident that the combustion and heat exchange phases are closely related to the DXN generation concentration.

The concentration of DXN generation is detected at the outlet of the economizers. By integrating the MSWI process, the available data encompasses process numerical data (PND), flame video data (FVD), and recorded operational data (ROD). Derived from the distributed control system (DCS), PND consists of data points recorded at intervals of seconds. Before the generation of DXN, it included 119 physical variables related to the feeding, combustion, and heat exchange stages. FVD is captured by industrial cameras installed inside the incinerator, with each one-second video containing 25 frames. Each image has a resolution of 576×720 pixels and three color channels. ROD is collected from manual statistics maintained by operational experts, with data recorded once per shift. It includes metrics such as feed rate, incineration efficiency, average load, steam production per ton of MSW, and calorific value of MSW.

To construct a model for DXN generation concentration, the training data encompasses three different adoption cycles, necessitating a design that incorporates multimodal attributes. Given the multimodal data, such as PND and FVD, it is evident that the input variables during the generation phase exhibit significantly high-dimensional properties. Consequently, the information available for modeling displays both multimodal and high-dimensional characteristics. In practical scenarios, however, the available data on DXN generation concentration is extremely limited due to constraints imposed by detection equipment and associated costs. As a result, the output DXN information suitable for modeling is characterized by a small sample size, which poses challenges for robust model development.

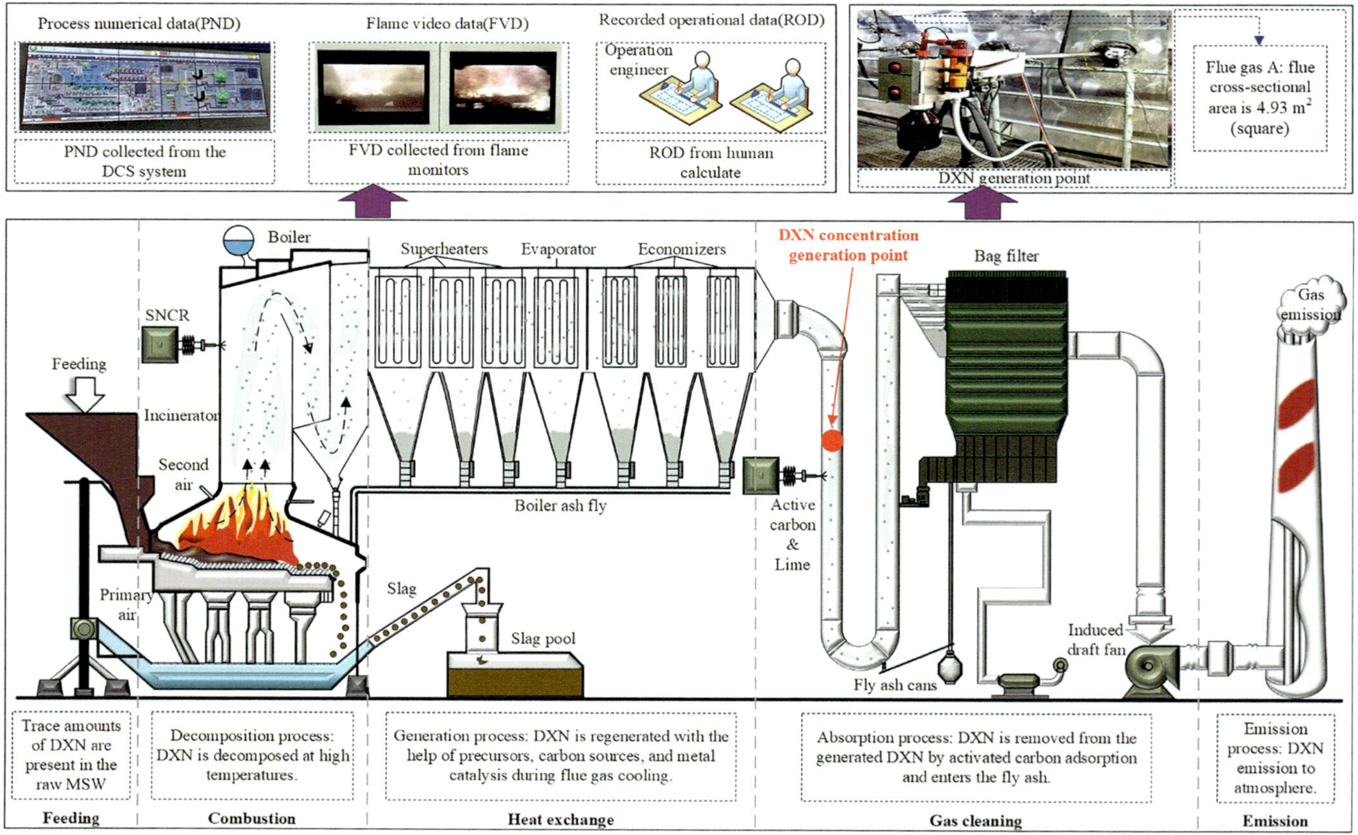

FIGURE 9.1 MSWI process in the view of the DXN generation concentration.

9.3 MODELING STRATEGY

This work introduces a multimodal data-driven tree-based deep learning (MDTDL) approach aimed at achieving online measurement of DXN generation concentration. The proposed modeling strategy consists of two main components: multimodal data preprocessing and tree-based DL modeling. In the first component, time scale matching of multimodal data is performed alongside statistical calculations based on mechanistic knowledge to generate the multimodal modeling dataset. In the tree-based DL modeling component, a novel adaptive DFR algorithm based on cross-layer full connection (ADFR-clfc) is proposed. This algorithm is applied to PND and ROD, while a CNN utilizing transfer learning, in combination with ADFR-clfc, is employed for modeling FVD. The final DXN generation concentration is obtained by calculating the arithmetic average of the outputs from both models. The structure of the MDTDL approach is illustrated in Figure 9.2.

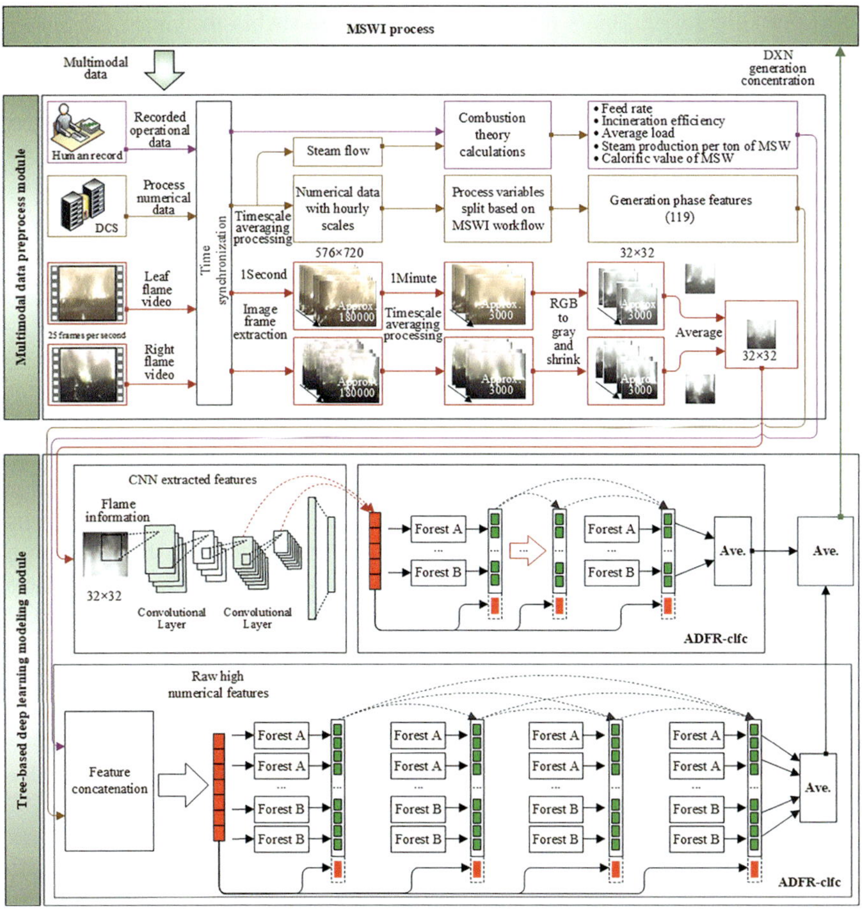

FIGURE 9.2 Modeling strategy of DXN generation concentration.

The main contributions of this study can be summarized as follows:

(1) Introduction of a Multimodal Data-Driven Modeling Strategy: This study is the first to introduce a multimodal data-driven modeling strategy for the soft sensor of DXN generation concentration. It utilizes PND from the DCS, FVD from the combustion monitoring system, and ROD from domain expert operation records of the MSWI power plant, enabling the construction of an accurate DXN generation concentration model.
(2) Addressing Data Limitations: The availability of DXN generation concentration data is constrained due to the economic and labor costs, as well as the complexity of laboratory detection methods, leading to small sample sizes. To tackle this challenge, a parallel fusion strategy based on transfer learning and explainable tree-based models is employed.
(3) Real-Time Measurement of DXN Generation Concentration: Driven by the need to minimize DXN generation concentration in the MSWI process, an explainable DL model that incorporates multimodal data is leveraged. To our knowledge, this is the first study to utilize such a technique for real-time measurement of DXN generation concentration. This model has the potential to provide engineers with valuable insights into DXN generation dynamics, facilitating more informed operational decisions and reducing emissions from pollution sources.
(4) Validation of the Proposed Method: The proposed modeling method is validated by comparing its performance against other approaches using nearly one year of actual data, measured in terms of accuracy and precision. This validation underscores its effectiveness for real-world applications.

9.4 ALGORITHM IMPLEMENTATION

9.4.1 Multimodal Data Preprocess Module

The multimodal data preprocessing module is illustrated in Figure 9.2. Different data types exhibit distinct time scales. For instance, DXN concentration is measured on an hourly basis, ROD is collected daily, and PND and FVD are recorded at a second-level frequency. To ensure coherence, the timing of the multimodal data is synchronized with the DXN sampling time. Specifically, the 2-hour average concentration of DXN corresponds to 2 hours of PND, 2 hours of flame video, and one set of ROD. An example of the time scale adjustment process for multimodal data is depicted in Figure 9.3.

One set of ROD includes the following parameters: feed rate (unit: tons/h), incineration efficiency (unit: %), average load (unit: tons/h), steam production per ton of MSW (unit: tons/h), and calorific value of MSW (unit: kJ). These values are calculated as Eqs. (9.1)–(9.5).

$$x_{\mathrm{FR}} = \frac{Q_2}{24} \tag{9.1}$$

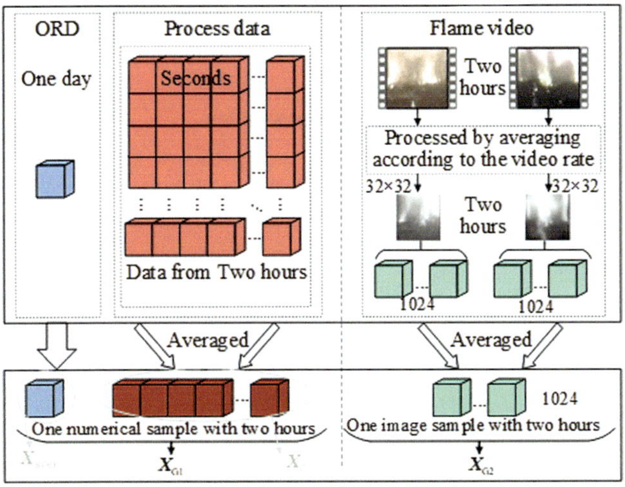

FIGURE 9.3 A specific case of data processing.

$$x_{\mathrm{IE}} = \frac{Q_2}{Q_1} \tag{9.2}$$

$$x_{\mathrm{AL}} = \frac{1}{T}\sum_{t=1}^{T} x_{87} \tag{9.3}$$

$$x_{\mathrm{SP}} = \frac{x_{\mathrm{AL}} \times 24}{Q_2} \tag{9.4}$$

$$x_{\mathrm{CV}} = \frac{x_{\mathrm{SP}} \times \left(H_{\mathrm{Steam}} - H_{\mathrm{Water}}\right)}{\eta_{\mathrm{Boiler}} \times 4.186} \tag{9.5}$$

where Q_1 represents the rated capacity of the MSWI process; Q_2 represents the actual volume handled in one day; x_{87} indicates the steam flow; H_{Steam} denotes the steam enthalpy value; H_{Water} expresses the water enthalpy value; η_{Boiler} represents the boiler thermal efficiency; x_{FR}, x_{IE}, x_{AL}, x_{SP}, and x_{CV} represent the feed rate, incineration efficiency, average load, steam production per ton of MSW, and calorific value of MSW, respectively.

The detailed interpretation of Eqs. (9.1)–(9.5) are shown in Table 9.1.

TABLE 9.1
Explanations of Eqs. (9.1)–(9.5)

Mark	Interpretation
x_{FR}	The feeding rate represents the amount of waste fed into the incinerator per hour, expressed in tons per hour. It is calculated by dividing the total daily treatment volume by 24 hours. This method ensures that the feeding rate accurately reflects the hourly input of waste into the incineration process.
x_{IE}	The combustion rate represents the ratio between the actual treatment capacity of the incinerator and its designed treatment capacity. It is calculated by dividing the total treatment volume by the designed treatment capacity. This calculation provides insight into the efficiency and effectiveness of the MSWI process compared to its intended design specifications.
x_{AL}	The calculation method for the average load is straightforward: it involves dividing the sum of all sample points by the total number of sample points.
x_{SP}	The steam production per ton of MSW indicates the amount of steam generated from the combustion of one ton of MSW. It is calculated by dividing the total steam volume produced over a 24-hour period by the actual treatment volume of MSW during that same period.
x_{CV}	The calculation principle is based on the inverse deduction of the calorific value of MSW by utilizing steam enthalpy, boiler efficiency, and the volume of MSW treated.

With the above calculations, a valid ROD can be obtained, denoted as

$$\boldsymbol{X}_{\text{ROD}} = \left\{\boldsymbol{x}_{\text{FR}}, \boldsymbol{x}_{\text{IE}}, \boldsymbol{x}_{\text{AL}}, \boldsymbol{x}_{\text{SP}}, \boldsymbol{x}_{\text{CV}}\right\} \in R^{N_G \times M_{\text{ROD}}}, \tag{9.6}$$

where N_G is the number of samples and M_{ROD} represents the dimensions of variables.

The processing of PND consists of two parts, i.e., data averaging and variable slicing. The specific method of data averaging is as follows:

$$x_i = \frac{1}{T}\sum_{t=1}^{T} x_i^t \tag{9.7}$$

where x_i is the ith process variables, T indicates the time scale. After data averaging, the obtained PND dataset is denoted as

$$\boldsymbol{X}_{\text{PND}} = \left\{\boldsymbol{x}_i\right\}_{i=1}^{N_G} \in R^{N_G \times M_{\text{PND}}} \tag{9.8}$$

where M_{PND} represents the dimensions of process numerical variables.

The original FVD consists of two parts, left and right, which share the same dimensional characteristics but contain different information. During the actual incineration process, flame changes occur relatively slowly. Therefore, video frames are extracted at one-second intervals, corresponding to a second-level time scale. Specifically, the first frame of the video is extracted every second as picture data. The flame image data from both the left and right sides are collected, resulting in approximately 180,000 images (size: 576 × 720 × 3 pixels) per side.

Next, the second-level image data to obtain minute-scale flame images (size: 576 × 720 × 3 pixels) are averaged, yielding around 3,000 images for each side. These images are then converted to grayscale, reducing the channels from three to one, and resized to 32 × 32 pixels. Consequently, flame image data from both the left and right sides are generated, consisting of approximately 3,000 images (size: 32 × 32 pixels) per side. The 3,000 images from each side are averaged to produce a single flame image of size 32 × 32 pixels.

Finally, the left and right flame images are averaged together to obtain the final image data, completing the fusion of the hourly scale flame video information from both sides. This process results in a single flame image dataset with a size of 32× 32 pixels.

Based on the above video preprocess, an FVD data denoted as is obtained, as follows:

$$\boldsymbol{X}_{\text{FVD}} = \left\{\boldsymbol{X}_i^{2\text{D}}\right\}_{i=1}^{N_F} \in R^{N_F \times (32\times32)} \tag{9.9}$$

9.4.2 Tree-Based Deep Learning Modeling Module

The input data for modeling the DXN generation concentration include $\boldsymbol{X}_{\text{ROD}}$, $\boldsymbol{X}_G$, and $\boldsymbol{X}_{\text{FVD}}$. The inputs of the PND and ROD-driven model and the FVD-driven model are denoted as follows:

$$\boldsymbol{X}_{\text{G1}} = \left\{\boldsymbol{X}_{\text{GPND}}, \boldsymbol{X}_{\text{ROD}}\right\} \in R^{N_G \times M_{\text{G1}}} \tag{9.10}$$

$$\boldsymbol{X}_{\text{G2}} = \boldsymbol{X}_{\text{FVD}} \in R^{N_{\text{F}} \times (32\times32)} \tag{9.11}$$

The specific modeling process is as follows.

9.4.2.1 PND and ROD-Driven Model Based on ADFR-clfc

The ADFR-clfc model consists of three components: the layer forest model, cross-layer full connection, and layer adaptability. Each layer forest model is composed of RFs and completely RFs (CRFs), with each forest method comprising multiple decision tree (DT) units.

The training dataset is denoted as

$$\boldsymbol{D}_{\text{G1}} = \left\{\boldsymbol{X}_{\text{G1}}, \boldsymbol{y}_G\right\} \in R^{N_G \times (M_{\text{G1}}+1)} \tag{9.12}$$

Taking the k_{G1}th layer as an example, the modeling process is described as follows. The input representation data of the k_{G1}th layer is denoted as,

$$\left.\begin{array}{l} X_{G1} = \left\{ x_{G1}^{i} \right\}_{i=1}^{N_G} \in R^{N_G \times M_{G1}} \\ \left\{ H_1^{\text{Map}}, H_2^{\text{Map}}, \ldots, H_{k_{G1}-1}^{\text{Map}} \right\} \\ N_G \\ \left(M_{G1} + \sum_{k=2}^{k_{G1}} J_{G1,\text{Forest}} \times \left(k_{G1} - 1 \right) \right) \\ y_G \end{array} \right\} \Rightarrow D_{G1}^{k_{G1}} = \left\{ x_{G1}^{i}, \ldots, h_{k_{G1}}^{\text{Map}}, \ldots, y_{G1}^{i} \right\}_{i=1}^{N_G} \in R^{N_G \times \left(M_{G1}^{k_{G1}} + 1 \right)} \tag{9.13}$$

where $H_{k_{G1}-1}^{\text{Map}}$ represents the output of the $\left(k_{G1}-1\right)$th layer, $M_{G1} + \sum_{k=2}^{k_{G1}} J_{G1,\text{Forest}} \times \left(k_{G1}-1\right)$ is the dimension of the feature, $J_{G1,\text{Forest}}$ is the number of the forest model, and $D_{G1}^{k_{G1}}$ indicates the input representation data to the k_{G1}th layer with dimension $M_{G1}^{k_{G1}}$; and $M_{G1}^{k_{G1}}$ is calculated as follows:

$$M_{G1}^{k_{G1}} = M_{G1} + \sum_{k=2}^{k_{G1}} J_{G1,\text{Forest}} \times \left(k_{G1} - 1 \right) \tag{9.14}$$

Subsequently, the input representation dataset $D_{G1}^{k_{G1}}$ is preprocessed using bootstrap and random subspace methods, which can be denoted as follows:

$$\left\{ D_{G1}^{k_{G1}, j_{G1,\text{DT}}} \right\}_{j_{G1,\text{DT}}=1}^{J_{G1,\text{DT}}} = f_{\text{RSM}} \left(f_{\text{Bootstrap}} \left(D_{G1}^{k_{G1}}, N_G, J_{G1,\text{DT}} \right), M_{\text{RSM}} \right) \tag{9.15}$$

where $f_{\text{Bootstrap}}(\cdot)$ is the bootstrap sampling function; $J_{G1,\text{DT}}$ expresses the times of bootstrap; $f_{\text{RSM}}(\cdot)$ represents the feature sampling function; $D_{G1}^{k_{G1}, j_{G1,\text{DT}}}$ denotes the $j_{G1,\text{DT}}$th subset from input representation data $D_{G1}^{k_{G1}}$; and M_{RSM} is the number of features in $D_{G1}^{k_{G1}, j_{G1,\text{DT}}}$, where $M_{\text{RSM}} << M_{G1}^{k_{G1}}$.

The dataset is then utilized to construct the DT model within the RF framework. In this process, each DT employs mean squared error to identify the optimal cut-off point for tree growth. This can be represented as follows:

$$\left(m_{G1}^{\text{bset}}, x_{G1}^{m_{G1}^{\text{bset}}, s} \right) = \min \left(\sum_{x_{G1}^{i} \in R_L} \left(y_{G1}^{i} - c_L \right)^2 + \sum_{x_{G1}^{i} \in R_R} \left(y_{G1}^{i} - c_R \right)^2 \right) \tag{9.16}$$

$$\begin{cases} R_{\text{L}} = \{ x_{G1} \mid x_{G1}^{i} \le x_{G1}^{m_{G1}^{\text{bset}}, s} \} \\ R_{\text{R}} = \left\{ x_{G1} \middle| x_{G1}^{i} \rangle x_{G1}^{m_{G1}^{\text{bset}}, s} \right\} \end{cases} \tag{9.17}$$

where $m_{\mathrm{G1}}^{\mathrm{bset}}$ is the serial number of the best features; $x_{\mathrm{G1}}^{m_{\mathrm{G1}}^{\mathrm{bset}},s}$ represents the value of the best features; $\boldsymbol{R}_L$ and $\boldsymbol{R}_R$ are the datasets of the left sub-node and the right sub-node, respectively; c_L and c_R denotes the average of the output in $\boldsymbol{R}_L$ and $\boldsymbol{R}_R$, respectively.

By using the limit of minimum samples in the leaf node, the $j_{\mathrm{G1,DT}}$th DT model can be written as follows,

$$\Gamma_{\mathrm{RF}}^{j_{\mathrm{G1,DT}}}(\cdot)=\sum_{q=1}^{Q} c_{\mathrm{RF}}^{j_{\mathrm{G1,DT}},q} I\left(\boldsymbol{x}_{\mathrm{G1}}^{i}\in \boldsymbol{R}_q\right) \tag{9.18}$$

where $I(\cdot)$ is a symbolic function.

Consequently, the RF model, comprising $J_{\mathrm{G1,DT}}$ DT model, can be expressed as follows,

$$f_{\mathrm{G1,RF}}^{k_{\mathrm{G1}}}(\cdot)=\frac{1}{J_{\mathrm{G1,DT}}}\sum_{j_{\mathrm{G1,DT}}=1}^{J_{\mathrm{G1,DT}}}\Gamma_{\mathrm{RF}}^{j_{\mathrm{G1,DT}}}(\cdot) \tag{9.19}$$

In the CRF, the growth of each complete DT is based on a random split point, while the remaining strategies are consistent with those used in standard DTs. Consequently, the description of this process can be simplified by representing the CRF model as $f_{\mathrm{G1,CRF}}^{k_{\mathrm{G1}}}(\cdot)$.

Given the number of trees in both the RF and the CRF, the layer forest model can be expressed as follows,

$$f_{\mathrm{G1}}^{k_{\mathrm{G1}}}(\cdot)=\left\{f_{\mathrm{G1,RF}}^{k_{\mathrm{G1}},j_{\mathrm{G1,Forest}}}(\cdot),f_{\mathrm{G1,CRF}}^{k_{\mathrm{G1}},j_{\mathrm{G1,Forest}}}(\cdot)\right\}_{j_{\mathrm{G1,Forest}}=1}^{J_{\mathrm{G1,Forest}}} \tag{9.20}$$

The structure of the ADFR-clfc model is illustrated in Figure 9.2. The layer regression vector serves as the output of the middle layer forest model. The layer regression vector for the k_{G1}th layer can be expressed as follows:

$$\boldsymbol{h}_{\mathrm{G1}}^{k_{\mathrm{G1}},i}=f_{\mathrm{G1}}^{k_{\mathrm{G1}}}\left(\boldsymbol{x}_{\mathrm{G1}}^{i}\right)\in R^{1\times J_{\mathrm{G1,Forest}}} \tag{9.21}$$

During the learning process of the ADFR-clfc model, a cross-layer full connection strategy is employed to enhance the representation performance of the layer regression vector. Consequently, the input feature information for the $(k_{\mathrm{G1}}+1)$th layer can be expressed as follows:

$$\boldsymbol{X}_{\mathrm{G1}}^{k_{\mathrm{G1}}+1}=\left[\boldsymbol{X}_{\mathrm{G1}},\boldsymbol{H}_{\mathrm{G1}}^{1},\boldsymbol{H}_{\mathrm{G1}}^{2},\ldots,\boldsymbol{H}_{\mathrm{G1}}^{k_{\mathrm{G1}}}\right]R^{N_G\times\left(M_{\mathrm{G1}}+J_{\mathrm{G1,Forest}}\times k_{\mathrm{G1}}\right)} \tag{9.22}$$

where

$$\boldsymbol{H}_{\mathrm{G1}}^{k_{\mathrm{G1}}}=\begin{bmatrix}\boldsymbol{h}_{\mathrm{G1}}^{k_{\mathrm{G1}},1}\\ \boldsymbol{h}_{\mathrm{G1}}^{k_{\mathrm{G1}},2}\\ \vdots\\ \boldsymbol{h}_{\mathrm{G1}}^{k_{\mathrm{G1}},N}\end{bmatrix}\in R^{N_G\times J_{\mathrm{G1,Forest}}} \tag{9.23}$$

where $\boldsymbol{H}_{\mathrm{G1}}^{k_{\mathrm{G1}}}$ is the output of the k_{G1}th layer forest model.

Compared to the DFR-clfc method, the ADFR-clfc model incorporates a layer adaptation mechanism to enhance learning performance. When the error during the layer-by-layer learning process converges, the learning process is automatically halted. This can be described as follows:

$$\theta_{\mathrm{G1}}=\frac{\left(\sqrt{\left(f_{\mathrm{G1}}^{k_{\mathrm{G1}}}\left(\boldsymbol{X}_{\mathrm{G1}}\right)-\boldsymbol{y}_G\right)^2}-\sqrt{\left(f_{\mathrm{G1}}^{k_{\mathrm{G1}}-1}\left(\boldsymbol{X}_{\mathrm{G1}}\right)-\boldsymbol{y}_G\right)^2}\right)}{N} \tag{9.24}$$

where

$$f_{\mathrm{G1}}^{k_{\mathrm{G1}}}=\frac{1}{J_{\mathrm{G1,Forest}}}\sum_{j_{\mathrm{G1,Forest}}=1}^{J_{\mathrm{G1,Forest}}}f_{\mathrm{G1,RF}}^{k_{\mathrm{G1}},j_{\mathrm{G1,Forest}}}(\cdot)+f_{\mathrm{G1,CRF}}^{k_{\mathrm{G1}},j_{\mathrm{G1,Forest}}}(\cdot) \tag{9.25}$$

where θ_{G1} is a threshold set based on empirical experience.

The ADFR-clfc model continues to increase the number of layers to achieve better learning outcomes when the error does not meet the specified criteria. If the error falls within acceptable limits after the ADFR-clfc model has learned up to K_{G1} layers, the final DXN generation model based on PND and ROD is obtained, which is denoted as $F_{\mathrm{G1}}(\cdot)$.

Therefore, the ADFR-clfc model consists of a multi-layer structure, where each layer incorporates two types of forest models (RF and CRF). Within each forest model, there are multiple tree models. Thus, the formal representation can be expressed as follows:

$$f_{\mathrm{G1}}^{(k-1)_{\mathrm{G1}}}\xrightarrow{\boldsymbol{X}_{\mathrm{G1}}^{k_{\mathrm{G1}}},y_G}\left\{\overbrace{\left[\begin{matrix}\begin{bmatrix}\Gamma_{\mathrm{RF}}^1\\ \vdots\\ \Gamma_{\mathrm{RF}}^{J_{\mathrm{G1,DT}}}\end{bmatrix}=f_{\mathrm{G1,RF}}^{k_{\mathrm{G1}},j_{\mathrm{G1,Forest}}}\\ \vdots\\ \begin{bmatrix}\Gamma_{\mathrm{CRF}}^1\\ \vdots\\ \Gamma_{\mathrm{CRF}}^{J_{\mathrm{G1,DT}}}\end{bmatrix}=f_{\mathrm{G1,CRF}}^{k_{\mathrm{G1}},j_{\mathrm{G1,Forest}}}\end{matrix}\right]}^{L_{k\mathrm{G1}}}=f_{\mathrm{G1}}^{k_{\mathrm{G1}}}\right\}\xrightarrow{\boldsymbol{X}_{\mathrm{G1}}^{k_{\mathrm{G1}}+1},y_G}f_{\mathrm{G1}}^{(k+1)_{\mathrm{G1}}}. \tag{9.26}$$

9.4.2.2 FVD-Driven Model Based on CNN and ADFR-clfc

A transferred CNN is employed to extract deep features from the image data. This transferred CNN is based on the flame status recognition model proposed in [56]. For the sake of completeness, a brief description of the model is provided below.

The CNN model features an 8-layer structure that includes an input layer, convolutional layers, pooling layers, fully connected layers, and an output layer. Given the input flame image dataset $\boldsymbol{X}_i^{2\text{D}}$, the CNN model extracts deep features as follows:

$$\boldsymbol{A}_{j\text{CNN}}^{\text{L1}} = f_{\tanh}\left(\boldsymbol{X}_i^{2\text{D}} * \boldsymbol{K}_{1,j\text{CNN}} + \boldsymbol{B}_1\right), \quad \boldsymbol{K}_{1,j\text{CNN}} \in R^{5\times5\times6} \;\; \boldsymbol{B}_1 \in R^{6\times6} \tag{9.27}$$

$$\boldsymbol{A}_{j\text{CNN}}^{\text{L2}} = \text{Mean}\left(\boldsymbol{A}_{j\text{CNN}}^{\text{L1}}, \boldsymbol{K}_2\right), \quad \boldsymbol{K}_2 \in R^{2\times2\times6} \tag{9.28}$$

$$\boldsymbol{A}_{j\text{CNN}}^{\text{L3}} = f_{\tanh}\left(\boldsymbol{A}_{j\text{CNN}}^{\text{L2}} \otimes \boldsymbol{K}_{3,m}\right) + \boldsymbol{B}_2, \quad \boldsymbol{K}_{3,m} \in R^{5\times5\times6\times16} \;\; \boldsymbol{B}_2 \in R^{16\times16} \tag{9.29}$$

$$\boldsymbol{A}_{j\text{CNN}}^{\text{L4}} = \text{Mean}\left(\boldsymbol{A}_{j\text{CNN}}^{\text{L3}}, \boldsymbol{K}_4\right), \quad \boldsymbol{K}_4 \in R^{2\times2\times16} \tag{9.30}$$

$$\boldsymbol{a}_{j\text{CNN}}^{\text{L5}} = f_{\tanh}\left(\boldsymbol{A}_{j\text{CNN}}^{\text{L4}} \otimes \boldsymbol{K}_5\right) + \boldsymbol{b}_3, \quad \boldsymbol{K}_5 \in R^{5\times5\times16\times120} \;\; \boldsymbol{b}_3 \in R^{1\times120} \tag{9.31}$$

$$\boldsymbol{a}_{j\text{CNN}}^{\text{L6}} = f_{\tanh}\left(\boldsymbol{a}_{j\text{CNN}}^{\text{L5}} \boldsymbol{k}_6 + \boldsymbol{b}_4\right), \quad \boldsymbol{k}_6 \in R^{120\times4} \;\; \boldsymbol{b}_4 \in R^{1\times4} \tag{9.32}$$

where $\boldsymbol{A}_{j\text{CNN}}^{\text{L1}} \in R^{28\times28\times6}$, $\boldsymbol{A}_{j\text{CNN}}^{\text{L2}} \in R^{14\times14\times6}$, $\boldsymbol{A}_{j\text{CNN}}^{\text{L3}} \in R^{10\times10\times16}$, $\boldsymbol{A}_{j\text{CNN}}^{\text{L4}} \in R^{5\times5\times16}$, $\boldsymbol{a}_{j\text{CNN}}^{\text{L5}} \in R^{1\times120}$, and $\boldsymbol{a}_{j\text{CNN}}^{\text{L6}} \in R^{1\times4}$ are the output feature maps of the convolution layer 1, layer 2, layer 3, layer 4, layer 5, and layer 6, respectively; $*$ represents the convolution process; $\otimes$ denotes the multi-channel convolution operation; $f_{\tanh}(\cdot)$ indicates the Tanh activation function; $\text{Mean}(\cdot)$ is the mean function of the matrix; and $\boldsymbol{K}_i$ is the core matrix.

The output feature maps from convolution layer 5 and layer 6 are utilized as inputs for the ADFR-clfc model. This can be expressed as follows:

$$\boldsymbol{x}_{\text{G2}}^{i} = \left[\boldsymbol{a}_{j\text{CNN}}^{\text{L5}}, \boldsymbol{a}_{j\text{CNN}}^{\text{L6}}\right] = F_{\text{CNN}}\left(\boldsymbol{X}_i^{2\text{D}}\right) \in R^{1\times124} \tag{9.33}$$

where $F_{\text{CNN}}(\cdot)$ represents the CNN model.

As a result, the acquired deep features of FVD can be denoted as

$$\boldsymbol{X}_{\text{G2}} = \left\{\boldsymbol{x}_{\text{G2}}^{i}\right\}_{i=1}^{N_A} \in R^{N_A\times124} \tag{9.34}$$

Subsequently, an ADFR-clfc model is trained using the deep features extracted from the CNN and the DXN concentration values. Based on the aforementioned process, the resulting model developed from the FVD data utilizing CNN and ADFR-clfc is denoted as $F_{\text{G2}}(\cdot)$.

Finally, the DXN generation concentration model is denoted as follows,

$$\hat{y}^{\text{Generation}} = \frac{1}{2}\left(F_{\text{G1}}(\cdot) + F_{\text{G2}}(\cdot)\right) \tag{9.35}$$

The above process involves the application of transfer learning, which leverages the transferability of the model. The classification model proposed in [56] for flame combustion state identification acts as the source domain model and is directly reused for the target domain's flame images, with no additional design or manipulation required during the transfer process. Since both the literature [56] and the data utilized in this chapter originate from the same data source, the source domain model can be seamlessly reused. Moreover, as illustrated in Figure 9.3, after preprocessing, each image is resized to 32 × 32 pixels, resulting in an input dimension of 1,024. The transferred CNN model is employed for feature extraction from these images, effectively capturing deep features and reducing the learning cost for downstream models. The dimensionality of the extracted deep features is 124, significantly lower than the original 1,024 dimensions. Thus, utilizing the transferred CNN model for feature extraction is expected to enhance the performance of the predictive model.

9.4.3 Comprehensive Flow Implementation

The specific implementation process of the proposed method is illustrated in Figure 9.4.

As illustrated in Figure 9.4, the modeling process is divided into two parts: numerical data modeling and image data modeling. For numerical data modeling, the obtained data features are utilized as input to create the first ADFR-clfc prediction model. In the image data modeling phase, the CNN model is first employed to extract features from the image data. These extracted features are then used for modeling to generate the second ADFR-clfc prediction model. In this context, the CNN model acts as a classification model for combustion conditions; however, in this chapter, it is directly transferred and utilized solely as a feature extractor. Consequently, the detailed structure of the CNN model is not elaborated upon here. The primary modeling algorithm utilized throughout this process remains the ADFR-clfc.

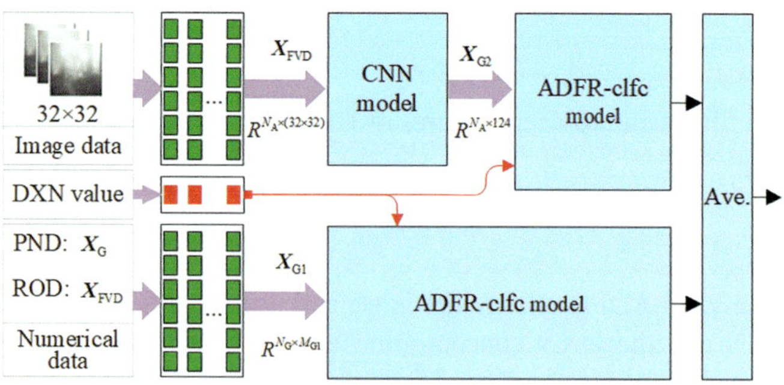

FIGURE 9.4 Implementation of the modeling flow.

The motivation for using the arithmetic mean method to obtain the final DXN generation concentration is based on the following points:

(1) Modeling DXN generation concentration is a typical regression problem, with the primary objective being to predict a specific numerical value. The arithmetic mean effectively reduces fluctuations in the predictions of individual DTs, thereby enhancing the stability and reliability of the overall prediction.
(2) The arithmetic mean strategy is a simple yet effective data processing method that has been widely adopted in ensemble learning algorithms such as RF. It has been demonstrated to be an effective approach for addressing regression problems.
(3) From a practical engineering perspective, the DXN generation concentration predicted using the arithmetic mean strategy exhibits better generality and reference value. This makes it more applicable in real-world scenarios.

However, it is important to acknowledge certain drawbacks associated with the arithmetic mean approach. These include the inability to accurately delineate the individual contributions of each model to the final result, the potential for reduced predictive accuracy, and increased susceptibility to outliers. Nonetheless, the currently designed model retains good explorability and adaptability in addressing these challenges.

9.5 CASE STUDY

9.5.1 Data Description and Evaluation Index

To obtain the DXN generation concentration from the MSWI process, detection samples were collected and analyzed from November 2020 to January 2022 using laboratory assays in accordance with the National Environmental Protection Standard (HJ 77.2-2008). As illustrated in Figure 9.1, isokinetic tracking sampling was employed to collect flue gas samples at the time of DXN generation, with each sampling session exceeding 2 hours. In total, 47 valid samples were collected from the MSWI plant. The distribution of sampling dates and times is depicted in Figure 9.5.

The DXN concentration obtained from the HRGC/HRMS equipment is denoted as c_s. This concentration requires further transformation to obtain the TEQ (Toxic Equivalent) concentration under dry air conditions at 101.325 kPa and 237 K (with 11% oxygen). The transformation process can be described as follows:

$$c = \frac{21\% - 11\%}{21\% - \varphi(O_2)} \times \sum_{i=1}^{17} c_s^i \cdot \text{TEF}_i \tag{9.36}$$

where c indicates the TEQ concentration with unit $\text{ng I} - \text{TEQ}/\text{Nm}^3$; $\varphi(O_2)$ represents the oxygen content at the stack; and TEF_i is the typical DXN congener concentration.

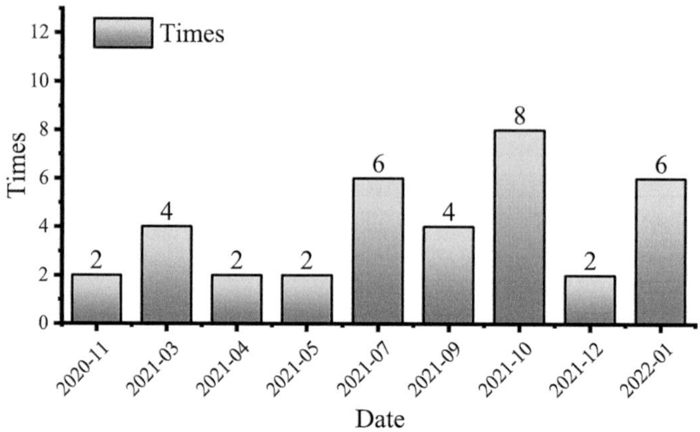

FIGURE 9.5 DXN sample sampling date and times.

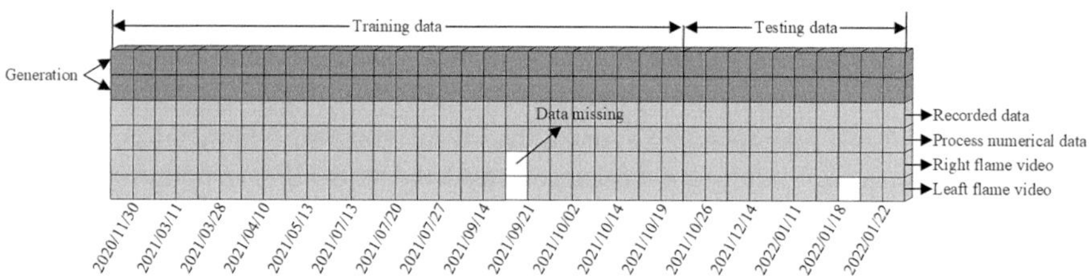

FIGURE 9.6 Distribution of the modeling data.

The input and output data for the MDTDL model are illustrated in Figure 9.6.

In this study, data collected before October 19, 2021, is utilized as the training set, while data acquired after that date serves as the testing set. As shown in Figure 9.6, the acquired data contains missing values, which necessitates careful handling. To address the issue of missing ground truth for flame videos, the data from September 21, 2021, is excluded during the flame data modeling process. For the remaining flame video from January 18, 2022, missing values were replaced with random noise to facilitate processing. This approach ensures that the model can still operate effectively despite the presence of incomplete data.

9.5.2 Modeling Results

The hyperparameter settings for the ADFR-clfc model are as follows: minimum sample in leaf node is 3, number of DT in each forest is 5, number of RF is 2, number of CRF is 2, selected features is $\sqrt{M}$, and threshold is 1×10^{-4}. The hyperparameter settings for the CNN model are as follows: structure of CNN is input-three convolution layer-fully connected layer -output, size and number of three convolution kernel are set to $\left[5\times5\times6\right]-\left[5\times5\times16\right]-\left[5\times5\times16\right]$, the epoch is 200, and the learning rate is 0.01.

With the aforementioned hyperparameter settings, the results of the model training process are presented in Figure 9.7.

Figure 9.7 shows the following observations:

(1) In the PND modeling process, the training error gradually converges as the number of layers increases, as shown in Figure 9.7(a). The ADFR-clfc model ultimately converges with a final layer count reaching 538.

(2) The CNN image combustion status recognition model demonstrates classification accuracies exceeding 90% across training, validation, and testing

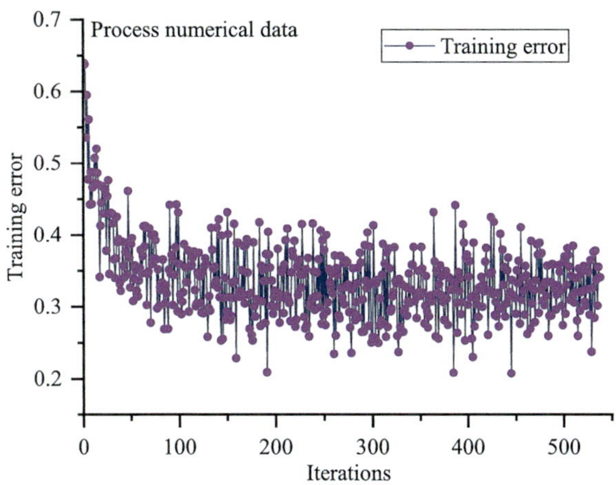

(a) Training error curves for PND

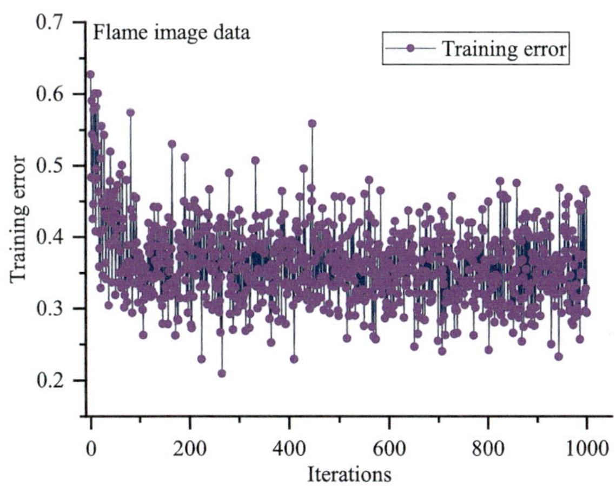

(b) Training error curves for FVD

FIGURE 9.7 Training results of ADFR-clfc model and CNN-ADFR-clfc model. (a) Training error curves for PND. (b) Training error curves for FVD.

(Continued)

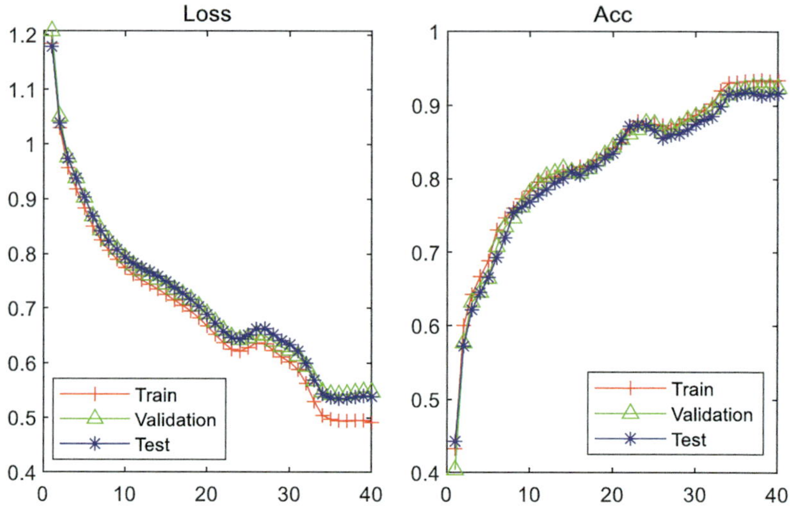

(c) Training results of the CNN model

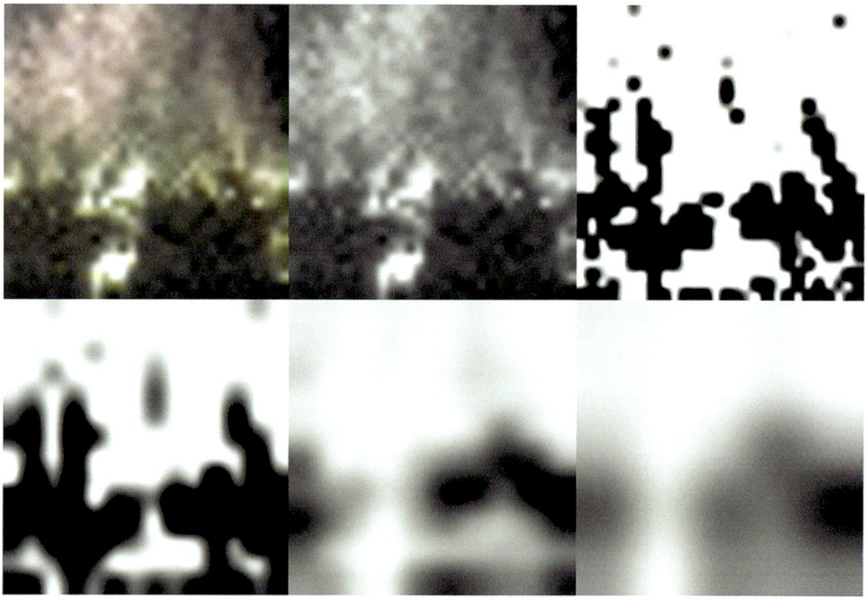

(d) Image extract results from the CNN model

FIGURE 9.7 (Continued) (c) Training results of the CNN model. (d) Image extract results from the CNN model.

phases, as illustrated in Figure 9.7(c). This underscores the effectiveness of the CNN model in learning from flame images.

(3) Figure 9.7(d) showcases partial results of the transfer CNN model for flame image extraction in this study, highlighting the effective feature extraction capabilities of the CNN model.

(4) Figure 9.7(b) presents the training error curve of the flame image data-driven ADFR-clfc model, indicating that the ADFR-clfc model effectively converges during training, even when dealing with high-dimensional data.

9.5.3 Comparison Results

9.5.3.1 Comparison Results of Model Based on PND and ROD

The comparative methods employed in this phase include DT, RF, SVR, BPNN, Multi-layer back propagation neural network (MLBPNN), and DFR. The hyperparameter settings for these models are as follows:

(1) DT: Minimum samples in a leaf node are set to 3.
(2) RF: Minimum samples in a leaf node are set to 3, the number of DT is 500, and selected features are $\sqrt{M}$.
(3) SVR: The kernel function is Gaussian, and the number of iterations is set to 1,000.
(4) BPNN: The structure is input-hidden-output, with the number of hidden neurons set to $(2\times M)-1$, the number of epochs is 1,500, and the learning rate is 0.001.
(5) MLBPNN: The structure is input-hidden1-hidden2-hidden3-output, with the number of neurons in the three hidden layers set to $(2\times M)-(4\times M)-(2\times M)$, the number of epochs is 1,500, and the learning rate is 0.001.
(6) DFR: The number of layers is 50, the minimum samples in a leaf node are set to 3, the number of DTs in each forest is 500, the number of RF is 2, the number of CRF is 2, and selected features are $\sqrt{M}$.

These hyperparameter settings were chosen to optimize the performance of each model in the comparative analysis.

Additionally, several advanced modeling methods have been included in the experimental comparison section, namely TTBL [57], RSWFNN [58], and SM-RSC [59]. Each method underwent 20 experiments, and the statistically best results are presented in Table 9.2. The corresponding fitting curves are depicted in Figure 9.8.

Based on the experimental results presented in Table 9.2 and Figure 9.8, the following conclusions can be drawn:

(1) Overfitting is observed in the BPNN and SM-RSC methods, likely due to the high-dimensional characteristics of the training data combined with a limited number of samples. While SM-RSC demonstrates slightly better generalization performance under overfitting conditions compared to BPNN, the latter exhibits a significant deviation between the measured values in the testing set and the actual situation.
(2) A comparison between DL and machine learning methods reveals a notable difference in stability, with DL methods showing superior consistency. For instance, the MLBPNN approach displays greater stability when contrasted with the BPNN method. Experimental results indicate that deep models are

TABLE 9.2
Statistical Results of Different Methods in DXN Generation Model by Using PND and ROD

	Training Set		Testing Set	
Method	**RMSE**	**MAE**	**RMSE**	**MAE**
DT	0.0366	0.0167	0.0940	0.0681
RF	0.0531	0.0355	0.0948	0.0708
SVR	0.0605	0.0229	0.0980	0.0704
BPNN	0.0031	0.0025	0.2476	0.2051
MLBPNN	0.0185	0.0137	0.0980	0.0674
DFR	0.0561	0.0376	0.0938	0.0704
TTBL [57]	0.1015	0.0730	0.1028	0.0783
RSWFNN [58]	0.0616	0.0476	0.0844	0.0689
SM-RSC [59]	0.0001E−01	0.0002E−02	0.0965	0.0891
ADFR-clfc	0.0361	0.0204	0.0838	0.0652

more effective in representing high-dimensional feature spaces and complex scenarios, suggesting that increasing model depth contributes positively to stability.

(3) The proposed ADFR-clfc method outperforms DT, RF, and DFR in terms of learning efficiency and generalization capability. Its measurement trend within the testing set closely aligns with the actual values, indicating high accuracy. Unlike neural network-based methods, the ADFR-clfc method does not demonstrate overfitting in the training set fitting. Although it shows weaker performance compared to BPNN, MLBPNN, and SM-RSC, it outperforms the TTBL and RSWFNN methods. The prediction results from the testing set indicate that the ADFR-clfc method achieves better generalization performance, benefiting from the incorporation of DTs as stable learning units within the proposed model.

9.5.3.2 Comparison Results of Model Based on FVD

The comparative methods employed in this model include MLBPNN, CNN for regression (CNNR), DFR, and CNN-DFR. The hyperparameter settings for each method are provided as follows:

(1) MLBPNN: The structure consists of input-hidden1-hidden2-hidden3-output layers, with the number of neurons in the three hidden layers set to $(2\times M)-(4\times M)-(2\times M)$, an epoch count of 2,500, and a learning rate of 0.001.

(2) CNNR: The structure comprises input followed by three convolutional layers, a fully connected layer, and an output layer. The size and number of kernels for

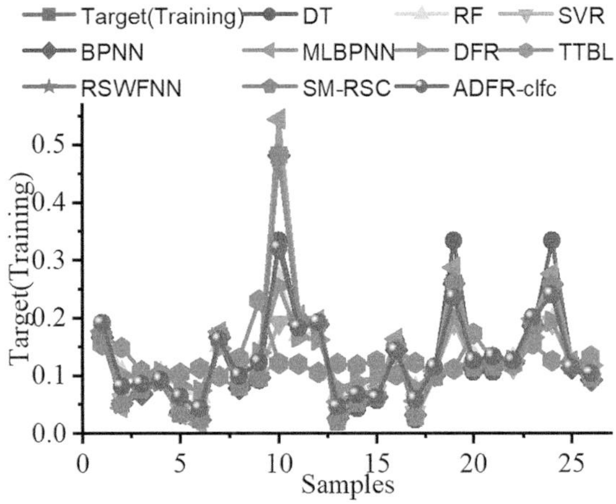

(a) Training curves by using PND and ROD

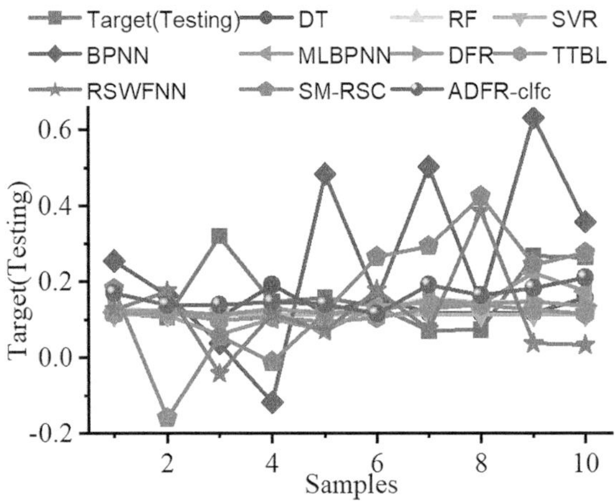

(b) Testing curves by using PND and ROD

FIGURE 9.8 Fitting curves for different methods in the DXN generation model by using PND and ROD. (a) Training curves by using PND and ROD. (b) Testing curves by using PND and ROD.

the three convolutional layers are set to $[5\times5\times8]-[5\times5\times16]-[5\times5\times32]$, with an epoch count of 2,500 and a learning rate of 0.001.

(3) DFR: This model features 50 layers, with a minimum sample size in the leaf node set to 3. It includes 500 DTs in each forest, 2 RFs, and 2 CRFs, alongside selected features of $\sqrt{M}$.

(4) CNN-DFR: The CNN structure consists of input followed by three convolutional layers, a fully connected layer, and an output layer. The size and number of kernels for the three convolutional layers are set to $[5\times5\times6]-[5\times5\times16]-[5\times5\times16]$, with an epoch count of 200 and a

learning rate of 0.01. In addition, it has 50 layers, a minimum sample size in the leaf node of 3, 500 DTs in each forest, 2 RFs, and 2 completely randomized forests, along with selected features of $\sqrt{M}$.

Similar to the results obtained from the models based on PND and ROD, the statistically best results are presented in Table 9.3. The corresponding fitting curves are illustrated in Figure 9.9.

By combining the information from Table 9.3 and Figure 9.9, several observations can be made:

(1) With the exception of SM-RSC, all methods have demonstrated satisfactory learning outcomes on the training set, indicating that the trained models are adequately qualified. Notably, compared to PND and ROD data, the feature information in image data may exhibit smaller differences.
(2) The experimental results for MLBPNN and CNNR indicate that modeling with image features extracted by convolutional networks outperforms directly inputting the images into the network. This suggests that convolutional networks are more effective at extracting critical feature information from images. This conclusion is further supported by the experimental results of DFR and CNN-DFR.
(3) The method proposed in this chapter employs CNN for feature extraction, followed by modeling using the ADFR-clfc approach. The experimental results presented in this section demonstrate the effectiveness of the proposed method. Compared to classical methods (MLBPNN, CNNR, DFR, DFR-clfc), CNN-DFR exhibits improved performance in both training and testing phases. Furthermore, relative to existing high-performance methods (TTBL, RSWFNN, SM-RSC), CNN-DFR shows superior generalization performance.

TABLE 9.3
Statistical Results of Different Methods in DXN Generation Model Using FCD

	Training Set		Testing Set	
Method	RMSE	MAE	RMSE	MAE
MLBPNN	0.0792	0.0481	0.1041	0.0790
CNNR	0.0659	0.0413	0.1072	0.0786
DFR	0.0587	0.0382	0.1072	0.0803
DFR-clfc	0.0436	0.0236	0.0976	0.0785
CNN-DFR	0.0552	0.0372	0.1016	0.0770
TTBL [57]	0.0944	0.0662	0.1025	0.0769
RSWFNN [58]	0.0649	0.0428	0.0963	0.0782
SM-RSC [59]	0.0002E−11	0.0001E−11	0.1027	0.0896
CNN-ADFR-clfc	0.0350	0.0171	0.0889	0.0762

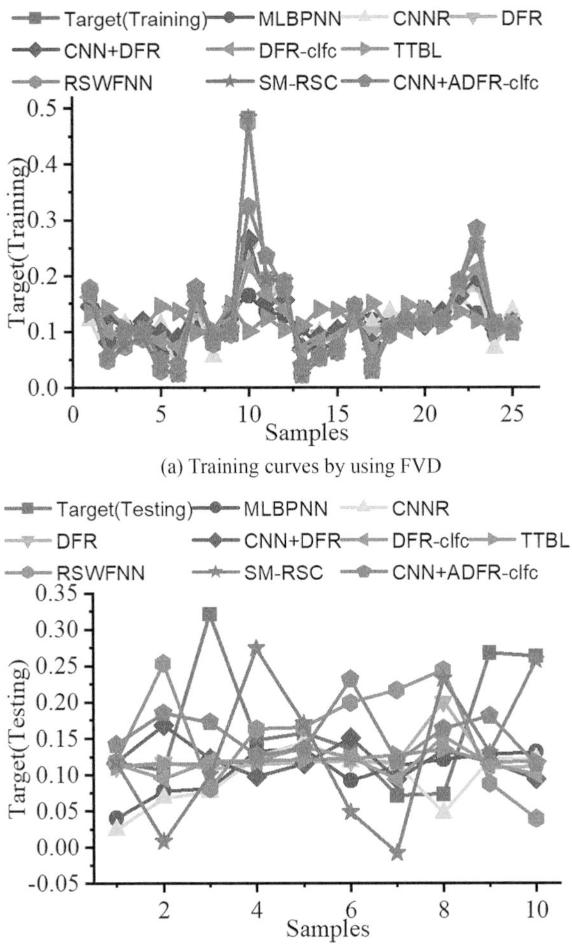

(a) Training curves by using FVD

(b) Testing curves by using FVD

FIGURE 9.9 Fitting curves for different methods in the DXN generation model by using FVD. (a) Training curves using FVD. (b) Testing curves using FVD.

9.5.4 Results and Discussion

To further analyze the performance of the proposed approach and strategy, the aforementioned statistical results of various methods are evaluated using the average strategy. The results of this evaluation are presented in Table 9.3.

In the methods outlined in Table 9.4, MLBPNN, TTBL, RSWFNN, and SM-RSC are classified as neural network-based methods, while DFR is categorized as a non-neural network-based method. The experiments reveal that neural network-based methods exhibit stronger learning capabilities on the training set compared to their non-neural counterparts. However, the advantage of non-neural network-based

TABLE 9.4
Statistical Results of Different Methods by Average Strategy

	Training Set		Testing Set	
Method	RMSE	MAE	RMSE	MAE
MLBPNN	0.0489	0.0309	0.1011	0.0732
DFR	0.0574	0.0379	0.1005	0.0754
TTBL [57]	0.0980	0.0696	0.1027	0.0776
RSWFNN [58]	0.0633	0.0452	0.0904	0.0736
SM-RSC [59]	5.0E–06	1.0E–06	0.0996	0.0894
Our work	0.0356	0.0188	0.0864	0.0707

methods lies in their relatively stable models, a characteristic that is also reflected in their performance on the testing data.

This chapter has improved upon the DFR method by proposing a modeling approach specifically designed for multimodal data. The training results indicate that the enhanced ADFR-clfc method significantly boosts the fitting performance of the model on the training data, demonstrating the effectiveness of the proposed approach.

Based on the experimental results presented in Table 9.4, the average fusion strategy proposed in this chapter integrates existing methods into the new strategy, revealing that the modeling method and strategy designed herein are optimal for both training and testing phases. Notably, compared to the original DFR method, the enhancements in training and testing performance in terms of the RMSE metric are 38% and 14%, respectively. When compared to other neural network-based methods, such as MLBPNN, the improvements in training and testing performance in the RMSE metric are 27% and 15%, respectively. Additionally, relative to SM-RSC, there is a 99% decrease in training performance and a 13% increase in testing performance in the RMSE metric.

The proposed model is specifically tailored to the characteristics of multimodal data and employs an average fusion modeling strategy that leverages the advantages of both neural network and non-neural network models. The experimental results substantiate the effectiveness of the model in simulating DXN generation concentrations.

9.5.5 Sensitivity Analyses

The hyperparameters of the proposed DXN generation concentration model are as follows: minimum samples in a leaf node, number of DTs in each forest, number of RF, number of CRF, selected features, and threshold. To assess the model's sensitivity to these hyperparameters, a single-factor analysis method is employed.

The details of each hyperparameter setting are provided in Table 9.5, which outlines the specific values used for testing. Additionally, Figure 9.10 illustrates the response curves for the proposed DXN generation concentration model, showcasing how variations in each hyperparameter influence the model's performance.

TABLE 9.5
Details of Each Hyperparameter Setting

Hyperparameter	Test Interval
Minimum sample in leaf node	Min: 1; Max: 20; Step: 1
Number of DT in each forest	Min: 5; Max: 100; Step: 5
Number of RF	Min: 1; Max: 20; Step: 1
Number of CRF	Min: 1; Max: 20; Step: 1
Selected features	Min: 5; Max: 50; Step: 2
Threshold	Min: 1×10^{-5}; Max: 1×10^{-3}; Step: 5×10^{-5}

Figure 9.10 demonstrates the following observations regarding the hyperparameters of the proposed DXN generation concentration model:

(1) Minimum Sample in Leaf Node: As the number of samples in the leaf nodes increases, the training effect of the model tends to deteriorate, while generalization performance improves. The experimental results indicate that the model achieves optimal performance overall when there are six samples in the leaf nodes.
(2) Number of DT in Each Forest: The number of decision trees has a significant impact on the model's testing performance, with optimal points observed at 25 and 80 trees. These points suggest a balance between complexity and model accuracy.
(3) Number of RF: Generally, increasing the number of RFs leads to reduced learning performance of the model. However, when 13 RFs are used for modeling, the model achieves its best testing set accuracy.
(4) Number of CRF: Increasing the number of CRFs typically enhances the model's learning performance; however, this improvement comes at the cost of significant decreases in testing set accuracy.
(5) Selected Features: Variations in the number of selected features have a minimal impact on both training and testing performance, as indicated by the response curve, which shows fluctuations within a small range.
(6) Threshold: An increase in the threshold reduces the model's layer-by-layer learning process, resulting in fewer layers in the model. Experimental results show that reducing the number of layers significantly decreases the model's learning capability.

Based on the analysis above, it is clear that the minimum sample in leaf nodes and the threshold are the most critical hyperparameters affecting the performance of the fusion model. These findings underscore the importance of carefully tuning these parameters to optimize model performance.

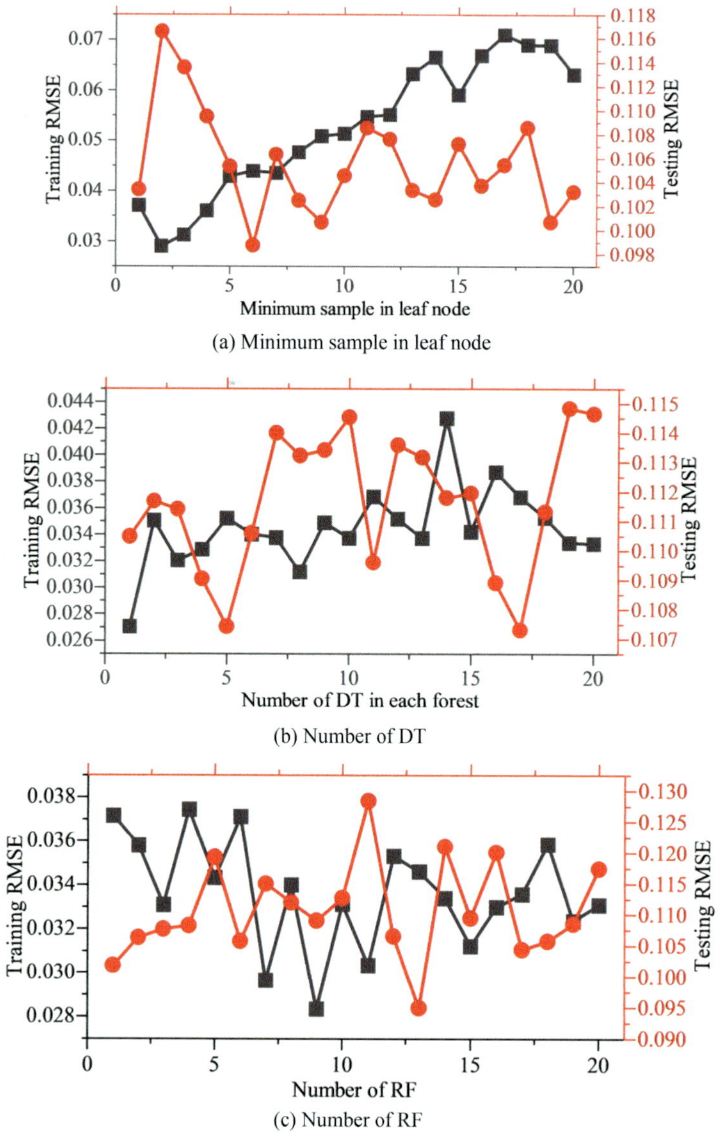

FIGURE 9.10 Sensitivity results of the proposed DXN generation concentration model with various settings of each hyperparameter. (a) Minimum sample in leaf node. (b) Number of DT. (c) Number of RF.

(Continued)

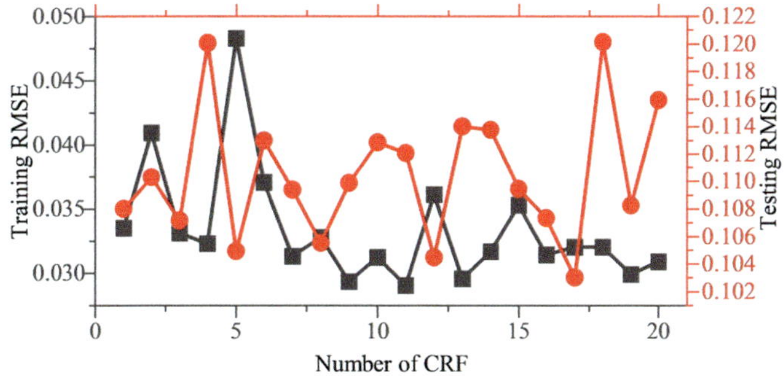

(d) Number of CRF

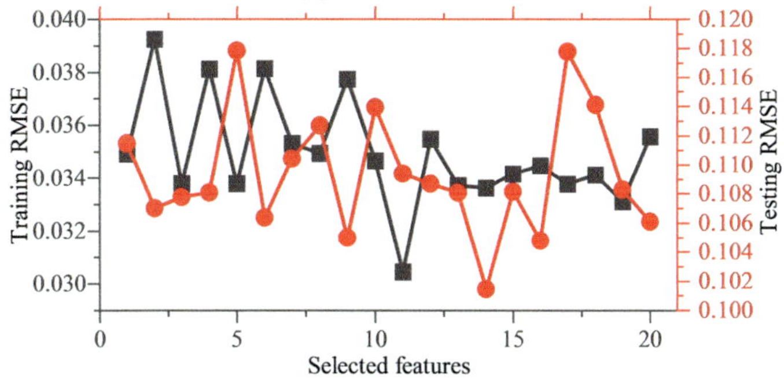

(e) Selected features

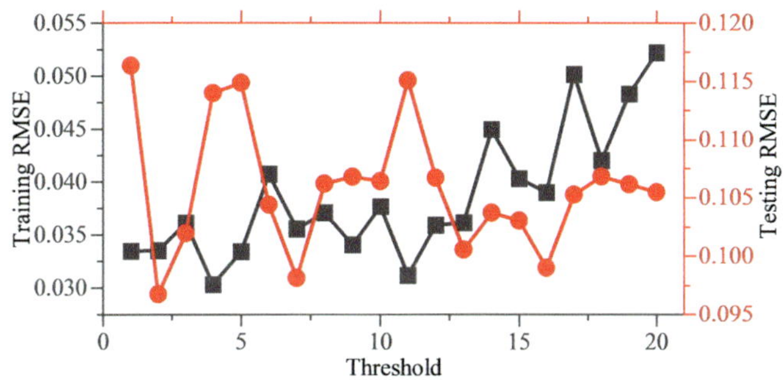

(f) Threshold

FIGURE 9.10 (Continued) (d) Number of CRF. (e) Selected features. (f) Threshold.

9.6 SUMMARY

Based on the multimodal data from the MSWI process, this chapter introduces an interpretable DL modeling method called MDTDL for modeling DXN generation concentrations. The experimental results indicate that the proposed method offers several advantages over existing modeling approaches, particularly in terms of modeling accuracy and interpretability.

The potential implications of this work are as follows: (1) Novel hybrid-driven strategy: This is the first instance of constructing a DXN model utilizing a hybrid-driven strategy that combines both numerical and image data, paving the way for new avenues in data-driven DXN modeling research; (2) First report on DXN modeling in MSWI: The modeling of DXN generation concentration within the MSWI process is reported for the first time, which could facilitate advancements in research focused on DXN emission reduction and control; (3) Validation of MDTDL effectiveness: The experimental results substantiate the modeling performance of MDTDL, providing a foundational model and theoretical support for potential online engineering applications in the future.

However, the current model does have limitations that require attention: (1) Overall interpretability: While individual DTs are interpretable, the presence of multiple layers and numerous DTs in MDTDL may diminish the model's overall interpretability; (2) Impact of outliers: The arithmetic mean strategy employed in the model can lead to decreased prediction accuracy when significant outliers are present in the predictions of single models; (3) Online deployment challenges: The current model design does not adequately address the challenges associated with online deployment processes.

Based on the analysis of these advantages and limitations, future research directions and scopes can be outlined as follows: (1) Exploration of advanced modeling methods: Investigating modeling methods that demonstrate superior generalization performance by leveraging deeper features that capture different hidden information, along with advanced fusion strategies; (2) Development of online dynamic modeling strategies: Focusing on strategies for online dynamic modeling that enhance the adaptability of the model in practical applications.

REFERENCES

1. Guo J, Bo X, Xie Y, et al. Health effects of future dioxins emission mitigation from Chinese municipal solid waste incinerators. *Journal of Environmental Management*, 2023, 345: 118805.
2. Ministry of Ecology Environment. Ambient air and waste gas determination of polychlorinated dibenzo-p-dioxins (PCDDs) and polychlorinated dibenzofurans (PCDFs) isotope dilution HRGC/HRMS. [Online]. April 1, 2009. Available: https://english.mee.gov.cn/Resources/standards/Air_Environment/air_method/201102/t20110215_200696.shtml
3. Xiong S, Lu S, Shang F, et al. Online predicting PCDD/F emission by formation pathway identification clustering and box-cox transformation. *Chemosphere*, 2021, 274: 129780.
4. Chen K, Peng Y, Lu S, et al. Bagging based ensemble learning approaches for modeling the emission of PCDD/Fs from municipal solid waste incinerators. *Chemosphere*, 2021, 274: 129802.
5. Ying Y, Lin X, Wu A, Li X. Review and outlook on municipal solid waste smart incineration. *CIESC Journal*, 2021, 72(2): 886–900.
6. Chang N, Huang S. Statistical modelling for the prediction and control of PCDDs and PCDFs emissions from municipal solid waste incinerators. *Waste Management & Research*, 1995, 13: 379–400.

7. Ishikawa R, Buekens A, Huang H, Watanabe K. Influence of combustion conditions on dioxin in an industrial-scale fluidized-bed incinerator: experimental study and statistical modelling. *Chemosphere*, 1997, 35(3): 465–477.
8. Bunsan S, Chen W, Chen H, Chuang Y, Grisdanurak N. Modeling the dioxin emission of a municipal solid waste incinerator using neural networks. *Chemosphere*, 2013, 92(3): 258–264.
9. Tang J, Xia H, Aljerf L, et al. Prediction of dioxin emission from municipal solid waste incineration based on expansion, interpolation, and selection for small samples. *Journal of Environmental Chemical Engineering*, 2022, 10(5): 108314.
10. Tang J, Xia H, Zhang J, et al. Deep forest regression based on cross-layer full connection. *Neural Computing and Applications*, 2021, 33(15): 9307–9328.
11. Xia H, Tang J, Aljerf L. Dioxin emission prediction based on improved deep forest regression for municipal solid waste incineration process. *Chemosphere*, 2022, 294: 133716.
12. Xia H, Tang J, Aljerf L, et al. Dioxin emission modeling using feature selection and simplified DFR with residual error fitting for the grate-based MSWI process. *Waste Management*, 2023, 168: 256–271.
13. Xia H, Tang J, Cong Q, et al. Dioxin emission concentration forecasting model for MSWI process with random forest-based transfer learning. In: *2020 39th Chinese Control Conference*, Shenyang, China, 27–29 July 2020. 2020, 5724–5729.
14. Zhang J S, Li X, Tian J, et al. A variational local weighted deep sub-domain adaptation network for remaining useful life prediction facing cross-domain condition. *Reliability Engineering & System Safety*, 2023, 231: 108986.
15. Zhang J S, Tian J, Yan P, et al. Multi-hop graph pooling adversarial network for cross-domain remaining useful life prediction: a distributed federated learning perspective. *Reliability Engineering & System Safety*, 2024, 244: 109950.
16. Song Q, Zheng Y J, Sheng W G, Yang J. Tridirectional transfer learning for predicting gastric cancer morbidity. *IEEE Transactions on Neural Networks and Learning Systems*, 2021, 32(2): 561–574.
17. Jin H, Xiong Q, Wang B, et al. Soft sensor development based on unsupervised dynamic weighted domain adaptation for quality prediction of batch processes. *IEEE Transactions on Instrumentation and Measurement*, 2024, 73: 1–19.
18. Wei J, Li H, Liu J. Fate of dioxins in a municipal solid waste incinerator with state-of-the-art air pollution control devices in China. *Environmental Pollution*, 2021, 289, 1–10.
19. Fiedle H. Thermal formation of PCDD/PCDF: a survey. *Environmental Engineering Science*, 1998, 159(1): 49–58.
20. Zhang R, Zhuang T, Zhang Q, Wang W. Mechanistic studies on the dibenzofuran and dibenzo-p-dioxin formation reactions from anthracene. *Science of the Total Environment*, 2019, 662: 41–47.
21. Kim K, Hong K, Ko Y, Kim M. Emission characteristics of PCDD/Fs, PCBs, chlorobenzenes, chlorophenols, and PAHs from polyvinylchloride combustion at various temperatures. *Journal of the Air & Waste Management Association*, 2004, 54: 555–562.
22. Altwicker E. Some laboratory experimental designs for obtaining dynamic property data on dioxins. *The Science of the Total Environment*, 1991, 104: 47–72.
23. Altarawneh M, Dlugogorski B, Kennedy E, Mackie J. Mechanisms for formation, chlorination, dechlorination and destruction of polychlorinated dibenzo-p-dioxins and dibenzofurans (PCDD/Fs). *Progress in Energy and Combustion Science*, 2009, 35(3): 245–274.
24. Peng Y, Chen J, Lu S, et al. Chlorophenols in municipal solid waste incineration: a review. *Chemical Engineering Journal*, 2016, 292, 398–414.

25. Sun Q, Ge Z. A survey on deep learning for data-driven soft sensors. *IEEE Transactions on Industrial Informatics*, 2021, 17(9): 5853–5866.
26. He Y L, Wang P F, Zhu Q X. Improved bi-lstm with distributed nonlinear extensions and parallel inputs for soft sensing. *IEEE Transactions on Industrial Informatics*, 2024, 20(3): 3748–3755.
27. Yao L, Shen B, Cui L, et al. Semi-supervised deep dynamic probabilistic latent variable model for multimode process soft sensor application. *IEEE Transactions on Industrial Informatics*, 2023, 19(4): 6056–6068.
28. Guo R, Liu H, Xie G, et al. A self-interpretable soft sensor based on deep learning and multiple attention mechanism: from data selection to sensor modeling. *IEEE Transactions on Industrial Informatics*, 2023, 19(5): 6859–6871.
29. Kontschieder P, Fiterau M, Criminisi A, Bulò S R. Deep neural decision forests. In: *2015 IEEE International Conference on Computer Vision*, Santiago, Chile, 7–13 December 2015. 2015, 1467–1475.
30. Shen W, Shen W, Guo Y, et al. Deep regression forests for age estimation. In: *2018 IEEE/CVF Conference on Computer Vision and Pattern Recognition*. 2018, 2304–2313.
31. Kuang H L, Huang X, Ma X L, Liu X H. EfficientRF: facial age estimation based on efficientnet and random forest. In: *2023 IEEE 3rd International Conference on Information Technology, Big Data and Artificial Intelligence*, Chongqing, China, 26–28 May 2023. 2023, 3, 196–200.
32. Zhang L., Chen Z H, Cui W, et al. WiFi-based indoor robot positioning using deep fuzzy forests. *IEEE Internet of Things Journal*, 2020, 7(11): 10773–10781.
33. Cui W, Zhang L, Li B, et al. Semi-supervised deep adversarial forest for cross-environment localization. *IEEE Transactions on Vehicular Technology*, 2022, 71(9): 10215–10219.
34. Zhou Z H, Feng J. Deep forest. *National Science Review*, 2019, 6(1): 74–86.
35. Jia Z, Liu Z, Gan Y. A deep forest based fault diagnosis scheme for electronics-rich analog circuit systems. *IEEE Transactions on Industrial Electronics*, 2021, 68(10): 10087–10096.
36. Zhi Y J, Yang T, Fu D M. An improved deep forest model for forecast the outdoor atmospheric corrosion rate of low-alloy steels. *Journal of Materials Science & Technology*, 2020, 49: 202–210.
37. Sun L, Mo Z H, Yan F H, et al. Adaptive feature selection guided deep forest for covid-19 classification with chest CT. *IEEE Journal of Biomedical and Health Informatics*, 2020, 24(10): 2798–2805.
38. Liu B, Guo W Y, Chen X, et al. Morphological attribute profile cube and deep random forest for small sample classification of hyperspectral image. *IEEE Access*, 2020, 8: 117096–117108.
39. Huang J, Liu H, Wang Q, Cai W W. Limited-projection volumetric tomography for time-resolved turbulent combustion diagnostics via deep learning. *Aerospace Science and Technology*, 2020, 106: 106123.
40. Maged A, Nour M. Prediction of combustion pressure with deep learning using flame images. *Fuel*, 2025, 380: 133203.
41. Pan X, Tang J, Xia H. Flame combustion state identification based on CNN in municipal solid waste incineration process. In: *2023 5th International Conference on Industrial Artificial Intelligence (IAI)*, Shenyang, China, 21–24 August 2023. 2023, 1–4.
42. Lyu Z, Jia X, Yang Y, et al. A comprehensive investigation of LSTM-CNN deep learning model for fast detection of combustion instability. *Fuel*, 2021, 303: 121300.
43. Yu Y, Han Z, Xu C. NO_x concentration prediction based on deep convolution neural network and support vector machine. *Proceedings of the CSEE*, 2022, 42(1): 238–248.

44. Qiu T, Liu M, Zhou G, et al. An unsupervised classification method for flame image of pulverized coal combustion based on convolutional auto-encoder and hidden markov model. *Energies*, 2019, 12(13): 2585.
45. Han Z, Huang Y, Li J, Zhang B, et al. A hybrid deep neural network based prediction of 300 MW coalfired boiler combustion operation condition. *Science China Technological Sciences*, 2021, 64(10): 2300–2311.
46. Han Z, Xie Y, Hossain M M, Xu C. A hybrid deep neural network model for NO_x emission prediction of heavy oil-fired boiler flames. *Fuel*, 2023, 333(2): 126419.
47. Wang D, Zhang X, Chen H, et al. A sintering state recognition framework to integrate prior knowledge and hidden information considering class imbalance. *IEEE Transactions on Industrial Electronics*, 2021, 68(8): 7400–7411.
48. Han Z, Li J, Zhang B, et al. Prediction of combustion state through a semi-supervised learning model and flame imaging. *Fuel*, 2021, 289: 119745.
49. Lv Y, Qi X, Zheng X, et al. Unsupervised quantitative judgment of furnace combustion state with CBAM-SCAE-based flame feature extraction. *Journal of the Energy Institute*, 2024, 116: 101733.
50. Zhu Q, Wang Y, Li W, et al. Advancing multi-modal beam prediction with cross-modal feature enhancement and dynamic fusion mechanism. *IEEE Transactions on Communications*, 2025, 73(9): 7931–7940.
51. Xu P, Zhu X T, Clifton D A. Multimodal learning with transformers: a survey. *IEEE Transactions on Pattern Analysis and Machine Intelligence*, 2023, 45(10): 12113–12132.
52. Akbari H, Yuan L Z, Qian R, et al. VATT: transformers for multimodal self-supervised learning from raw video, audio and text. *Advances in Neural Information Processing Systems*, 2021, 34: 24206–24221.
53. Lin B, Jorgensen S B. Soft sensor design by multivariate fusion of image features and process measurements. *Journal of Process Control*, 2011, 21(4): 547–553.
54. Tóth P, Garami A, Csordás B. Image-based deep neural network prediction of the heat output of a step-grate biomass boiler. *Applied Energy*, 2017, 200: 155–169.
55. Zhou P, Gao B H, Wang S, Chai T Y. Identification of abnormal conditions for fused magnesium melting process based on deep learning and multisource information fusion. *IEEE Transactions on Industrial Electronics*, 2022, 69(3): 3017–3026.
56. Pan X, Tang J, Xia H, Wang T. Online combustion status recognition of municipal solid waste incineration process using DFC based on convolutional multi-layer feature fusion. *Sustainability*, 2023, 15(23): 16473.
57. Zhu J L, Jia M W, Zhang Y, et al. Transductive transfer broad learning for cross-domain information exploration and multigrade soft sensor application. *Chemometrics and Intelligent Laboratory Systems*, 2023, 235: 1–9.
58. Li W, Ding C, Qiao J. Robust neural network modeling with small-worldness for effluent total phosphorus prediction in wastewater treatment process. *IEEE Transactions on Reliability*, 2024, 74(1): 2473–2486.
59. Yan A J, Guo J C, Wang D H. Robust stochastic configuration networks for industrial data modelling with Student's-t mixture distribution. *Information Sciences*, 2022, 607: 493–505.

10 AI-Based Tree Modeling for Multi-Point Dioxin Concentrations

10.1 RELATED WORKS

10.1.1 Multi-Point Dioxin Concentrations Modeling Description

The rapid development of MSWI plants has raised concerns about exhaust emissions contributing to secondary environmental pollution, particularly DXN emissions [1, 2]. DXN is a byproduct of the MSWI process and consists of unintentionally produced persistent organic pollutants that can adversely affect both the surrounding environment and human health [3]. DXNs are a group of highly toxic organic compounds that present significant risks to both human health and the environment. These persistent pollutants can cause a variety of health issues, including carcinogenicity, endocrine disruption, immune system impairment, and neurological damage. Their environmental impacts are equally concerning, leading to ecosystem pollution, bioaccumulation, and disruption of ecological balance. While advancements in air pollution control technologies have significantly reduced pollutant emissions from the MSWI process [4, 5], it remains a primary source of DXN emissions [6, 7]. Consequently, the discharge of DXNs from the MSWI process warrants increased global attention. Due to the inherent delays associated with physical detection techniques, controlling and minimizing DXN emissions from existing MSWI processes presents significant challenges [8]. As a result, online measurements of DXN concentrations are essential for reconciling the need for sustainable development of MSWI with the negative environmental impacts of DXN emissions [9].

Artificial intelligence (AI)-based data-driven modeling algorithms can effectively facilitate real-time measurement of DXN concentrations and support its reduction control [10, 11]. Two distinct categories of methods can be employed for online DXN concentration measurements in the MSWI process: AI-based physical indirect online detection [12, 13] and AI-based soft measurement.

Both approaches utilize AI models to correlate easily measurable process variables with DXN emission concentrations. The primary distinction between these two methods lies in their input features. In AI-based physical indirect online detection methods, some researchers focus on using organic compounds, such as chlorobenzene and chlorophenols, as input features [14]. In their respective studies, Cao et al. employed 1,2,4-trichlorobenzenes as input features and utilized a linear model to measure DXN concentration in a 400 tons/d MSWI plant [15]. Following this, Xiong et al. also used 1,2,4-trichlorobenzenes as input features and developed a least-squares mapping

DOI: 10.1201/9781003634195-13

model for a 750 t/d full-scale grate furnace [16]. The results of these experiments indicate that measuring trace amounts of 1,2,4-trichlorobenzenes in the stack gas using TD-GC-REMPI-TOFMS serves as an effective indicator for DXN concentrations. Additionally, other compounds such as 1,3-dichlorobenzene, 1,4-dichlorobenzene, 1,2-dichlorobenzene, 1,3,5-trichlorobenzene, 1,2,4-trichlorobenzene, and 1,2,3-trichlorobenzene were employed as input variables in a bagging-based ensemble neural network mapping model specifically designed for four MSWI plants.

Furthermore, Xiong et al. adopted 1,2,4-trichlorobenzene, carbon monoxide, sulfur dioxide, oxynitride, particulate matter, fluoride, and hydrogen chloride as input features to develop a Bayesian support vector regression mapping model for detecting DXN emission concentrations [17]. The results from using multiple inputs demonstrate that machine learning models outperform the linear models proposed in previous studies. These research efforts validate the effectiveness of indirectly mapping DXN emission concentrations by utilizing indicators present in flue gas. As a result, AI-based physical indirect online detection methods can be employed to evaluate trends in DXN emission concentrations online. However, these indicators require real-time online detection equipment, which not only increases testing costs but also limits the testing environment.

AI-based soft measurement technology utilizes process data from the distributed control system (DCS) as input features to map DXN emission concentrations, distinguishing it from the first detection method. A soft measuring model for DXN concentration, constructed using readily measurable process data, facilitates efficient and real-time detection of DXN emissions. Current research on DXN emission models encompasses various approaches, including linear regression analysis, classical machine learning, and deep learning. The linear regression analysis approach focuses on elucidating the relationship between process variables and DXN emission concentrations [18]. In contrast, classical machine learning methods address the limitations of linear regression in nonlinear systems. Prominent techniques in this category include artificial neural networks [19], support vector machines [20], and decision trees (DT) [21].

In recent years, there has been significant progress in modeling DXN emission concentrations using deep learning approaches. These studies include methods such as deep belief networks [22], deep forest regression (DFR) [23], and improved DFR [24]. The findings from these studies demonstrate the feasibility of constructing a DXN emission concentration mapping model that utilizes process data as inputs. The numerical analyses have confirmed that AI-based soft measurement technology is effective for measuring DXN emission concentrations, with deep learning models outperforming both traditional machine learning and linear models. However, current AI-based modeling approaches primarily focus on DXN emission concentrations at the stack level, often overlooking DXN concentrations during the generation and adsorption stages in their model designs.

As mentioned above, it can be concluded that the AI-based process data-driven modeling method offers advantages such as low cost and ease of development. However, existing research has not yet developed models in terms of multi-point DXN concentration that incorporate the characteristics of the MSWI process along with the mechanisms of DXN generation, adsorption, and emission.

10.1.2 DXN Concentration Distribution

The studied MSWI process consists of several stages, including feeding, incineration, heat exchange, gas cleaning, and emission, as illustrated in Figure 10.1.

The depicted system in Figure 10.1 utilizes a typical grate-type incinerator with a daily capacity of 800 tons of municipal solid waste (MSW). The flue gas cleaning stage is equipped with advanced air pollution control devices (APCDs), including selective non-catalytic reduction, activated carbon injection, semi-dry deacidification, and a bag filter system.

The multiple stages of the MSWI process exhibit a clear cascade relationship with a back-and-forth progression. The raw MSW, initially containing trace amounts of dioxins (approximately 0.8 ng TEQ/kg [5]), is collected and stored for about one week to undergo fermentation, which enhances its calorific value. Subsequently, the MSW is fed into the furnace via a feeder and subjected to a series of processes, including drying, combustion, and burnout, transforming it into high-temperature flue gas that exceeds 850°C. At this elevated temperature, the decomposition of dioxins can be completed. The process then enters the heat recovery stage, where the high-temperature flue gas must be cooled down to 200°C. Numerous studies have identified three pathways for DXN generation during this stage: homogeneous synthesis (500°C–800°C) [25, 26], precursor synthesis (200°C–400°C) [27, 28], and de novo synthesis (200°C–400°C) [29, 30]. Consequently, the DXN generation point

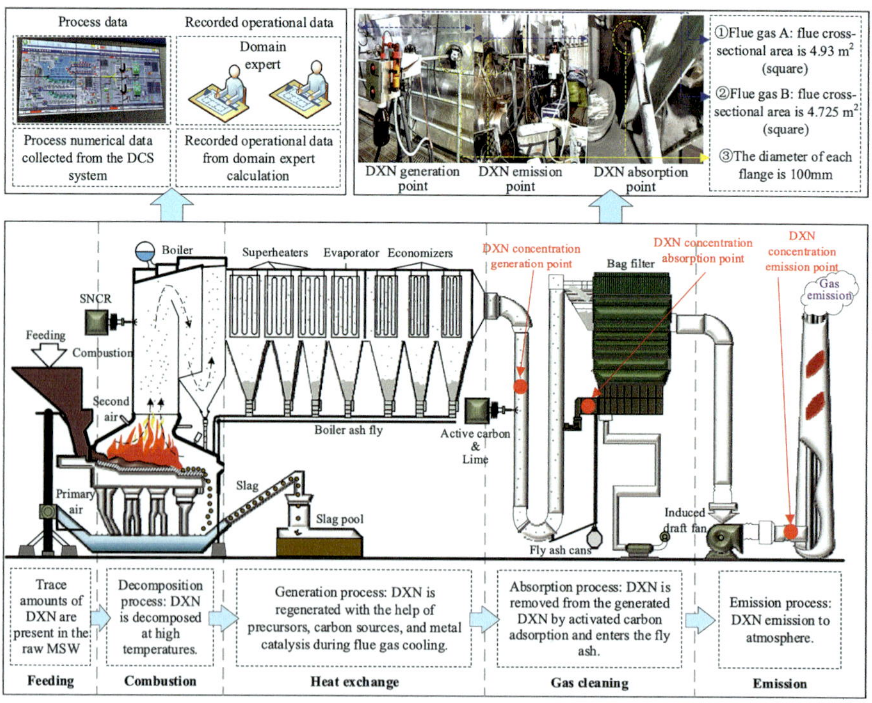

FIGURE 10.1 MSWI process with DXN generation, adsorption, and emission phases.

for online sampling and offline analysis is identified as the outlet of the economizer, as illustrated in Figure 10.1.

Next, the cooled flue gas is directed to the gas cleaning stage, where APCDs are employed to eliminate toxic and hazardous substances from the flue gas. The regenerated DXN is adsorbed by the activated carbon present in the fly ash, which is subsequently separated from the flue gas using a bag filter. Therefore, the DXN adsorption point for online sampling and offline analysis is located at the bottom of the bag filter, as illustrated in Figure 10.1.

After the flue gas cleaning process, the exhaust gas is discharged from the stack by the induced draft fan. Consequently, the outlet of the induced draft fan is considered the DXN emission point for both online sampling and offline analysis, serving as a regular location for DXN emission detection.

10.1.3 DXN Detection Data (Model Output)

To obtain a comprehensive understanding of DXN concentration throughout the MSWI process, detection samples were collected and analyzed between November 2020 and January 2022 using laboratory assays in accordance with the National Environmental Protection Standard (HJ 77.2-2008) of China. As illustrated in Figure 10.1, isokinetic tracking sampling was employed to sample flue gas at both the DXN generation and emission points, with each sampling session lasting over 2 hours.

During this 2-hour period, fly ash sampling was conducted using an equal time interval strategy to ensure the real-time collection of fly ash samples. The volume of each flue gas sample collected was approximately 3–4 m^3, while the weight of each fly ash sample was around 100 g. In total, 47 sample sets were collected from the studied MSWI power plant, comprising 47 generation samples, 47 emission samples, and 16 adsorption samples. The distribution of sampling dates and times is depicted in Figure 10.2.

Subsequently, the quartz filters, XAD-2 resin, and condensate water from the samples are combined for sample extraction and purification. During the preprocessing operation, 13C12-labeled compounds are used as internal standards for the samples.

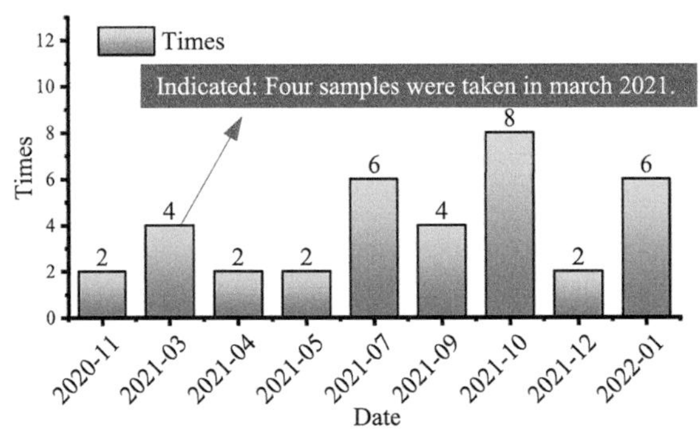

FIGURE 10.2 DXN sample sampling date and times.

In the instrument analysis phase, the temperature programming for high-resolution gas chromatography coupled with high-resolution mass spectrometry (HRGC/HRMS) is as follows: (1) the column temperature is initially maintained at 150°C for 3 minutes; (2) it is increased to 230°C at a rate of 20°C/min and held for 18 minutes; (3) then, the temperature is raised to 235°C at a rate of 5°C/min and maintained for 10 minutes; (4) finally, the column temperature is increased to 320°C at a rate of 4°C/min and held for 3 minutes.

The DXN concentrations measured using HRGC/HRMS equipment require further transformation to obtain the toxic equivalent quantity (TEQ) concentration under dry air conditions at 101.325 kPa and 237 K (with 11% oxygen). Based on our sampling strategy, it is noted that the time scale of the DXN concentration data represents a 2-hour average emission concentration during both the generation and emission phases, while a 4-hour average is used for the emission concentration in the adsorption phase. The distribution of the TEQ concentrations of DXN at the generation, adsorption, and emission sampling points is illustrated in Figure 10.3.

Figure 10.3 illustrates the following observations: (1) The DXN data from the generation phase (sample size of 36), adsorption phase (sample size of 16), and emission phase (sample size of 36) are characterized by relatively small sample sizes; (2) The number of DXN samples in the adsorption phase can be considered an ultra-small sample size when compared to the generation and emission phases; (3) A review of the DXN concentration distributions reveals that the adsorbed DXN data exhibit high variability. This variability indicates more uncertainty in the data during the adsorption phase.

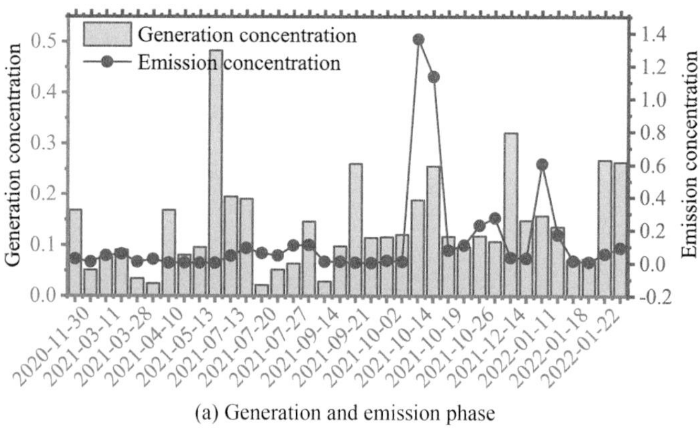

FIGURE 10.3 Distribution of the DXN concentration at generation, adsorption, and emission phases. (a) Generation and emission phase.

(Continued)

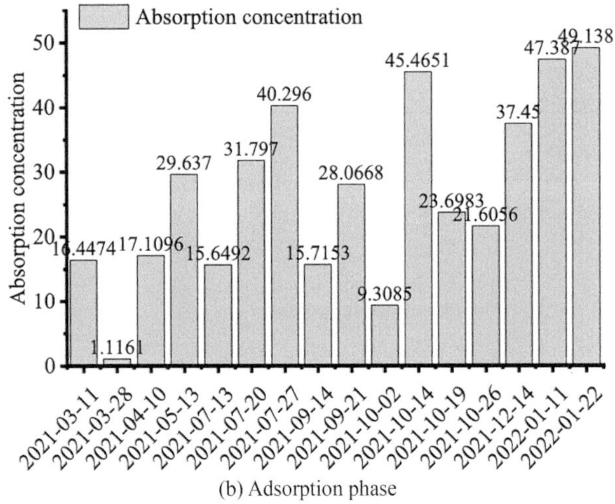

FIGURE 10.3 (Continued) (b) Adsorption phase.

10.1.4 Operational Data (Model Input)

The operational data generated by the MSWI process can be categorized into two types: process data (second-level scale) and operational data (ORD, day-level scale), as illustrated in Figure 10.1. Considering the actual sampling points for DXN generation, adsorption, and emissions, the operational data can be segmented based on the following guidelines:

(1) DXN Generation Phase: This phase includes data from two modes: process data and ORD. The process variables in this phase correspond to the feeding, combustion, and hcat exchange stages, with specific information on these variables provided in Table 10.1. The ORD comprises feed rate, incineration efficiency, average load, steam production per ton of MSW, and calorific value of MSW;
(2) DXN Adsorption Phase: The data used during this phase encompass both process data and DXN concentrations from the generation phase. The process variables primarily focus on activated carbon and lime, with detailed information on these variables presented in Table 10.1.
(3) DXN Emission Phase: In this phase, the input data consist of process variables and the DXN concentration from the generation and adsorption phases. The gas cleaning and emission stages supply the numerical process variables for this phase, with specific details provided in Tables 10.1 and 10.2.

In Tables 10.1 and 10.2, the dimensions of the process numerical variables at different phases are 119, 5, and 34, respectively. Based on the available data, it is evident that the input variables in the generation phase exhibit significantly high-dimensional

TABLE 10.1
Process Variables of Different Phases in Terms of DXN Generation, Adsorption, and Emissions

Phase	Number	Variables	Number	Variables
Generation	1	1st furnace gas temperature	62	2nd superheater outlet steam temperature
	2	2nd furnace gas temperature	64	3rd superheater outlet steam temperature
	3	3rd furnace gas temperature	65	Economizer outlet water temperature
	4	4th furnace gas temperature	66	1st dry grate airflow
	5	5th furnace gas temperature	67	2nd dry grate airflow
	6	6th furnace gas temperature	68	3rd dry grate airflow
	7	7th furnace gas temperature	69	4th dry grate airflow
	8	8th furnace gas temperature	70	1st burn1 grate airflow
	9	9th furnace gas temperature	71	2nd burn1 grate airflow
	10	1st top of burnout grate temperature	72	3rd burn1 grate airflow
	11	2nd top of burnout grate temperature	73	4th burn1 grate airflow
	12	Right temperature of the furnace	74	1st burn2 grate airflow
	13	Left temperature of the furnace	75	2nd burn2 grate airflow
	14	Average temperature of the furnace	76	3rd burn2 grate airflow
	15	1st dry grate temperature	77	4th burn2 grate airflow
	16	2nd dry grate temperature	78	1st burnout grate airflow
	17	3rd dry grate temperature	79	2nd burnout grate airflow
	18	4th dry grate temperature	80	1st secondary airflow
	19	1st dry and burn1 grate temperature	81	2nd secondary airflow
	20	2nd dry and burn1 grate temperature	82	Total economizer feedwater
	21	3rd dry and burn1 grate temperature	83	1st economizer feedwater
	22	4th dry and burn1 grate temperature	84	2nd economizer feedwater
	23	1st burn1 grate temperature	85	1st superheater cooling water
	24	2nd burn1 grate temperature	86	2nd superheater cooling water
	25	3rd burn1 grate temperature	87	Boiler steam flow
	26	4th burn1 grate temperature	88	Flue gas flow
	27	5th burn1 grate temperature	89	1st urea flow
	28	6th burn1 grate temperature	90	2nd urea flow
	29	7th burn1 grate temperature	91	Primary fan pressure
	30	8th burn1 grate temperature	92	Secondary fan pressure

(*Continued*)

TABLE 10.1 (Continued)
Process Variables of Different Phases in Terms of DXN Generation, Adsorption, and Emissions

Phase	Number	Variables	Number	Variables
	31	1st burn2 grate temperature	93	Negative pressure in furnace
	32	2nd burn2 grate temperature	94	Economizer outlet gas pressure
	33	3rd burn2 grate temperature	95	Boiler Tank Pressure
	34	4th burn2 grate temperature	96	3-stage superheater outlet steam pressure
	35	5th burn2 grate temperature	97	1st feed grate speed
	36	6th burn2 grate temperature	98	2nd feed grate speed
	37	7th burn2 grate temperature	99	3rd feed grate speed
	38	8th burn2 grate temperature	100	4th feed grate speed
	39	Left slag outlet temperature	101	1st dry grate speed
	40	Right slag outlet temperature	102	2nd dry grate speed
	41	1st first channel temperature	103	3rd dry grate speed
	42	2nd first channel temperature	104	4th dry grate speed
	43	1st protection tube temperature	105	1st burn1 grate speed
	44	2nd protection tube temperature	106	2nd burn1 grate speed
	45	1st 3-stage superheater temperature	107	3rd burn1 grate speed
	46	2nd 2-stage superheater temperature	108	4th burn1 grate speed
	47	1st evaporators temperature	109	1st burn2 grate speed
	48	2nd evaporators temperature	110	2nd burn2 grate speed
	49	1st economizer temperature	111	3rd burn2 grate speed
	50	2nd economizer temperature	112	4th burn2 grate speed
	51	Economizer outlet gas temperature	113	1st burnout grate speed
	52	1st primary airflow temperature	114	2nd burnout grate speed
	53	2nd primary airflow temperature	115	Primary fan frequency
	54	3rd primary airflow temperature	116	Secondary fan frequency
	55	Secondary airflow temperature	117	Boiler water level
	56	1st wall cooling air outlet temperature	118	1st oxygen concentration at economizer
	57	2nd wall cooling air outlet temperature	119	2nd oxygen concentration at economizer
	58	1st grate cooling air outlet temperature	120	Feed rate (ORD)

(*Continued*)

TABLE 10.1 (Continued)
Process Variables of Different Phases in Terms of DXN Generation, Adsorption, and Emissions

Phase	Number	Variables	Number	Variables
	59	2nd grate cooling air outlet temperature	121	Incineration efficiency (ORD)
	60	Dry grate outlet air temperature	122	Average load (ORD)
	61	Burn grate outlet air temperature	123	Steam production per ton of MSW (ORD)
	61	1st superheater outlet steam temperature	124	Calorific value of MSW(ORD)
Adsorption	88	Flue gas flow	127	Lime feeder speed
	125	1st mixer feedwater flow	128	Active carbon feeder speed
	126	2nd mixer feedwater flow		
Emission	88	Flue gas flow	145	Gas cleaning outlet gas pressure
	129	1st inlet flue gas temperature	146	1st reactor differential pressure
	130	2nd inlet flue gas temperature	147	2nd reactor differential pressure
	131	3rd inlet flue gas temperature	148	3rd reactor differential pressure
	132	1st bag filter inlet temperature	149	4th reactor differential pressure
	133	2nd bag filter inlet temperature	150	1st bag filter differential pressure
	134	Fluidizing fan heater temperature	151	2nd bag filter differential pressure
	135	1st outlet flue gas temperature	152	Induced draft fan speed
	136	2nd outlet flue gas temperature	153	Oxygen concentration of emissions
	137	3rd outlet flue gas temperature	154	Dust concentration of emissions
	138	4th outlet flue gas temperature	155	NO_x concentration of emissions
	139	5th outlet flue gas temperature	156	SO_2 concentration of emissions
	140	6th outlet flue gas temperature	157	HCl concentration of emissions
	141	Induced draft fan outlet gas temperature	158	CO concentration of emissions
	142	1st reactor inlet flue gas flow	159	CO_2 concentration of emissions
	143	2nd reactor inlet flue gas flow	160	HF concentration of emissions
	144	Induced draft fan outlet gas pressure	161	Light permeability

TABLE 10.2
Data Characteristics of Different Phases

Phase	Types	Dimension	Data Characteristics	Physical Property
Generation	Process data	119	High-dimensional	Sensor measurements: Temperature, flow, pressure, level, concentration, etc. Operational parameters: Airflow, speed, frequency
	ORD	5	—	—
Adsorption	Process data	5	Low-dimensional	Sensor measurements: Flow Operational parameters: Flow and speed
Emission	Process data	34	High-dimensional	Sensor measurements: Flow, temperature, pressure, and concentration Operational parameters: Speed

properties, while those in the adsorption phase demonstrate low-dimensional properties, and the variables in the emission phase show medium-dimensional characteristics. Consequently, different AI-based modeling algorithms should be employed to address these varying characteristics.

10.2 MODELING STRATEGY

Given the data characteristics of DXN in the generation, adsorption, and emission phases, this chapter introduces a comprehensive modeling strategy known as the whole-process tree-based model (WPTM). The WPTM method consists of several key modules, including data preprocessing, DXN generation model construction, DXN adsorption model construction, and DXN emission model construction. In the structure of WPTM, the soft measuring outputs from each phase are cascaded to achieve a holistic model for the whole lifecycle of DXN concentration, as illustrated in Figure 10.4.

The functions of the four modules of the WPTM are described as follows:

(1) Data Preprocessing Module: The collected data consists of process data and ORD data. The data is sliced according to the detection points for the generation, adsorption, and emission phases and is preprocessed based on the time scale of the online sampling of DXN data;
(2) DXN Generation Model Construction Module: A novel method known as ADFR-clfc is proposed for the DXN generation model. This method is specifically designed for handling high-dimensional, small-sample process data;
(3) DXN Adsorption Model Construction Module: Addressing ultra-small samples and data uncertainty are critical challenges for the adsorption model. To tackle the issue of ultra-small samples, a latent variable virtual sample

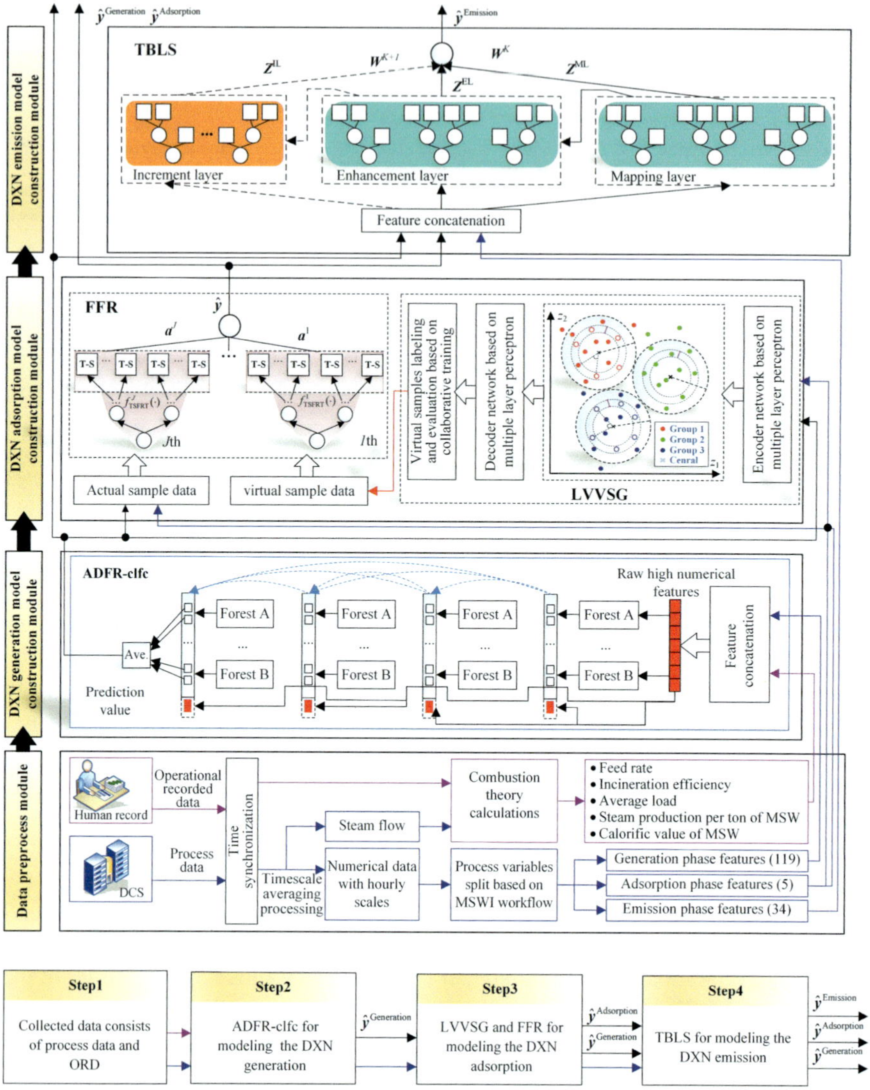

FIGURE 10.4 WPTM strategy of DXN generation, adsorption, and emission.

generation (LVVSG) method is employed. Additionally, a fuzzy learning method, referred to as Fuzzy forest regression (FFR), is utilized for modeling DXN adsorption.

(4) DXN Emission Model Construction Module: The tree broad learning system (TBLS) method is employed to rapidly and accurately learn from medium-dimensional, small sample size process data, using the DXN generation concentration, DXN adsorption concentration, and process numerical variables as inputs.

The key contributions of our work can be summarized as follows:

(1) Introduction of WPTM Method: The WPTM, grounded in data-driven principles, is introduced as the first approach to model the mechanistic processes of DXN concentration. To the best of our knowledge, this is the first AI-based model designed for soft measurement of the whole lifecycle of DXN concentration across multiple phases, including generation, adsorption, and emission;
(2) Detailed Delineation of DXN Processes: This study focuses on the MSWI process, providing an in-depth analysis of the processes of DXN generation, adsorption, and emission. The determination of data characteristics at different phases is based on a comprehensive understanding of the DXN mechanism, offering a novel framework for DXN modeling and future optimization for its reduction;
(3) Application of Advanced Modeling Techniques: State-of-the-art tree-based models, including adaptive DFR based on cross-layer full connection (ADFR-clfc), TBLS, and FFR, along with supporting modeling technologies, are employed to address diverse data characteristics. These characteristics encompass high-dimensional small samples, low-dimensional ultra-small samples, and medium-dimensional small samples across the various phases related to DXN;
(4) Validation of the WPTM Model: The proposed WPTM model is validated using nearly one year of authentic data pertaining to DXN generation, adsorption, and emission phases. The case study conclusively demonstrates that WPTM provides a robust and comprehensive modeling framework for DXN throughout the whole MSWI process.

The rationale for selecting specific machine learning techniques (e.g., ADFR-clfc, FFR, and TBLS) for the comprehensive modeling strategy known as the WPTM is based on the distinct characteristics of the modeling data across the three phases:

(1) Generation phase exhibits high-dimensional features with a limited number of samples. To address this, Tang et al. proposed the DFR-clfc [31]. Experimental results show that DFR-clfc performs exceptionally well when dealing with small sample sizes that have high-dimensional features. Consequently, this chapter proposes ADFR-clfc, an extension of the DFR-clfc method, for modeling tasks in the DXN generation phase;
(2) Adsorption phase presents low-dimensional features, ultra-small sample sizes, and inherent uncertainty. Xia et al. introduced a Takagi–Sugeno fuzzy regression trees ensemble approach for modeling complex industrial processes [32]. Their findings indicate that Takagi–Sugeno fuzzy regression trees provide superior representational performance compared to classification DT when handling uncertain data. Therefore, an ensemble approach utilizing FFR is implemented within the WPTM framework to effectively model the adsorption phase;

(3) Emission phase features medium-dimensional data with a limited number of samples. Recognizing the strengths of ensemble learning and broad learning systems (BLS), Xia et al. developed a TBLS specifically for modeling small datasets [33]. Given its demonstrated advantages with medium-sized datasets, the TBLS approach is employed for modeling the emission phase.

By tailoring the modeling algorithms to the specific data characteristics of each stage, the WPTM effectively addresses the challenges associated with DXN concentration modeling throughout the whole process.

10.3 ALGORITHM IMPLEMENTATION

10.3.1 Data Preprocess Module

The methodology of data synchronization is illustrated in Figure 10.5.

Figure 10.5 provides a comprehensive visual representation of the data synchronization process. Different data types operate on distinct time scales: DXN concentration is measured on an hourly basis, ORD is tracked daily, and process data is recorded every second. The synchronization of these data types aligns with the DXN sampling time; for instance, the two-hour average concentration of DXN corresponds to two hours of process data and one set of ORD. The details of this synchronization are outlined as follows.

One set of ORD data comprises the following parameters: feed rate (unit: tons/h), incineration efficiency (unit: %), average load (unit: tons/h), steam production per ton of MSW (unit: tons/h), and calorific value of MSW (unit: kJ). These values are calculated as follows:

$$x_{FR} = \frac{Q_2}{24} \tag{10.1}$$

$$x_{IE} = \frac{Q_2}{Q_1} \tag{10.2}$$

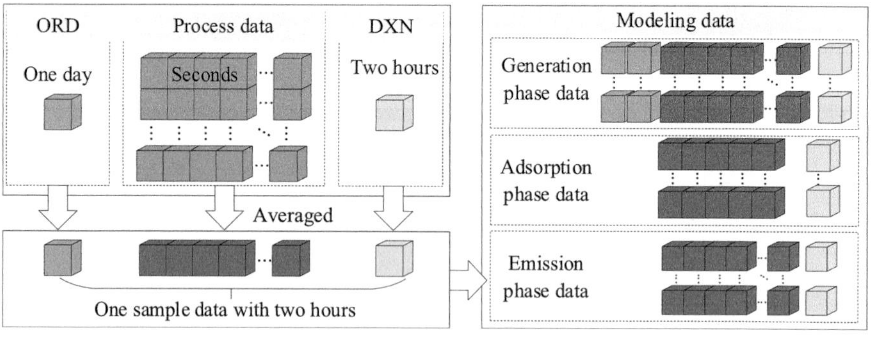

FIGURE 10.5 Methodology of data synchronization of the proposed method.

$$x_{\mathrm{AL}} = \frac{1}{T}\sum_{t=1}^{T} x_{87} \tag{10.3}$$

$$x_{\mathrm{SP}} = \frac{x_{\mathrm{AL}} \times 24}{Q_2} \tag{10.4}$$

$$x_{\mathrm{CV}} = \frac{x_{\mathrm{SP}} \times \left(H_{\mathrm{Steam}} - H_{\mathrm{Water}}\right)}{\eta_{\mathrm{Boiler}} \times 4.186} \tag{10.5}$$

where x_{FR}, x_{IE}, x_{AL}, x_{SP}, and x_{CV} are the feed rate, incineration efficiency, average load, steam production per ton of MSW, and calorific value of MSW, respectively; Q_1 represents the rated capacity of the MSWI process; Q_2 is the actual volume handled in one day; x_{87} indicates the steam flow; H_{Steam} denotes the steam enthalpy value; H_{Water} expresses the water enthalpy value; and η_{Boiler} is boiler thermal efficiency.

With the above calculations, a valid ORD can be obtained, denoted as

$$\boldsymbol{X}_{\mathrm{ORD}} = \left\{\boldsymbol{x}_{\mathrm{FR}}, \boldsymbol{x}_{\mathrm{IE}}, \boldsymbol{x}_{\mathrm{AL}}, \boldsymbol{x}_{\mathrm{SP}}, \boldsymbol{x}_{\mathrm{CV}}\right\} \in R^{N_G \times M_{\mathrm{ORD}}} \tag{10.6}$$

where N_{G} is the number of samples and M_{ORD} represents the dimensions of variables.

The processing of process data consists of two parts, i.e., data averaging and variable slicing. The specific method of data averaging is as follows:

$$x_i = \frac{1}{T}\sum_{t=1}^{T} x_i^t \tag{10.7}$$

where x_i is the ith process variables, and T indicates the time scale.

After data averaging, the processed data can be obtained, denoted as

$$\boldsymbol{X}_{\mathrm{PND}} = \left\{\boldsymbol{x}_i\right\}_{i=1}^{N_G} \in R^{N_G \times M_{\mathrm{PND}}} \tag{10.8}$$

where M_{PND} represents the dimensions of process variables.

The three key points in the processes of DXN generation, adsorption, and emission are identified as the economizer outlet, the bottom of the bag filter, and the induced draft fan outlet, respectively. Correspondingly, process variables are categorized into three distinct sets, as detailed in Table 10.1. As a result, the input feature dimensions for the generation, adsorption, and emission models are 124, 5, and 34, respectively. These three datasets are preprocessed using standardized methods to produce DXN generation process data, DXN adsorption process data, and DXN emission process data, as follows:

$$\boldsymbol{X}_{\mathrm{GPND}} = \left\{\boldsymbol{x}_G^i\right\}_{i=1}^{N_G} \in R^{N_G \times M_{\mathrm{GPND}}} \tag{10.9}$$

$$\boldsymbol{X}_{\text{APND}} = \left\{ \boldsymbol{x}_A^i \right\}_{i=1}^{N_A} \in R^{N_A \times M_{\text{APND}}} \tag{10.10}$$

$$\boldsymbol{X}_{\text{EPND}} = \left\{ \boldsymbol{x}_E^i \right\}_{i=1}^{N_G} \in R^{N_G \times M_{\text{EPND}}} \tag{10.11}$$

In summary, data preprocessing serves several crucial functions:

(1) Time Scale Unification: Preprocessing aligns the time scales of process data and ORD data, facilitating the acquisition of synchronized modeling input information;
(2) Feature Clarification and Alignment: Preprocessing clarifies the input feature information for each modeling stage and aligns the time scales using averaging techniques;
(3) Dimensionality Reduction: Preprocessing addresses dimensional differences in characteristic information, ultimately enhancing model accuracy.

10.3.2 DXN Generation Model Construction Module

In the DXN generation model construction module, the input data $\boldsymbol{X}_{\text{G1}}$ includes the ORD dataset and the DXN generation process data, while the output data $\boldsymbol{y}_G$ consists of the DXN generation concentration.

The ADFR-clfc model consists of three components: the layer forest model, cross-layer full connection, and layer adaptation. Each layer forest model is composed of a random forest (RF) and a completely random forest (CRF), with each forest approach consisting of multiple DT units.

The ADFR-clfc model comprises three components: the layer forest model, cross-layer full connection, and layer adaptive. Each layer forest model consists of RF, CRF, and each forest approach comprises multiple DT units.

The training dataset is denoted as $\boldsymbol{D}_{\text{G1}} = \left\{ \boldsymbol{X}_{\text{G1}}, \boldsymbol{y}_G \right\}$, and by taking the k_{G1}th layer as an example, and the modeling process is described as follows.

The production process of the input representation data $\boldsymbol{D}_{\text{G1}}^{k_{\text{G1}}}$ of the k_{G1}th layer is denoted as,

$$\left.\begin{array}{l} \boldsymbol{X}_{\text{G1}} = \left\{ \boldsymbol{x}_{\text{G1}}^i \right\}_{i=1}^{N_G} \in R^{N_G \times M_{\text{G1}}} \\ \left\{ \boldsymbol{H}_1^{\text{Map}}, \boldsymbol{H}_2^{\text{Map}}, \ldots, \boldsymbol{H}_{k_{\text{G1}}-1}^{\text{Map}} \right\} \\ N_G \\ M_{\text{G1}} + \sum_{k=2}^{k_{\text{G1}}} J_{\text{G1,Forest}} \times \left(k_{\text{G1}} - 1 \right) \\ \boldsymbol{y}_G \end{array}\right\} \Rightarrow \boldsymbol{D}_{\text{G1}}^{k_{\text{G1}}} = \left\{ \boldsymbol{x}_{\text{G1}}^i, \ldots, \boldsymbol{h}_{k_{\text{G1}}}^{\text{Map}}, \ldots, y_{\text{G1}}^i \right\}_{i=1}^{N_G} \in R^{N_G \times \left(M_{\text{G1}}^{k_{\text{G1}}} + 1 \right)} \tag{10.12}$$

where $\boldsymbol{H}_{k_{\mathrm{G1}}-1}^{\mathrm{Map}}$ represents the output of the $(k_{\mathrm{G1}}-1)$th layer; $J_{\mathrm{G1,Forest}}$ is the number of the forest model; $M_{\mathrm{G1}}+\sum_{k=2}^{k_{\mathrm{G1}}} J_{\mathrm{G1,Forest}}\times(k_{\mathrm{G1}}-1)$ is the dimension of the input feature; and its dimensions are represented as follows:

$$\boldsymbol{D}_{\mathrm{G1}}^{k_{\mathrm{G1}}}=\left\{\boldsymbol{x}_{\mathrm{G1}}^{i},y_{\mathrm{G1}}^{i}\right\}_{i=1}^{N_G}\in R^{N_G\times\left(M_{\mathrm{G1}}^{k_{\mathrm{G1}}}+1\right)} \tag{10.13}$$

$$M_{\mathrm{G1}}^{k_{\mathrm{G1}}}=M_{\mathrm{G1}}+\sum_{k=2}^{k_{\mathrm{G1}}} J_{\mathrm{G1,Forest}}\times(k_{\mathrm{G1}}-1) \tag{10.14}$$

The $\boldsymbol{D}_{\mathrm{G1}}^{k_{\mathrm{G1}}}$ is preprocessed using the bootstrap and random subspace method, it can be denoted as,

$$\left\{\boldsymbol{D}_{\mathrm{G1}}^{k_{\mathrm{G1}},j_{\mathrm{G1,DT}}}\right\}_{j_{\mathrm{G1,DT}}=1}^{J_{\mathrm{G1,DT}}}=f_{\mathrm{RSM}}\left(f_{\mathrm{Bootstrap}}\left(\boldsymbol{D}_{\mathrm{G1}}^{k_{\mathrm{G1}}},N_G,J_{\mathrm{G1,DT}}\right),M_{\mathrm{RSM}}\right) \tag{10.15}$$

where $f_{\mathrm{Bootstrap}}(\cdot)$ is the bootstrap sampling function; $J_{\mathrm{G1,DT}}$ expresses the times of bootstrap; $f_{\mathrm{RSM}}(\cdot)$ represents the feature sampling function; $\boldsymbol{D}_{\mathrm{G1}}^{k_{\mathrm{G1}},j_{\mathrm{G1,DT}}}$ denotes the $j_{\mathrm{G1,DT}}$th subset from input representation data $\boldsymbol{D}_{\mathrm{G1}}^{k_{\mathrm{G1}}}$; M_{RSM} is the number of features in $\boldsymbol{D}_{\mathrm{G1}}^{k_{\mathrm{G1}},j_{\mathrm{G1,DT}}}$, where $M_{\mathrm{RSM}} << M_{\mathrm{G1}}^{k_{\mathrm{G1}}}$.

The $\left\{\boldsymbol{D}_{\mathrm{G1}}^{k_{\mathrm{G1}},j_{\mathrm{G1,DT}}}\right\}_{j_{\mathrm{G1,DT}}=1}^{J_{\mathrm{G1,DT}}}$ is then employed to build the DT models in the forest model. In the RF model, each DT utilizes mean squared error to identify the optimal cut-off point for completing tree growth. This process is represented as follows:

$$\left(m_{\mathrm{G1}}^{\mathrm{bset}},x_{\mathrm{G1}}^{m_{\mathrm{G1}}^{\mathrm{bset}},s}\right)=\min\left(\sum_{\boldsymbol{x}_{\mathrm{G1}}^{i}\in\boldsymbol{R}_L}\left(y_{\mathrm{G1}}^{i}-c_L\right)^2+\sum_{\boldsymbol{x}_{\mathrm{G1}}^{i}\in\boldsymbol{R}_R}\left(y_{\mathrm{G1}}^{i}-c_R\right)^2\right) \tag{10.16}$$

$$\begin{cases}\boldsymbol{R}_L=\{\boldsymbol{x}_{\mathrm{G1}}\mid x_{\mathrm{G1}}^{i}\le x_{\mathrm{G1}}^{m_{\mathrm{G1}}^{\mathrm{bset}},s}\}\\ \boldsymbol{R}_R=\{\boldsymbol{x}_{\mathrm{G1}}\mid x_{\mathrm{G1}}^{i}> x_{\mathrm{G1}}^{m_{\mathrm{G1}}^{\mathrm{bset}},s}\}\end{cases} \tag{10.17}$$

where $m_{\mathrm{G1}}^{\mathrm{bset}}$ is the serial number of the best features; $x_{\mathrm{G1}}^{m_{\mathrm{G1}}^{\mathrm{bset}},s}$ represents the value of the best features; $\boldsymbol{R}_L$ and $\boldsymbol{R}_R$ are the datasets of the left sub-node and the right sub-node, respectively; c_L and c_R denotes the average of the output in $\boldsymbol{R}_L$ and $\boldsymbol{R}_R$, respectively.

By applying the minimum sample limit in the leaf node, the $j_{\mathrm{G1,DT}}$th DT model can be expressed as follows,

$$\Gamma_{\mathrm{RF}}^{j_{\mathrm{G1,DT}}}(\cdot)=\sum_{q=1}^{Q} c_{\mathrm{RF}}^{j_{\mathrm{G1,DT}},q} I\left(\boldsymbol{x}_{\mathrm{G1}}^{i}\in\boldsymbol{R}_q\right) \tag{10.18}$$

where $I(\cdot)$ is a symbolic function.

Consequently, the RF model, comprising $J_{\text{G1,DT}}$ DT model, can be expressed as,

$$f_{\text{G1,RF}}^{k_{\text{G1}}}(\cdot) = \frac{1}{J_{\text{G1,DT}}} \sum_{j_{\text{G1,DT}}=1}^{J_{\text{G1,DT}}} \Gamma_{\text{RF}}^{j_{\text{G1,DT}}}(\cdot) \tag{10.19}$$

In the CRF model, the growth of each DT is based on a random split point, while the remaining strategies are consistent with those used in traditional decision trees. Therefore, the description of this process is simplified by representing the CRF model as follows: $f_{\text{G1,CRF}}^{k_{\text{G1}}}(\cdot)$. Given the number of RF and CRF models, the layer forest model is written as,

$$f_{\text{G1}}^{k_{\text{G1}}}(\cdot) = \left\{ f_{\text{G1,RF}}^{k_{\text{G1}},j_{\text{G1,Forest}}}(\cdot), f_{\text{G1,CRF}}^{k_{\text{G1}},j_{\text{G1,Forest}}}(\cdot) \right\}_{j_{\text{G1,Forest}}=1}^{J_{\text{G1,Forest}}} \tag{10.20}$$

The structure of the ADFR-clfc model is illustrated in Figure 10.4, which shows that the layer regression vector is the output of the middle layer forest model. The layer regression vector of the k_{G1}th layer is expressed as follows,

$$\boldsymbol{h}_{\text{G1}}^{k_{\text{G1}},i} = f_{\text{G1}}^{k_{\text{G1}}}\left(\boldsymbol{x}_{\text{G1}}^{i}\right) \in R^{1 \times J_{\text{G1,Forest}}} \tag{10.21}$$

During the learning process of the ADFR-clfc model, the cross-layer full connection strategy is employed to enhance the representation performance of the layer regression vector. Consequently, the input feature information for the $\left(k_{\text{G1}}+1\right)$th layer can be expressed as follows:

$$\boldsymbol{X}_{\text{G1}}^{k_{\text{G1}}+1} = \left[\boldsymbol{X}_{\text{G1}}, \boldsymbol{H}_{\text{G1}}^{1}, \boldsymbol{H}_{\text{G1}}^{2}, \ldots, \boldsymbol{H}_{\text{G1}}^{k_{\text{G1}}} \right] R^{N_G \times \left(M_{\text{G1}} + J_{\text{G1,Forest}} \times k_{\text{G1}}\right)} \tag{10.22}$$

where

$$\boldsymbol{H}_{\text{G1}}^{k_{\text{G1}}} = \begin{bmatrix} \boldsymbol{h}_{\text{G1}}^{k_{\text{G1}},1} \\ \boldsymbol{h}_{\text{G1}}^{k_{\text{G1}},2} \\ \vdots \\ \boldsymbol{h}_{\text{G1}}^{k_{\text{G1}},N} \end{bmatrix} \in R^{N_G \times J_{\text{G1,Forest}}} \tag{10.23}$$

where $\boldsymbol{H}_{\text{G1}}^{k_{\text{G1}}}$ is the output of the k_{G1}th layer forest model.

Compared to the DFR method based on cross-layer full connection, the ADFR-clfc model incorporates a layer adaptation mechanism to enhance learning performance. When the error in the layer-by-layer learning process converges, the learning process is automatically halted. This can be described as follows:

$$\theta_{G1} = \frac{\left(\sqrt{\left(f_{G1}^{k_{G1}}\left(X_{G1}\right)-y_G\right)^2}-\sqrt{\left(f_{G1}^{k_{G1}-1}\left(X_{G1}\right)-y_G\right)^2}\right)}{N} \quad (10.24)$$

where

$$f_{G1}^{k_{G1}} = \frac{1}{J_{G1,\text{Forest}}} \sum_{j_{G1,\text{Forest}}=1}^{J_{G1,\text{Forest}}} f_{G1,\text{RF}}^{k_{G1},j_{G1,\text{Forest}}}\left(\cdot\right) + f_{G1,\text{CRF}}^{k_{G1},j_{G1,\text{Forest}}}\left(\cdot\right) \quad (10.25)$$

where θ_{G1} is a user-defined threshold value.

The ADFR-clfc model continues to increase the number of layers to achieve better learning outcomes until the error criteria are met. Assuming that the error requirements are satisfied when the ADFR-clfc model has been trained up to K_{G1} layers, the resulting model can be denoted as $F_{G1}\left(\cdot\right)$. Consequently, the DXN generation model is expressed as follows:

$$\hat{y}^{\text{Generation}} = F_{G1}\left(\cdot\right) \quad (10.26)$$

10.3.3 DXN Adsorption Model Construction Module

The modeling of the adsorption model consists of two steps. Firstly, the modeling dataset based on the original dataset is expanded using the VSG technique. Secondly, the virtual sample data with the real data are combined for fuzzy modeling.

In the DXN adsorption model construction module, the input data includes the DXN generation concentration and the DXN generation process data, while the output data are the DXN adsorption concentration. Therefore, the original dataset available for VSG in the DXN adsorption model is denoted as follows:

$$\boldsymbol{D}_{\text{APND}} = \left\{\left\{\boldsymbol{X}_{\text{APND}}, \boldsymbol{y}_G\right\}, \boldsymbol{y}_A\right\} = \left\{\boldsymbol{x}_{\text{A}}^1, \ldots, \boldsymbol{x}_A^{M_{\text{APND}}+1}, \boldsymbol{y}_A\right\} \in R^{N_A \times \left(M_{\text{APND}}+2\right)} \quad (10.27)$$

10.3.3.1 VSG Algorithm

As shown in Figure 10.4, the latent variable space is represented as a 2-dimensional coordinate system, resulting in an output layer of the encoder network consisting of four units. The encoder network discussed in this chapter is designed with multiple hidden layers. Therefore, the output of the encoder network can be expressed as follows:

$$\left[\mu_i, \sigma_i\right]^{\text{T}} = f_{\text{Encoder}}\left(\boldsymbol{x}_A^i; \boldsymbol{W}_{\text{Encoder}}, \boldsymbol{b}_{\text{Encoder}}\right) \quad (10.28)$$

where $\mu_i \in R^{1\times 2}$ is the two-dimensional mean vector; $\sigma_i \in R^{1\times 2}$ represents the two-dimensional variance vector; $f_{\text{Encoder}}(\cdot)$ denotes the encoder network model; and $\boldsymbol{W}_{\text{Encoder}}$ and $\boldsymbol{b}_{\text{Encoder}}$ are the weight and bias of the encoder network model, respectively.

By utilizing the mean and variance information, the latent vector z^i_{Encoder} can be expressed as follows:

$$z^i_{\text{Encoder}} = \mu_i + \exp(\sigma_i / 2) \odot \varepsilon_i \in \mathbb{R}^{2\times 1} \tag{10.29}$$

where ε_i represents a standard normal distribution random number.

Furthermore, the latent vector can be represented as follows:

$$z^i_{\text{Encoder}} = \left[z^{i,1}_{\text{Encoder}}, z^{i,2}_{\text{Encoder}} \right]^{\text{T}} \tag{10.30}$$

By inputting the process data $\boldsymbol{X}_{\text{APND}}$ into the encoder network model $f_{\text{Encoder}}(\cdot)$, the latent variables matrix can be represented as follows:

$$\boldsymbol{Z}_{\text{Encoder}} = \left\{ z^i_{\text{Encoder}} \right\}_{i=1}^{N_A} \in \mathbb{R}^{N_A \times 2} \tag{10.31}$$

Consequently, the 2-dimensional coordinate system can be constructed according to $\boldsymbol{Z}_{\text{Encoder}}$. In this two-dimensional coordinate system, the VSG method is applied based on clustering and geometric rotation to generate virtual samples.

Firstly, the latent variable samples are clustered into $K_{\text{VSG}} = 3$ clusters using the K-means algorithm, as follows:

$$\begin{aligned} \boldsymbol{C}_k &= \boldsymbol{C}_k \cup \min \left\{ \sqrt{\left\| z^i_{\text{Encoder}} - \frac{1}{N_{C_k}} \sum_{z^i_{\text{Encoder}} \in \boldsymbol{C}_k} z^i_{\text{Encoder}} \right\|^2} \right\}_{i=1}^{N_A} \\ &= \boldsymbol{C}_k \cup \min \left\{ \sqrt{\left\| z^i_{\text{Encoder}} - \boldsymbol{c}^{k_{\text{VSG}}}_{\text{VSG}} \right\|^2} \right\}_{i=1}^{N_A} \end{aligned} \tag{10.32}$$

where $\boldsymbol{C}_k$ represents the data set of the k_{VSG}th cluster, N_{C_k} indicates the number of latent variable samples in the k_{VSG}th cluster $\boldsymbol{C}_k$, and $\boldsymbol{c}^{k_{\text{VSG}}}_{\text{VSG}}$ is the centroid of the k_{VSG}th cluster $\boldsymbol{C}_k$.

Based on the three obtained cluster datasets and their corresponding centroids $\left[\boldsymbol{c}^1_{\text{VSG}}, \boldsymbol{c}^2_{\text{VSG}}, \boldsymbol{c}^3_{\text{VSG}} \right]$, three circular regions $H^1_{\text{Ring}}, H^2_{\text{Ring}}$, and H^3_{Ring} that are tangent to each other can be identified in the two-dimensional coordinate system (as illustrated in Figure 10.4). The radii of these three circular regions are denoted as r_1, r_2, and r_3. With the distribution of latent variable samples in each circular region, each region can be divided into multiple concentric rings, as follows:

$$\varsigma^{k_{\text{VSG}}}_{\text{Ring}} = \frac{1}{K_{k_{\text{VSG}}}} \left(r_{k_{\text{VSG}}} - d^{\min}_{k_{\text{VSG}}} \right) \tag{10.33}$$

where $\zeta_{\text{Ring}}^{k_{\text{VSG}}}$ is the radius difference of each concentric ring in the k_{VSG}th cluster, $K_{k_{\text{VSG}}}$ represents the number of concentric rings in the k_{VSG}th cluster, and $d_{k_{\text{VSG}}}^{\min}$ denotes the minimum distance between latent variable samples and the k_{VSG}th centroid.

Therefore, each circle consists of multiple concentric rings with the same radius difference $\zeta_{\text{Ring}}^{k_{\text{VSG}}}$ in the 2-dimensional coordinate system. Subsequently, a rotation weight matrix method based on rule information is employed to adaptively generate virtual samples. The detailed process for generating these virtual samples can be found in the [34]. Finally, the resulting latent variable virtual samples dataset is represented as follows:

$$\tilde{Z}_{\text{Encoder}}^{\text{VSG}} = \left\{ \left\{ \left\{ \ldots, \tilde{z}_1^i, \ldots \right\}_{i=1}^{N_k^{\text{VSG}}} \right\}, \left\{ \left\{ \ldots, \tilde{z}_2^i, \ldots \right\}_{i=1}^{N_k^{\text{VSG}}} \right\}, \left\{ \left\{ \ldots, \tilde{z}_3^i, \ldots \right\}_{i=1}^{N_k^{\text{VSG}}} \right\} \right\}_{k_{\text{VSG}}=1}^{3} \quad (10.34)$$

After obtaining the latent variable virtual samples $\tilde{Z}_{\text{Encoder}}^{\text{VSG}}$, the decoding network, which shares the same structure as the encoder network, is utilized to map these latent variable virtual samples back to the original space with physically meaningful process numerical variables. Therefore, the output of the decoder network can be expressed as follows:

$$\tilde{\boldsymbol{x}}_{k_{\text{VSG}}}^i = f_{\text{Decoder}}\left(\tilde{z}_{k_{\text{VSG}}}^i ; \boldsymbol{W}_{\text{Decoder}}, \boldsymbol{b}_{\text{Decoder}} \right) \quad (10.35)$$

where $\tilde{\boldsymbol{x}}_{k_{\text{VSG}}}^i \in R^{1\times(M_{\text{APND}}+1)}$ is the virtual process numerical samples; $f_{\text{Decoder}}(\cdot)$ denotes the decoder network model; and $\boldsymbol{W}_{\text{Decoder}}$ and $\boldsymbol{b}_{\text{Decoder}}$ are the weight and bias of the decoder network model, respectively.

By using the decoder network model ($f_{\text{Decoder}}(\cdot)$), the unlabeled virtual process numerical samples can be calculated by inputting the latent variable virtual samples $\tilde{Z}_{\text{Encoder}}^{\text{VSG}}$. They are denoted as follows:

$$\tilde{\boldsymbol{X}}_{\text{Encoder}}^{\text{VSG}} = \left\{ \left\{ \left\{ \ldots, \tilde{\boldsymbol{x}}_1^i, \ldots \right\}_{i=1}^{N_1^{\text{VSG}}} \right\}, \left\{ \left\{ \ldots, \tilde{\boldsymbol{x}}_2^i, \ldots \right\}_{i=1}^{N_2^{\text{VSG}}} \right\}, \left\{ \left\{ \ldots, \tilde{\boldsymbol{x}}_3^i, \ldots \right\}_{i=1}^{N_3^{\text{VSG}}} \right\} \right\}_{k_{\text{VSG}}=1}^{3} \quad (10.36)$$

Subsequently, a multi-tree cooperative method is proposed for labeling and evaluating these unlabeled virtual samples. This method encompasses three steps: training the linear tree model, labeling and evaluating the unlabeled virtual samples, and performing collaborative iteration.

The linear tree model shares the same structure and growth method as a DT, with the primary difference being in the calculation method for the leaf nodes. After obtaining the linear tree structure, the leaf nodes are calculated as follows:

$$c_{\text{LinearTree}}^{k_{\text{VSG}},q} = \frac{1}{2}\left(\boldsymbol{X}_A^{k_{\text{VSG}},q} \boldsymbol{w}^{k_{\text{VSG}},q} + \frac{1}{N^{k_{\text{VSG}},q}} \sum y_A^{k_{\text{VSG}},q} \right) \quad (10.37)$$

where $c_{\text{LinearTree}}^{k_{\text{VSG}},q}$ is the output of the qth leaf node in the k_{VSG}th linear tree, $\boldsymbol{X}_{\text{A}}^{k_{\text{VSG}},q}$ indicates the feature space data of the qth leaf node, $\boldsymbol{w}^{k_{\text{VSG}},q}$ represents the weight vector (a ridge regression method is used to solve the weight vector in this chapter), $N^{k_{\text{VSG}},q}$ denotes the number of samples in $\boldsymbol{X}_{A}^{k_{\text{VSG}},q}$, and $y_{A}^{k_{\text{VSG}},q}$ expresses the target value in the qth leaf node.

Therefore, the k_{VSG}th linear tree model can be expressed as,

$$\Gamma_{\text{LinearTree}}^{k_{\text{VSG}}}(\cdot)=\sum_{q=1}^{Q} c_{\text{LinearTree}}^{k_{\text{VSG}},q} I\left(\boldsymbol{x}_{A}^{i}\in\boldsymbol{R}_{q}\right) \tag{10.38}$$

To label and evaluate the unlabeled virtual samples, three linear tree models, as denoted $\Gamma_{\text{LinearTree}}^{1}(\cdot)$, $\Gamma_{\text{LinearTree}}^{2}(\cdot)$ and $\Gamma_{\text{LinearTree}}^{3}(\cdot)$ are constructed, each corresponding to one of the three categories of virtual samples. Subsequently, these virtual sample data are fed into their respective linear tree models, and they are labeled using the leaf nodes. The calculation process is outlined as follows:

$$\begin{aligned}\tilde{y}_{k_{\text{VSG}}}^{i} &= \Gamma_{\text{LinearTree}}^{k_{\text{VSG}}}\left(\tilde{\boldsymbol{x}}_{k_{\text{VSG}}}^{i},\boldsymbol{w}^{k_{\text{VSG}},q},\lambda\right)\\ &=\frac{1}{2}\left(\tilde{\boldsymbol{x}}_{k_{\text{VSG}}}^{i}\boldsymbol{w}^{k_{\text{VSG}},q}+\frac{1}{N^{k_{\text{VSG}},q}}\sum y_{A}^{k_{\text{VSG}},q}\right)\end{aligned} \tag{10.39}$$

where $\tilde{y}_{k_{\text{VSG}}}^{i}$ is the label of ith virtual samples in k_{VSG}th group, and λ represents the regular term parameter.

Consequently, the pseudo-labeled virtual sample data can be denoted as follows:

$$\tilde{\boldsymbol{D}}_{\text{Encoder}}^{\text{VSG}}=\left\{\left\{\left\{\tilde{\boldsymbol{x}}_{1}^{i},\tilde{y}_{1}^{i}\right\}_{i=1}^{N_{1}^{\text{VSG}}}\right\},\left\{\left\{\tilde{\boldsymbol{x}}_{2}^{i},\tilde{y}_{2}^{i}\right\}_{i=1}^{N_{2}^{\text{VSG}}}\right\},\left\{\left\{\tilde{\boldsymbol{x}}_{3}^{i},\tilde{y}_{3}^{i}\right\}_{i=1}^{N_{3}^{\text{VSG}}}\right\}\right\}_{k_{\text{VSG}}=1}^{3} \tag{10.40}$$

Furthermore, each pseudo-labeled virtual sample is used to update the corresponding linear DT model without altering the structure of the tree. The evaluation index for the pseudo-labeled virtual samples is determined as follows:

$$\delta_{k_{\text{VSG}}}^{i}=\sum_{\boldsymbol{x}_{A}^{i}\in\boldsymbol{R}_{q}}\left(\left(y_{A}^{i,q}-\boldsymbol{x}_{A}^{i}\hat{\boldsymbol{w}}^{k_{\text{VSG}},q}\right)^{2}-\left(y_{A}^{i,q}-\boldsymbol{x}_{A}^{i}\boldsymbol{w}^{k_{\text{VSG}},q}\right)^{2}\right) \tag{10.41}$$

where $\delta_{k_{\text{VSG}}}^{i}$ is the evaluation index of the ith pseudo-labeled virtual samples in the k_{VSG}th group, and $\hat{\boldsymbol{w}}^{k_{\text{VSG}},q}$ represents the updated weight of the qth leaf node in the k_{VSG}th linear tree.

The evaluation set for the three groups of pseudo-labeled virtual samples is obtained as follows:

$$\tilde{\boldsymbol{D}}_{\text{Evaluation}}^{\text{VSG}} = \left\{ \left\{ \ldots, \delta_1^i, \ldots \right\}_{i=1}^{N_1^{\text{VSG}}}, \left\{ \ldots, \delta_2^i, \ldots \right\}_{i=1}^{N_2^{\text{VSG}}}, \left\{ \ldots, \delta_3^i, \ldots \right\}_{i=1}^{N_3^{\text{VSG}}} \right\} \tag{10.42}$$

Subsequently, a collaborative strategy is employed to select the positive pseudo-labeled virtual samples. This collaborative process involves the following steps:

(1) Define a threshold θ_{Percent} to select positive pseudo-labeled virtual samples from the evaluation set during each iteration;
(2) Add the positive pseudo-labeled virtual samples to the training set of the other linear tree models and update the leaf nodes of these models;
(3) Input the remaining pseudo-labeled virtual samples into the updated linear tree models to obtain the evaluation index set.

Consequently, a set of positive pseudo-labeled virtual sample data is acquired, which is denoted as follows:

$$\tilde{\boldsymbol{D}}^{\text{VSG}} = \left\{ \tilde{\boldsymbol{X}}^{\text{VSG}}, \tilde{\boldsymbol{y}}^{\text{VSG}} \right\} \tag{10.43}$$

10.3.3.2 FFR Algorithm

The dataset for fuzzy modeling now comprises actual data and positive pseudo-labeled virtual sample data. The combination of these two datasets is denoted as follows:

$$\boldsymbol{D}_A = \left\{ \boldsymbol{D}_{\text{APND}}; \tilde{\boldsymbol{D}}^{\text{VSG}} \right\} = \left\{ \boldsymbol{x}_A^1, \ldots, \boldsymbol{x}_A^{M_{\text{APND}}+1}, \boldsymbol{y}_A \right\} \in R^{\left(N_A + \tilde{N}^{\text{VSG}}\right) \times \left(M_{\text{APND}}+2\right)} \tag{10.44}$$

The structure of the FFR model, as illustrated in Figure 10.4, consists of several fuzzy tree models organized in an ensemble through weight assignment. Each fuzzy tree is constructed based on the training dataset $\boldsymbol{D}_A$. The structure of the fuzzy trees is the same as that of DT. Subsequently, a fuzzy system is constructed at the leaf nodes to complete the formation of the fuzzy tree. Assuming the input data for the qth leaf node is represented as follows:

$$\boldsymbol{D}_A^{j\text{FDT}} = \left\{ \boldsymbol{x}_A^i, \boldsymbol{y}_A \right\}_{i=1}^{M_A^{j\text{FDT},q}} \in R^{N_A^{j\text{FDT},q} \times \left(M_A^{j\text{FDT},q} + 1 \right)} \tag{10.45}$$

where $M_A^{j\text{FDT},q} \leq M_{\text{APND}}$ is the feature dimension.

In the premise part of the fuzzy system, each input feature is processed fuzzily as follows,

$$o_{k^{\text{Fuzzy}}} = \prod_{m=1}^{M_A^{j\text{FDT},q}} \mu^{k^{\text{Fuzzy}}}_{A^{k^{\text{Fuzzy}}}}\left(\boldsymbol{x}_A^i\right) = \mu^{k^{\text{Fuzzy}}}_{A_1^{k^{\text{Fuzzy}}}}\left(x_A^{i,1}\right) \wedge \mu^{k^{\text{Fuzzy}}}_{A_2^{k^{\text{Fuzzy}}}}\left(x_A^{i,2}\right) \wedge \ldots \wedge \mu^{k^{\text{Fuzzy}}}_{A^{k^{\text{Fuzzy}}}_{M_A^{j\text{FDT},q}}}\left(x_A^{i,M_A^{j\text{FDT},q}}\right)$$
$$= \exp\left[-\frac{\left\|x_A^{i,1} - c^{\text{Fuzzy}}_{1,k^{\text{Fuzzy}}}\right\|^2}{\left(\sigma^{\text{Fuzzy}}_{1,k^{\text{Fuzzy}}}\right)^2}\right] \wedge \exp\left[-\frac{\left\|x_A^{i,2} - c^{\text{Fuzzy}}_{2,k^{\text{Fuzzy}}}\right\|^2}{\left(\sigma^{\text{Fuzzy}}_{2,k^{\text{Fuzzy}}}\right)^2}\right] \wedge \ldots \wedge$$
$$\exp\left[-\frac{\left\|x_A^{i,M_A^{j\text{FDT},q}} - c^{\text{Fuzzy}}_{3,k^{\text{Fuzzy}}}\right\|^2}{\left(\sigma^{\text{Fuzzy}}_{3,k^{\text{Fuzzy}}}\right)^2}\right] \tag{10.46}$$

where $o_{k^{\text{Fuzzy}}}$ is the output of the k^{Fuzzy}th fuzzy rule.

Given K^{Fuzzy} fuzzy rules in the premise part, normalization is employed to obtain the output weight of the k^{Fuzzy}th fuzzy rule as follows:

$$\bar{o}_{k^{\text{Fuzzy}}} = \frac{o_{k^{\text{Fuzzy}}}}{\sum_{k^{\text{Fuzzy}}=1}^{K^{\text{Fuzzy}}} o_{k^{\text{Fuzzy}}}} \tag{10.47}$$

In the consequent part, a single hidden-layer neural network without an activation function (containing a number of neurons equal to K^{Fuzzy}) is used to map each input feature, denoted as follows:

$$\phi_{k^{\text{Fuzzy}}} = \omega^1_{k^{\text{Fuzzy}}} x_A^{i,1} + \omega^2_{k^{\text{Fuzzy}}} x_A^{i,2} + \cdots + \omega^{M_A^{j\text{FDT},q}}_{k^{\text{Fuzzy}}} x_A^{i,M_A^{j\text{FDT},q}} \tag{10.48}$$

Finally, the leaf node output of all the K^{Fuzzy} fuzzy rules is expressed as follows:

$$c_{\text{FDT}}^{j\text{FDT},q} = \sum_{k^{\text{Fuzzy}}=1}^{K^{\text{Fuzzy}}} \bar{o}_{k^{\text{Fuzzy}}} \phi_{k^{\text{Fuzzy}}}$$
$$= \sum_{k^{\text{Fuzzy}}=1}^{K^{\text{Fuzzy}}} \left\{\left(\frac{o_{k^{\text{Fuzzy}}}}{\sum_{k^{\text{Fuzzy}}=1}^{K^{\text{Fuzzy}}} o_{k^{\text{Fuzzy}}}}\right)\left(\omega^1_{k^{\text{Fuzzy}}} x_A^{i,1} + \omega^2_{k^{\text{Fuzzy}}} x_A^{i,2} + \cdots + \omega^{M_A^{j\text{FDT},q}}_{k^{\text{Fuzzy}}} x_A^{i,M_A^{j\text{FDT},q}}\right)\right\} \tag{10.49}$$

where $c_{\text{FDT}}^{j\text{FDT},q}$ is the output of the qth leaf node in the j_{FDT}th fuzzy tree model.

In each fuzzy tree, the parameters of the premise part are updated using a gradient descent approach, while the parameters of the consequent part are updated using least-squares methods. Therefore, the j_{FDT}th fuzzy tree model can be expressed as follows:

$$\Gamma_{\text{FDT}}^{j_{\text{FDT}}}\left(\cdot\right)=\sum_{q=1}^{Q} c_{\text{FDT}}^{j_{\text{FDT}},q} \boldsymbol{I}\left(\boldsymbol{x}_A^i \in \boldsymbol{R}_q\right) \tag{10.50}$$

After obtaining all the J_{FDT} fuzzy trees, a ridge regression method is used to assign the weight to each fuzzy tree, as follows,

$$\begin{aligned}\omega_{\text{FFR}} = & \left[\ldots,\Gamma_{\text{FDT}}^{j_{\text{FDT}}}\left(\boldsymbol{x}_A^1,\ldots,\boldsymbol{x}_A^{M_{\text{APND}}+1}\right),\ldots\right]^{\text{T}}\left[\ldots,\Gamma_{\text{FDT}}^{j_{\text{FDT}}}\left(\boldsymbol{x}_A^1,\ldots,\boldsymbol{x}_A^{M_{\text{APND}}+1}\right),\ldots\right] \\ & +\lambda \boldsymbol{I})^{-1}\left[\ldots,\Gamma_{\text{FDT}}^{j_{\text{FDT}}}\left(\boldsymbol{x}_A^1,\ldots,\boldsymbol{x}_A^{M_{\text{APND}}+1}\right),\ldots\right]^{\text{T}} y_A\end{aligned} \tag{10.51}$$

Finally, the DXN absorption model based on the FFR algorithm is expressed as follows:

$$\hat{\boldsymbol{y}}^{\text{Adsorption}} = F_{\text{FFR}}\left(\cdot\right)=\sum_{j_{\text{FDT}}=1}^{J_{\text{FDT}}} \Gamma_{\text{FDT}}^{j_{\text{FDT}}}\left(\cdot\right) w_{\text{FFR}} \tag{10.52}$$

10.3.4 DXN Emission Model Construction Module

In the DXN emission model construction module, the input data include DXN generation concentration, DXN absorption concentration, and DXN emission process data, while the output data are the DXN emission concentration. Therefore, by combining the aforementioned datasets, the dataset for modeling the DXN emission concentration is denoted as follows:

$$\boldsymbol{D}_E=\left\{\left\{\boldsymbol{X}_{\text{APND}},\boldsymbol{y}_G,\boldsymbol{y}_A\right\},\boldsymbol{y}_E\right\}=\left\{\boldsymbol{x}_E^1,\ldots,\boldsymbol{x}_E^{M_{\text{APND}}+2},\boldsymbol{y}_E\right\}\in R^{N_G\times\left(M_{\text{EPND}}+3\right)} \tag{10.53}$$

The TBLS algorithm is employed to construct the DXN emission model. This model consists of mapping, enhancement, and increment layers, as illustrated in Figure 10.4. In each layer of the TBLS, the basic unit is a DT model, and the number of DTs can be determined through experimentation.

The DT model of the TBLS is expressed as follows.

$$\Gamma_{\text{TBLS}}^{j_{\text{Map}}}\left(\cdot\right)=\sum_{q=1}^{Q} c_{\text{DT}}^{j_{\text{Map}},q} \boldsymbol{I}\left(\boldsymbol{x}_{\text{E}}^i \in \boldsymbol{R}_q\right) \tag{10.54}$$

where

$$c_{\text{DT}}^{j_{\text{Map}},q}=\frac{1}{N^{j_{\text{Map}},q}}\sum y_E^{j_{\text{Map}},q} \tag{10.55}$$

where $\Gamma_{\text{TBLS}}^{j_{\text{Map}}}(\cdot)$ is the j_{Map}th DT model of the mapping layer.

Therefore, the output of the j_{Map}th DT unit in the mapping layer is denoted as follows:

$$z_{\text{Map}}^{j_{\text{Map}}} = \Gamma_{\text{TBLS}}^{j_{\text{Map}}}\left(\left\{\boldsymbol{x}_E^1,\ldots,\boldsymbol{x}_E^{M_{\text{APND}}+2}\right\}\right) \in \boldsymbol{R}^{N_G \times 1} \quad (10.56)$$

As a result, the output of the mapping layer can be expressed as follows:

$$\boldsymbol{Z}_{\text{Map}} = \left[\ldots, z_{\text{Map}}^{j_{\text{Map}}}, \ldots\right] \in \boldsymbol{R}^{N_G \times J_{\text{Map}}} \quad (10.57)$$

Simultaneously, the input of the enhancement layer is obtained by combining $\boldsymbol{D}_E$ and the output of the mapping layer $\boldsymbol{Z}_{\text{Map}}$. After this combination, the input data is represented as follows:

$$\boldsymbol{D}_E^{\text{Enh}} = \left\{\left\{\boldsymbol{X}_{\text{APND}}, \boldsymbol{y}_G, \boldsymbol{y}_A, \boldsymbol{Z}_{\text{Map}}\right\}, \boldsymbol{y}_E\right\} \in R^{N_G \times \left(M_{\text{EPND}}+2+J_{\text{Map}}+1\right)} \quad (10.58)$$

Subsequently, the output of the j_{Enh}th DT unit in the enhancement layer is denoted as follows:

$$z_{\text{Enh}}^{j_{\text{Enh}}} = \Gamma_{\text{TBLS}}^{j_{\text{Enh}}}\left(\left\{\boldsymbol{X}_{\text{APND}}, \boldsymbol{y}_G, \boldsymbol{y}_A, \boldsymbol{Z}_{\text{Map}}\right\}\right) \in \boldsymbol{R}^{N_G \times 1} \quad (10.59)$$

The output of the enhancement layer is represented as follows:

$$\boldsymbol{Z}_{\text{Enh}} = \left[\ldots, z_{\text{Enh}}^{j_{\text{Enh}}}, \ldots\right] \in \boldsymbol{R}^{N_G \times J_{\text{Enh}}} \quad (10.60)$$

Further, the combined output of the mapping and enhancement layers can be denoted as follows:

$$\boldsymbol{G}^{J_{\text{TBLS}}} = [\boldsymbol{Z}_{\text{Map}} \mid \boldsymbol{Z}_{\text{Enh}}] \in \boldsymbol{R}^{N_G \times \left(J_{\text{Map}}+J_{\text{Enh}}\right)} \quad (10.61)$$

Consequently, the static part of the TBLS model is represented as follows:

$$\begin{aligned} f_{\text{TBLS}}^1(\cdot) &= \boldsymbol{G}^{J_{\text{TBLS}}} \omega_{\text{TBLS}}^{J_{\text{TBLS}}} \\ &= [z_{\text{Map}}^1, z_{\text{Map}}^2, \ldots, z_{\text{Map}}^{J_{\text{Map}}} \mid z_{\text{Enh}}^1, z_{\text{Enh}}^2, \ldots, z_{\text{Enh}}^{J_{\text{Enh}}}] \omega_{\text{TBLS}}^{J_{\text{TBLS}}} \end{aligned} \quad (10.62)$$

where $\omega_{\text{TBLS}}^{J_{\text{TBLS}}} \in \boldsymbol{R}^{\left(J_{\text{Map}}+J_{\text{Enh}}\right)\times 1}$ is a weight vector.

By connecting the mapping layer, enhancement layer, and output layer, the process of weight calculation is expressed as follows:

$$\begin{aligned}\omega_{\text{TBLS}}^{J_{\text{TBLS}}} &= \left(\left(\boldsymbol{G}^{J_{\text{TBLS}}}\right)^{\text{T}}\boldsymbol{G}^{J_{\text{TBLS}}}+\lambda\boldsymbol{I}\right)^{-1}\left(\boldsymbol{G}^{J_{\text{TBLS}}}\right)^{\text{T}}\boldsymbol{y}_E \\ &= \left(\boldsymbol{G}^{J_{\text{TBLS}}}\right)^{*}\boldsymbol{y}_E\end{aligned} \tag{10.63}$$

where $\left(\boldsymbol{G}^{J_{\text{TBLS}}}\right)^{*}$ represents the pseudoinverse matrix of $\boldsymbol{G}^{J_{\text{TBLS}}}$, λ denotes the regularization coefficient, and $\boldsymbol{I}$ is the identity matrix.

When the static part of the TBLS model encounters challenges in achieving the desired learning effect, TBLS adapts by quickly adding DT models. By incorporating a DT model in the increment layer and inputting the outputs of the mapping layer, enhancement layer, and raw data, the output of the added DT is denoted as follows:

$$z_{\text{Inc}}^{1} = \Gamma_{\text{TBLS}}^{1}\left(\left\{\boldsymbol{X}_{\text{APND}},\boldsymbol{y}_G,\boldsymbol{y}_A,\boldsymbol{Z}_{\text{Map}},\boldsymbol{Z}_{\text{Enh}}\right\}\right)\in\boldsymbol{R}^{N_G\times 1} \tag{10.64}$$

Consequently, the new TBLS model is expressed as follows:

$$\begin{aligned}\hat{\boldsymbol{y}}^{\text{Emission\_static}} &= F_{\text{TBLS}}^{\text{static}}(\cdot) = \left[z_{\text{Map}}^{1},z_{\text{Map}}^{2},\ldots,z_{\text{Map}}^{J_{\text{Map}}}\middle| z_{\text{Enh}}^{1},z_{\text{Enh}}^{2},\ldots,z_{\text{Enh}}^{J_{\text{Enh}}}\middle| z_{\text{Inc}}^{1}\right]\omega_{\text{TBLS}}^{J_{\text{TBLS}}+1} \\ &= \left[z_{\text{Map}}^{1},z_{\text{Map}}^{2},\ldots,z_{\text{Map}}^{J_{\text{Map}}}\middle| z_{\text{Enh}}^{1},z_{\text{Enh}}^{2},\ldots,z_{\text{Enh}}^{J_{\text{Enh}}}\middle| z_{\text{Inc}}^{1}\right]\left(\boldsymbol{G}^{J_{\text{TBLS}}+1}\right)^{*}\boldsymbol{y}_E\end{aligned} \tag{10.65}$$

Accordingly, the update procedure of the pseudoinverse matrix is shown as follows:

$$\left[\boldsymbol{G}^{J_{\text{TBLS}}+1}\right]^{*} = \begin{bmatrix}\left[\boldsymbol{G}^{J_{\text{TBLS}}+1}\right]^{*}-\boldsymbol{D}\boldsymbol{B}^{\text{T}} \\ \boldsymbol{B}^{\text{T}}\end{bmatrix} \tag{10.66}$$

where

$$\begin{cases}\boldsymbol{D} = \left[\boldsymbol{G}^{J_{\text{TBLS}}}\right]^{*} z_{\text{Inc}}^{1} \\ \boldsymbol{B}^{\text{T}} = \begin{cases}\left[\boldsymbol{C}\right]^{*}, & \text{if } C\neq 0 \\ \left[\boldsymbol{I}+\boldsymbol{D}^{\text{T}}\boldsymbol{D}\right]^{-1}\boldsymbol{D}^{\text{T}}\left[\boldsymbol{G}^{J_{\text{TBLS}}}\right]^{*}, & \text{if } C = 0\end{cases} \\ \boldsymbol{C} = z_{\text{Inc}}^{1} - \boldsymbol{G}^{J_{\text{TBLS}}}\boldsymbol{D}\end{cases} \tag{10.67}$$

The incremental process automatically stops growing when the model error reaches convergence. The final DXN emission model based on the TBLS algorithm is denoted as follows:

$$\hat{y}^{\text{Emission}} = F_{\text{TBLS}}(\cdot) \tag{10.68}$$

10.4 CASE STUDY

10.4.1 Results and Discussion

10.4.1.1 Modeling Results

This section provides a detailed description of the hyperparameter settings and the results of the training process for each model within the WPTM method.

10.4.1.1.1 Generation Model Results

The hyperparameter settings of the ADFR-clfc model are as follows: minimum sample in leaf node is 3, number of DT in each forest is 5, number of RF is 2, number of CRF is 2, selected features is $\sqrt{M}$, and threshold is 1×10^{-4}. With these hyperparameter settings, the results of the model training process for the DXN generation model are illustrated in Figure 10.6.

Figure 10.6 demonstrates that the training error gradually converges as the number of layers increases. The ADFR-clfc model ultimately achieves convergence with a final count of 538 layers. This indicates that the ADFR-clfc model effectively converges during training, even when dealing with high-dimensional data.

10.4.1.1.2 Adsorption Model Results

The hyperparameter settings of the VSG model are as follows: structure of VAE network is input-two hidden layers-encode output-decode input -two hidden layers-output, the number of neurons in each layer is set to 6-512-256-4-2-256-512-6, the epoch

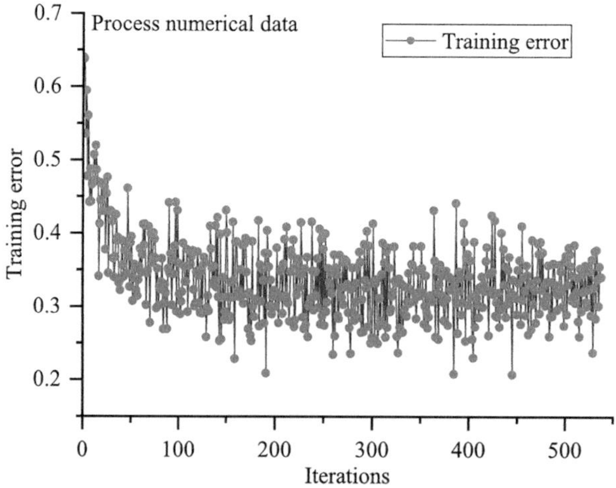

FIGURE 10.6 Training results of the ADFR-clfc model.

of VAE is 1,500, the number of central is 3, the number of rings is 8, the percent of co-training is 0.95, the regularization term of co-training is 0.5, the iterations of co-training is 5. Under the hyperparameter settings of the VSG-FFR model, the results of the model training process for DXN absorption are presented in Figure 10.7.

Figure 10.7(a) reveals that the error of the encoder and decoder network effectively converges during the training process, suggesting that the encoder and decoder exhibit strong characterization performance for low-dimensional process data in the adsorption phase. Figure 10.7(b) illustrates the distribution of the raw ultra-small samples in the latent variable space, while Figure 10.7(c) displays the distribution of the virtual samples generated by the VSG method. The results of this experiment demonstrate that the adopted VSG approach can generate a substantial number of similar virtual samples.

After labeling and evaluating the collaboration strategy, the distribution between the obtained pseudo-labels and the true values is shown in Figure 10.7(d). The pseudo-labels closely align with the distribution of the true value labeling, demonstrating the effectiveness of the collaboration strategy. Subsequently, the virtual samples are added to the original ultra-small samples to assess their impact on the FFR model. The experimental results illustrated in Figure 10.7(e) indicate that selecting 20 virtual samples based on the performance curve maximizes the FFR model's performance.

10.4.1.1.3 Emission Model Results

The hyperparameter settings of the TBLS model are as follows: the minimum sample in a leaf node is set to 3, the number of selected features is $\sqrt{M}$, the number of DT in the mapping and enhancement layer is 10, and the regularization term is 0.01. With the hyperparameter settings of the TBLS model, the results of the training process for the DXN emission model are illustrated in Figure 10.8.

Figure 10.8 illustrates that the learning performance of the static part of the TBLS model is limited. However, the training error of the TBLS model gradually decreases when the DT model is added online. As a result, the TBLS model ultimately converges for the emission phase when the number of incremental DTs exceeds 30.

10.4.1.2 Comparison Results

Considering the model characteristics established in this chapter, several classical and state-of-the-art methods are employed for experimental comparison with the trained WPTM model.

10.4.1.2.1 Generation Model Comparison Based on Process Data

The sizes of the training dataset and the testing dataset are 26 and 10, respectively. The comparative methods employed in this phase include DT, RF, support vector regression (SVR), back propagation neural network (BPNN), multi-layer back propagation neural network (MLBPNN), and DFR. The hyperparameter settings for the aforementioned methods are provided as follows:

(1) DT: minimum sample in leaf node is set to 3;
(2) RF: minimum sample in leaf node is set to 3, number of DT is 500, and selected features are $\sqrt{M}$. SVR: the kernel function is Gaussian, the number of iterations is 1,000;

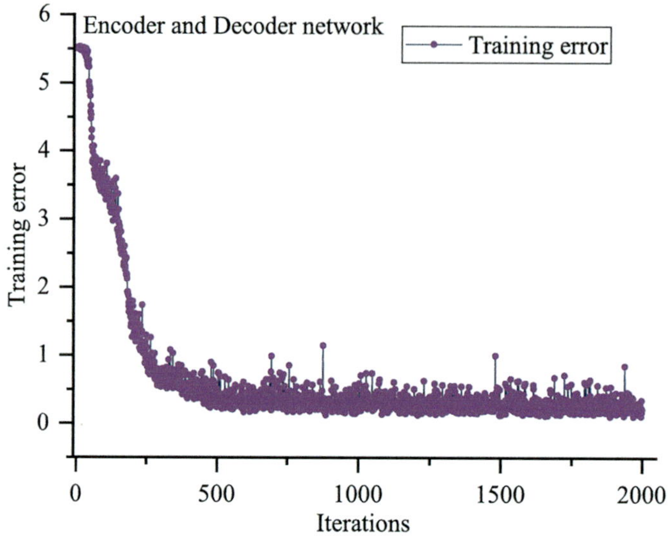

(a) Training error curves for encoder and decoder

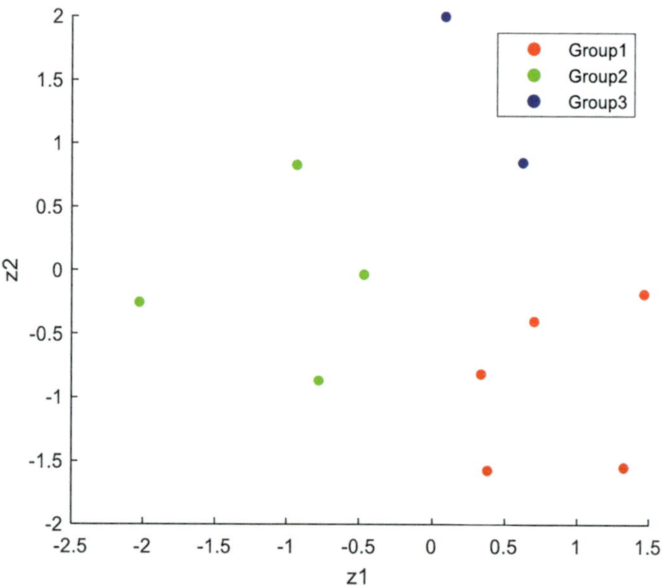

(b) Distribution of real adsorption truth samples

FIGURE 10.7 Training results of the VSG-FFR-based DXN absorption model. (a) Training error curves for encoder and decoder. (b) Distribution of real adsorption truth samples.

(Continued)

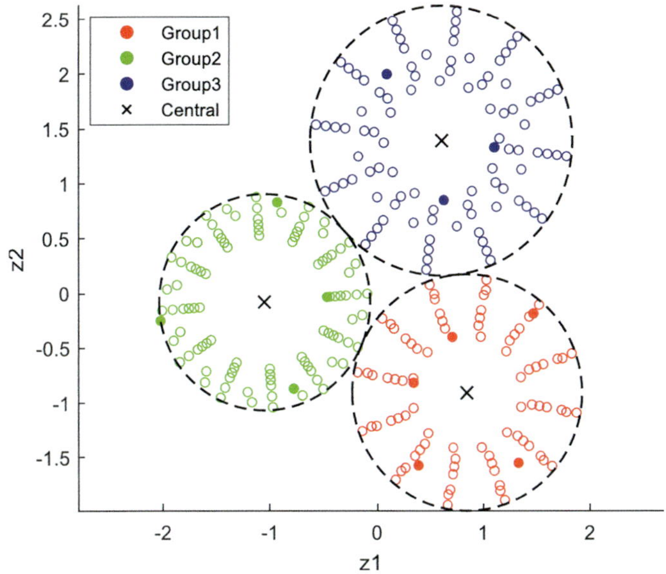

(c) Distribution of virtual and real samples

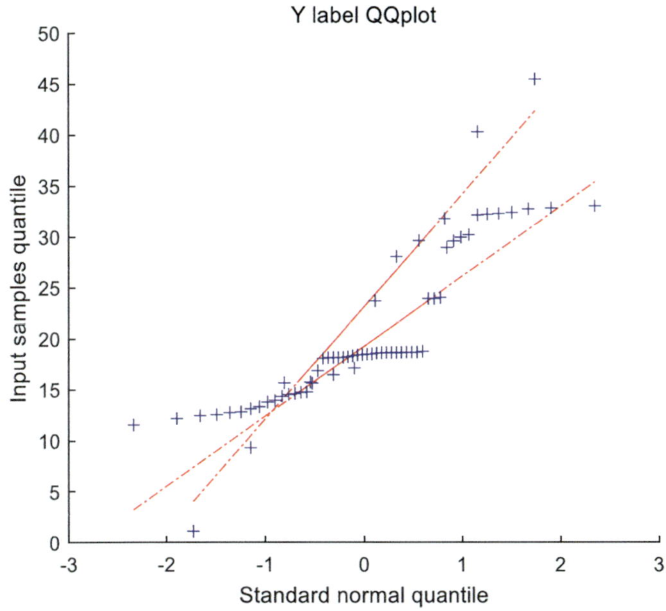

(d) Distribution of truth and pseudo-labeled value

FIGURE 10.7 (Continued) (c) Distribution of virtual and real samples. (d) Distribution of truth and pseudo-labeled value.

(*Continued*)

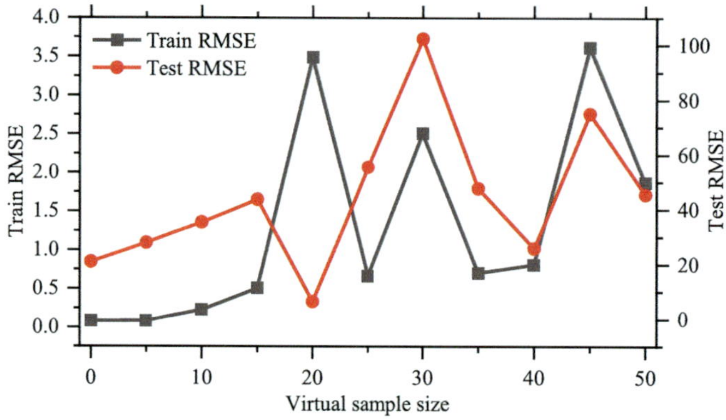

(e) The effect of the number of virtual samples

FIGURE 10.7 (Continued) (e) The effect of the number of virtual samples.

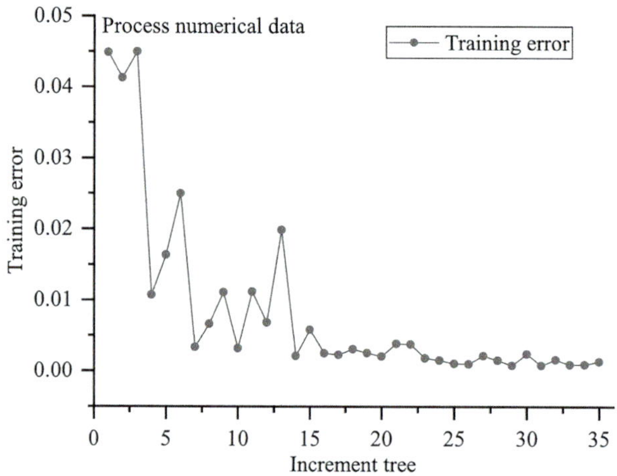

FIGURE 10.8 Training error curves of the TBLS-based DXN emission model.

(3) BPNN: structure is input-hidden-output, number of hidden neurons is set $(2\times M)-1$, epoch is 1,500, and learning rate is 0.001;
(4) MLBPNN: structure is input-hidden1-hidden2-hidden3-output, the number of three hidden neurons is set to $(2\times M)$-$(4\times M)$-$(2\times M)$, the epoch is 1,500, and the learning rate is 0.001.
(5) DFR: number of layers is 50, minimum sample in leaf node is 3, number of DT in each forest is 500, number of RF is 2, number of completely RF is 2, and selected features is $\sqrt{M}$.

Each method is subjected to 20 experiments, and the statistically best results are presented in Table 10.3. The corresponding fitting curves are illustrated in Figure 10.9.

TABLE 10.3
Statistical Results of Different Methods in DXN Generation Model by Using Process Data

	Training Set		Testing Set	
Method	**RMSE**	**MAE**	**RMSE**	**MAE**
DT	0.0366	0.0167	0.0940	0.0681
RF	0.0531	0.0355	0.0948	0.0708
SVR	0.0605	0.0229	0.0980	0.0704
BPNN	0.0031	0.0025	0.2476	0.2051
MLBPNN	0.0185	0.0137	0.0980	0.0674
DFR	0.0561	0.0376	0.0938	0.0704
ADFR-clfc	0.0361	0.0204	0.0838	0.0652

Based on the experimental results presented in Table 10.3 and Figure 10.9, the following conclusions can be drawn:

(1) Overfitting is observed in the BPNN method, resulting in a significant deviation of its measured values in the testing set from the actual conditions.
(2) A comparison between deep learning methods and traditional machine learning methods reveals that deep learning approaches demonstrate greater stability; for instance, the MLBPNN method outperforms the standard BPNN.
(3) The proposed ADFR-clfc method exhibits superior learning and generalization performance compared to DT, RF, and DFR. Its detection trend in the testing set closely aligns with the actual values. Furthermore, compared to MLBPNN, ADFR-clfc shows greater stability in measurements, making it more suitable for practical applications.

10.4.1.2.2 Absorption Model Comparison Based on Process Data

The sizes of the training dataset and the testing dataset are 12 and 4, respectively. The comparative methods employed in this phase include BPNN, SVR, RF, Fuzzy neural network (FNN), and FFR. The hyperparameter settings for the aforementioned methods are provided as follows:

(1) BPNN: Structure is input-hidden-output, number of hidden neurons is set $(2 \times M) - 1$, the epoch is 1,500, and the learning rate is 0.001.
(2) SVR: Kernel function is Gaussian, number of iterations is 1,000.
(3) RF: Minimum sample in leaf node is set to 3, number of DT is 500, and selected features are $\sqrt{M}$.
(4) FNN: Number of fuzzy rules is M, learning rate is 0.01, and iterations are 1000.

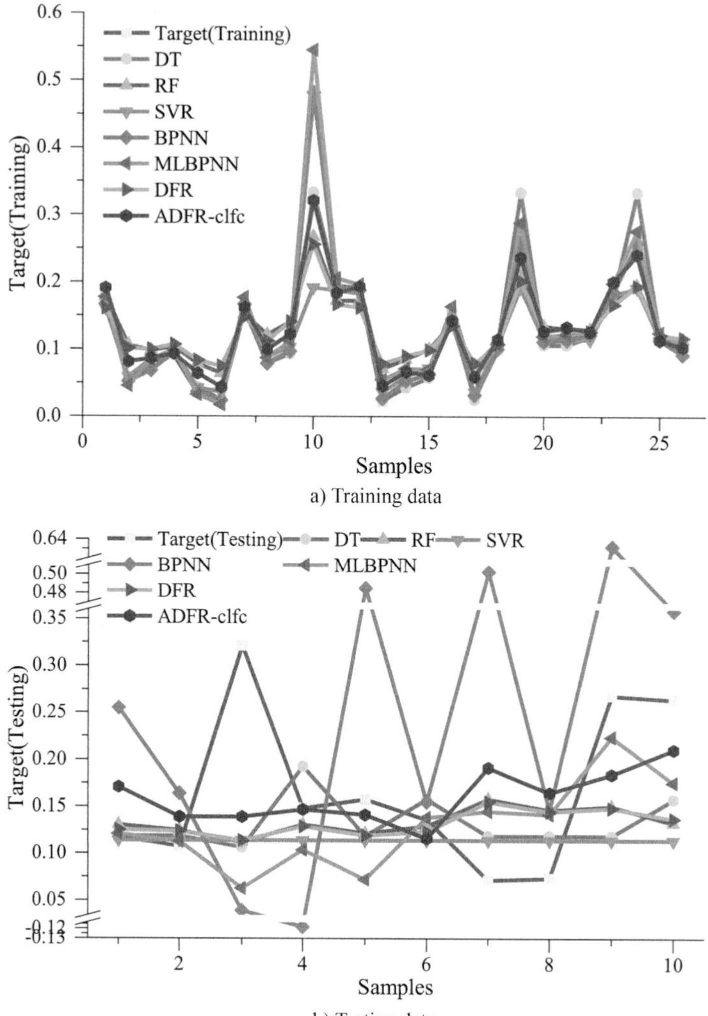

FIGURE 10.9 Fitting curves for different methods in the DXN generation model by using process data. (a) Training data. (b) Testing data.

(5) MLBPNN: Structure is input-hidden1-hidden2-hidden3-output, the number of three hidden neurons is set to $(2\times M)$-$(4\times M)$-$(2\times M)$, the epoch is 1,500, and the learning rate is 0.001.

(6) FFR: Minimum sample in leaf node is 3, number of fuzzy DT is 20, number of fuzzy rules is M'(feature selected), learning rate is 0.01, iterations are 10, and the regularization term is 0.01.

Each method is subjected to 20 experiments, and the statistically best results are presented in Table 10.4. The corresponding fitting curves are illustrated in Figure 10.10.

TABLE 10.4
Statistical Results of Different Methods in DXN Adsorption Model by Using Process Data

	Training Set		Testing Set	
Method	**RMSE**	**MAE**	**RMSE**	**MAE**
BPNN	7.3493	5.9939	11.0984	9.5908
SVR	5.5138	3.8129	19.3716	16.5794
RF	8.1289	6.8939	16.8517	14.3371
FNN	6.7141	4.9369	15.2474	12.7090
MLBPNN	0.3684	0.1985	11.3819	10.0082
FRF	0.0990	0.0777	12.7826	9.4559
VSG+FRF	5.7131	4.5671	9.4500	8.9695

Table 10.4 and Figure 10.10 indicate that MLBPNN and FFR exhibit the best learning ability in the training set. This suggests that highly complex models are prone to overfitting when working with low-dimensional, small-sample data; for example, the generalization performance of MLBPNN is weaker than that of the BPNN model.

In comparison to SVR and RF, the FNN demonstrates better measurement accuracy in the testing set. This implies that the introduction of fuzzy rules in the model aids in characterizing the uncertainty of the data, thereby enhancing the model's generalization performance.

The results comparing VSG-FFR and FFR reveal that the addition of virtual samples serves as a regularizer for the FFR model. In other words, it enhances the model's generalization performance by mitigating potential overfitting. Overall, these findings demonstrate that the proposed VSG and FFR methods are effective in the DXN adsorption model.

10.4.1.2.3 Emission Model Comparison Based on Process Data

In this phase, the sizes of the training dataset and the testing dataset are 26 and 10, respectively. The comparative methods employed include DT, BPNN, SVR, RF, MLBPNN, and Broad learning system neural network (BLSNN). The hyperparameter settings of the above algorithms are provided as follows:

(1) DT: Minimum sample in leaf node is set to 3;
(2) BPNN: Structure is input-hidden-output, number of hidden neurons is set $(2\times M)-1$, epoch is 1,500, and learning rate is 0.001;
(3) SVR: Kernel function is Gaussian, number of iterations is 1,000;
(4) RF: Minimum sample in leaf node is set to 3, number of DT is 500, and selected features is $\sqrt{M}$;
(5) MLBPNN: Structure is input-hidden1-hidden2-hidden3-output, the number of three hidden neurons is set to $(2\times M)$-$(4\times M)$-$(2\times M)$, epoch is 1,500, and learning rate is 0.001;

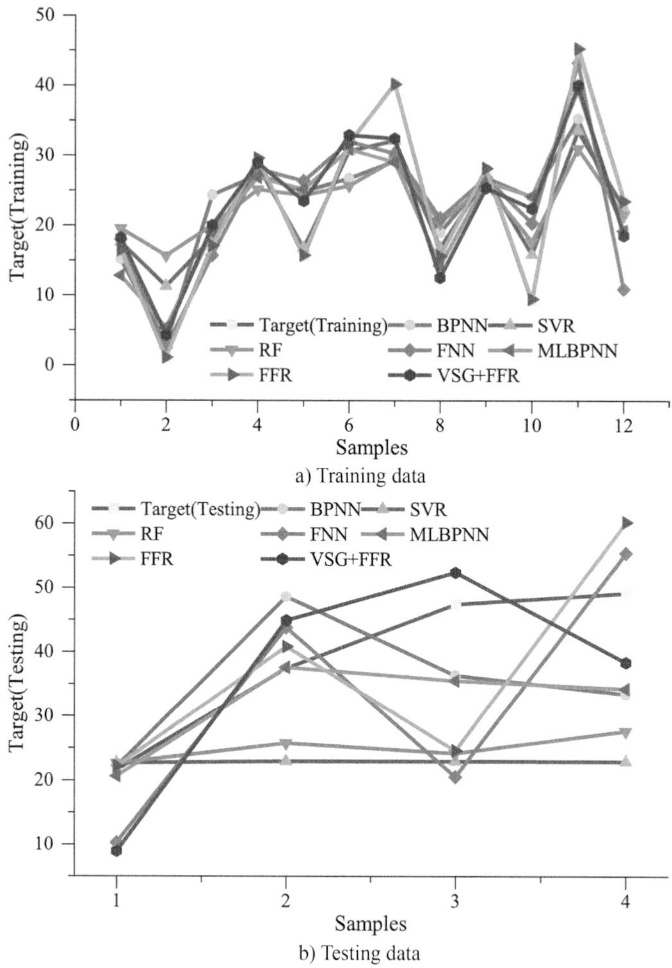

FIGURE 10.10 Fitting curves for different methods in the DXN adsorption model using process data. (a) Training data. (b) Testing data.

(6) BLSNN: Number of feature nodes is 9, mapping groups is 5, enhancement nodes is 41, and regularization term is 0.5.

Each method is subjected to 20 experiments, and the statistically best results are presented in Table 10.5. The corresponding fitting curves are illustrated in Figure 10.11.

Based on the results presented in Table 10.5 and Figure 10.11, the following observations can be made:

(1) There are significant distributional differences between the training and testing data, resulting in considerably higher detection rates in the testing set for most methods, such as DT, BPNN, and BLSNN.

TABLE 10.5
Statistical Results of Different Methods in DXN Emission Model by Using Process Data

	Training Set		Testing Set	
Method	**RMSE**	**MAE**	**RMSE**	**MAE**
DT	0.0323	0.0124	0.6694	0.4837
BPNN	0.0228	0.0171	0.5324	0.4252
SVR	0.3227	0.0953	0.2084	0.1266
RF	0.1684	0.0900	0.1611	0.1424
MLBPNN	0.0185	0.0057	0.1668	0.1381
BLSNN	0.0357	0.0257	0.1980	0.1604
TBLS	0.0004	0.0003	0.1053	0.0937

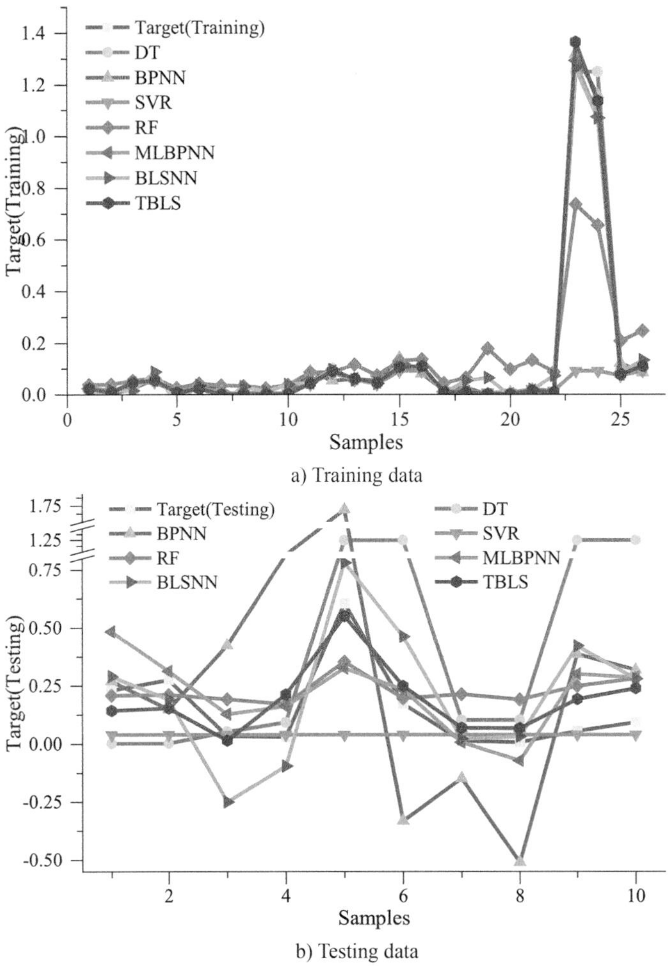

FIGURE 10.11 Fitting curves for different methods in the DXN emission model using process data. (a) Training data. (b) Testing data.

(2) The DT model produces inaccurate output values in the testing set; however, the trend of its curves closely aligns with the actual values. Additionally, the experimental results for RF indicate that aggregating multiple DT models enhances the stability and generalization of the model.
(3) Analyzing the performance of tree-based models (DT, RF, and TBLS) versus neural network-based models (BPNN, MLBPNN, and BLSNN), it is observed that tree-based models generally exhibit better stability, particularly RF and TBLS.
(4) Compared to BLSNN, the TBLS model demonstrates superior learning ability in the training set, achieving stable and highly accurate measurements.

In summary, the experiments conducted validate that the proposed WPTM model is effective in measuring DNX generation, adsorption, and emission at multiple points. This effectiveness is evidenced by comparisons with classical methods and current state-of-the-art approaches. Consequently, the WPTM model, which includes ADFR-clfc, VSG-FFR, and TBLS, successfully achieves comprehensive process modeling for multi-point DXN concentration in the MSWI process.

10.4.1.3 Sensitivity Analysis of Model Input Variables

Utilizing the WPTM model established for the comprehensive process of DXN, this section analyzes the significance of input variables in the application of the model. The objective is to provide additional insights into the proposed framework. The approach begins by setting upper and lower limits for the input variables based on expert knowledge. Subsequently, 100 values are selected at uniform intervals within this range and fed into the WPTM. The importance of the input variables is then computed by examining the variations in the soft measuring values of the DXN generation, adsorption, and emission models, as illustrated in Figure 10.12.

10.4.1.3.1 Generation Phase Input Variables Analysis

Figure 10.12(a)–(g) depicts the impact of input variables associated with the DXN generation process, encompassing process data from the DXN generation phase and ORD, on the WPTM model.

(1) Overall, the preceding 124 variables show minimal influence on the DXN emission concentration. This suggests that the control measures implemented during the combustion and heat recovery stages have a limited effect on the DXN emission concentration in the exhaust gas.
(2) When comparing the adsorption and generation phases, the scores for the adsorption phase are higher than those for the generation phase. This indicates that operational control of the combustion process has a more pronounced impact on DXN adsorption concentration. The most influential variables (with scores exceeding 0.2) include the 4th furnace gas temperature (x4), the 2nd burn1 grate temperature (x24), the 3rd burn1 grate temperature (x25), the 8th burn1 grate temperature (x30), the 2nd evaporators

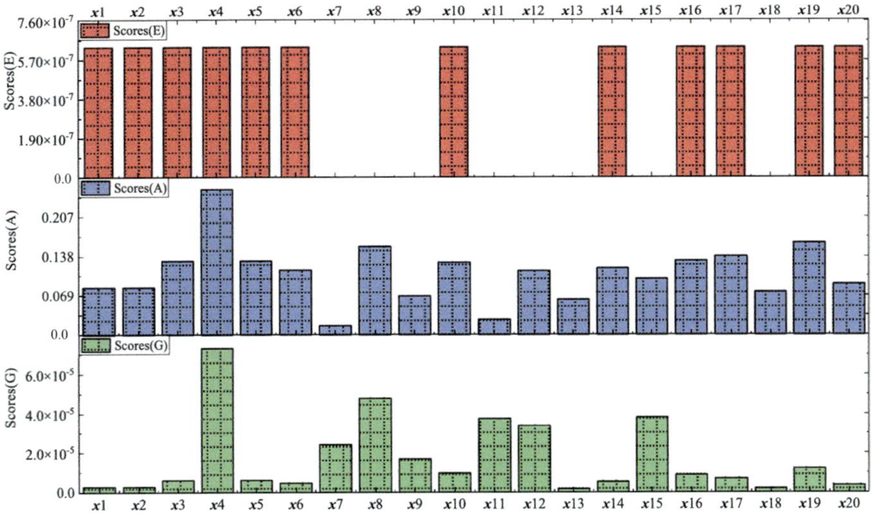

(a) Input variables 1-20 from DXN generation phase

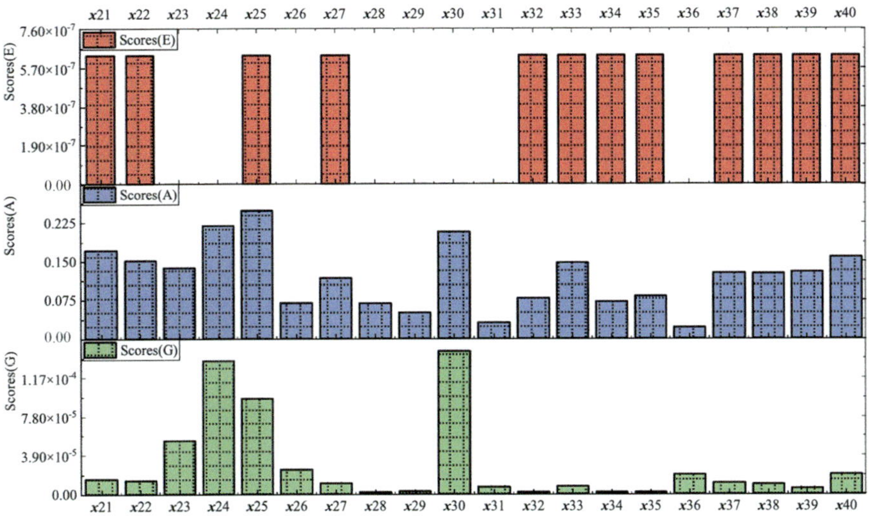

(b) Input variables 21-40 from DXN generation phase

FIGURE 10.12 Impact of input variables from process data of multi-point DXN concentration on the WPTM model. (a) Input variables 1–20 from the DXN generation phase. (b) Input variables 21–40 from the DXN generation phase.

(Continued)

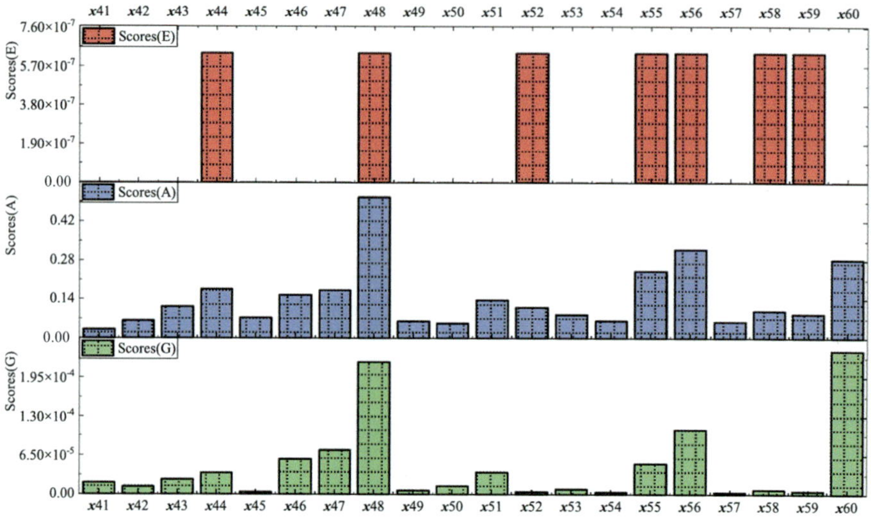

(c) Input variables 41-60 from DXN generation phase

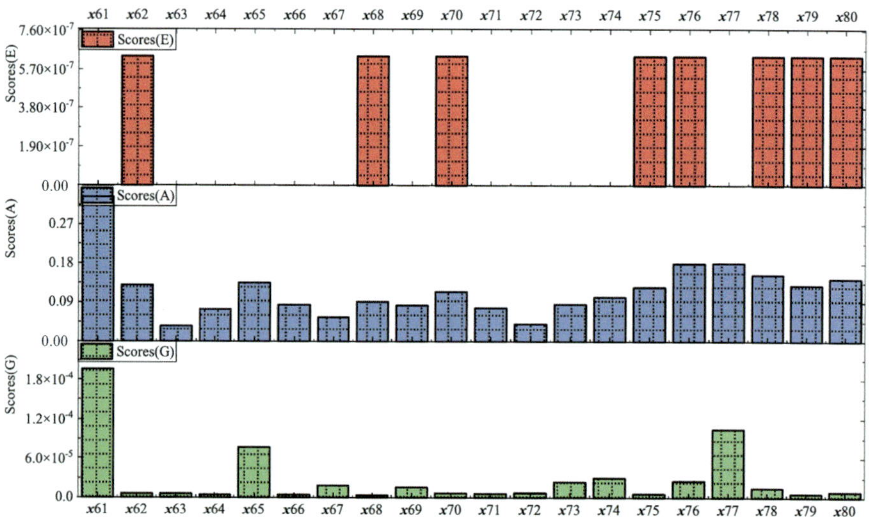

(d) Input variables 61-80 from DXN generation phase

FIGURE 10.12 (Continued) (c) Input variables 41–60 from the DXN generation phase. (d) Input variables 61–80 from the DXN generation phase.

(Continued)

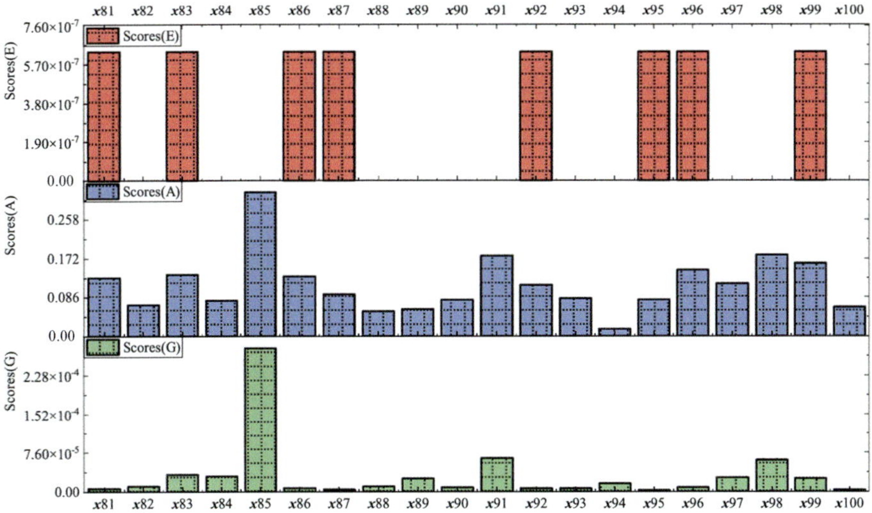

(e) Input variables 81-100 from DXN generation phase

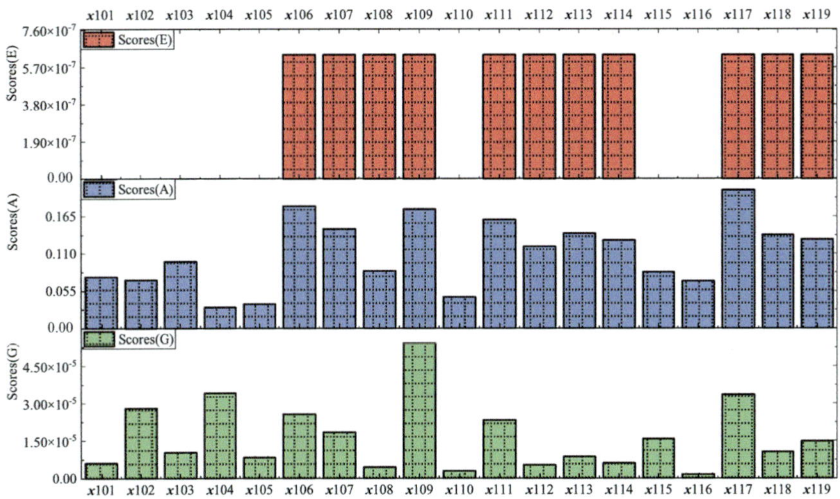

(f) Input variables 101-119 from DXN generation phase

FIGURE 10.12 (Continued) (e) Input variables 81–100 from the DXN generation phase. (f) Input variables 101–119 from the DXN generation phase.

(Continued)

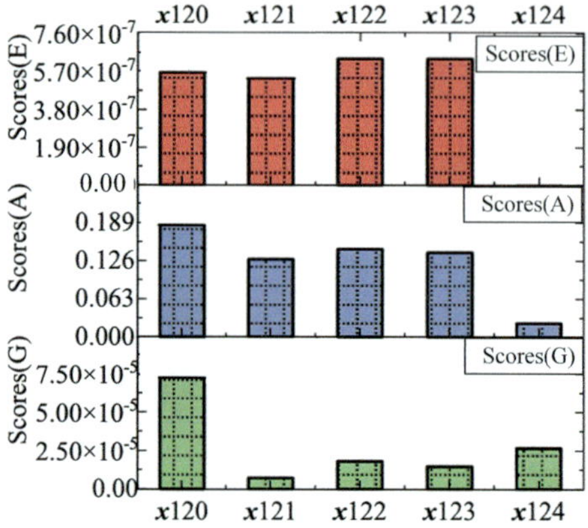

(g) Input variables 120-124 from DXN generation phase

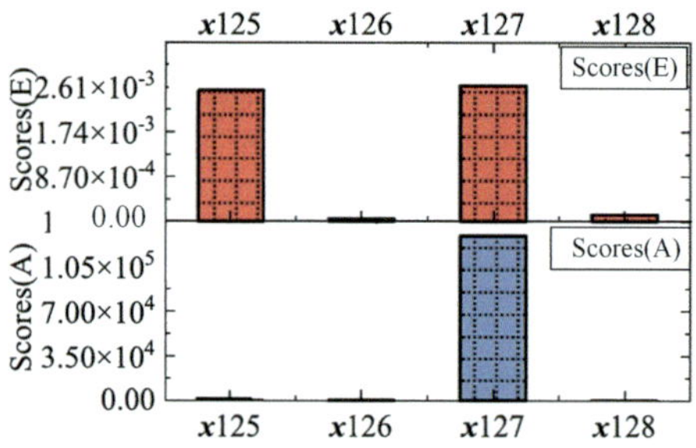

(h) Input variables 125-128 from DXN adsorption phase

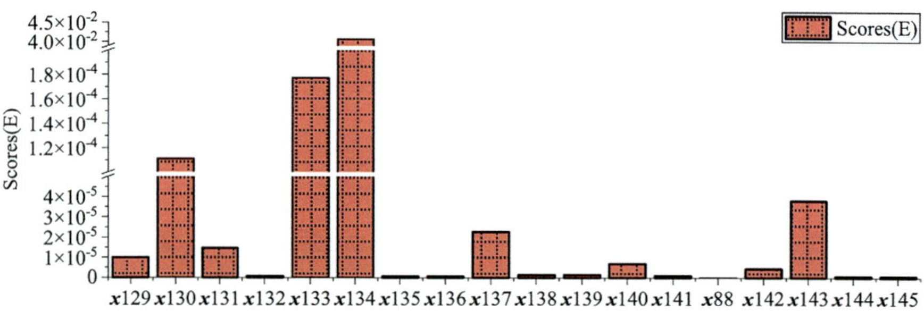

(i) Input variables 129-145 from DXN emission phase

FIGURE 10.12 (Continued) (g) Input variables 120–124 from the DXN generation phase. (h) Input variables 125–128 from the DXN adsorption phase. (i) Input variables 129–145 from the DXN emission phase.

(Continued)

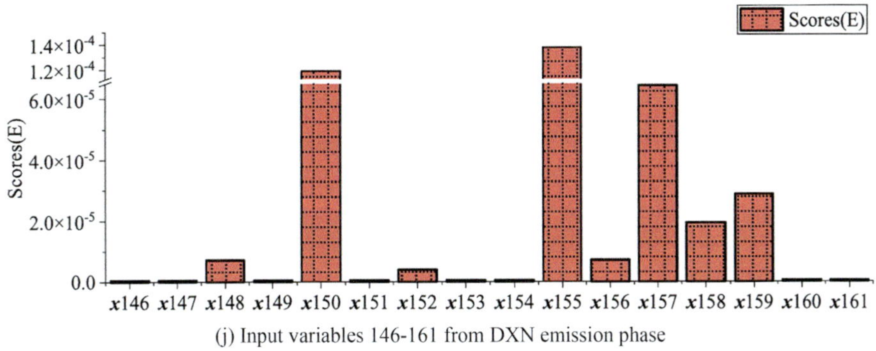

FIGURE 10.12 (Continued) (j) Input variables 146–161 from the DXN emission phase.

temperature (x48), the secondary airflow temperature (x55), the 1st wall cooling air outlet temperature (x56), the dry grate outlet air temperature (x60), the burn grate outlet air temperature (x61), the 1st superheater cooling water (x85), and the boiler water level (x117).

(3) For DXN generation concentration, significant impacting variables include the 2nd burn1 grate temperature (x24), the 8th burn1 grate temperature (x30), the 2nd evaporators temperature (x48), the 1st wall cooling air outlet temperature (x56), the dry grate outlet air temperature (x60), the burn grate outlet air temperature (x61), the 4th burn2 grate airflow (x77), and the 1st superheater cooling water (x85).

(4) Among the recorded operational variables, the feed rate (x120) exhibits a high impact on both DXN generation and adsorption concentrations.

10.4.1.3.2 Absorption Phase Input Variables Analysis

Figure 10.12(h) illustrates the impact of input variables related to the DXN adsorption process on the WPTM model. The observations are as follows:

(1) For DXN adsorption concentration, the lime feeder speed (x127) has a more significant impact than other variables.

(2) The numerical process variables in the adsorption phase exert distinct programmable effects on DXN emission concentrations, with the most pronounced effects observed for the 1st mixer feedwater flow (x125) and the B. lime feeder speed (x127).

10.4.1.3.3 Emission Phase Input Variables Analysis

Figure 10.12(i) and (j) present the impact of input variables from the DXN emission phase on the WPTM model. The results indicate that:

(1) Variables 129 through 152 are related to the gas cleaning stage, while variables 153 through 161 pertain to conventional pollutants. The experimental results reveal that the WPTM model is more sensitive to the variables in the flue gas cleaning stage compared to those associated with conventional pollutants.
(2) Among the variables from 129 to 152, the most influential on DXN emission concentrations are the 2nd inlet flue gas temperature (x130), the 2nd bag filter inlet temperature (x133), the fluidizing fan heater temperature (x134), and the 1st bag filter differential pressure (x150).
(3) Among the conventional pollutants, the concentrations of NO_x (x155), SO_2 (x 156), HCl (x157), CO (x158), and CO_2 (x159) have a primary impact on DXN emission concentration.

10.4.1.3.4 Summary Analysis

This chapter presents a comprehensive framework for developing an AI-based soft measuring model for DXN concentration across the whole MSWI process. It employs various tree-based modeling approaches tailored for each phase. In the generation phase, a novel ADFR-clfc method is introduced to manage the high-dimensional feature data associated with DXN generation. The adsorption phase addresses the challenges posed by ultra-small sample sizes and inherent uncertainties through the utilization of VSG and FFR methods. For the emission phase, characterized by medium-dimensional small-sample data, the TBLS approach is implemented to construct the emission model. The proposed WPTM approach utilizes a hierarchical structure with DT as fundamental units, leveraging deep ensemble techniques, broad ensemble methods, fuzzy operators, and VSG technology. The effectiveness of the WPTM approach is demonstrated through experimental results.

The distribution concentrations of the DXN generation, adsorption, and emission phases under different input data in the MSWI process are illustrated in Figure 10.13.

This result demonstrates the effectiveness of the proposed WPTM model in soft measuring DXN across multiple phases using multimodal data for the MSWI process. Consequently, this framework can be utilized to explore mechanism characterization and support studies aimed at optimizing DXN reduction control.

10.4.2 Strengths

For further investigation and engineering applications, the following strengths of the WPTM approach are highlighted:

(1) Comprehensive Understanding: The WPTM model provides a thorough understanding of the complex interplay between DXN generation, adsorption, and emission within the MSWI process by leveraging readily available process data. This deep understanding enhances informed decision-making

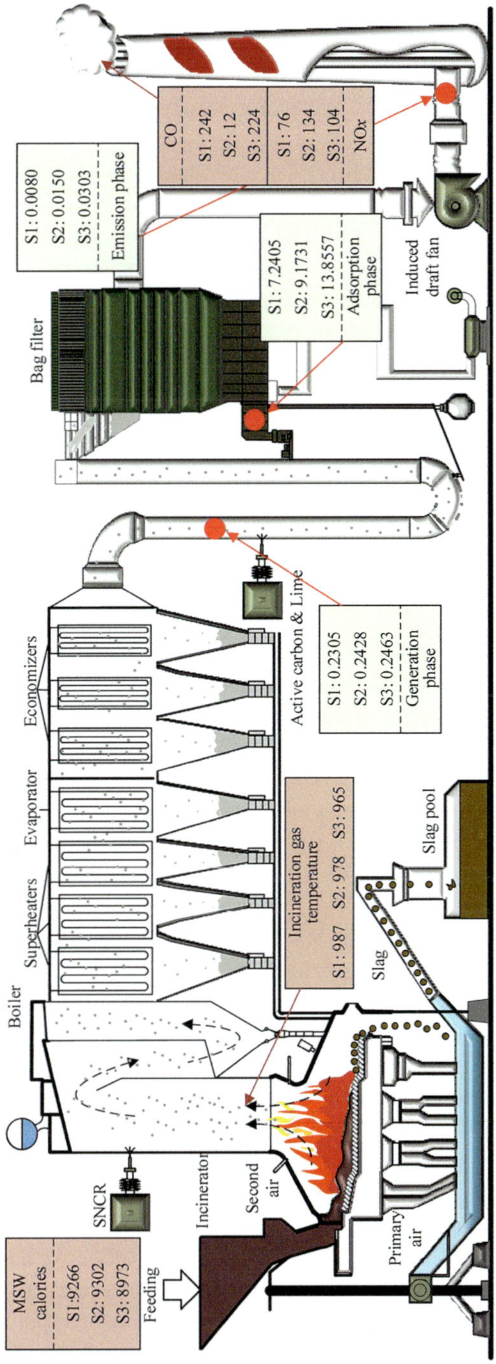

FIGURE 10.13 Distribution concentration of DXN generation, adsorption, and emission phases with different input data.

for optimizing process control and reducing DXN emissions. Notably, the WPTM approach is adaptable to other waste-to-energy plants that involve similar stages, such as incineration, heat exchange, gas cleaning, and gas emission. This versatility makes the proposed WPTM method a valuable tool for achieving environmental and energy sustainability by facilitating effective DXN reduction and emission control across a wide range of waste-to-energy facilities.

(2) Robustness and Interpretability: The foundation of the WPTM approach lies in the use of tree-based models, including ADFR-clfc for generation, VSG-FFR for adsorption, and TBLS for emission. Tree models are recognized for their robustness to noise and outliers, ensuring model stability even amid data imperfections. Their transparent and easily interpretable computational process aids in understanding and analyzing the model's predictions. Experimental results demonstrate the interpretability of the VSG technique, further enhancing the WPTM model's capability to trace and explain anomalous prediction outcomes. This combination of robustness and interpretability makes the WPTM model particularly well-suited for accurately predicting DXN generation, adsorption, and emission concentrations in MSWI processes, ultimately contributing to improved process control and environmental performance.

(3) Transferability and Engineering Applications: The WPTM framework, encompassing the ADFR-clfc generation model, VSG-FFR adsorption model, and TBLS emission model, exhibits significant transferability beyond MSWI. These individual components can be readily adapted for DXN concentration prediction in various industrial settings, including those handling industrial or medical waste. By incorporating relevant process data and potentially adjusting model parameters or structures, the WPTM approach can effectively address the specific requirements of different waste types and industrial processes, fostering broader applicability and promoting environmental sustainability across a range of industries.

From an engineering perspective, the WPTM model offers valuable insights for optimizing MSWI processes and minimizing DXN emissions. It can identify key process variables, such as the 2nd inlet flue gas temperature (x130), 2nd bag filter inlet temperature (x133), fluidizing fan heater temperature (x134), and 1st bag filter differential pressure (x150), that significantly impact DXN concentration. This knowledge enables targeted adjustments to operating conditions to minimize emissions. Furthermore, the WPTM can facilitate automated control systems by providing real-time predictions of DXN concentration, allowing for dynamic adjustments to process parameters, optimization of combustion conditions, and informed implementation of specific air pollution control technologies.

In conclusion, the WPTM approach presents a highly effective and versatile modeling framework designed to meet the complex demands of predicting DXN concentration in the MSWI process. Its ability to provide interpretable predictions makes it a promising tool for advancing our understanding and control of DXN emissions. By enabling informed decision-making for process optimization and emission reduction, the WPTM significantly contributes to improved environmental performance and the advancement of sustainable waste management practices.

10.4.3 Limitations

Assuming that the model's predictions for DXN generation, adsorption, and emission concentrations in the MSWI process meet actual needs, the following three scenarios illustrate the role of the model:

> Scenario 1: When the model predicts low DXN generation and emission concentrations, the recommended optimized guidance, such as "maintain the current control strategy," is provided to the operating engineer;
> Scenario 2: When the model predicts high DXN generation and emission concentrations, the optimized guidance recommendation is to "slowly control the key process variables under the precondition of stable MSW combustion" with the aim of reducing DXN emission concentrations;
> Scenario 3: When the model predicts low DXN generation concentration but high emission concentration, the recommended guidance, such as "check the operating status of the bag filter," is provided to the operating engineer.

Despite its strengths, the current version of the WPTM model has several limitations that need to be addressed in future research. These limitations include:

(1) Handling Data Bias: The training data for the WPTM model spans a single year with a limited sample size, which restricts the model's ability to capture the full range of variations in the target phenomenon. This limitation introduces biases into the model's detection processes for generation, adsorption, and emission concentrations, ultimately impacting overall accuracy. Future research should explore the use of unsupervised data or VSG technology to enhance the WPTM model's performance and mitigate the effects of limited training data.
(2) Hyperparameter Optimization: The WPTM model incorporates ADFR-clfc, VSG-FFR, and TBLS approaches for predicting DXN concentrations within the MSWI process. To ensure a fair and objective comparison, all methods are standardized with similar hyperparameters, adhering to common best practices whenever possible. Although the WPTM model demonstrates superior performance based on expert-determined hyperparameter settings, future research should prioritize optimizing hyperparameters specifically for the DXN generation, adsorption, and emission models within the WPTM framework. This optimization will further enhance the model's accuracy and effectiveness in predicting DXN concentrations in the MSWI process.
(3) Condition Adaptability: The current modeling data represents specific operating conditions, and unseen data can be categorized into two scenarios: similar and dissimilar. The WPTM model has shown robustness and effectiveness under similar conditions. However, dissimilar scenarios represent

concept drift in operating conditions and data distribution, making accurate measurement of DXN generation, adsorption, and emission concentrations challenging. To address this issue in engineering applications, where MSW composition is inherently uncertain, the WPTM model should be adaptable to varying operating conditions.

10.5 SUMMARY

This chapter introduces a multi-point DXN concentration modeling method using an AI-based advanced tree algorithm, referred to as WPTM, for the comprehensive assessment of DXN concentrations in the whole MSWI process. The proposed WPTM model is validated against historical data from an actual MSWI power plant in Beijing, and its validity is demonstrated through comparisons with other state-of-the-art methods.

The main contributions of this work are summarized as follows: (1) A soft sensor for measuring DXN concentration during the generation, adsorption, and emission phases is developed using an advanced tree-based deep and broad learning approach; (2) Distinct modeling strategies are designed for each phase, ensuring that they align with the specific characteristics of each stage; (3) These results lay the groundwork for the soft measurement of DXN concentration at each phase, thereby enhancing our understanding of the incineration process.

As demonstrated by experimental results, the proposed method outperforms current modeling algorithms. However, the WPTM model has some limitations, including: (1) The current version is a static model, which may lead to significant deviations in measuring DXN concentrations under unknown operating conditions; (2) The small sample size of the modeling data poses a risk to the generalization performance of the WPTM; (3) Operational condition drift is not fully accounted for during the modeling and design process.

To address these limitations, future research should focus on the following areas: (1) Exploring online updates of model components based on dynamic operating conditions to ensure effective deployment of the WPTM model in real-world scenarios; (2) Integrating mechanistic knowledge and data into the WPTM model to enhance generalization performance and better align with actual scenario requirements; (3) Developing condition drift monitoring methods for the MSWI process to provide a basis for the online updating of the WPTM model.

APPENDIX

TABLE 10A.1
Explanation of Abbreviations

Abbreviations	Explanation
MSW	Municipal solid waste
MSWI	MSW incineration
DXN	Dioxins
AI	Artificial intelligence

(Continued)

TABLE 10A.1 (Continued)
Explanation of Abbreviations

Abbreviations	Explanation
ADFR-clfc	Adaptive DFR based on cross-layer full connection
TBLS	Tree broad learning system
FFR	Fuzzy forest regression
DCS	Distribution control system
DT	Decision Tree
DFR	Deep forest regression
WPTM	Whole-process tree-based model
ORD	operational recorded data
VSG	Virtual sample generation
APCDs	Air Pollution Control Devices
HRGC/HRMS	High-resolution gas chromatography, high-resolution mass spectroscopy
LVVSG	Latent variable VSG
RF	Random forest
CRF	Completely RF
MAE	Mean absolute error
RMSE	Root mean square error
SVR	Support vector regression
BPNN	Back propagation neural network
MLBPNN	Multiple-layer backpropagation neural network
FNN	Fuzzy neural network
BLSNN	Broad learning system neural network

REFERENCES

1. McKay G. Dioxin characterisation, formation and minimisation during municipal solid waste (MSW) incineration: review. *Chemical Engineering Journal*, 2002, 86(3): 343–368.
2. Altarawneh M, Saeed A, Al-Harahsheh M, et al. Thermal decomposition of brominated flame retardants (BFRs): products and mechanisms. *Progress in Energy and Combustion Science*, 2019, 70: 212–259.
3. Lin B C, Yang L L, Zheng M H, et al. Synergetic promoting/inhibiting mechanisms of copper/calcium compounds in the formation of persistent organic pollutants and environmentally persistent free radicals from anthracene. *Chemical Engineering Journal*, 2022, 441: 1–11.
4. Fu Z, Lin S, Tian H, et al. A comprehensive emission inventory of hazardous air pollutants from municipal solid waste incineration in China. *Science of the Total Environment*, 2022, 826: 1–11.
5. Wei J, Li H, Liu J. Fate of dioxins in a municipal solid waste incinerator with state-of-the-art air pollution control devices in China. *Environmental Pollution*, 2021, 289: 1–10.
6. Lei R, Xu Z, Xing Y, et al. Global status of dioxin emission and China's role in reducing the emission. *Journal of Hazardous Materials*, 2021, 418: 1–10.
7. Xia H, Tang J, Aljerf L, et al. Assessment of PCDD/Fs formation and emission characteristics at a municipal solid waste incinerator for one year. *Science of the Total Environment*, 2023, 883: 1–15.

8. Eftekhari M, Gheibi M, Azizi-Toupkanloo H, et al. Statistical optimization, soft computing prediction, mechanistic and empirical evaluation for fundamental appraisal of copper, lead and malachite green adsorption. *Journal of Industrial Information Integration*, 2021, 23: 1–21.
9. Qiao J F, Guo Z H, Tang J. Dioxin emission concentration measurement approaches for municipal solid wastes incineration process: a survey. *Acta Automatica Sinica*, 2020, 46(6): 1063–1089.
10. Munir M T, Li B, Naqvi M. Revolutionizing municipal solid waste management (MSWM) with machine learning as a clean resource: opportunities, challenges and solutions. *Fuel*, 2023, 348: 1–19.
11. Tang J, Wang T Z, Xia H, et al. An overview of artificial intelligence application for optimal control of municipal solid waste incineration process. *Sustainability*, 2024, 16(5): 1–41.
12. Cao X, Lu S, Stevens W R, et al. Fast indirect measurement of PCDD/F TEQ emission from municipal solid waste incineration: a review. *Waste Disposal & Sustainable Energy*, 2019, 1: 39–51.
13. Chen K, Peng Y, Lu S, et al. Bagging based ensemble learning approaches for modeling the emission of PCDD/Fs from municipal solid waste incinerators. *Chemosphere*, 2021, 274:1–8.
14. Blumenstock M, Zimmermann R, Schramm K W, et al. Identification of surrogate compounds for the emission of PCDD/F (I-TEQ value) and evaluation of their on-line realtime detectability in flue gases of waste incineration plants by REMPI–TOFMS mass spectrometry. *Chemosphere*, 2021, 42(5–7): 507–518.
15. Cao X, Stevens W R, Tang S, et al. Atline measurement of 1, 2, 4-trichlorobenzene for polychlorinated dibenzo-p-dioxin and dibenzofuran International Toxic Equivalent Quantity prediction in the stack gas. *Environmental Pollution*, 2019, 244: 202–208.
16. Xiong S, Lu S, Shang F, et al. Online predicting PCDD/F emission by formation pathway identification clustering and Box-Cox Transformation. *Chemosphere*, 2021, 274: 1–9.
17. Xiong S, Peng Y, Lu S, et al. Generalized prediction and optimal operating parameters of PCDD/F emissions by explainable Bayesian support vector regression. *Waste Management*, 2021, 135: 437–447.
18. Hasberg W, May H, Dorn I. Description of the residence-time behaviour and burnout of PCDD, PCDF and other higher chlorinated aromatic hydrocarbons in industrial waste incineration plants. *Chemosphere*, 1989, 19(1–6): 565–571.
19. Tang J, Xia H, Aljerf L, et al. Prediction of dioxin emission from municipal solid waste incineration based on expansion, interpolation, and selection for small samples. *Journal of Environmental Chemical Engineering*, 2022, 10(5): 1–13.
20. Tang J, Qiao J F. Dioxin emission concentration soft measuring approach of municipal solid waste incineration based on selective ensemble kernel learning algorithm. *CIESC Journal*, 2019, 70(2): 696–706.
21. Heng X, Tang J, Cong Q M, et al. Dioxin emission concentration forecasting model for MSWI process with random forest-based transfer learning. In: *39th Chinese Control Conference*, Shenyang, China, 27–29 July 2020. 2020, 5724–5729.
22. Guo Z H, Tang J, Qiao J F, et al. Dioxin emission concentration soft measurement model of MSWI process based on unmarked samples and improved deep belief network. In: *39th Chinese Control Conference*, Shenyang, China, 27–29 July 2020. 2020, 5784–5789.

23. Tang, J, Xia H, Qiao J F, et al. Soft measurement of dioxin emission concentration based on deep forest regression algorithm. *International Journal of System Control and Information Processing*, 2021, 3(3): 208–228.
24. Xia H, Tang J, Aljerf L. Dioxin emission prediction based on improved deep forest regression for municipal solid waste incineration process. *Chemosphere*, 2022, 294: 1–13.
25. Kim K S, Hong K H, Ko Y H, et al. Emission characteristics of PCDD/Fs, PCBs, chlorobenzenes, chlorophenols, and PAHs from polyvinylchloride combustion at various temperatures. *Journal of the Air & Waste Management Association*, 2004, 54(5): 555–562.
26. Altwicker E R. Some laboratory experimental designs for obtaining dynamic property data on dioxins. *The Science of the Total Environment*, 1991, 104(1–2): 47–72.
27. Altarawneh M, Dlugogorski B Z, Kennedy E M, et al. Mechanisms for formation, chlorination, dechlorination and destruction of polychlorinated dibenzo-p-dioxins and dibenzofurans (PCDD/Fs). *Progress in Energy and Combustion Science*, 2009, 35(3): 245–274.
28. Peng Y, Chen J, Lu S, et al. Chlorophenols in municipal solid waste incineration: a review. *Chemical Engineering Journal*, 2016, 292: 398–414.
29. Stieglitz L. Selected topics on the de novo synthesis of PCDD/PCDF on fly ash. *Environmental Engineering Science*, 1998, 15(1): 5–18.
30. Zhang M M, Buekens A. De novo synthesis of dioxins: a review. *International Journal of Environment and Pollution*, 2016, 60(1–4): 63–110.
31. Tang J, Xia H, Zhang, J. et al. Deep forest regression based on cross-layer full connection. *Neural Computing and Applications*, 2021, 33: 9307–9328.
32. Xia H, Tang J, Yu W, et al. Takagi–Sugeno fuzzy regression trees with application to complex industrial modeling. *IEEE Transactions on Fuzzy Systems*, 2023, 31(7): 2210–2224.
33. Xia H, Tang J, Yu W, et al. Tree broad learning system for small data modeling. *IEEE Transactions on Neural Networks and Learning Systems*, 2024, 35(7): 8909–8923.
34. Xia H, Tang J, and Cui C L. Virtual sample generation method based on feature scaling and co-training label for industrial data modeling. In: *5th International Conference on Industrial Artificial Intelligence*, Shenyang, China, 21–24 August 2023. 2023, 1–6.

11 Virtual Sample Generation Method Based on Generative Adversarial Fuzzy Neural Network for Dioxin Emission Risk Warning

11.1 RELATED WORKS

11.1.1 Dioxin Emission Risk Warning Description

Measuring the Key Performance Indicator (KPI) of complex industrial processes online is challenging due to limitations in detection technology and high economic costs [1]. Additionally, industrial processes often operate in a steady state for extended periods, resulting in an unbalanced distribution of abnormal conditions and a limited number of fault samples [2]. Consequently, KPI modeling for industrial processes must address the issue of small sample sizes. The complexity of industrial processes arises from their mechanisms and structures, driven by increasing demands for quality, performance, and cost-effectiveness. Advances in automation and information technologies have enabled the use of a growing number of instruments and devices to collect operational data from various perspectives. As a result, industrial process data exhibit high dimensionality and significant uncertainty [3, 4]. In summary, KPI modeling samples in industrial contexts are characterized by high dimensionality, considerable uncertainty, and limited sample sizes.

Various small sample modeling methods have been proposed to address the challenge of limited modeling samples, including support vector machine (SVM) [5], Bayesian network [6], and kernel regression [7]. With advancements in hardware performance and neural network techniques, deep learning (DL) has rapidly evolved and become closely integrated with industry, demonstrating a strong capability for handling complex modeling tasks [8]. Deep neural networks (DNN), as a typical DL model, are widely employed in regression prediction [9, 10], image recognition [11, 12], and fault diagnosis [13, 14]. However, they struggle to effectively tackle KPI modeling problems based on small sample datasets. To address this limitation, deep forest classification (DFC), which is based on non-neural network models and incorporates concepts from DNNs and ensemble forests, has been proposed for small sample scenarios [15, 16]. Nevertheless, DFC may not guarantee robust

DOI: 10.1201/9781003634195-14

generalization performance due to the limited sample sizes available for KPI modeling [17, 18]. Virtual sample generation (VSG) has been introduced to mitigate these challenges and is widely utilized in complex industrial processes such as petroleum, chemical, and mechanical manufacturing [19, 20].

In nature, VSG relies on original data to create virtual samples that help bridge the gap between existing data [21]. This approach can partially address issues related to small sample sizes and class imbalance [22]. Currently, there are three main VSG methods: those based on sampling, information diffusion, and generative adversarial networks (GANs). The sampling-based VSG generates new samples by resampling from the original dataset or its distribution. For instance, Bennin et al. employed resampling techniques on minority classes within software defect data to balance sample sizes [23]; however, this method does not introduce any new samples. Similarly, Xie et al. proposed a VSG method based on Gaussian distribution sampling [24]. Nonetheless, this approach is contingent upon the characteristics of the original sample distribution, making it challenging to apply in high-dimensional contexts. Subsequently, high-dimensional samples were projected into low-dimensional space using *t*-distributed Stochastic Neighbor Embedding (*t*-SNE) to generate virtual samples [25]. However, this approach can lead to information loss and the creation of redundant samples. On the other hand, information diffusion-based VSG generates virtual samples by leveraging the theoretical diffusion range of characteristics. This includes methods such as mega-trend diffusion [26] and tree-based trend diffusion [27]. However, these techniques are vulnerable to the influence of outlier data, which can introduce bias into the diffusion range.

In contrast to the previously mentioned methods, GAN-based VSG enhances the realism of generated virtual samples through the adversarial interplay between the generator and the discriminator [28]. In the realm of image generation, Lian et al. successfully used GANs to create defect images [29], thereby increasing the number of training samples. Their experimental results indicated a significant improvement in defect detection accuracy. When it comes to generating industrial process data, GAN-based VSG is primarily utilized for producing fault signal samples. For instance, Wang et al. proposed a data rebalancing mechanism leveraging Wasserstein GAN [30]. They employed a long short-term memory full convolutional network along with high-dimensional vibration frequency spectrums to facilitate fault diagnosis after generating fault samples. Similarly, Cui et al. introduced a DXN model based on active learning GAN (AL-GAN) [31]. However, this approach is subject to human interference and neglects the inherent uncertainty of DXN samples. As a result, research into GAN-based VSG for industrial process data remains an open and evolving area.

Feature reduction is a common strategy that encompasses both feature extraction and selection to address the challenges posed by high-dimensional modeling samples [32]. Feature extraction employs either linear or nonlinear methods to derive low-dimensional data that replaces the original high-dimensional data [33]. However, the extracted features often lack physical significance, and features with low contribution rates can destabilize the model. In contrast, feature selection involves discarding certain features while retaining those with clear physical meanings, making it particularly suitable for industrial processes where causal relationships exist between inputs and outputs [34]. For instance, Xia et al. proposed a feature selection algorithm based on random forests (RF) [35]. However, this adaptive selection does not take into account the characteristics of the data.

Fuzzy neural networks (FNN) have been introduced to address the challenges associated with uncertainty in modeling samples. FNN is a versatile modeling algorithm that combines the nonlinear processing and analytical capabilities of fuzzy systems with the parameter learning and optimization functions of neural networks [36, 37]. For example, Zhang et al. proposed an FNN framework for nonlinear system identification in industrial coke furnaces [38], and their simulation results demonstrated the advantages of this approach. Similarly, Wang et al. utilized an adaptive FNN to connect multiple parameters and properties of high-performance gray cast iron [39], with sensitivity test results indicating strong performance. An error probability density function FNN, updated via an adaptive learning rate gradient descent, was introduced to predict the effluent ammonia nitrogen levels in sewage treatment plants [40]. This method exhibited superior prediction accuracy and model stability compared to other techniques. Juang et al. used FNN to estimate the automatic hitting duration for rechargeable impact wrenches [41], allowing for precise torque control. Additionally, Adeleke et al. combined FNN with evolutionary algorithms to predict the lower heating value of waste [42], achieving optimal accuracy. However, there are limited reports on the application of FNN for generating virtual samples.

11.1.2 Random Forest (RF)

RF is an ensemble algorithm based on decision trees (DT), which utilizes the bagging method to construct various DT models. As a result, RF exhibits strong adaptability to noise and effective capabilities for processing nonlinear data [15]. The process of establishing the RF is outlined as follows.

Let the data be represented as $\left\{\left(\boldsymbol{x}_i, y_i\right)\right\}_{i=1}^{N} \in R^{N\times(D+1)}$, where $\boldsymbol{x}_i$ is the input and y_i is the output. The number of samples in $\left\{\left(\boldsymbol{x}_i, y_i\right)\right\}_{i=1}^{N}$ is N, and the feature dimension is D. The bagging method is employed to randomly draw $n\,(n<N)$ samples from $\left\{\left(\boldsymbol{x}_i, y_i\right)\right\}_{i=1}^{N}$ with replacement. The dataset is then divided into in-bag samples and out-of-bag (oob) samples.

For in-bag samples, $m\,(m<D)$ features are selected randomly to construct a DT. The dth variable is designated as the splitting variable, and s represents the splitting point. The sample space is divided into two regions: the left space R_{Left} and the right space R_{Right}, based on the chosen split point. The optimal splitting point is determined using the following calculation:

$$\min_{i,y_i}\left[\min_{c_{\text{Left}}}\sum_{\boldsymbol{x}\in R_{\text{Left}}}\left(y_a-c_{\text{Left}}\right)^2+\min_{c_{\text{Right}}}\sum_{\boldsymbol{x}\in R_{\text{Right}}}\left(y_b-c_{\text{Right}}\right)^2\right] \tag{11.1}$$

where c_{Left} and c_{Right} are the means of R_{Left} and R_{Right} outputs; and y_a and y_b are the outputs in R_{Left} and R_{Right}.

The aforementioned process is repeated until a stopping criterion is met. The output of the DT is then calculated as follows:

$$\hat{y}_{DT} = \sum_{l_{DT}=1}^{L_{DT}} c_{l_{DT}} I\left(\boldsymbol{x} \in R_{l_{DT}}\right) \tag{11.2}$$

where L_{DT} is the number of leaf nodes; $c_{l_{DT}}$ and $R_{l_{DT}}$ represent the output and space of the l_{DT}th leaf node; and $I(\cdot)$ denotes a symbolic function.

The aforementioned process is repeated to establish N_{RF} DTs. The output of the RF is obtained by taking the mean of the outputs from the N_{RF} DTs.

11.1.3 Fuzzy Neural Network (FNN)

FNN are neural network algorithms grounded in fuzzy set theory, offering remarkable capabilities for processing nonlinear and uncertain data [43]. The Takagi–Sugeno fuzzy neural network (TS-FNN) model is one of the most widely utilized forms of FNN [44]. Its structure is illustrated in Figure 11.1.

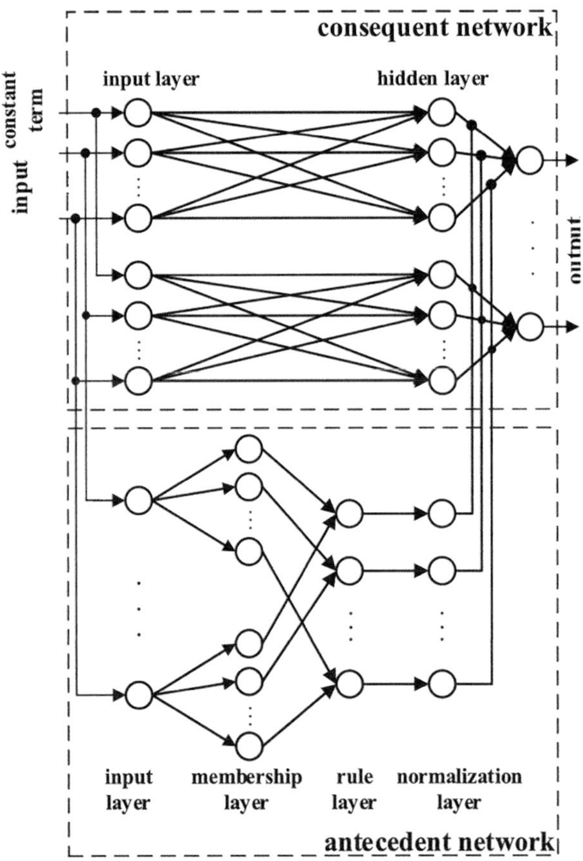

FIGURE 11.1 TS-FNN model structure.

The TS-FNN comprises an antecedent network and a consequent network. The antecedent network is responsible for matching the antecedents of the fuzzy rules, while the consequent network generates the consequents of these rules.

The antecedent network consists of several layers: the input layer, membership layer, rule layer, and normalization layer. Initially, the input data passes through the input layer, where fuzzy membership is determined by the membership layer. Next, the rule layer evaluates the applicability of each antecedent of the fuzzy rules based on the membership values. Finally, the output of the antecedent network is obtained through the normalization layer.

The consequent network includes an input layer and a hidden layer. The output of the hidden layer is calculated by multiplying the input data by the corresponding weights and adding a constant term. The overall output of the TS-FNN is derived from the weighted sum of the outputs from both the antecedent network and the hidden layer.

11.1.4 Deep Forest Classification (DFC)

In 2017, Zhou et al. introduced a novel forest structure for small sample classification problems [45], known as DFC. This framework first establishes a depth structure using a non-neural network composed of forest algorithms. In the multi-grained scanning module, a class distribution vector serves as the feature representation between layers. The cascade forest module employs both CRF and RF as basic learners [46]. The DFC framework for a three-class problem is illustrated in Figure 11.2.

Figure 11.2 illustrates three sizes of sliding windows that are employed to scan the raw input features. The features scanned from the same window serve as inputs to both the CRF and RF. Subsequently, class vectors are generated and concatenated to create enhanced features.

The first set of enhanced features is utilized to train the first layer of the cascade forest. Similarly, other sliding windows produce corresponding enhanced features, which are combined with the class vector generated by the previous layer to train the second and third layers of the cascade forest. This process is repeated until convergence is achieved.

11.2 ALGORITHM STRATEGY

The proposed VSG method based on GAFNN comprises four modules: RF-based feature adaptive selection, GAFNN-based candidate VSG, multiple constraint-based virtual sample selection, and mixed sample-based classification modules. The overall framework of the proposed method is illustrated in Figure 11.3.

In Figure 11.3, $\boldsymbol{X}_{\text{org}}$ and $\boldsymbol{Y}_{\text{real}}$ denote the input and output of the original real samples from the dataset, respectively; $\boldsymbol{X}_{\text{noise}}$ represents random noise; $\boldsymbol{X}_{\text{real}}$ indicates

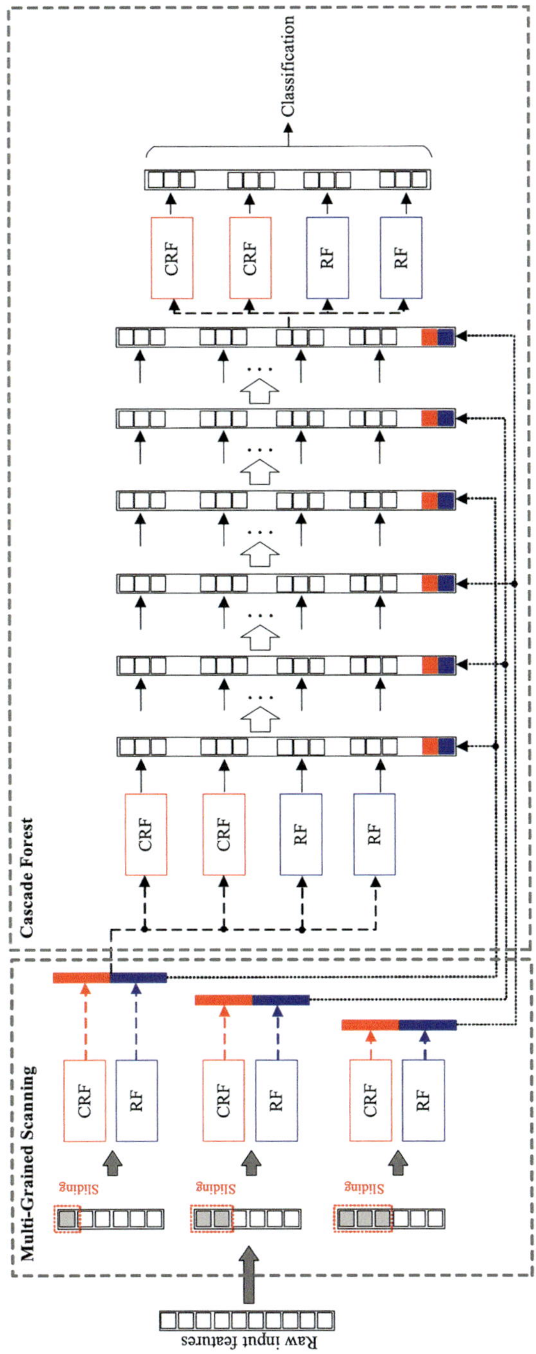

FIGURE 11.2 DFC framework of a three-class problem.

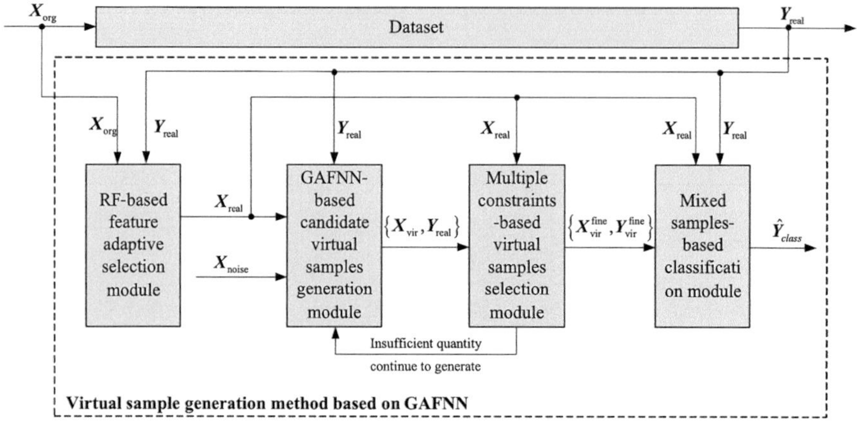

FIGURE 11.3 Strategy of VSG method based on GAFNN.

selected feature; $\{X_{vir}, Y_{real}\}$ denotes generated candidate virtual samples through GAFNN; $\{X_{vir}^{fine}, Y_{vir}^{fine}\}$ indicates qualified virtual samples after screening; X_{vir}^{fine} and Y_{vir}^{fine} represent the input and output, respectively; and $\hat{Y}_{class}$ is the predicted value vector generated by the classification model.

Figure 11.3 illustrates the steps of the proposed method as follows. First, X_{org} and Y_{real} are utilized as inputs to the RF-based feature adaptive selection module, producing the final selected feature X_{real}. Second, GAFNN is trained using X_{real}, X_{noise}, X_{org} and Y_{real}, to generate candidate virtual samples $\{X_{vir}^{fine}, Y_{vir}^{fine}\}$. Third, the multiple constraint-based virtual sample selection module selects $\{X_{vir}^{fine}, Y_{vir}^{fine}\}$ to ensure the quality of the virtual samples. Finally, the qualified virtual samples $\{X_{vir}^{fine}, Y_{vir}^{fine}\}$ and the original samples $\{X_{real}, Y_{real}\}$ are mixed and used to build the DFC in the mixed sample-based classification module.

The main contributions of this chapter are summarized as follows: (1) A novel GAFNN method is proposed for VSG in the context of KPI modeling, incorporating adaptive feature selection, optimized VSG, and DFC model construction; (2) The RF-based feature selection algorithm is enhanced to adaptively select important inputs; (3) Class-specific virtual samples are generated through GAFNN for the first time; and (4) A multi-constrained selection mechanism is introduced, which includes discriminant probability, maximum mean difference (MMD), and nearest neighbor class criteria to ensure the acquisition of high-quality virtual samples.

11.3 ALGORITHM IMPLEMENTATION

11.3.1 RF-Based Feature Adaptive Selection Module

RF is employed to assess the importance of input features, allowing for the adaptive selection of the most relevant inputs.

The input and output of the original real samples from the dataset are assumed to be as follows:

$$R_{\text{org}} = \left\{ \boldsymbol{X}_{\text{org}}, \boldsymbol{Y}_{\text{real}} \right\} = \left[\left(\boldsymbol{x}_{\text{org}}^{1}, y_{\text{real}}^{1} \right), \left(\boldsymbol{x}_{\text{org}}^{2}, y_{\text{real}}^{2} \right), \ldots, \left(\boldsymbol{x}_{\text{org}}^{N_{\text{org}}}, y_{\text{real}}^{N_{\text{org}}} \right) \right] \tag{11.3}$$

where N_{org} is the number of original real samples.

Then, the Bootstrap resampling technique is applied to randomly sample data from $\left\{ \boldsymbol{X}_{\text{org}}, \boldsymbol{Y}_{\text{real}} \right\}$ to construct R group training subsets, as follows:

$$R_{\text{org}}^{\text{sel}} = \left\{ \left\{ \boldsymbol{X}_1, \boldsymbol{Y}_1 \right\}, \left\{ \boldsymbol{X}_2, \boldsymbol{Y}_2 \right\}, \ldots, \left\{ \boldsymbol{X}_R, \boldsymbol{Y}_R \right\} \right\} \tag{11.4}$$

Furthermore, the R groups of oob data are represented as follows:

$$R_{\text{org}}^{\text{obb}} = \left\{ \left\{ \boldsymbol{X}_1^{\text{oob}}, \boldsymbol{Y}_1^{\text{oob}} \right\}, \left\{ \boldsymbol{X}_2^{\text{oob}}, \boldsymbol{Y}_2^{\text{oob}} \right\}, \ldots, \left\{ \boldsymbol{X}_R^{\text{oob}}, \boldsymbol{Y}_R^{\text{oob}} \right\} \right\} \tag{11.5}$$

Next, the R group DTs are constructed using the R group training subsets. Subsequently, the R groups of out-of-bag (OOB) data are fed into the DTs to generate R group prediction vectors, as follows:

$$\hat{\boldsymbol{Y}}_{\text{org}}^{\text{pre}} = \left[\hat{\boldsymbol{Y}}_1^{\text{pre}}, \hat{\boldsymbol{Y}}_2^{\text{pre}}, \ldots, \hat{\boldsymbol{Y}}_R^{\text{pre}} \right] \tag{11.6}$$

Thus, the rth obb error is calculated by

$$E_r = \frac{1}{2} \sum_{s=1}^{N_{\text{oob}}} \left[y_{r,s}^{\text{oob}} - y_{r,s}^{\text{pre}} \right]^2 \tag{11.7}$$

where N_{oob} is the number of obb data; $y_{r,s}^{\text{pre}}$ is the sth value of the predicted vector; $y_{r,s}^{\text{oob}}$ is the sth value of the rth true value of obb data with $r = 1, 2, \ldots, R$ and $s = 1, 2, \ldots, N_{\text{oob}}$.

Noise is added to the input features of the obb data for prediction, and the R group noise predictions are expressed as follows:

$$\hat{\boldsymbol{Y}}_{\text{org}}^{\text{noise}} = \left[\hat{\boldsymbol{Y}}_1^{\text{nosie},h}, \hat{\boldsymbol{Y}}_2^{\text{nosie},h}, \ldots, \hat{\boldsymbol{Y}}_R^{\text{nosie},h} \right] \tag{11.8}$$

Thus, the rth noise obb error is calculated by

$$E_r^{\text{noise},h} = \frac{1}{2} \sum_{s=1}^{N_{\text{oob}}} \left[y_{r,s}^{\text{oob}} - \hat{y}_{r,s}^{\text{noise},h} \right]^2 \tag{11.9}$$

where $\hat{y}_{r,s}^{\text{noise},h}$ is the sth value of rth group noise prediction of the hth input feature.

The importance score of the hth feature is calculated as follows:

$$\gamma_h = \frac{1}{R}\sum_{r=1}^{R}\left|E_r - E_r^{\text{noise},h}\right| \tag{11.10}$$

After the aforementioned operations are repeated to calculate the importance scores for all features, we proceed to normalize these scores. The normalized importance score of the hth feature is calculated by:

$$\gamma_h^{\text{scale}} = \frac{\gamma_h - \min(\gamma)}{\max(\gamma) - \min(\gamma)} \tag{11.11}$$

where $\max(\gamma)$ and $\min(\gamma)$ represent the maximum and minimum of all the calculated scores.

The normalized importance scores are sorted in descending order, as follows:

$$\left[\gamma_1^{\text{descend}}, \gamma_2^{\text{descend}}, \ldots, \gamma_L^{\text{descend}}, \ldots, \gamma_{D_{\text{org}}}^{\text{descend}}\right] = f_{\text{descend}}\left(\left\{\gamma_h^{\text{scale}}\right\}_{h=1}^{D_{\text{org}}}\right) \tag{11.12}$$

where D_{org} is the number of features in the original sample, and $\gamma_L^{\text{descend}}$ represents the importance score of the Lth feature in the descending order of scores.

When the feature corresponding to the $(L+1)$th descending importance score increases by less than 5%, the features associated with the first L descending importance scores are adaptively selected as input features. This approach eliminates the need to manually set thresholds. The increase is calculated by:

When the feature of the descending importance score increases by less than 5%, the features corresponding to the first descending importance scores are adaptively selected as input features. Thus, the need to manually set thresholds is avoided. The increase ξ_{L+1} is calculated by

$$\xi_{L+1} = \frac{\gamma_{L+1}^{\text{descend}}}{\sum_{l=1}^{L}\gamma_l^{\text{descend}}} \tag{11.13}$$

where $\gamma_{L+1}^{\text{descend}}$ denotes the importance score of the corresponding feature of the $(L+1)$th descending score.

If $\xi_{L+1} \le 0.05$, then the addition of features is halted, and the final selected features are denoted as:

$$\boldsymbol{X}_{\text{real}} = \left[\boldsymbol{x}_1^{\text{descend}}, \ldots, \boldsymbol{x}_l^{\text{descend}}, \ldots, \boldsymbol{x}_L^{\text{descend}}\right] \tag{11.14}$$

where $\boldsymbol{x}_l^{\text{descend}}$ is the lth selected feature.

11.3.2 GAFNN-Based Candidate VSG Module

The backpropagation neural network is employed to enhance antagonism in the original GAN [28]. This module utilizes a TS-FNN as both the generator and discriminator, referred to as GAFNN, to improve the capability of handling nonlinearity and uncertainty. Additionally, sample labels are used as conditional information to control the type of generated virtual samples. The proposed framework for generating candidate virtual samples based on GAFNN is illustrated in Figure 11.4.

The learning rates for the generator and discriminator are denoted as $\alpha_{\text{lr}}^{\text{G}}$ and $\alpha_{\text{lr}}^{\text{D}}$, respectively, while the training epoch is represented as N_e. The input noise is defined as follows:

$$\{\boldsymbol{X}_{\text{noise}}, \boldsymbol{Y}_{\text{real}}\} = \left[\left(\boldsymbol{x}_{\text{noise}}^{1}, y_{\text{real}}^{1}\right), \ldots, \left(\boldsymbol{x}_{\text{noise}}^{s}, y_{\text{real}}^{s}\right), \ldots, \left(\boldsymbol{x}_{\text{noise}}^{N_{\text{vir}}}, y_{\text{real}}^{N_{\text{vir}}}\right)\right] \in R^{(D_{\text{vir}}+1) \times N_{\text{vir}}} \tag{11.15}$$

where N_{vir} and D_{vir} represent the number and dimension of candidate virtual samples to be generated.

For conciseness, $N_{\text{vir}} = N_{\text{org}}$ is set accordingly. The generator's antecedent network comprises an input layer, a membership layer, a rule layer, and a normalization layer.

Taking the sth input $\left(\boldsymbol{x}_{\text{noise}}^{s}, y_{\text{real}}^{s}\right)$ as an example, it is first input into the membership layer of the generator's antecedent network. The calculation of the membership of the input to each fuzzy set is shown as follows:

$$\mu_{i,j}^{G} = \exp\left\{-\frac{\left(x_{\text{noise}}^{s,i} - c_{i,j}^{G}\right)^2}{\left(\sigma_{i,j}^{G}\right)^2}\right\} \tag{11.16}$$

where $x_{\text{noise}}^{s,i}$ is the ith element of $\left(\boldsymbol{x}_{\text{noise}}^{s}, y_{\text{real}}^{s}\right)$, $i = 1, 2, \ldots, D_{\text{vir}} + 1$; $\mu_{i,j}^{G}$ is jth membership of fuzzy sets, $j = 1, 2, \ldots, M_G$, and M_G is the number of fuzzy sets; $c_{i,j}^{G}$ and $\sigma_{i,j}^{G}$ are the center and width of the membership function.

Next, the fuzzy logic rules in the rule layer are calculated by:

$$\eta_j^{G} = \mu_{1,j}^{G} \mu_{2,j}^{G} \ldots \mu_{D_{\text{vir}},j}^{G} \tag{11.17}$$

Finally, the output of the generator's antecedent network is produced through the normalization layer, as follows:

$$\bar{\eta}_j^{G} = \frac{\eta_j^{G}}{\sum_{j=1}^{M_G} \eta_j^{G}} \tag{11.18}$$

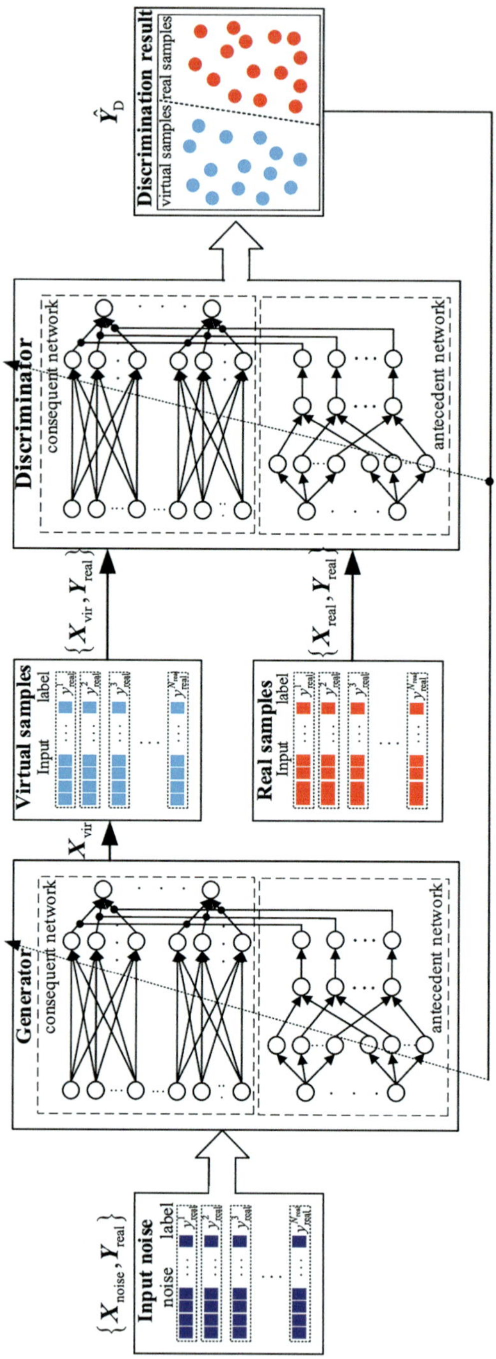

FIGURE 11.4 Framework for generating candidate virtual samples based on GAFNN.

The generator's consequent network consists of D_{vir} parallel sub-networks, each with the same structure, and each producing an output. The input to the generator's consequent network is $\left(\boldsymbol{x}_{\text{noise}}^{s}, y_{\text{real}}^{s}\right)$. The output of the hidden layer is generated as follows:

$$y_{k,j}^{\text{G}} = p_{0,j}^{\text{G}} + p_{1,j}^{\text{G}} x_{\text{noise}}^{s,1} + \cdots + p_{D_{\text{vir}}+1,j}^{\text{G}} x_{\text{noise}}^{s,D_{\text{vir}}+1} \tag{11.19}$$

where $k = 1,2,\ldots,D_{\text{vir}}$; $y_{k,j}^{\text{G}}$ is the jth output of the kth sub-network.

The ith output of the generator is calculated by

$$Y_k^G = \sum_{j=1}^{M_G} \overline{\eta}_j^G y_{k,j}^G \tag{11.20}$$

Therefore, the candidate's virtual sample was generated based on $\left(\boldsymbol{x}_{\text{noise}}^{s}, y_{\text{real}}^{s}\right)$ is denoted as follows:

$$\boldsymbol{x}_{\text{vir}}^{s} = \left[Y_1^G, Y_2^G, \ldots, Y_{D_{\text{vir}}}^G\right]^{\text{T}} \tag{11.21}$$

Therefore, the generated candidate virtual samples are denoted as follows:

$$\boldsymbol{X}_{\text{vir}} = \left[\boldsymbol{x}_{\text{vir}}^{1}, \ldots, \boldsymbol{x}_{\text{vir}}^{2}, \ldots, \boldsymbol{x}_{\text{vir}}^{N_{\text{vir}}}\right] \tag{11.22}$$

Furthermore, $\left\{\boldsymbol{X}_{\text{vir}}, \boldsymbol{Y}_{\text{real}}\right\}$ and $\left\{\boldsymbol{X}_{\text{real}}, \boldsymbol{Y}_{\text{real}}\right\}$ are input to the discriminator simultaneously, and its implementation process is similar to that of the generator. The key difference is that a sigmoid layer is added to the output layer of the discriminator to produce the discrimination result $\hat{\boldsymbol{Y}}_D$. Consequently, the loss function of GAFNN is calculated by:

$$\min_G \max_D V(D,G) = E_{P(\boldsymbol{X}_{\text{real}})}\left(\log \boldsymbol{Y}_D^{\text{real}}\right) + E_{P(\boldsymbol{X}_{\text{noise}})}\left[\log\left(1 - \boldsymbol{Y}_D^{\text{vir}}\right)\right] \tag{11.23}$$

where $P\left(\boldsymbol{X}_{\text{real}}\right)$ and $P\left(\boldsymbol{X}_{\text{noise}}\right)$ are the distribution of $\boldsymbol{X}_{\text{real}}$ and $\boldsymbol{X}_{\text{noise}}$; $\boldsymbol{Y}_D^{\text{real}}$ and $\boldsymbol{Y}_D^{\text{vir}}$ are the discriminator output for $\left\{\boldsymbol{X}_{\text{real}}, \boldsymbol{Y}_{\text{real}}\right\}$ and $\left\{\boldsymbol{X}_{\text{vir}}, \boldsymbol{Y}_{\text{real}}\right\}$, respectively.

GAFNN employs the discriminator to calculate the probability that the input sample originates from $P\left(\boldsymbol{X}_{\text{real}}\right)$ or $P\left(\boldsymbol{X}_{\text{noise}}\right)$. The generator learns the distribution $P\left(\boldsymbol{X}_{\text{real}}\right)$ of $\boldsymbol{X}_{\text{real}}$ based on the discriminant results to minimize the loss $\log\left(1 - \boldsymbol{Y}_D^{\text{vir}}\right)$. The adversarial game between the generator and the discriminator enables the candidate virtual samples to increasingly resemble the real samples.

11.3.3 Multiple Constraint-Based Virtual Sample Selection Module

Typically, establishing comprehensive evaluation indicators to assess and select virtual samples is crucial for ensuring their quality. The proposed multiple constraint-based virtual sample selection module incorporates three metrics: MMD, discriminator probability, and nearest neighbor class.

First, generators with numbers N_{MMD} are selected during the stabilization stage, and each generator produces a set of candidate virtual samples as follows:

$$R_{\mathrm{cand}} = \left\{ \left\{ \boldsymbol{X}_{\mathrm{cand}}^{1}, \boldsymbol{Y}_{\mathrm{cand}} \right\}, \left\{ \boldsymbol{X}_{\mathrm{cand}}^{2}, \boldsymbol{Y}_{\mathrm{cand}} \right\}, \ldots, \left\{ \boldsymbol{X}_{\mathrm{cand}}^{N_{\mathrm{MMD}}}, \boldsymbol{Y}_{\mathrm{cand}} \right\} \right\} \tag{11.24}$$

The MMD value between the virtual and real samples is calculated as follows:

$$f_{\mathrm{MMD}}\left(\boldsymbol{X}_{\mathrm{cand}}^{t}, \boldsymbol{X}_{\mathrm{real}} \right) = \sup_{\phi \in H} \left\| E_{p}\left[\phi\left(\boldsymbol{X}_{\mathrm{cand}}^{t} \right) \right] - E_{q}\left[\phi\left(\boldsymbol{X}_{\mathrm{real}} \right) \right] \right\|_{H} \tag{11.25}$$

where $\boldsymbol{X}_{\mathrm{cand}}^{t}$ denotes the input of tth group candidate virtual samples; H represents reproducing kernel Hilbert space (RKHS); $\phi(\cdot)$ denotes the mapping of samples to high-dimensional RKHS; $E_{p}\left[\phi\left(\boldsymbol{X}_{\mathrm{vir}}^{\mathrm{ori}} \right) \right]$ and $E_{q}\left[\phi\left(\boldsymbol{X}_{\mathrm{real}} \right) \right]$ are the expected value of the sample mapped to the RKHS.

The generator with the lowest MMD value is selected as the filter generator, which is represented as follows:

$$\phi_{\mathrm{MMD}} = \min_{\mathrm{MMD}}\left(f_{\mathrm{MMD}}\left(\boldsymbol{X}_{\mathrm{cand}}^{1}, \boldsymbol{X}_{\mathrm{real}} \right), \ldots, f_{\mathrm{MMD}}\left(\boldsymbol{X}_{\mathrm{cand}}^{t}, \boldsymbol{X}_{\mathrm{real}} \right), \ldots, f_{\mathrm{MMD}}\left(\boldsymbol{X}_{\mathrm{cand}}^{N_{\mathrm{MMD}}}, \boldsymbol{X}_{\mathrm{real}} \right) \right) \tag{11.26}$$

where $\min_{\mathrm{MMD}}(\cdot)$ is the generator number corresponding to the virtual samples with the smallest MMD value

Next, a group of virtual samples $\left\{ \boldsymbol{X}_{\mathrm{filter}}, \boldsymbol{Y}_{\mathrm{filter}} \right\}$ is generated from $\mathrm{G}_{\phi_{\mathrm{MMD}}}$, and their discriminator probability is calculated. Virtual samples with a discriminator probability greater than the threshold θ_{disc} are selected as follows:

$$P\left(\left(\boldsymbol{x}_{\mathrm{filter}}^{i}, y_{\mathrm{filter}}^{i} \right) = \mathrm{True} \right) \geq \theta_{\mathrm{disc}} \tag{11.27}$$

where $\left(\boldsymbol{x}_{\mathrm{filter}}^{i}, y_{\mathrm{filter}}^{i} \right)$ represent the ith virtual sample of $\left\{ \boldsymbol{X}_{\mathrm{filter}}, \boldsymbol{Y}_{\mathrm{filter}} \right\}$.

Then, $\boldsymbol{X}_{\mathrm{filter}}$ and $\boldsymbol{X}_{\mathrm{real}}$ are denoted as follows:

$$\boldsymbol{X}_{\mathrm{filter}} = \left[\boldsymbol{x}_{\mathrm{filter}}^{1}, \boldsymbol{x}_{\mathrm{filter}}^{2}, \ldots, \boldsymbol{x}_{\mathrm{filter}}^{N_{\mathrm{filter}}} \right] \in R^{D_{\mathrm{real}} \times N_{\mathrm{filter}}} \tag{11.28}$$

$$\boldsymbol{X}_{\mathrm{real}} = \left[\boldsymbol{x}_{\mathrm{real}}^{1}, \boldsymbol{x}_{\mathrm{real}}^{2}, \ldots, \boldsymbol{x}_{\mathrm{real}}^{N_{\mathrm{real}}} \right] \in R^{D_{\mathrm{real}} \times N_{\mathrm{real}}} \tag{11.29}$$

where D_{real} represents the dimension of $\boldsymbol{X}_{\mathrm{real}}$; N_{filter} and N_{real} represent the sample size of $\boldsymbol{X}_{\mathrm{filter}}$ and $\boldsymbol{X}_{\mathrm{real}}$, respectively.

The Euclidean distance of each sample in $\left\{ \boldsymbol{X}_{\mathrm{filter}}, \boldsymbol{Y}_{\mathrm{filter}} \right\}$ to each sample in $\left\{ \boldsymbol{X}_{\mathrm{real}}, \boldsymbol{Y}_{\mathrm{real}} \right\}$ is calculated by

$$d_{i,j} = \mathrm{dist}\left(\boldsymbol{x}_{\mathrm{filter}}^{i}, \boldsymbol{x}_{\mathrm{real}}^{j} \right) = \sqrt{\sum_{a=1}^{D_{\mathrm{real}}} \left(\boldsymbol{x}_{\mathrm{filter}}^{a,i} - \boldsymbol{x}_{\mathrm{real}}^{a,j} \right)^{2}} \tag{11.30}$$

where $\text{dist}\left(\boldsymbol{x}_{\text{filter}}^{i}, \boldsymbol{x}_{\text{real}}^{j}\right)$ represents the Euclidean distance between the ith sample $\boldsymbol{x}_{\text{filter}}^{i}$ of $\boldsymbol{X}_{\text{filter}}$ and the jth sample $\boldsymbol{x}_{\text{real}}^{j}$ of $\boldsymbol{X}_{\text{real}}$; $\boldsymbol{x}_{\text{filter}}^{a,i}$ and $\boldsymbol{x}_{\text{real}}^{a,j}$ represent the ath element of $\boldsymbol{x}_{\text{filter}}^{i}$ and $\boldsymbol{x}_{\text{real}}^{a,j}$, respectively.

The K minimum distances are selected, and the majority class among these is designated as the nearest neighbor class, as follows:

$$\left\{d_{i,1}^{\min}, d_{i,2}^{\min}, \ldots, d_{i,K}^{\min}\right\} = \min_K\left(d_{i,1}, d_{i,2}, \ldots, d_{i,N_{\text{real}}}\right) \tag{11.31}$$

$$y_i^{\text{knn}} = \max_{\text{knn}}\left(d_{i,1}^{\min}, d_{i,2}^{\min}, \ldots, d_{i,K}^{\min}\right) \tag{11.32}$$

where $\min_K(\cdot)$ denotes that the K minimum distances are selected from all distances; $\left\{d_{i,1}^{\min}, d_{i,2}^{\min}, \ldots, d_{i,K}^{\min}\right\}$ are K minimum distances of $\left(\boldsymbol{x}_{\text{filter}}^{i}, y_{\text{filter}}^{i}\right)$; $\max_{\text{knn}}(\cdot)$ denotes the majority class in $\left\{d_{i,1}^{\min}, d_{i,2}^{\min}, \ldots, d_{i,K}^{\min}\right\}$; and y_i^{knn} is the nearest neighbor class of $\left(\boldsymbol{x}_{\text{filter}}^{i}, y_{\text{filter}}^{i}\right)$.

The virtual samples are retained when the values of y_i^{knn} and y_{filter}^{i} are consistent, as follows:

$$\xi_i^{\text{sel}} = \begin{cases} 1, & y_i^{\text{knn}} = y_{\text{filter}}^{i} \\ 0, & y_i^{\text{knn}} \neq y_{\text{filter}}^{i} \end{cases} \tag{11.33}$$

If the condition $\xi_i^{\text{sel}} = 1$ holds, then we keep $\left(\boldsymbol{x}_{\text{filter}}^{i}, y_{\text{filter}}^{i}\right)$; otherwise, we discard $\left(\boldsymbol{x}_{\text{filter}}^{i}, y_{\text{filter}}^{i}\right)$.

Finally, the virtual samples that meet the aforementioned requirements are considered qualified virtual samples $\left\{\boldsymbol{X}_{\text{vir}}^{\text{fine}}, \boldsymbol{Y}_{\text{vir}}^{\text{fine}}\right\}$.

In this module, MMD ensures the diversity of virtual samples, the discriminator probability threshold guarantees the accuracy of the selected virtual samples, and the nearest neighbor category maintains the distinction between the selected virtual samples and those from other categories. The parameters θ_{disc} and K also need to be determined based on the quality requirements of the virtual samples and the characteristics of the dataset.

11.3.4 Mixed Sample-Based Classification Module

The qualified virtual samples $\left\{\boldsymbol{X}_{\text{vir}}^{\text{fine}}, \boldsymbol{Y}_{\text{vir}}^{\text{fine}}\right\}$ and the original samples $\left\{\boldsymbol{X}_{\text{real}}, \boldsymbol{Y}_{\text{real}}\right\}$ are mixed to generate mixed virtual samples R_{mix}, as follows:

$$R_{\text{mix}} = \left\{\boldsymbol{X}_{\text{mix}}, \boldsymbol{Y}_{\text{mix}}\right\} = \left\{\left\{\boldsymbol{X}_{\text{vir}}^{\text{fine}}, \boldsymbol{Y}_{\text{vir}}^{\text{fine}}\right\}, \left\{\boldsymbol{X}_{\text{real}}, \boldsymbol{Y}_{\text{real}}\right\}\right\} \tag{11.34}$$

The hyperparameters of the DFC model include the number of forest layers N_{FL}, the number of cascade forests N_{CF}, the number of DTs N_{DT}, and the number of samples per leaf node N_{leaf}. The construction steps of the DFC model are as follows:

> First, R_{mix} is used as the training data for the DFC, with two forest algorithms (RF and CFR) employed as the basic learners, each consisting of N_{DT} DTs.

Second, a N_{class}-dimensional estimated class distribution vector (which depends on the classification task) is calculated for each of the RFs and CRFs.

Third, the class distribution vectors are concatenated and used as input for the next layer.

Finally, the aforementioned operations are repeated until a N_{FL}-layer DFC is constructed.

11.3.5 Algorithm Pseudocode

The pseudocode of the GAFNN algorithm is demonstrated in Table 11.1.

TABLE 11.1
GAFNN Algorithm Pseudocode

Input:	Original real sample $\{\boldsymbol{X}_{\text{org}}, \boldsymbol{Y}_{\text{real}}\}$; Random noise $\boldsymbol{X}_{\text{noise}}$; Training epoch N_e; Number of generated virtual samples N_{vir}; Number of generator choices N_{MMD}; Threshold for the discriminator probability θ_{disc}; Number of nearest neighbor samples K
Output:	Mixed virtual samples R_{mix}
Step 1:	For $h = 1 : D_{\text{org}}$
Step 2:	Feature importance scores γ_h are calculated;
Step 3:	Normalized importance scores γ_h^{scale} are calculated;
Step 4:	End
Step 5:	The normalized importance scores in descending order $\left[\gamma_1^{\text{descend}}, \gamma_2^{\text{descend}}, \ldots, \gamma_l^{\text{descend}}, \ldots, \gamma_{D_{\text{org}}}^{\text{descend}}\right]$ are obtained;
Step 6:	For $L = 1 : D_{\text{org}}$
Step 7:	RF is constructed based on the first l importance scores in descending order corresponding to the features;
Step 8:	$\gamma_{L+1}^{\text{descend}} \Big/ \sum_{l=1}^{L} \gamma_l^{\text{descend}}$ is calculated;
Step 9:	If $\gamma_{L+1}^{\text{descend}} \Big/ \sum_{l=1}^{L} \gamma_l^{\text{descend}} < 0.05$
Step 10:	break;
Step 11:	end
Step 12:	end
Step 13:	The real sample input after feature selection $\boldsymbol{X}_{\text{real}} = \left[\boldsymbol{x}_1^{\text{descend}}, \boldsymbol{x}_2^{\text{descend}}, \ldots, \boldsymbol{x}_{l-1}^{\text{descend}}\right]$ is obtained;
Step 14:	For $e = 1 : N_e$
Step 15:	The generator and discriminator are trained;
Step 16:	end

(Continued)

TABLE 11.1 (Continued)
GAFNN Algorithm Pseudocode

Step 17:	While true
Step 18:	Choose N_{MMD} generators and generate N_{MMD} groups of candidate virtual sample $\left\{\boldsymbol{X}_{\text{cand}}^{l}, \boldsymbol{Y}_{\text{cand}}\right\}_{l=1}^{N_{\text{MMD}}}$;
Step 19:	The virtual samples are selected according to θ_{disc} and K criteria;
Step 20:	The virtual samples that meet the requirements are added to the qualified virtual samples $\left\{\boldsymbol{X}_{\text{vir}}^{\text{fine}}, \boldsymbol{Y}_{\text{vir}}^{\text{fine}}\right\}$, and the sample size of $\left\{\boldsymbol{X}_{\text{vir}}^{\text{fine}}, \boldsymbol{Y}_{\text{vir}}^{\text{fine}}\right\}$ is $N_{\text{vir}}^{\text{fine}}$;
Step 21:	If $N_{\text{vir}}^{\text{fine}} = N_{\text{vir}}$
Step 22:	Break;
Step 23:	End
Step 24:	End
Step 25:	$\left\{\boldsymbol{X}_{\text{vir}}^{\text{fine}}, \boldsymbol{Y}_{\text{vir}}^{\text{fine}}\right\}$ and $\left\{\boldsymbol{X}_{\text{real}}, \boldsymbol{Y}_{\text{real}}\right\}$ are mixed to generate a mixed sample R_{mix}

TABLE 11.2
Statistical Results of Benchmark Datasets

Dataset	Size of Original Samples	Size of Selected Samples	Number of Input Features
SC	21,263	107	81
RB	372	124	103

11.4 EXPERIMENTAL STUDY

11.4.1 Benchmark Data

11.4.1.1 Dataset Description

The benchmark data includes superconductivity (SC) and residential building (RB) datasets sourced from the University of California, Irvine (UCI) Machine Learning Repository (https://archive.ics.uci.edu/ml/datasets.php). The SC dataset consists of 21,263 samples and features 81 input variables, with critical temperature as the output. In contrast, the RB dataset contains 372 samples and 103 input variables, with construction cost as the output. To emphasize the small sample characteristic of the modeling data, only 1/200 of the original SC samples and 1/3 of the original RB samples were randomly selected. These selected samples are further divided into three subsets: training, validation, and testing, comprising 50%, 25%, and 25% of the selected samples, respectively. The statistical results are presented in Table 11.2. The outputs of the SC and RB datasets are categorized into three levels: low, medium, and high, as shown in Table 11.3.

TABLE 11.3
Division Types of Benchmark Datasets

Dataset Name	Standard	Type	Number of Samples
SC	$0 \le y_{SC} < 20$	Low	43
	$20 \le y_{SC} < 60$	Middle	33
	$60 \le y_{SC}$	High	31
RB	$0 \le y_{RB} < 200$	Low	60
	$200 \le y_{RB} < 400$	Middle	41
	$400 \le y_{RB}$	High	23

11.4.1.2 Modeling Results

The number of DT in the RF-based feature adaptive selection module is set to 500, with a leaf node sample size of 1. The results of the RF-based adaptive feature selection for the SC and RB datasets are presented in Figures 11.5 and 11.6.

Figures 11.5 and 11.6 illustrate that the importance scores of the 10th and 6th features increase by less than 5% in descending order. Therefore, the first nine features are adaptively selected for the SC dataset, while the first ten features are selected for the RB dataset. The parameters of the GAFNN based on the SC and RB datasets are presented in Tables 11.4 and 11.5.

The training epoch N_e is set to 300, with both the generator and discriminator learning rates $\alpha_{\mathrm{lr}}^{\mathrm{G}}$ and $\alpha_{\mathrm{lr}}^{\mathrm{D}}$ at 0.0001. The MMD values for the SC and RB datasets, which represent the differences between the virtual and real samples generated during training, are displayed in Figure 11.7.

Figure 11.7 demonstrates that the MMD values for the SC and RB datasets stabilize when the training epochs reach 200 and 250, respectively. The process of selecting qualified virtual samples based on multiple constraints involves generating high-, medium-, and low-level samples for each dataset.

Initially, five generators are selected at intervals of 20 epochs from the 200th to the 280th epoch for the SC dataset, and at intervals of 10 epochs from the 250th to the 290th epoch for the RB dataset. Each generator produces a group of virtual samples equal in number to the training samples. The MMD values for these five groups are presented in Table 11.6.

Table 11.6 indicates that the MMD value for the 220th generator is the lowest for the SC dataset, while the 280th generator exhibits the lowest MMD value for the RB dataset. Consequently, these generators are selected as the screening generators.

Next, six candidate virtual samples are generated using the screening generators, and the discriminator probabilities are calculated, as shown in Table 11.7.

Table 11.7 demonstrates that the 3rd and 6th virtual samples meet the requirement with a discriminator probability of 0.8 for the SC dataset. Similarly, the 1st, 3rd, and 4th virtual samples satisfy the requirement with a discriminator probability of 0.6 for the RB dataset.

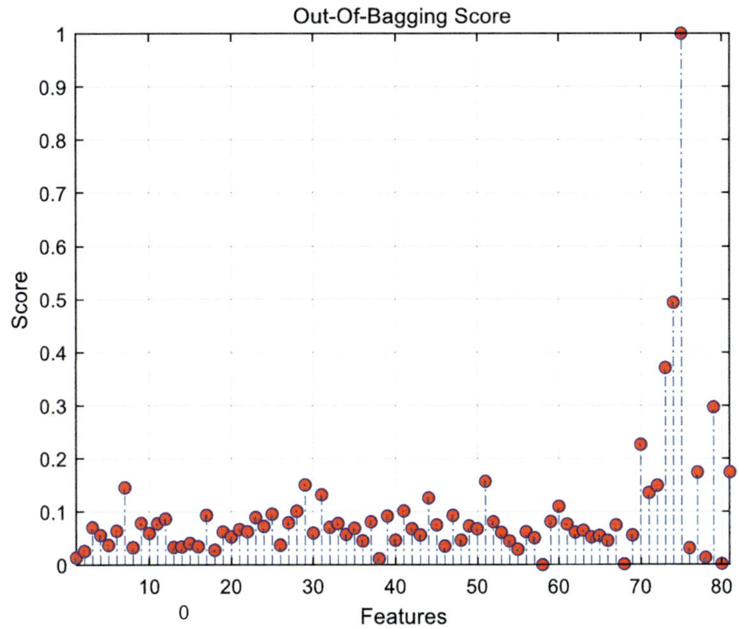

(a) Importance score of features

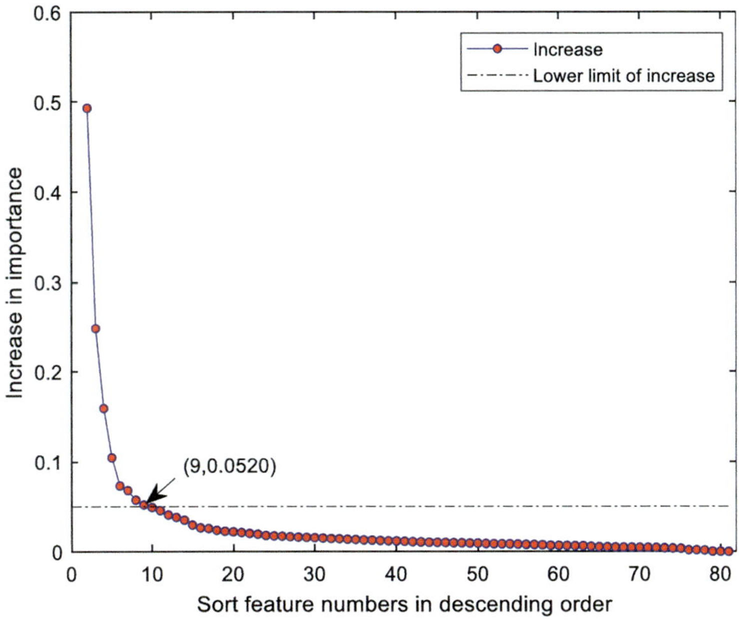

(b) Importance score increase of feature

FIGURE 11.5 Results of RF-based adaptive feature selection for the SC dataset. (a) Importance score of features. (b) Importance score increase of feature.

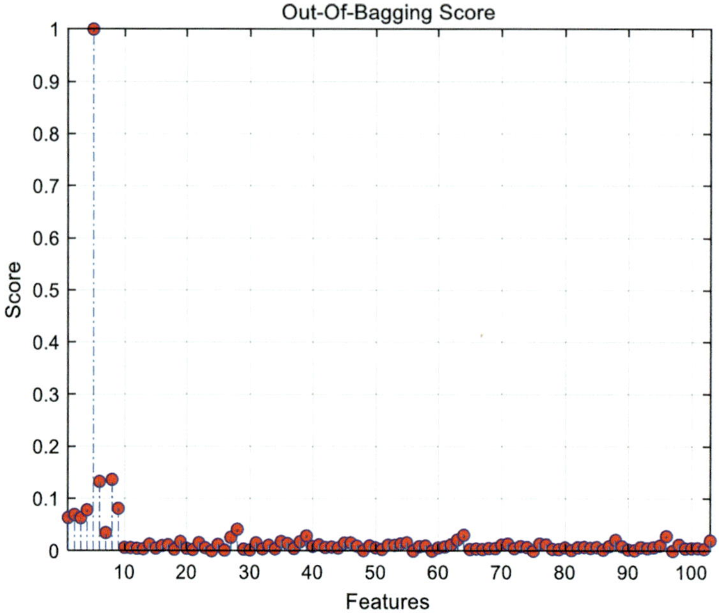

(a) Importance score of features

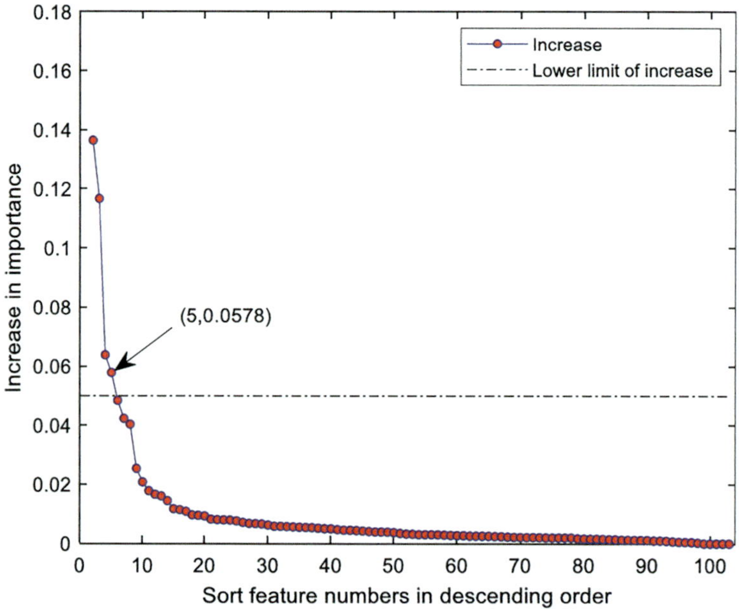

(b) Importance score increase of feature

FIGURE 11.6 Results of RF-based adaptive feature selection for the RB dataset. (a) Importance score of features. (b) Importance score increase of feature.

TABLE 11.4
Parameters of GAFNN for SC Data

Structure	Network		Number of Neurons
Generator	Antecedent network	Input layer	10
		Membership layer	60 × 10
		Rule layer	60
		Normalization layer	60
	Consequent network	Hidden layer	60 × 9
		Output layer	9
Discriminator	Antecedent network	Input layer	10
		Membership layer	60 × 10
		Rule layer	60
		Normalization layer	60
	Consequent network	Hidden layer	60 × 1
		Output layer	1

TABLE 11.5
Parameters of GAFNN for RB Data

Structure	Network		The Number of Neurons
Generator	Antecedent network	Input layer	6
		Membership layer	36 × 6
		Rule layer	36
		Normalization layer	36
	Consequent network	Hidden layer	36 × 5
		Output layer	5
Discriminator	Antecedent network	Input layer	6
		Membership layer	36 × 6
		Rule layer	36
		Normalization layer	36
	Consequent network	Hidden layer	36 × 1
		Output layer	1

Subsequently, the threshold is set to five for all datasets. The nearest neighbor classifications of the aforementioned virtual samples are presented in Table 11.8.

Finally, the aforementioned process is repeated to generate qualified virtual samples that match the size of the training samples. The virtual and training samples are then reduced to three dimensions using t-SNE to illustrate their relationship. The results are presented in Figure 11.8.

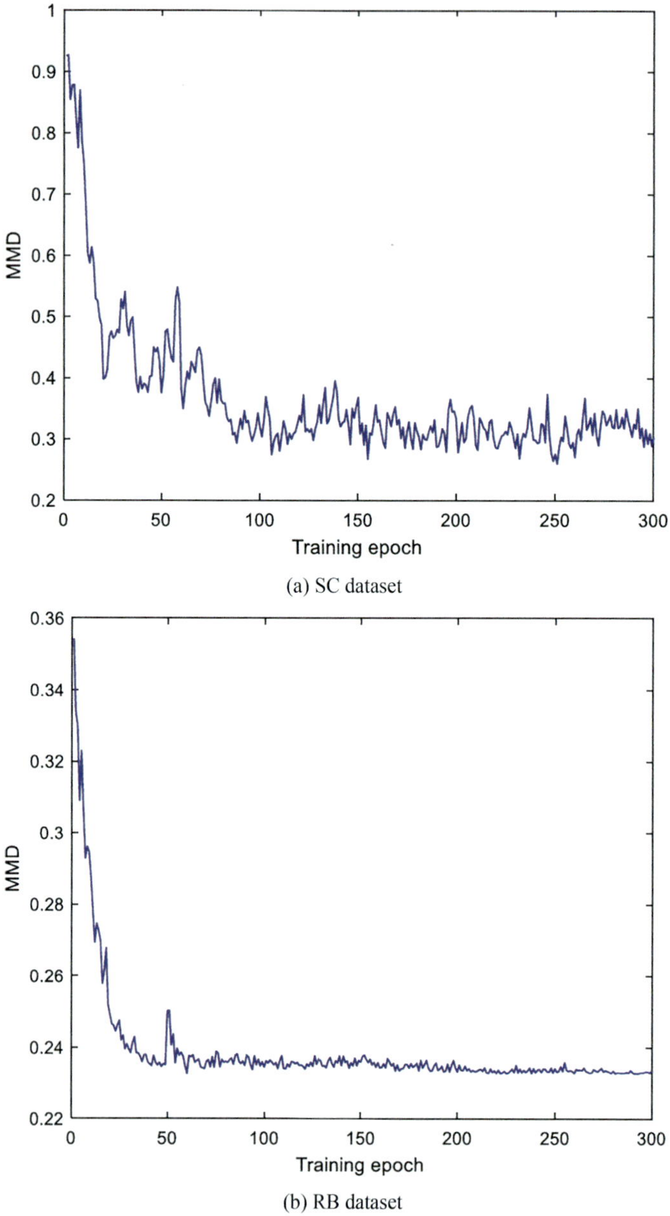

FIGURE 11.7 Relationship between the MMD value and the training algorithm for benchmark datasets. (a) SC dataset. (b) RB dataset.

TABLE 11.6
MMD-Based Generator Initial Screening for Benchmark Datasets

Dataset Name	Training Epoch	MMD Values
	200	0.2841
	220	0.2883
SC	240	0.3098
	260	0.2916
	280	0.3076
	250	0.2437
	260	0.2357
RB	270	0.2397
	280	0.2378
	290	0.2387

TABLE 11.7
Discriminator Probability of Candidate Virtual Samples for Benchmark Datasets

Dataset Name	No.	Level	Discriminator Probability
SC	1	Low	0.494055
	2	Low	0.568173
	3	Middle	0.914151
	4	Middle	0.628289
	5	High	0.723913
	6	High	0.865288
RB	1	Low	0.770992
	2	Low	0.594538
	3	Middle	0.756663
	4	Middle	0.649891
	5	High	0.506117
	6	High	0.542996

TABLE 11.8
Nearest Neighbor Class of Candidate Virtual Samples for Benchmark Datasets

Dataset Name	No.	Level	*K* Class with the Smallest Distance					Nearest Neighbor Class
SC	3	Middle	Middle	Middle	High	High	High	High
	6	High	High	Middle	High	High	High	High
RB	1	Low	Middle	Middle	Low	Middle	Low	Middle
	3	Middle	Low	Middle	Middle	Middle	Low	Middle
	4	Middle	Middle	Middle	Middle	Middle	Middle	Middle

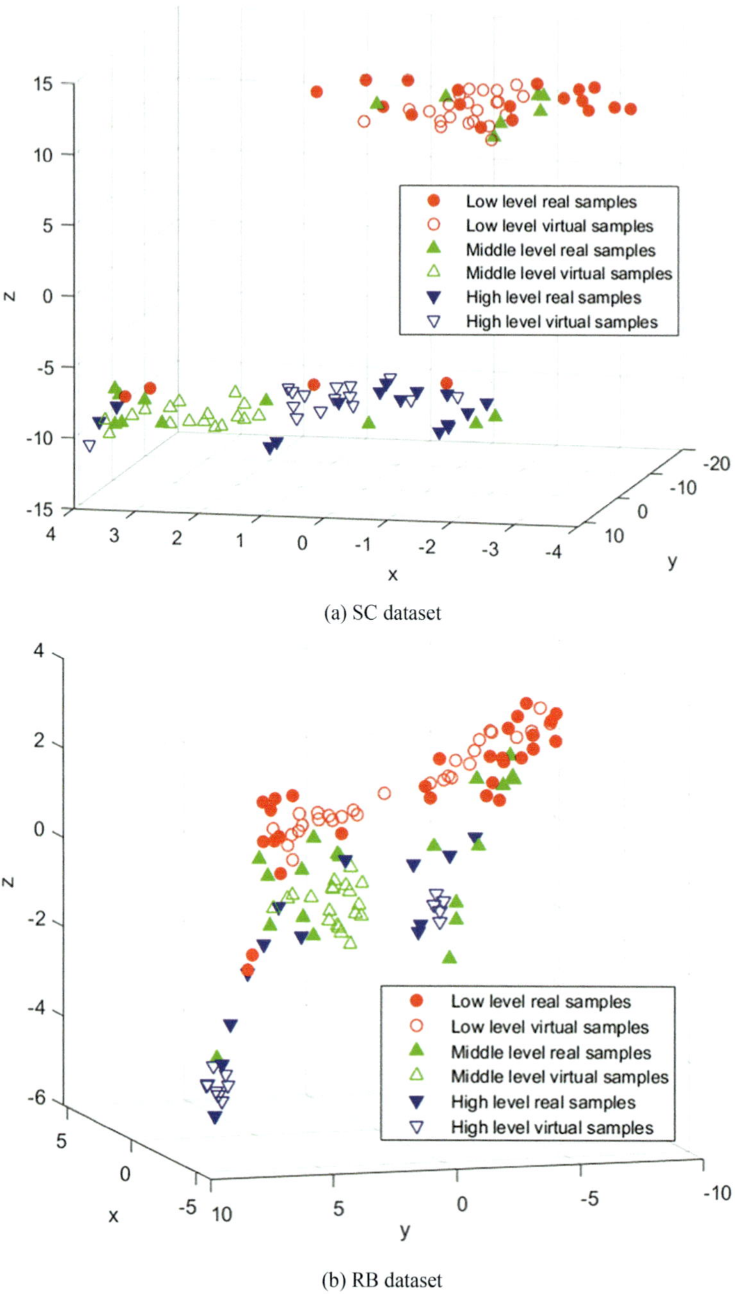

FIGURE 11.8 Display of virtual samples and training samples for benchmark datasets. (a) SC dataset. (b) RB dataset.

The parameters for the DF-based classification model are configured as follows: N_{FL} is set to 50, N_{CF} is 4 (comprising 2 CRF and 2 RF), N_{DT} is 50, and N_{leaf} is 1/10 of the training samples. Mixed samples consist of qualified virtual and real samples generated after the previously mentioned screening. Each time, a group of virtual samples (e.g., one high-level, one medium-level, and one low-level sample) is added to the real samples as training samples, resulting in a total of 20 groups. Additionally, each experiment is repeated 30 times to account for the randomness inherent in DFC. The experimental results for the benchmark datasets are presented in Figures 11.9 and 11.10, as well as Tables 11.9 and 11.10.

Figure 11.9 and Table 11.9 indicate that the accuracy and stability of the classification model are enhanced by adding virtual samples for the SC dataset. However, the improvement becomes minimal after more than 16 groups of virtual samples are added. Conversely, Figure 11.10 and Table 11.10 demonstrate that virtual samples can significantly improve model accuracy for the RB dataset. Based on the validation results, 15 and 19 groups of virtual samples are added in the testing experiments for the SC and RB datasets, respectively, and the results are presented in Table 11.11.

Table 11.11 presents the average accuracy and standard deviation for the testing dataset, revealing values of 0.741 and 0.02175 for the SC dataset, and 0.928 and 0.03134 for the RB dataset.

11.4.1.3 Comparative Results

The DFC is compared with classic algorithms, specifically RF and SVM, using the raw small samples. To account for randomness, the experiments for both DFC and RF are repeated 30 times, and the results are averaged. The outcomes of the comparative experiments involving the classic algorithms are presented in Table 11.12.

Table 11.12 indicates that the DFC achieves better accuracy than both RF and SVM on the benchmark datasets. Following this, the GAFNN is compared with the original GAN and AL-GAN [31]. The parameters for both GAN and AL-GAN are detailed in Table 11.13. The MMD values of the virtual and training samples across the three methods are presented in Table 11.14. In the comparative experiments for the SC and RB datasets, 15 and 19 groups of virtual samples are added, respectively, and the results are shown in Tables 11.15 and 11.16.

Table 11.15 concludes the following: (1) The GAN achieves the highest accuracy, while the GAFNN exhibits the best stability in the training dataset; (2) The accuracy and stability of AL-GAN are lower than those of the GAN for both the training and validation phases of the RB dataset, mainly due to human interference in the virtual samples; (3) The proposed GAFNN method demonstrates superior accuracy and stability compared to the other methods in the validation and testing datasets.

Table 11.16 further concludes that: (1) The GAN maintains the best accuracy, whereas the GAFNN shows the greatest stability in the training of the RB dataset; (2) Although the proposed GAFNN method outperforms the comparison methods in terms of accuracy during validation and testing datasets, its stability does not match that of the GAN. Therefore, the stability of the GAFNN for the RB dataset requires improvement.

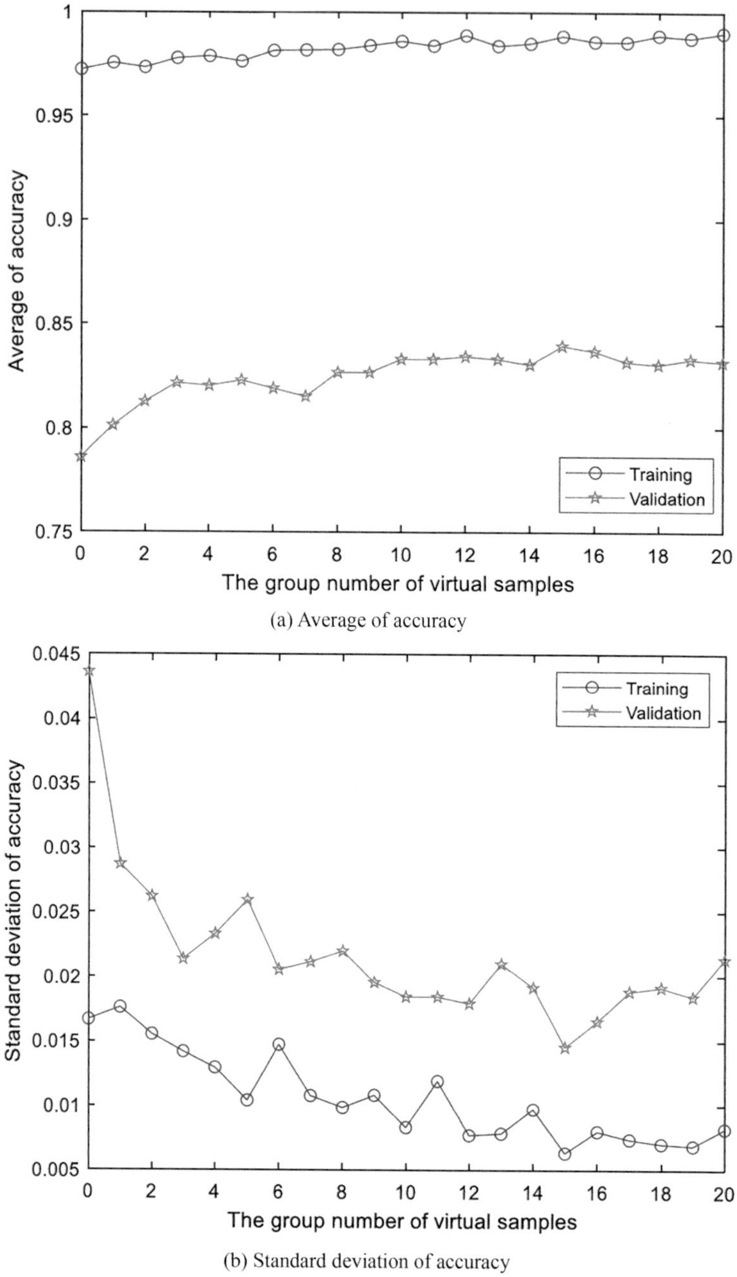

FIGURE 11.9 Experimental results of mixed samples for the SC dataset. (a) Average of accuracy. (b) Standard deviation of accuracy.

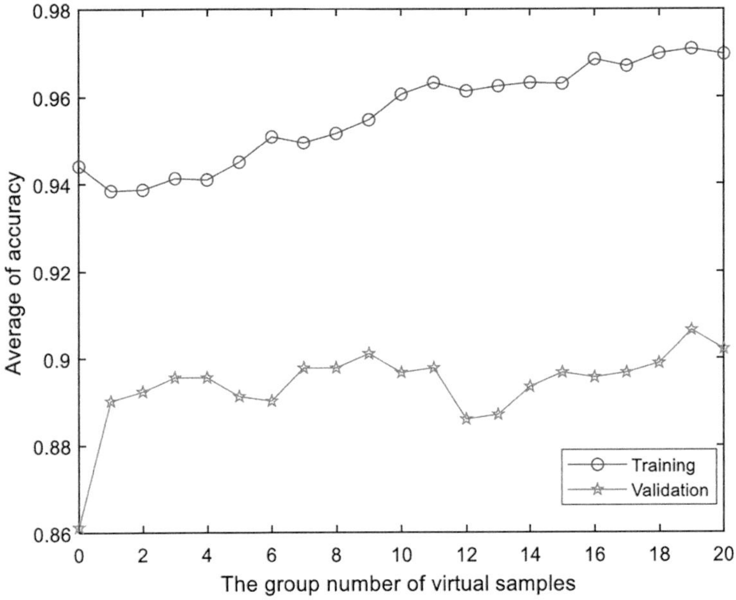

(a) Average of accuracy

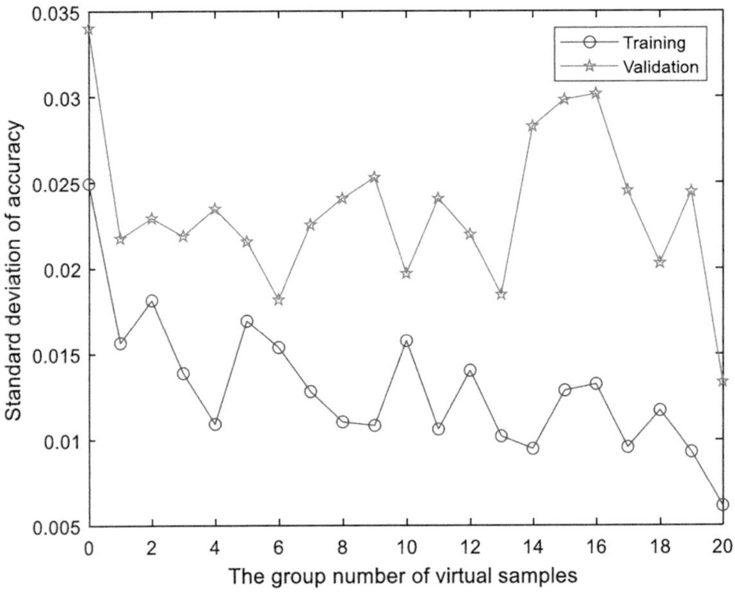

(b) Standard deviation of accuracy

FIGURE 11.10 Experimental results of mixed samples for the RB dataset. (a) Average of accuracy. (b) Standard deviation of accuracy.

TABLE 11.9
Statistical Results of Mixed Samples for SC Dataset

Dataset	Average of Accuracy		Standard Deviation of Accuracy	
	Best	Group Number	Best	Group Number
Training	0.990	20	6.352E–03	15
Validation	0.840	15	1.458E–02	15

TABLE 11.10
Statistical Results of Mixed Samples for RB Dataset

Dataset	Average of Accuracy		Standard Deviation of Accuracy	
	Best	Group Number	Best	Group Number
Training	0.971	19	6.145E–03	20
Validation	0.906	19	1.335E–02	20

TABLE 11.11
Statistical Results of Testing Experiments for Benchmark Datasets

Dataset	Average of Accuracy	Standard Deviation of Accuracy
SC	0.741	2.175E–02
RB	0.928	3.134E–02

TABLE 11.12
Results of Comparative Experiment of Classic Algorithms for Benchmark Datasets

Dataset	Method	Training Accuracy	Validation Accuracy	Testing Accuracy
SC	DFC	0.986	0.829	0.677
	RF	0.918	0.8333	0.627
	SVM	0.870	0.808	0.667
RB	DFC	0.960	0.856	0.823
	RF	0.953	0.693	0.746
	SVM	0.952	0.742	0.774

TABLE 11.13
Comparative Experimental Parameters for Benchmark Datasets

Dataset	Generator			Discriminator			Learning Rate	Training Epoch
	Input	Hidden	Output	Input	Hidden	Output		
SC	10	60	9	10	60	1	0.0001	300
RB	6	36	5	6	36	1	0.0001	300

TABLE 11.14
Comparative Experimental MMD for Benchmark Datasets

Dataset	Method	MMD		
		Training	Validation	Testing
	GAFNN	0.2902	0.5038	0.5671
SC	GAN	0.3551	0.6808	0.6692
	AL-GAN	0.3314	0.6271	0.6437
	GAFNN	0.2340	0.4512	0.4772
RB	GAN	0.2399	0.5061	0.5337
	AL-GAN	0.2380	0.4631	0.4817

TABLE 11.15
Comparative Experimental Results for SC Dataset

Method	Dataset	Average of Accuracy	Standard Deviation of Accuracy
	Training	0.989	6.352E−03
GAFNN	Validation	0.840	1.458E−02
	Testing	0.741	2.175E−02
	Training	0.991	7.928E−03
GAN	Validation	0.801	2.041E−02
	Testing	0.722	2.877E−02
	Training	0.990	8.589E−03
AL-GAN	Validation	0.799	2.800E−02
	Testing	0.722	2.708E−02

TABLE 11.16
Comparative Experimental Results for RB Dataset

Method	Dataset	Average of Accuracy	Standard Deviation of Accuracy
	Training	0.971	9.293E–03
GAFNN	Validation	0.906	2.448E–02
	Testing	0.928	3.134E–02
	Training	0.973	1.112E–02
GAN	Validation	0.899	2.902E–02
	Testing	0.899	2.775E–02
	Training	0.966	1.058E–02
AL-GAN	Validation	0.904	2.317E–02
	Testing	0.925	3.207E–02

TABLE 11.17
Division Standard for DXN Dataset

Dataset Name	Standard	Risking Level	Number of Samples
DXN	$0 \le y_{DXN} < 0.02$	Low	27
	$0.02 \le y_{DXN} < 0.05$	Middle	17
	$0.05 \le y_{DXN}$	High	24

11.4.2 DXN Data from Actual Industrial Process

11.4.2.1 Dataset Description

The DXN data is derived from the actual MSWI power plant in Beijing, covering the years 2012–2018 [47, 48]. This dataset comprises 67 samples and includes 127 input features, with the DXN emission concentration serving as the output variable. Following the same partition strategy used for the benchmark datasets, the DXN dataset is divided into training, validation, and testing sets in the proportions of 1/2, 1/4, and 1/4, respectively.

Additionally, the DXN dataset is categorized into three risk levels, with the division criteria presented in Table 11.17.

11.4.2.2 Modeling Results

The number of DT in the RF model is set to 500, with a minimum sample size of 1 for each leaf node. The results of the RF-based adaptive feature selection are illustrated in Figure 11.11.

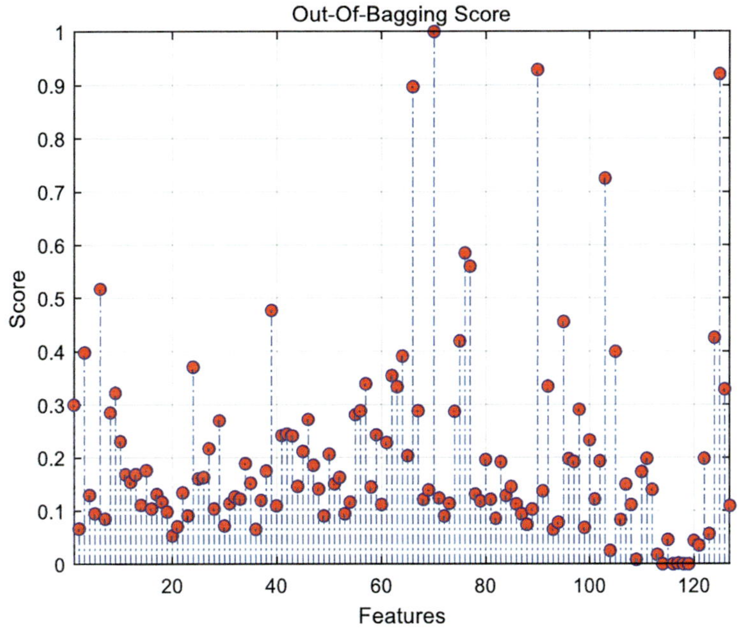

(a) Importance score of features

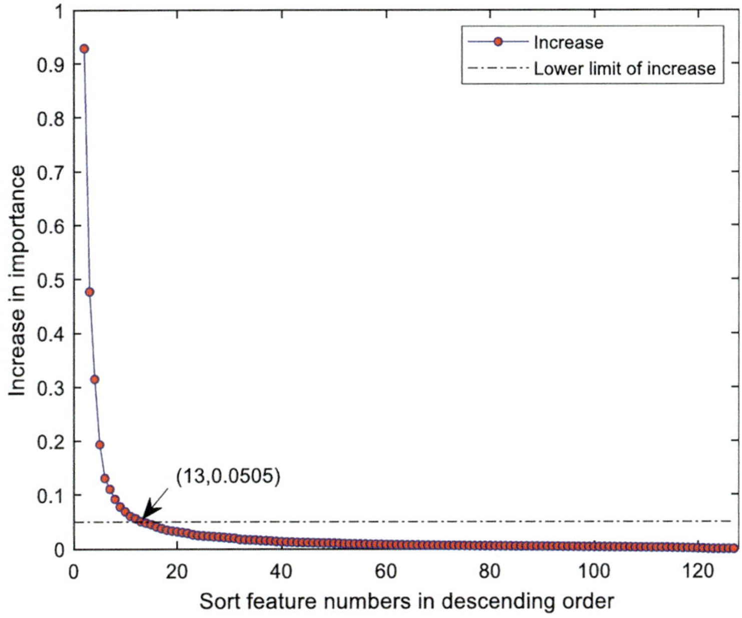

(b) Importance score increase of feature

FIGURE 11.11 Results of RF-based adaptive feature selection for the DXN dataset. (a) Importance score of features. (b) Importance score increase of feature.

Figure 11.11 demonstrates that the importance scores for the 14th feature increase by less than 5% in descending order. Consequently, the first 13 features are adaptively selected. The parameters of the GAFNN model based on the DXN dataset are presented in Table 11.18.

The training epoch N_e is set to 500, with the learning rates α_{lr}^{G} and α_{lr}^{D} for both the generator and discriminator set at 0.0001. The MMD values between the virtual and real samples generated during training are illustrated in Figure 11.12.

Figure 11.12 shows that the MMD values gradually decrease as the training epochs increase, stabilizing once the training epochs reach 400.

Similar to the benchmark datasets, 10 generators are selected at intervals of 10 from the 400th epoch to the 490th epoch. Each generator produces a group of virtual samples equal in number to the training samples. The MMD values for these 10 groups are calculated and presented in Table 11.19.

Table 11.19 indicates that the MMD value of the 470th generator is the lowest; therefore, it is selected as the screening generator.

Subsequently, nine candidate virtual samples are generated using the screening generator, and the discriminator probabilities are calculated, as shown in Table 11.20.

Table 11.20 reveals that the 1st, 6th, 7th, and 8th virtual samples meet the requirement with a discriminator probability of 0.8.

Next, the nearest neighbor classification is employed to further filter the virtual samples using the specified criteria. The nearest neighbor classifications for the aforementioned virtual samples are presented in Table 11.21.

Table 11.21 presents the risk levels and nearest neighbor classifications for the 1st, 7th, and 8th virtual samples, resulting in three qualified virtual samples.

Finally, the aforementioned process is repeated to generate a total of 34 qualified virtual samples, matching the size of the training samples. These virtual and training samples are then reduced to three dimensions using t-SNE to illustrate their relationship, with the results displayed in Figure 11.13.

TABLE 11.18
Parameters of GAFNN for DXN Data

Structure	Network		Number of Neurons
Generator	Antecedent network	Input layer	14
		Membership layer	70 × 14
		Rule layer	70
		Normalization layer	70
	Consequent network	Hidden layer	70 × 13
		Output layer	13
Discriminator	Antecedent network	Input layer	14
		Membership layer	70 × 14
		Rule layer	70
		Normalization layer	70
	Consequent network	Hidden layer	70 × 1
		Output layer	1

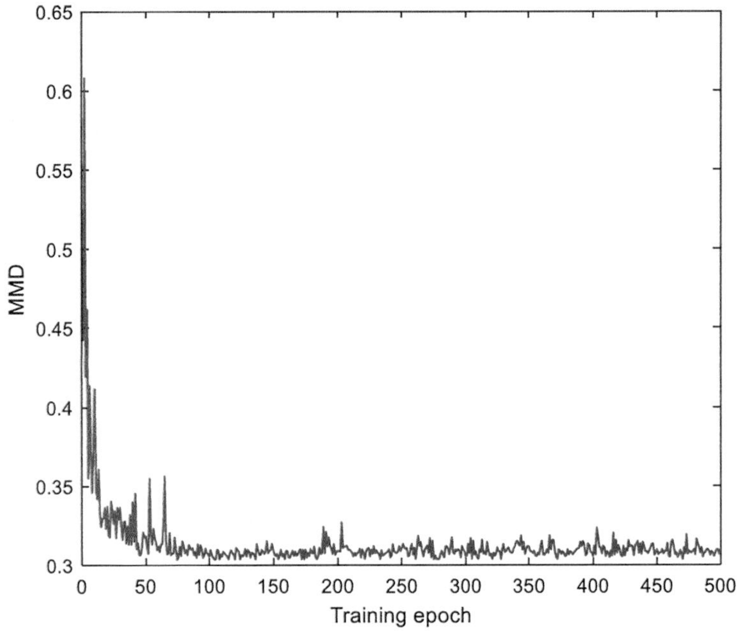

FIGURE 11.12 Relationship between the MMD Value and the Training Algebra for DXN Dataset.

TABLE 11.19
MMD-Based Generator Initial Screening Results for DXN Dataset

Training Epoch	MMD Values
400	0.3149
410	0.3169
420	0.3122
430	0.3109
440	0.3115
450	0.3144
460	0.3122
470	0.3094
480	0.3120
490	0.3145

The parameters of the DFC are configured as follows: $N_{FL} = 50$, $N_{CF} = 4$, $N_{DT} = 50$, and N_{leaf} as 1/10 of the training samples. The mixed samples consist of qualified virtual and real samples. For each iteration, a group of virtual samples is added to the real samples as training data, including one high-risk, one medium-risk, and one

TABLE 11.20
Discriminator Probability of Candidate Virtual Samples for DXN Dataset

Number of Virtual Samples	Level	Discriminator Probability
1	Low	0.8993
2	Low	0.5942
3	Low	0.7109
4	Middle	0.7877
5	Middle	0.6963
6	Middle	0.8392
7	High	0.8752
8	High	0.9197
9	High	0.6927

TABLE 11.21
Nearest Neighbor Class of Candidate Virtual Samples for DXN Dataset

Number of Virtual Samples	Level	*K* Class with the Smallest Distance					Nearest Neighbor Class
1	Low	Middle	Low	Low	Low	Low	Low
6	Middle	Middle	Middle	Low	Low	Low	Low
7	High	High	High	Low	High	Low	High
8	High	Low	Low	High	High	High	High

low-risk level virtual sample, resulting in a total of 20 groups added. Additionally, each experiment is repeated 30 times to account for the randomness inherent in the DFC. The experimental results are presented in Figure 11.14 and Table 11.22.

Figure 11.14 and Table 11.22 demonstrate that the inclusion of virtual samples enhances the accuracy of the risk warning model. In the testing experiments, 7 groups of virtual samples are added based on the validation results, resulting in an average accuracy of 0.857 and a standard deviation of 0.03343.

11.4.2.3 Comparative Results

The DFC is compared with classic algorithms, specifically RF and SVM, using the raw small samples. The experiments for both DFC and RF are repeated 30 times, and the results are averaged. The outcomes of the comparative experiments involving these classic algorithms are presented in Table 11.23.

Table 11.23 indicates that SVM outperforms the other methods in terms of training and validation, while DFC achieves the highest accuracy in testing.

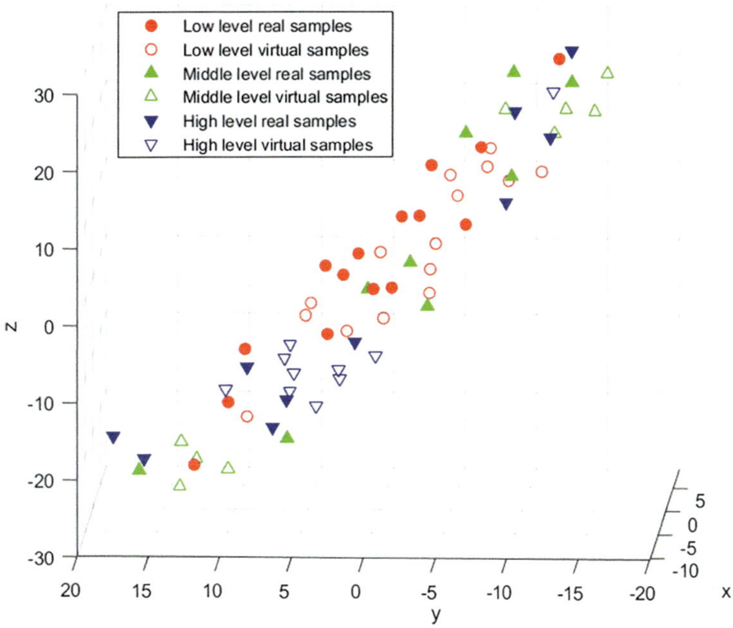

FIGURE 11.13 Display of virtual samples and training samples for the DXN dataset.

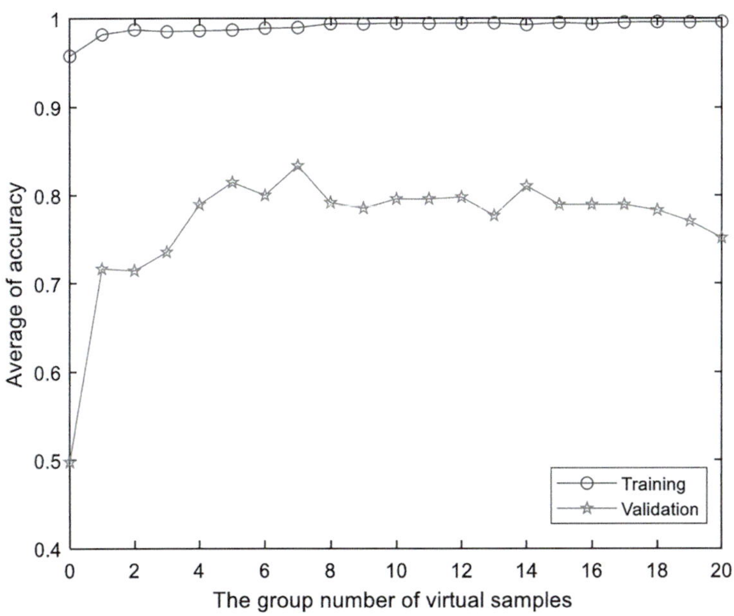

(a) Average of accuracy

FIGURE 11.14 Experimental results of mixed samples for the DXN dataset. (a) Average of accuracy.

(Continued)

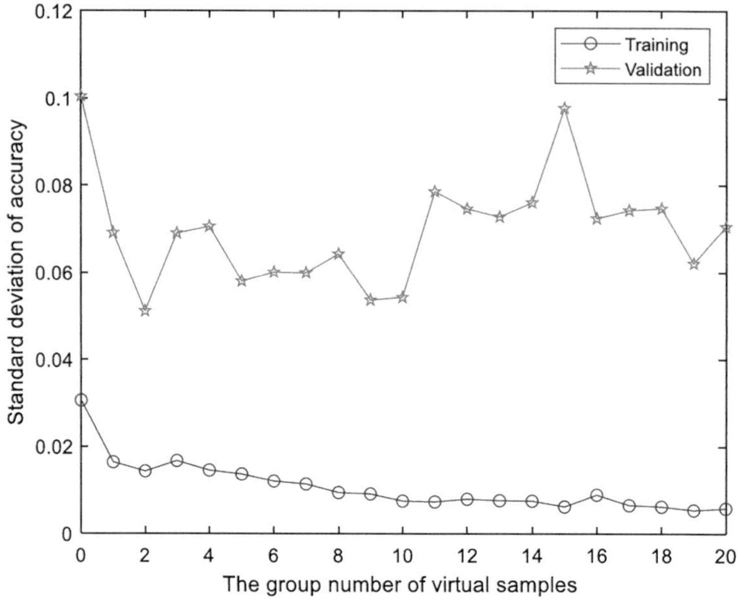

(b) Standard deviation of accuracy

FIGURE 11.14 (Continued) (b) Standard deviation of accuracy.

TABLE 11.22
Statistical Results of Mixed Samples for DXN Dataset

	Average of Accuracy		Standard Deviation of Accuracy	
Dataset	**Best**	**Group Number**	**Best**	**Best**
Training	0.996	20	5.386E−03	19
Validation	0.833	7	5.993E−02	2

TABLE 11.23
Results of Comparative Experiments of Classic Algorithms for Benchmark Datasets

Method	Training Accuracy	Validation Accuracy	Testing Accuracy
DFC	0.982	0.569	0.678
RF	0.990	0.5500	0.655
SVM	1	0.625	0.647

TABLE 11.24
Comparative Experimental MMD for DXN Dataset

		MMD		
Dataset	**Method**	**Training**	**Validation**	**Testing**
SC	GAFNN	0.3045	0.5699	0.3763
	GAN	0.3277	0.5983	0.4229
	AL-GAN	0.3208	0.5829	0.3979

TABLE 11.25
Comparative Experimental Results for DXN Dataset

Method	Dataset	Average of Accuracy	Standard Deviation of Accuracy
GAFNN	Training	0.990	1.138E–02
	Validation	0.833	5.993E–02
	Testing	0.857	3.343E–02
GAN	Training	0.981	1.306E–02
	Validation	0.788	5.085E–02
	Testing	0.837	5.049E–02
AL-GAN	Training	0.984	1.328E–02
	Validation	0.723	7.643E–02
	Testing	0.845	5.458E–02

The GAFNN is then compared with the original GAN and AL-GAN. The parameters for both GAN and AL-GAN are set as follows: the generator consists of 14 input neurons, 70 hidden layer neurons, and 13 output layer neurons; the discriminator comprises 14 input neurons, 70 hidden layer neurons, and 1 output layer neuron; the learning rate is set to 0.0001, and the training epoch is 500.

The MMD values of the virtual and training samples across the three methods are presented in Table 11.24. Additionally, seven groups of virtual samples are included in the comparative experiments, with the results shown in Table 11.25.

Table 11.25 reveals the following findings: (1) The GAFNN exhibits the highest accuracy and stability in both the training and testing datasets; (2) While the GAFNN demonstrates the best accuracy in the validation dataset, the GAN shows superior stability. Therefore, the proposed GAFNN method outperforms the others in terms of both accuracy and stability.

11.4.3 Analysis on Hyperparameter Sensitivity

In the proposed GAFNN, four hyperparameters are analyzed: the number of fuzzy sets for the generator (M_G), the learning rate for the generator (α_{lr}^{G}), the number of fuzzy sets for the discriminator (M_D), and the learning rate for the discriminator (α_{lr}^{D}). The intervals for these hyperparameters are presented in Table 11.26.

Twenty groups of virtual samples are generated and added to all datasets. The results of the hyperparameter analysis on the testing set are presented in Figures 11.15–11.18.

TABLE 11.26
Interval Setting of Hyperparameters

	Interval		
Hyperparameter	**SC**	**RB**	**DXN**
M_G	[30, 100]	[18, 60]	[56, 140]
$\alpha_{\rm lr}^{G}$	[1E–04, 1E–03]	[1E–04, 1E–03]	[1E–04, 1E–03]
M_D	[30, 100]	[18, 60]	[56, 140]
$\alpha_{\rm lr}^{D}$	[1E–04, 1E–03]	[1E–04, 1E–03]	[1E–04, 1E–03]

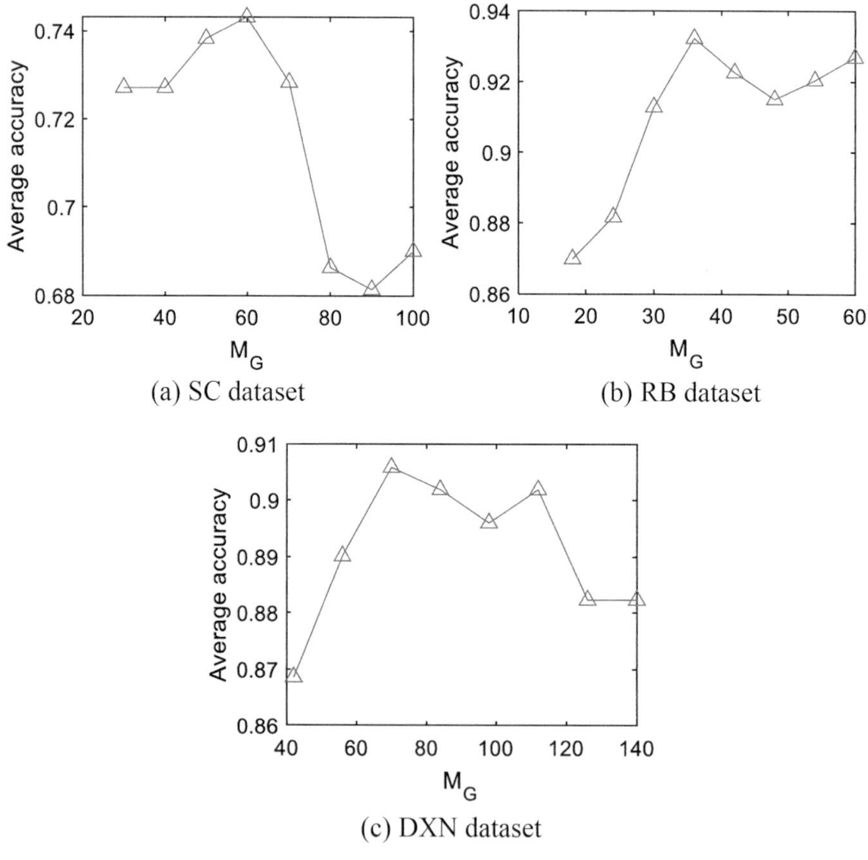

FIGURE 11.15 Average accuracy under M_G. (a) SC dataset. (b) RB dataset. (c) DXN dataset.

Figures 11.15–11.18 show the following:

(1) Regarding the number of fuzzy sets for the generator (M_G), Figure 11.15 illustrates that higher accuracy is achieved within the ranges of [50, 60], [36, 42], and [70, 84] for the SC, RB, and DXN datasets, respectively. Additionally, the accuracy tends to decrease as M_G increases beyond [70, 100] for the SC dataset. Therefore, M_G should not be excessively large. The generator performs optimally when its value is appropriately set. Thus, it is determined that M_G should be five to six times the dimension of the input, which is reasonable for the aforementioned modeling data.

(2) Regarding the learning rate for the generator $\left(\alpha_{\mathrm{lr}}^{G}\right)$, Figure 11.16 demonstrates that accuracy declines as α_{lr}^{G} increases when the training epochs are held constant. The highest accuracy is attained within the ranges of [1E–04, 3E–04], [1E–04, 4E–04], and [1E–04, 2E–04] for the SC, RB, and DXN datasets, respectively. Therefore, setting α_{lr}^{G} at [1E–04, 2E–04] is deemed reasonable for most datasets.

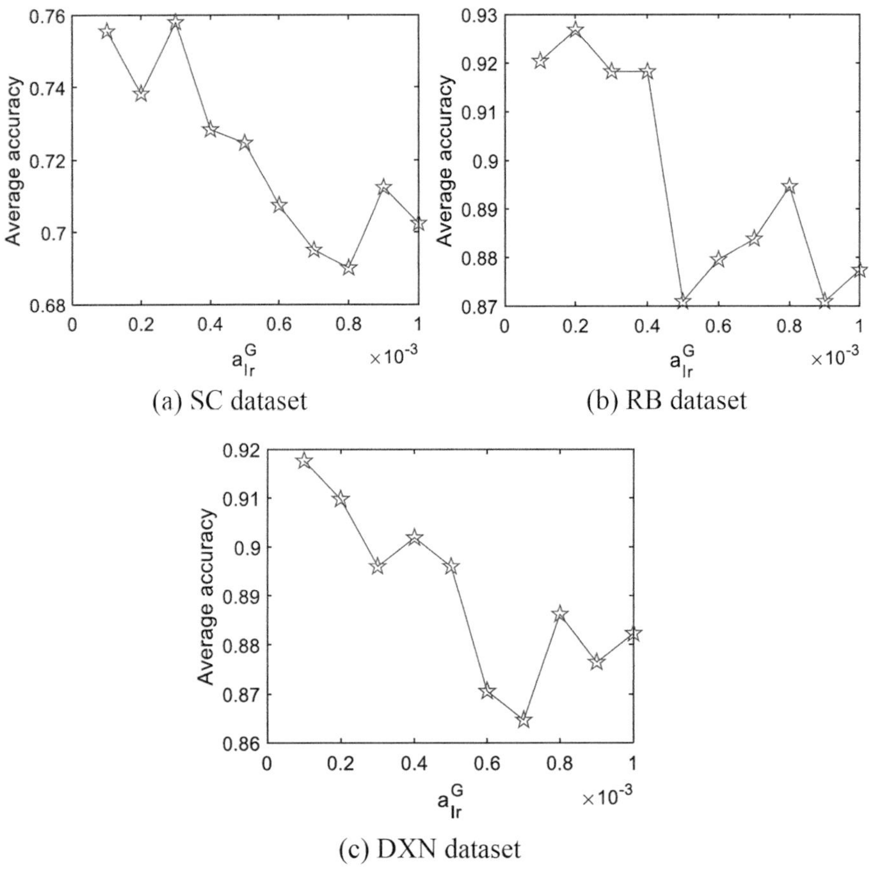

FIGURE 11.16 Average accuracy under α_{lr}^{G}. (a) SC dataset. (b) RB dataset. (c) DXN dataset.

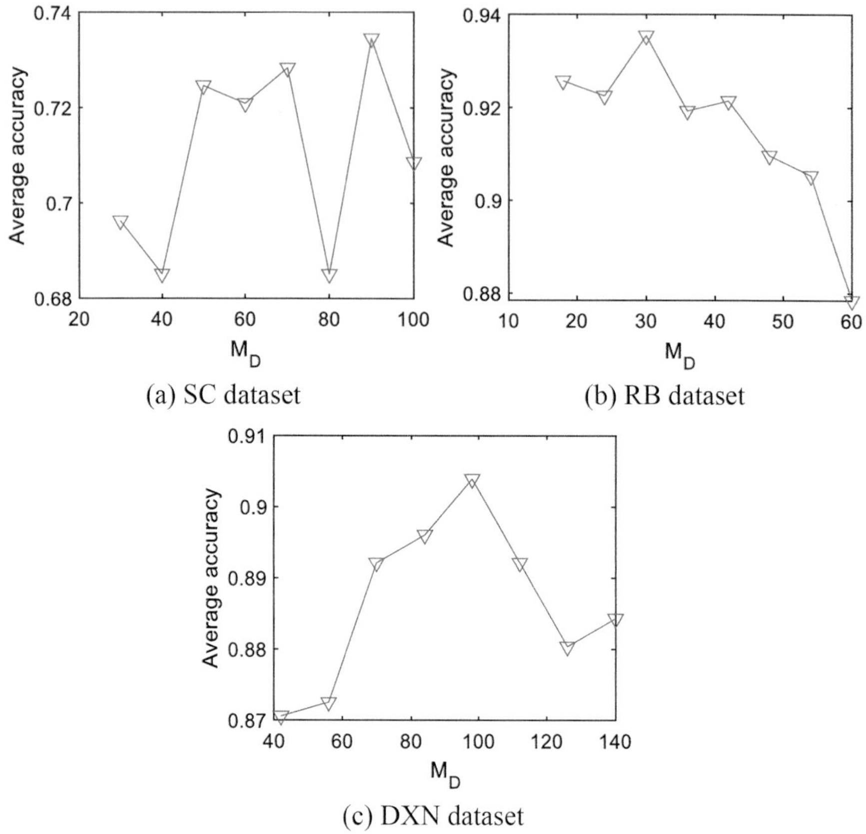

FIGURE 11.17 Average accuracy under M_D. (a) SC dataset. (b) RB dataset. (c) DXN dataset.

(3) In terms of the number of fuzzy sets for the discriminator (M_D), Figure 11.17 indicates that the highest accuracy is achieved at 90, 30, and 98 for the SC, RB, and DXN datasets, respectively. Therefore, $M_D \geq M_G$ is considered efficient within these ranges.
(4) Regarding the learning rate for the discriminator $\left(\alpha_{\text{lr}}^{D}\right)$, Figure 11.18 shows that the highest accuracy is achieved at 2E−04, 1E−04, and 2E−04 for the SC, RB, and DXN datasets, respectively. Additionally, the accuracy decreases as α_{lr}^{D} increases. Therefore, it can be concluded that $\alpha_{\text{lr}}^{D} = \alpha_{\text{lr}}^{G}$ is suitable for most of the modeling data.

In summary, different hyperparameters exert varying effects on distinct datasets. Given the interactions among these hyperparameters, global optimization is essential, taking into account the specific characteristics of each dataset.

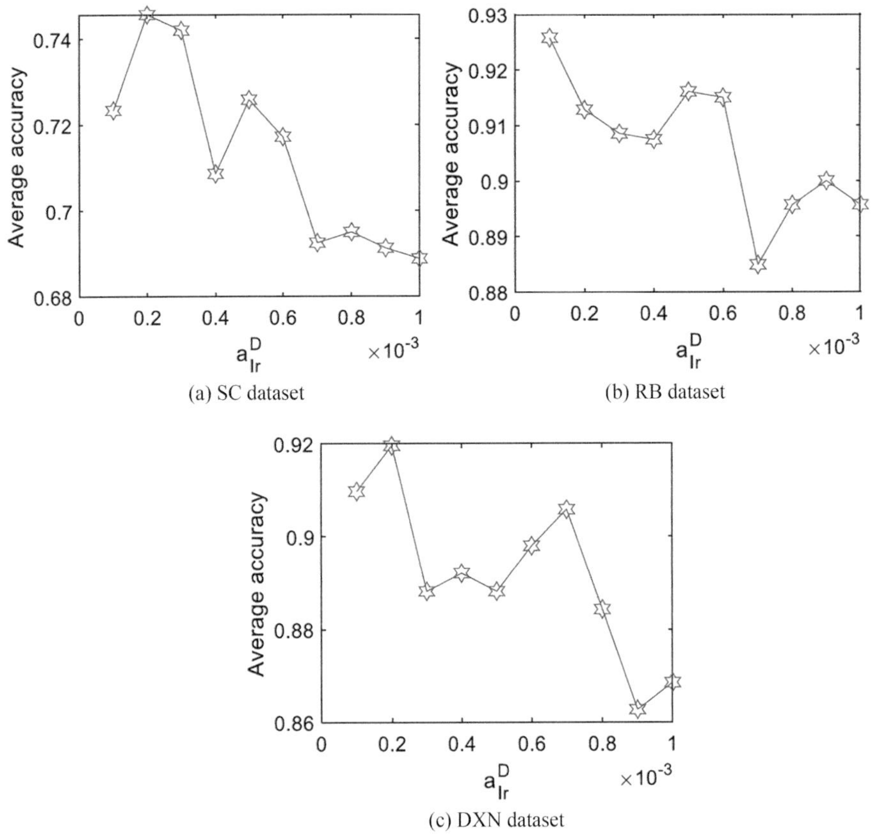

FIGURE 11.18 Average accuracy under α_{lr}^{D}. (a) SC dataset. (b) RB dataset. (c) DXN dataset.

11.4.4 Ablation Experiment

In the ablation study, experiments A–D are conducted to evaluate the impact of the RF-based feature adaptive selection module (Module 1) and the mixed sample-based classification module (Module 2) on the GAFNN. The setup for the ablation experiments is detailed in Table 11.27.

Twenty groups of virtual samples are added to all datasets, and each experiment is conducted 30 times. The results of the ablation experiments on the testing set are presented in Table 11.28.

Table 11.28 presents the following findings:

(1) The results of experiments B and C indicate that the RF-based feature adaptive selection module has a more significant impact on accuracy compared to the mixed sample-based classification module. However, the mixed sample-based classification module contributes more to stability.

TABLE 11.27
Ablation Experiment Setup

No.	Module 1	Module 2
A	√	√
B	×	√
C	√	×
D	×	×

TABLE 11.28
Statistical Results of Ablation Experiment

	SC		RB		DXN	
No.	Average of Accuracy	Standard Deviation of Accuracy	Average of Accuracy	Standard Deviation of Accuracy	Average of Accuracy	Standard Deviation of Accuracy
A	0.736	2.116E–02	0.938	3.274E–02	0.917	2.530E–02
B	0.710	2.193E–02	0.889	3.355E–02	0.878	2.646E–02
C	0.728	2.448E–02	0.923	3.452E–02	0.908	2.965E–02
D	0.701	2.740E–02	0.880	3.486E–02	0.847	3.656E–02

(2) The results of experiments A and D demonstrate that both the RF-based feature adaptive selection module and the mixed sample-based classification module assist the algorithm in generating high-quality virtual samples, which subsequently enhance model accuracy and stability.

11.5 SUMMARY

This study presents a novel VSG method based on GAFNN for modeling small sample data. The RF-based feature adaptive selection method is introduced to minimize randomness and human interference in the selection process. GAFNN is utilized to generate virtual samples, enhancing the training sample size and effectively addressing uncertainty in the training data. Additionally, virtual samples undergo screening through discriminant probability, MMD, and nearest neighbor classification within a multiple constraint strategy, ensuring the quality of the generated samples. Experimental results on benchmark and DXN datasets demonstrate that the GAFNN method outperforms other approaches in terms of accuracy and stability. Future work will focus on determining the optimal virtual sample size and evaluating the impact of the generative model on the quality of the generated virtual samples.

The explanation of abbreviations in this chapter is provided in Table 11.29.

TABLE 11.29
Explanation of Abbreviations

Abbreviation	Explanation
KPI	Key performance indicator
DL	Deep learning
DNN	Deep neural network
DFC	Deep forest classification
VSG	Virtual sample generation
GAN	Generative adversarial network
t-SNE	*t*-distributed stochastic neighbor embedding
RF	Random forest
DXN	Dioxin
FNN	Fuzzy neural network
GAFNN	Generative adversarial fuzzy neural network
MMD	Maximum mean difference
DT	Decision tree
Obb	out-of-bag
TS-FNN	Fuzzy neural network based on Takagi–Sugeno
CRF	Completely random forest
RKHS	Reproducing kernel Hilbert space
SC	Superconductivity
RB	Residential building
UCI	University of California, Irvine
SVM	Support vector machine
AL-GAN	Active learning generative adversarial network

REFERENCES

1. Zhang T, Chen J, Xie J, Pan T. SASLN: signals augmented self-taught learning networks for mechanical fault diagnosis under small sample condition. *IEEE Transactions on Instrumentation and Measurement*, 2021, 70: 1–11.
2. Pan T Y, Chen J L, Zhang T C, et al. Generative adversarial network in mechanical fault diagnosis under small sample: a systematic review on applications and future perspectives. *ISA Transactions*, 2021, 128(Part B): 1–11.
3. Yin S, Ding X S, Xie X C, Luo H. A review on basic data-driven approaches for industrial process monitoring. *IEEE Transactions on Industrial Electronics*, 2014, 61(11): 6418–6428.
4. Zhu J L, Ge Z Q, Song Z H, Gao F R. Review and big data perspectives on robust data mining approaches for industrial process modeling with outliers and missing data. *Annual Reviews in Control*, 2018, 46: 107–133.
5. Chen Y, Xu P, Chu Y, et al. Short-term electrical load forecasting using the Support Vector Regression (SVR) model to calculate the demand response baseline for office buildings. *Applied Energy*, 2017, 195: 659–670.
6. Pang T Y, Yu T X, Song B F. A Bayesian network model for fault diagnosis of a lock mechanism based on degradation data. *Engineering Failure Analysis*, 2021, 122: 1–21.

7. Ding Y F, Jia M P, Miao Q H, Huang P. Remaining useful life estimation using deep metric transfer learning for kernel regression. *Reliability Engineering & System Safety*, 2021, 212: 1–11.
8. Yuan X F, Gu Y J, Wang Y L, et al. A deep supervised learning framework for data-driven soft sensor modeling of industrial processes. *IEEE Transactions on Neural Networks and Learning Systems*, 2019, 31(11): 4737–4746.
9. Hu X B, Niu P F, Wang J M, Zhang X X. Multi-objective prediction of coal-fired boiler with a deep hybrid neural networks. *Atmospheric Pollution Research*, 2020, 11(7): 1084–1090.
10. Liu X J, Zhang H, Kong X B, Lee K Y. Wind speed forecasting using deep neural network with feature selection. *Neurocomputing*, 2020, 397: 393–403.
11. Hemanth D J, Deperlioglu O, Kose U. An enhanced diabetic retinopathy detection and classification approach using deep convolutional neural network. *Neural Computing and Applications*, 2020, 32(3): 707–721.
12. Kromp F, Fischer L, Bozsaky E, et al. Evaluation of deep learning architectures for complex immunofluorescence nuclear image segmentation. *IEEE Transactions on Medical Imaging*, 2021, 40(7): 1934–1949.
13. Liang Y, Li B B, Jiao B. A deep learning method for motor fault diagnosis based on a capsule network with gate-structure dilated convolutions. *Neural Computing and Applications*, 2020, 33(5): 1401–1418.
14. Wen L, Li X Y, Gao L. A transfer convolutional neural network for fault diagnosis based on ResNet-50. *Neural Computing and Applications*, 2019, 32(10): 6111–6124.
15. Tang J, Xia H, Zhang J, et al. Deep forest regression based on cross-layer full connection. *Neural Computing and Applications*, 2021, 33: 9307–9328.
16. Xia H, Tang J, Qiao J F, et al. DF classification algorithm for constructing a small sample size of data-oriented DF regression model. *Neural Computing and Applications*, 2022, 34: 2785–2811.
17. Gong H F, Chen Z S, Zhu Q X, He Y L. A Monte Carlo and PSO based virtual sample generation method for enhancing the energy prediction and energy optimization on small data problem: an empirical study of petrochemical industries. *Applied Energy*, 2017, 197: 405–455.
18. Zhu Q X, Chen Z S, Zhang H X, et al. Dealing with small sample size problems in process industry using virtual sample generation: a Kriging-based approach. *Soft Computing*, 2020, 24: 6889–6902.
19. Zhu Q X, Liu D P, Xu Y, He Y L. Novel space projection interpolation based virtual sample generation for solving the small data problem in developing soft sensor. *Chemometrics and Intelligent Laboratory Systems*, 2021, 217: 1–13.
20. Tang J, Qiao J F, Chai T Y, et al. Modeling multiple components mechanical signals by means of virtual sample generation technique. *Acta Automatica Sinica*, 2018, 44(9): 1569–1589.
21. Tang J, Xia H, Aljerf L, et al. Prediction of dioxin emission from municipal solid waste incineration based on expansion, interpolation, and selection for small samples. *Journal of Environmental Chemical Engineering*, 2022, 10: 108314.
22. Li C A, Lin L S. Generating information for small data sets with a multi-modal distribution. *Decision Support Systems*, 2014, 66: 71–81.
23. Bennin E K, Keung J W, Monden A. On the relative value of data resampling approaches for software defect prediction. *Empirical Software Engineering*, 2019, 24(2): 602–636.
24. Xie Y X, Qiu M, Zhang H B, Peng L Z, Chen Z X. Gaussian distribution based oversampling for imbalanced data classification. *IEEE Transactions on Knowledge and Data Engineering*, 2020, 34(2): 667–679.

25. He Y L, Hua Q, Zhu Q X, Lu S. Enhanced virtual sample generation based on manifold features: applications to developing soft sensor using small data. *ISA Transactions*, 2022, 126: 398–406.
26. Li D C, Wu S C, Tsai T I, Lina Y S. Using mega-trend-diffusion and artificial samples in small data set learning for early flexible manufacturing system scheduling knowledge. *Computers and Operations Research*, 2007, 34(4): 966–982.
27. Li D C, Chen C C, Chang C J, Lin W K. A tree-based-trend-diffusion prediction procedure for small sample sets in the early stages of manufacturing systems. *Expert Systems with Applications*, 2012, 39(1): 1575–1581.
28. Goodfellow I J, Pouget-Abadie J, Mirza M, et al. Generative adversarial networks. *Advances in Neural Information Processing Systems*, 2014, 3: 2672–2680.
29. Lian J, Jia W K, Zareapoor M, et al. Deep-learning-based small surface defect detection via an exaggerated local variation-based generative adversarial network. *IEEE Transactions on Industrial Informatics*, 2020, 16(2): 1343–1351.
30. Li Y B, Zou W T, Jiang L. Fault diagnosis of rotating machinery based on combination of Wasserstein generative adversarial networks and long short term memory fully convolutional network. *Measurement*, 2022, 191: 1–16.
31. Tang J, Cui C L, Xia H, et al. Dioxin emission risk warning model in MSWI process based on GAN with active learning mechanism. *Journal of Beijing University of Technology*, 2023, 49(5): 507–522.
32. Tang J, Liu Z, Zhang J, et al. Kernel latent features adaptive extraction and selection method for multi-component non-stationary signal of industrial mechanical device. *Neurocomputing*, 2016, 216: 296–309.
33. Wan Y T, Ma A L, Zhong Y F, et al. Multiobjective hyperspectral feature selection based on discrete sine cosine algorithm. *IEEE Transactions on Geoscience and Remote Sensing*, 2020, 58(5): 3601–3618.
34. Bao S, Zhang L, Han X S, et al. Feature selection method for nonintrusive load monitoring with balanced redundancy and relevancy. *IEEE Transactions on Industry Applications*, 2022, 58(1): 163–172.
35. Xia H, Tang J, Aljerf L. Dioxin emission prediction based on improved deep forest regression for municipal solid waste incineration process. *Chemosphere*, 2022, 294: 1–13.
36. Zadeh L A. Outline of a new approach to the analysis of complex systems and decision processes. *IEEE Transactions on Systems, Man, and Cybernetics*, 1973, 3(1): 28–44.
37. Jang J S R. ANFIS: adaptive-network-based fuzzy inference system. *IEEE Transactions on Systems, Man, and Cybernetics*, 1993, 23(3): 665–685.
38. Zhang R D, Tao J L. A nonlinear fuzzy neural network modeling approach using an improved genetic algorithm. *IEEE Transactions on Industrial Electronics*, 2018, 65(7): 5882–5892.
39. Wang G Q, Chen X, Li Y X. Fuzzy neural network analysis on gray cast iron with high tensile strength and thermal conductivity. *China Foundry*, 2019, 16(3): 190–197.
40. Qiao J F, Quan L M, Yang C L. Design of modeling error PDF based fuzzy neural network for effluent ammonia nitrogen prediction. *Applied Soft Computing Journal*, 2020, 91: 1–11.
41. Juang C F, Chen Y W. Automatic hitting-duration estimation of a rechargeable impact wrench using a fuzzy neural network to reach target toques. *International Journal of Fuzzy System*, 2022, 25: 29–41.
42. Adeleke O, Akinlabi S, Jen T C, et al. Evolutionary-based neuro-fuzzy modelling of combustion enthalpy of municipal solid waste. *Neural Computing and Applications*, 2022, 34: 7419–7436.

43. Lee S C, Lee E T. Fuzzy neural networks. *Mathematical Biosciences*, 1975, 23(1–2): 151–177.
44. Wan P, Sun D H, Zhao M, Huang S. Multistability for almost-periodic solutions of Takagi–Sugeno fuzzy neural networks with nonmonotonic discontinuous activation functions and time-varying delays. *IEEE Transactions on Fuzzy Systems*, 2019, 29(2): 400–414.
45. Zhou Z H, Feng J. Deep forest. *National Science Review*, 2019, 6: 74–86.
46. Fan W, Wang H. Is random model better? On its accuracy and efficiency. In: *ICDM'03: Proceedings of the Third IEEE International Conference on Data Mining*. 2013, 51–58.
47. Tang J, Guo Z H, Qiao J F. Dioxin emission concentration soft measurement based on multi-source latent feature selective ensemble modeling for municipal solid waste incineration process. *Acta Automatica Sinica*, 2022, 48(1): 223–238.
48. Tang J, Wang D D, Guo Z H, Qiao J F. Prediction of dioxin emission concentration in municipal solid waste incineration process based on optimal selection of virtual samples. *Journal of Beijing University of Technology*, 2021, 47(5): 431–443.

12 Summary and Outlook

12.1 SUMMARY OF KEY FINDINGS

- **Understanding DXN dynamics**: The book meticulously delves into the intricate mechanisms underlying dioxin (DXN) formation, adsorption, and emission within the municipal solid waste incineration (MSWI) process. This knowledge provides a foundation for developing sophisticated modeling strategies.
- **Numerical simulation modeling**: The authors successfully establish a comprehensive numerical simulation model (CNSM) by coupling FLIC and ASPEN software. This model effectively captures the evolution of DXN concentration throughout the MSWI process, providing a theoretical basis for understanding DXN dynamics.
- **AI-driven modeling**: Various advanced AI-based tree modeling techniques are introduced and applied to address different data characteristics in DXN modeling. These methods include the adaptive deep forest regression algorithm based on cross-layer full connection (ADFR-clfc), tree broad learning system (TBLS), fuzzy forest regression (FFR), and virtual sample generation (VSG) techniques.
- **Whole-process tree-based model (WPTM)**: The book proposes the WPTM framework, which integrates the aforementioned AI models into a hierarchical structure, effectively capturing the whole DXN concentration process within the MSWI plant.
- **Multimodal data-driven modeling**: The authors successfully implement multimodal data-driven approaches, leveraging process numerical data, flame video data, and operational recorded data. This approach provides a more comprehensive understanding of DXN generation and facilitates more accurate prediction models.
- **Multi-demand prediction**: The book demonstrates the capability of the developed models to address various prediction needs, including real-time prediction (second scale), hourly prediction, and daily prediction using both process and production data.

12.2 OUTLOOK AND FUTURE DIRECTIONS

- **Model enhancement**: The research highlights the need for further refinement and optimization of the AI models. This includes improving their ability to handle data bias, optimizing hyperparameters for increased accuracy, and enhancing their adaptability to dynamic operating conditions.
- **Dynamic updating**: Future research should explore dynamic updating mechanisms for the AI models, enabling them to adapt to changing conditions and maintain high prediction accuracy over time.

DOI: 10.1201/9781003634195-15

- **Real-time application**: The book highlights the potential for the developed models to be integrated into online control systems for real-time DXN concentration measurement. This would revolutionize emission control by enabling immediate adjustments to process parameters.
- **Multi-pollutant modeling**: Expanding the modeling framework to include other key pollutants, such as CO, CO_2, NOx, SO_2, and HCl, would provide a more holistic understanding of the MSWI process and facilitate comprehensive emission control strategies.

12.3 CONCLUSION

This book represents a significant contribution to the field of DXN modeling in MSWI processes. By utilizing both numerical simulation and AI-driven approaches, it provides invaluable insights into the complex dynamics of DXN and offers powerful tools for effective emission control. The research provides a strong foundation for continued advancements in online DXN concentration measurement and pollution reduction strategies, paving the way for a more sustainable and environmentally responsible MSWI industry.

Index

Pages in *italics* refer to figures and pages in **bold** refer to tables.

D

E

F

O

P

Q

R

S

T

For Product Safety Concerns and Information please contact our EU representative GPSR@taylorandfrancis.com Taylor & Francis Verlag GmbH, Kaufingerstraße 24, 80331 München, Germany